P9-DFJ-222

Student Edition

SpringBoard®

Mathematics

Algebra 1

CollegeBoard

About the College Board

The College Board is a mission-driven not-for-profit organization that connects students to college success and opportunity. Founded in 1900, the College Board was created to expand access to higher education. Today, the membership association is made up of over 6,000 of the world's leading educational institutions and is dedicated to promoting excellence and equity in education. Each year, the College Board helps more than seven million students prepare for a successful transition to college through programs and services in college readiness and college success—including the SAT® and the Advanced Placement Program®. The organization also serves the education community through research and advocacy on behalf of students, educators, and schools.

For further information, visit www.collegeboard.org.

ISBN: 1-4573-0151-2
ISBN: 978-1-4573-0151-3

4 5 6 7 8 9   16 17 18 19 20 21
Printed in the United States of America

Acknowledgments

The College Board gratefully acknowledges the outstanding work of the classroom teachers and writers who have been integral to the development of this revised program. The end product is testimony to their expertise, understanding of student learning needs, and dedication to rigorous but accessible mathematics instruction.

Michael Allwood
Brunswick School
Greenwich, Connecticut

Floyd Bullard
North Carolina School of Science and Mathematics
Durham, North Carolina

Marcia Chumas
East Mecklenburg High School
Charlotte, North Carolina

Kathy Fritz
Plano Independent School District
Plano, Texas

Shawn Harris
Ronan Middle School
Ronan, Montana

Marie Humphrey
David W. Butler High School
Charlotte, North Carolina

Brian Kotz
Montgomery College
Monrovia, Maryland

Chris Olsen
Prairie Lutheran School
Cedar Rapids, Iowa

Dr. Roxy Peck
California Polytechnic Institute
San Luis Obispo, California

Katie Sheets
Harrisburg School
Harrisburg, South Dakota

Andrea Sukow
Mathematics Consultant
Nashville, Tennessee

Stephanie Tate
Hillsborough School District
Tampa, Florida

SpringBoard Mathematics Product Development

Betty Barnett
Executive Director
Content Development

Kimberly Sadler, M.Ed.
Senior Math Product Manager

John Nelson
Mathematics Editor

Allen M. D. von Pallandt, Ph.D.
Senior Director
Mathematics Content Development

Judy Windle
Senior Math Instructional Specialist

Research and Planning Advisors

We also wish to thank the members of our SpringBoard Advisory Council and the many educators who gave generously of their time and their ideas as we conducted research for both the print and online programs. Your suggestions and reactions to ideas helped immeasurably as we planned the revisions. We gratefully acknowledge the teachers and administrators in the following districts.

ABC Unified
Cerritos, California

Albuquerque Public Schools
Albuquerque, New Mexico

Amarillo School District
Amarillo, Texas

Baltimore County Public Schools
Baltimore, Maryland

Bellevue School District 405
Bellevue, Washington

Charlotte Mecklenburg Schools
Charlotte, North Carolina

Clark County School District
Las Vegas, Nevada

Cypress Fairbanks ISD
Houston, Texas

District School Board of Collier County
Collier County, Florida

Denver Public Schools
Denver, Colorado

Frisco ISD
Frisco, Texas

Gilbert Unified School District
Gilbert, Arizona

Grand Prairie ISD
Grand Prairie, Texas

Hillsborough County Public Schools
Tampa, Florida

Houston Independent School District
Houston, Texas

Hobbs Municipal Schools
Hobbs, New Mexico

Irving Independent School District
Irving, Texas

Kenton County School District
Fort Wright, Kentucky

Lee County Public Schools
Fort Myers, Florida

Newton County Schools
Covington, Georgia

Noblesville Schools
Noblesville, Indiana

Oakland Unified School District
Oakland, California

Orange County Public Schools
Orlando, Florida

School District of Palm Beach County
Palm Beach, Florida

Peninsula School District
Gig Harbor, Washington

Polk County Public Schools
Bartow, Florida

Quakertown Community School District
Quakertown, Pennsylvania

Rio Rancho Public Schools
Rio Rancho, New Mexico

Ronan School District
Ronan, Montana

St. Vrain Valley School District
Longmont, Colorado

Scottsdale Public Schools
Phoenix, Arizona

Seminole County Public Schools
Sanford, Florida

Southwest ISD
San Antonio, Texas

Spokane Public Schools
Spokane, Washington

Volusia County Schools
DeLand, Florida

Contents

To the Student

Welcome to the SpringBoard program.

This program has been created with you in mind: the content you need to learn, the tools to help you learn, and the critical thinking skills that help you build confidence in your own knowledge of mathematics. The College Board publishes the SpringBoard program. It also publishes the PSAT/NMSQT, the SAT, and the Advanced Placement exams—all exams that you are likely to encounter in your student years. Preparing you to perform well on those exams and to develop the mathematics skills needed for high school success is the primary purpose of this program.

Standards-Based Mathematics Learning

The SpringBoard program is based on learning standards that identify the mathematics skills and knowledge that you should master to succeed in high school and in future college-level work. In this course, the standards follow these broad areas of mathematics knowledge:

- Mathematical practices
- Number and quantity
- Algebra
- Functions
- Modeling
- Statistics and probability

Mathematical practice standards guide your study of mathematics. They are actions you take to help you understand mathematical concepts rather than just mathematical procedures. For example, the mathematical practice standards state the following:

MP.1 Make sense of problems and persevere in solving them.

MP.2 Reason abstractly and quantitatively.

MP.3 Construct viable arguments and critique the reasoning of others.

MP.4 Model with mathematics.

MP.5 Use appropriate tools strategically.

MP.6 Attend to precision.

MP.7 Look for and make use of structure.

MP.8 Look for and express regularity in repeated reasoning.

As you continue your studies from middle school, you will examine expressions, equations, and functions, which will allow you to make comparisons between relations and functions. Expressions and equations connect with functions. Understanding the concept of functions is critical to future success in your study of algebra and the rest of the high school mathematics curriculum.

See pages xiii–xvi for a complete list of the Common Core State Standards for Mathematics for this course.

Strategies for Learning Mathematics

Some tools to help you learn are built into every activity. At the beginning of each activity, you will see suggested learning strategies. Each of these strategies is explained in full in the Resources section of your book. As you learn to use each strategy, you'll have the opportunity to decide which strategies work best for you. Suggested learning strategies include:

- Reading strategies
- Writing strategies
- Problem-solving strategies
- Collaborative strategies

Building Mathematics Knowledge and Skills

The SpringBoard program is built around the following.

Problem Solving Many of the problems in this book require you to ***analyze*** the situation and the information in a problem, ***make decisions, determine the strategies*** you'll use to solve the problem, and ***justify*** your solution.

Reasoning and Justification You will be asked to explain the reasoning behind how you solved problems, the mathematics concepts involved, and why your approach was appropriate.

Communication Communicating about mathematics, orally and in writing, with your classmates and teachers helps you organize your learning and explain mathematics concepts.

Mathematics Connections As you develop your mathematics knowledge, you will see the many connections between mathematics concepts and between mathematics and your own life.

Representations In mathematics, representations can take many forms, such as numeric, verbal, graphic, or symbolic. In this course, you are encouraged to use representations to organize problem information, present possible solutions, and communicate your reasoning.

We hope you enjoy your study of mathematics using the SpringBoard program.

Common Core State Standards for Mathematics

Algebra 1

HSN-RN The Real Number System

HSN-RN.A.1 Explain how the definition of the meaning of rational exponents follows from extending the properties of integer exponents to those values, allowing for a notation for radicals in terms of rational exponents. *For example, we define* $5^{\frac{1}{3}}$ *to be the cube root of 5 because we want* $\left(5^{\frac{1}{3}}\right)^3 = \left(5^{\frac{1}{3}}\right)^3$ *to hold, so* $\left(5^{\frac{1}{3}}\right)^3$ *must equal 5.*

HSN-RN.A.2 Rewrite expressions involving radicals and rational exponents using the properties of exponents.

HSN-RN.B.3 Explain why the sum or product of two rational numbers is rational; that the sum of a rational number and an irrational number is irrational; and that the product of a nonzero rational number and an irrational number is irrational.

HSN-Q Quantities

HSN-Q.A.1 Use units as a way to understand problems and to guide the solution of multi-step problems; choose and interpret units consistently in formulas; choose and interpret the scale and the origin in graphs and data displays.

HSN-Q.A.2 Define appropriate quantities for the purpose of descriptive modeling.

HSN-Q.A.3 Choose a level of accuracy appropriate to limitations on measurement when reporting quantities.

HSA-SSE Seeing Structure in Expressions

HSA-SSE.A.1 Interpret expressions that represent a quantity in terms of its context.*

HSA-SSE.A.1a Interpret parts of an expression, such as terms, factors, and coefficients.

HSA-SSE.A.1b Interpret complicated expressions by viewing one or more of their parts as a single entity. *For example, interpret $P(1 + r)^n$ as the product of P and a factor not depending on P.*

HSA-SSE.A.2 Use the structure of an expression to identify ways to rewrite it. *For example, see $x^4 - y^4$ as $(x^2)^2 - (y^2)^2$, thus recognizing it as a difference of squares that can be factored as $(x^2 - y^2)(x^2 + y^2)$.*

HSA-SSE.B.3 Choose and produce an equivalent form of an expression to reveal and explain properties of the quantity represented by the expression.*

HSA-SSE.B.3a Factor a quadratic expression to reveal the zeros of the function it defines.

HSA-SSE.B.3b Complete the square in a quadratic expression to reveal the maximum or minimum value of the function it defines.

HSA-SSE.B.3c Use the properties of exponents to transform expressions for exponential functions. *For example the expression 1.15^t can be rewritten as $\left(1.15^{\frac{1}{12}}\right)^{12t} \approx 1.012^{12t}$ to reveal the approximate equivalent monthly interest rate if the annual rate is 15%.*

HSA-APR Arithmetic with Polynomials & Rational Expressions

HSA-APR.A.1 Understand that polynomials form a system analogous to the integers, namely, they are closed under the operations of addition, subtraction, and multiplication; add, subtract, and multiply polynomials.

HSA-CED Creating Equations

HSA-CED.A.1 Create equations and inequalities in one variable and use them to solve problems. *Include equations arising from linear and quadratic functions, and simple rational and exponential functions.*

HSA-CED.A.2 Create equations in two or more variables to represent relationships between quantities; graph equations on coordinate axes with labels and scales.

HSA-CED.A.3 Represent constraints by equations or inequalities, and by systems of equations and/or inequalities, and interpret solutions as viable or nonviable options in a modeling context. *For example, represent inequalities describing nutritional and cost constraints on combinations of different foods.*

HSA-CED.A.4 Rearrange formulas to highlight a quantity of interest, using the same reasoning as in solving equations. *For example, rearrange Ohm's law $V = IR$ to highlight resistance R.*

HSA-REI Reasoning with Equations and Inequalities

HSA-REI.A.1 Explain each step in solving a simple equation as following from the equality of numbers asserted at the previous step, starting from the assumption that the original equation has a solution. Construct a viable argument to justify a solution method.

HSA-REI.B.3 Solve linear equations and inequalities in one variable, including equations with coefficients represented by letters.

HSA-REI.B.4 Solve quadratic equations in one variable.

HSA-REI.B.4a Use the method of completing the square to transform any quadratic equation in x into an equation of the form $(x - p)^2 = q$ that has the same solutions. Derive the quadratic formula from this form.

HSA-REI.B.4b Solve quadratic equations by inspection (e.g., for $x^2 = 49$), taking square roots, completing the square, the quadratic formula and factoring, as appropriate to the initial form of the equation. Recognize when the quadratic formula gives complex solutions and write them as $a \pm bi$ for real numbers a and b.

HSA-REI.C.5 Prove that, given a system of two equations in two variables, replacing one equation by the sum of that equation and a multiple of the other produces a system with the same solutions.

HSA-REI.C.6 Solve systems of linear equations exactly and approximately (e.g., with graphs), focusing on pairs of linear equations in two variables.

HSA-REI.C.7 Solve a simple system consisting of a linear equation and a quadratic equation in two variables algebraically and graphically. *For example, find the points of intersection between the line* $y = -3x$ *and the circle* $x^2 + y^2 = 3$.

HSA-REI.D.10 Understand that the graph of an equation in two variables is the set of all its solutions plotted in the coordinate plane, often forming a curve (which could be a line).

HSA-REI.D.11 Explain why the x-coordinates of the points where the graphs of the equations $y = f(x)$ and $y = g(x)$ intersect are the solutions of the equation $f(x) = g(x)$; find the solutions approximately, e.g., using technology to graph the functions, make tables of values, or find successive approximations. Include cases where $f(x)$ and/or $g(x)$ are linear, polynomial, rational, absolute value, exponential, and logarithmic functions.*

HSA-REI.D.12 Graph the solutions to a linear inequality in two variables as a half-plane (excluding the boundary in the case of a strict inequality), and graph the solution set to a system of linear inequalities in two variables as the intersection of the corresponding half-planes.

HSF-IF Interpreting Functions

HSF-IF.A.1 Understand that a function from one set (called the domain) to another set (called the range) assigns to each element of the domain exactly one element of the range. If f is a function and x is an element of its domain, then $f(x)$ denotes the output of f corresponding to the input x. The graph of f is the graph of the equation $y = f(x)$.

HSF-IF.A.2 Use function notation, evaluate functions for inputs in their domains, and interpret statements that use function notation in terms of a context.

HSF-IF.A.3 Recognize that sequences are functions, sometimes defined recursively, whose domain is a subset of the integers. *For example, the Fibonacci sequence is defined recursively by* $f(0) = f(1) = 1$, $f(n + 1) = f(n) + f(n - 1)$ *for* $n \geq 1$.

HSF-IF.B.4 For a function that models a relationship between two quantities, interpret key features of graphs and tables in terms of the quantities, and sketch graphs showing key features given a verbal description of the relationship. *Key features include: intercepts; intervals where the function is increasing, decreasing, positive, or negative; relative maximums and minimums; symmetries; end behavior; and periodicity.**

HSF-IF.B.5 Relate the domain of a function to its graph and, where applicable, to the quantitative relationship it describes. *For example, if the function h(n) gives the number of person-hours it takes to assemble n engines in a factory, then the positive integers would be an appropriate domain for the function.**

HSF-IF.B.6 Calculate and interpret the average rate of change of a function (presented symbolically or as a table) over a specified interval. Estimate the rate of change from a graph.*

HSF-IF.C.7 Graph functions expressed symbolically and show key features of the graph, by hand in simple cases and using technology for more complicated cases.*

HSF-IF.C.7a Graph linear and quadratic functions and show intercepts, maxima, and minima.

HSF-IF.C.7b Graph square root, cube root, and piecewise-defined functions, including step functions and absolute value functions.

HSF-IF.C.7c Graph exponential and logarithmic functions, showing intercepts and end behavior, and trigonometric functions, showing period, midline, and amplitude.

HSF-IF.C.8 Write a function defined by an expression in different but equivalent forms to reveal and explain different properties of the function.

HSF-IF.C.8a Use the process of factoring and completing the square in a quadratic function to show zeros, extreme values, and symmetry of the graph, and interpret these in terms of a context.

HSF-IF.C.8b Use the properties of exponents to interpret expressions for exponential functions. *For example, identify percent rate of change in functions such as* $y = (1.02)t$, $y = (0.97)t$, $y = (1.01)12t$, $y = (1.2)\frac{t}{10}$, *and classify them as representing exponential growth or decay.*

HSF-IF.C.9 Compare properties of two functions each represented in a different way (algebraically, graphically, numerically in tables, or by verbal descriptions). *For example, given a graph of one quadratic function and an algebraic expression for another, say which has the larger maximum.*

HSF-BF Building Functions

HSF-BF.A.1 Write a function that describes a relationship between two quantities.

HSF-BF.A.1a Determine an explicit expression, a recursive process, or steps for calculation from a context.

HSF-BF.A.1b Combine standard function types using arithmetic operations. *For example, build a function that models the temperature of a cooling body by adding a constant function to a decaying exponential, and relate these functions to the model.*

HSF-BF.A.2 Write arithmetic and geometric sequences both recursively and with an explicit formula, use them to model situations, and translate between the two forms.*

HSF-BF.B.3 Identify the effect on the graph of replacing $f(x)$ by $f(x) + k$, $k\,f(x)$, $f(kx)$, and $f(x + k)$ for specific values of k (both positive and negative); find the value of k given the graphs. Experiment with cases and illustrate an explanation of the effects on the graph using technology. Include recognizing even and odd functions from their graphs and algebraic expressions for them.

HSF-BF.B.4 Find inverse functions.

HSF-BF.B.4a Solve an equation of the form $f(x) = c$ for a simple function f [linear only] that has an inverse and write an expression for the inverse. *For example,* $f(x) = 2x^3$ *or* $f(x) = (x + 1)/(x - 1)$ *for* $x \neq 1$.

HSF-LE Linear, Quadratic, and Exponential Models

HSF-LE.A.1 Distinguish between situations that can be modeled with linear functions and with exponential functions.

HSF-LE.A.1a Prove that linear functions grow by equal differences over equal intervals, and that exponential functions grow by equal factors over equal intervals.

HSF-LE.A.1b Recognize situations in which one quantity changes at a constant rate per unit interval relative to another.

HSF-LE.A.1c Recognize situations in which a quantity grows or decays by a constant percent rate per unit interval relative to another.

HSF-LE.A.2 Construct linear and exponential functions, including arithmetic and geometric sequences, given a graph, a description of a relationship, or two input-output pairs (include reading these from a table).

HSF-LE.A.3 Observe using graphs and tables that a quantity increasing exponentially eventually exceeds a quantity increasing linearly, quadratically, or (more generally) as a polynomial function.

HSF-LE.B.5 Interpret the parameters in a linear or exponential function in terms of a context.

HSS-ID Interpreting Categorical and Quantitative Data

HSS-ID.A.1 Represent data with plots on the real number line (dot plots, histograms, and box plots).

HSS-ID.A.2 Use statistics appropriate to the shape of the data distribution to compare center (median, mean) and spread (interquartile range, standard deviation) of two or more different data sets.

HSS-ID.A.3 Interpret differences in shape, center, and spread in the context of the data sets, accounting for possible effects of extreme data points (outliers).

HSS-ID.B.5 Summarize categorical data for two categories in two-way frequency tables. Interpret relative frequencies in the context of the data (including joint, marginal, and conditional relative frequencies). Recognize possible associations and trends in the data.

HSS-ID.B.6 Represent data on two quantitative variables on a scatter plot, and describe how the variables are related.

HSS-ID.B.6a Fit a function to the data; use functions fitted to data to solve problems in the context of the data. Use given functions or choose a function suggested by the context. Emphasize linear, quadratic, and exponential models.

HSS-ID.B.6b Informally assess the fit of a function by plotting and analyzing residuals.

HSS-ID.B.6c Fit a linear function for a scatter plot that suggests a linear association.

HSS-ID.C.7 Interpret the slope (rate of change) and the intercept (constant term) of a linear model in the context of the data.

HSS-ID.C.8 Compute (using technology) and interpret the correlation coefficient of a linear fit.

HSS-ID.C.9 Distinguish between correlation and causation.

Equations and Inequalities

1

Unit Overview

Investigating patterns is a good foundation for studying Algebra 1. You will begin this unit by analyzing, describing, and generalizing patterns using tables, expressions, graphs, and words. You will then write and solve equations and inequalities in mathematical and real-world problems.

Key Terms

As you study this unit, add these and other terms to your math notebook. Include in your notes your prior knowledge of each word, as well as your experiences in using the word in different mathematical examples. If needed, ask for help in pronouncing new words and add information on pronunciation to your math notebook. It is important that you learn new terms and use them correctly in your class discussions and in your problem solutions.

Academic Vocabulary

- consecutive

Math Terms

- sequence
- common difference
- expression
- variable
- equilateral
- equation
- solution
- formula
- literal equation
- graph of an inequality
- solution of an inequality
- compound inequality
- conjunction
- disjunction
- absolute value
- absolute value notation
- absolute value equation
- absolute value inequality

ESSENTIAL QUESTIONS

? How can you represent patterns from everyday life by using tables, expressions, and graphs?

? How can you write and solve equations and inequalities?

EMBEDDED ASSESSMENTS

These assessments, following Activities 2 and 4, will give you an opportunity to demonstrate what you have learned about patterns, equations, and inequalities.

Embedded Assessment 1:

Patterns and Equations p. 33

Embedded Assessment 2:

Inequalities and Absolute Value p. 61

UNIT 1

Getting Ready

Write your answers on notebook paper.
Show your work.

1. What is $\frac{2}{3}+\frac{4}{5}$?
2. What condition must be met before you can add or subtract fractions?
3. Jennifer is checking Megan's homework. They disagree on the answer to this problem: $4^2 \times 2^2$. Jennifer says the product is 32 and Megan says it is 64. Who has the correct answer? Explain how she arrived at that correct product.
4. A piece of lumber $2\frac{1}{4}$ feet long is to be cut into 3 equal pieces. How long will each piece of cut wood be? Give the measurement in feet and in inches.
5. Arrange the following expressions in order of their value from least to greatest.
   a. $4-6$ **b.** $-4+6$
   c. $-4-6$
6. Which expression has the greater value? Justify your answer.
   A. $-8+3$ **B.** -8×3
7. Which of the following are equal to 14.95?
   A. 2.3×6.5 **B.** $21.45-6.5$
   C. $8.32+6.63$
8. Which equation has the least solution?
   A. $x+5=13$ **B.** $-6x=-30$
   C. $\frac{x}{4}=18$ **D.** $x-2=-11$
9. Combine like terms in the following expressions.
   a. $10x-4x$ **b.** $-15n+3n$
   c. $7.5y+1.6y-2$ **d.** $m+4-2m$
10. The Venn diagram below provides a visual representation of the students in Mr. Griffin's class who participate in music programs after school. What does the diagram tell you about the musical involvement of Student B and Student G? Explain how you reached your conclusion.

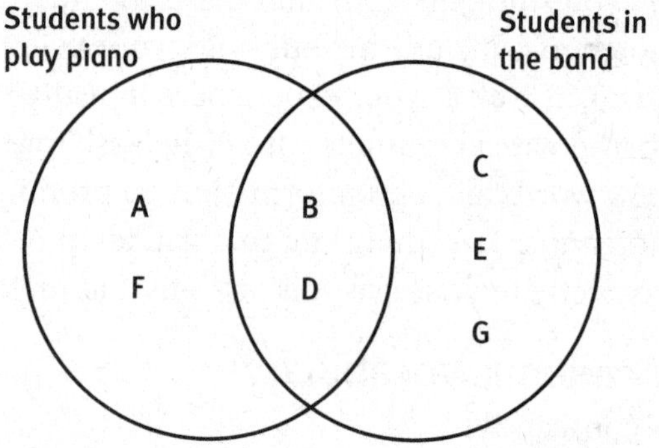

Investigating Patterns

Cross-Country Adventures

Lesson 1-1 Numeric and Graphic Representations of Data

My Notes

Learning Targets:

- Identify patterns in data.
- Use tables, graphs, and expressions to model situations.
- Use expressions to make predictions.

SUGGESTED LEARNING STRATEGIES: Sharing and Responding, Create Representations, Discussion Groups, Look for a Pattern, Interactive Word Wall

Mizing spent his summer vacation traveling cross-country with his family. Their first stop was Yellowstone National Park in Wyoming and Montana. Yellowstone is famous for its geysers, especially one commonly referred to as Old Faithful. A geyser is a spring that erupts intermittently, forcing a fountain of water and steam from a hole in the ground. Old Faithful can have particularly long and fairly predictable eruptions. As a matter of fact, park rangers have observed the geyser over many years and have developed patterns they use to predict the timing of the next eruption.

Park rangers have recorded the information in the table below.

Length of Eruption (in minutes)	Approximate Time Until Next Eruption (in minutes)
1	46
2	58
3	70
4	82

1. Describe any patterns you see in the table.

2. Why might it be important for park rangers to be able to predict the timing of Old Faithful's eruptions?

3. If an eruption lasts 8 minutes, about how long must park visitors wait to see the next eruption? Explain your reasoning using the patterns you identified in the table.

CONNECT TO HISTORY

Yellowstone National Park was the first National Park. The park was established by Congress on March 1, 1872. President Woodrow Wilson signed the act creating the National Park Service on August 25, 1916.

DISCUSSION GROUP TIPS

Work with your peers to set rules for:

- discussions and decision-making
- clear goals and deadlines
- individual roles as needed

My Notes

4. Graph the data from the table on the grid below.

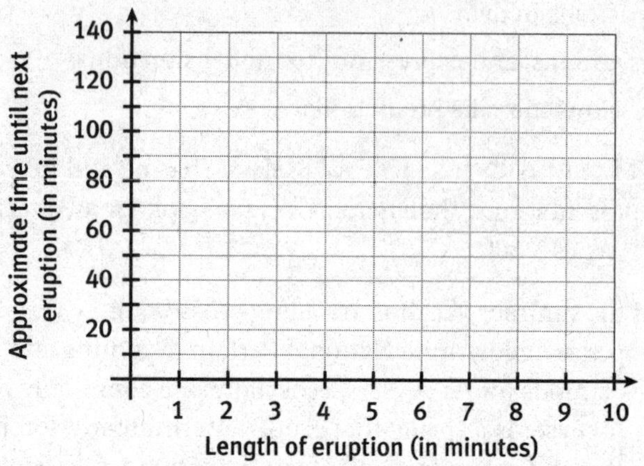

5. **Reason quantitatively.** Mizing and his family arrived at Old Faithful to find a sign indicating they had just missed an eruption and that it would be approximately 2 hours before the next one. How long was the eruption they missed? Explain how you determined your answer.

Patterns can be written as ***sequences.***

6. Using the table or graph above, write the approximate times until the next Old Faithful eruption as a sequence.

7. How would you describe this sequence of numbers?

MATH TERMS

A **sequence** is a list of numbers, and each number is called a *term* of the sequence. For example: 2, 4, 6, 8, … and 2, 5, 10, 17, … are sequences.

In the table below, 5 and 8 are ***consecutive*** terms. Some sequences have a ***common difference*** between consecutive terms. The common difference between the terms in the table below is 3.

Sequence: 5, 8, 11, 14...

Term number	Term
1	5
2	8
3	11
4	14

8. Identify two consecutive terms in the sequence of next eruption times that you created in Item 6.

9. The sequence of next eruption times has a common difference. Identify the common difference.

10. Each term in the sequence above can be written using the first term and repeated addition of the common difference. For example, the first term is 5, the second term is $5 + 3$, and the third term can be expressed as $5 + 3 + 3$ or $5 + 2(3)$. Similarly, the terms in the sequence of next eruption times can also be written using repeated addition of the common difference.

a. Write the approximate waiting time for the next eruption after eruptions lasting 4 and 5 minutes using repeated addition of the common difference.

b. Model with mathematics. Let n represent the number of minutes an eruption lasts. Write an ***expression*** using the ***variable*** n that could be used to determine the waiting time until the next eruption.

c. Check the accuracy of your expression by evaluating it when $n = 2$.

d. Use your expression to determine the number of minutes a visitor to the park must wait to see another eruption of Old Faithful after a 12-minute eruption.

My Notes

ACADEMIC VOCABULARY

Consecutive refers to items that follow each other in order.

MATH TIP

A common difference is also called a **constant difference.**

MATH TERMS

An **expression** may consist of numbers, variables, and operations. A **variable** is a letter or symbol used to represent an unknown quantity.

My Notes

Check Your Understanding

SB-Mobile charges $20 for each gigabyte of data used on any of its smartphone plans.

11. Copy and complete the table showing the charges for data based on the number of gigabytes used.

Number of Gigabytes Used	Total Data Charge
1	
2	
3	
4	
5	

12. Graph the data from the table. Be sure to label your axes.

13. Write a sequence to represent the total price of a data plan.

14. The sequence you wrote in Item 13 has a common difference. Identify the common difference.

15. Let *n* represent the number of gigabytes used. Write an expression that can be used to determine the total data charge for the phone plan.

16. Use your expression to calculate the total data charge if 10 gigabytes of data are used.

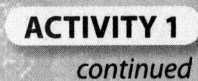

LESSON 1-1 PRACTICE

Travis owns stock in the SBO Company. After the first year of ownership the stock is worth $45 per share. Travis estimates that the value of a share will increase by $2.80 per year.

17. Copy and complete the table showing the value of the stock over the course of several years.

Year	Share Value
1	$45
2	
3	
4	
5	

18. Write a sequence to show the increase in the stock value over the course of several years.

19. **Make use of structure.** The sequence you wrote in Item 18 has a common difference. Identify the common difference.

20. Let n represent the number of years that have passed. Write an expression that can be used to determine the value of one share of SBO stock.

21. Use your expression to calculate the value of one share of stock after 20 years.

My Notes

CONNECT TO HISTORY

Mesa Verde National Park was created by President Theodore Roosevelt in 1906 as the first National Park designated to preserve "the works of man." The park protects nearly 5000 known archeological sites and 600 cliff dwellings, offering a look into the lives of the ancestral Pueblo people who lived there from 600–1300 AD.

Learning Targets:

- Use patterns to write expressions.
- Use tables, graphs, and expressions to model situations.

SUGGESTED LEARNING STRATEGIES: Look for a Pattern, Create Representations, Think-Pair-Share, Discussion Groups, Sharing and Responding

Mizing and his family also visited Mesa Verde National Park in Colorado. As Mizing investigated the artifacts on display from the ancestral Pueblo people who once called the area home, Mizing began to notice that the patterns used to decorate pottery, baskets, and textiles were geometric.

Mizing found a pattern similar to the one below particularly interesting.

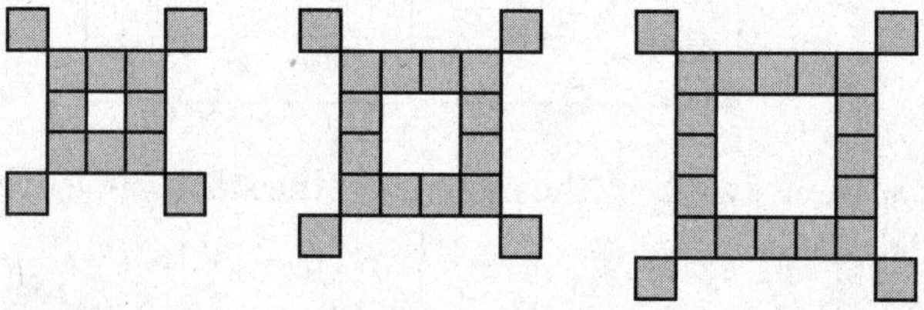

1. **Reason abstractly.** Draw the next two figures in the pattern.

2. Create a table showing the relationship between the figure number and the number of small squares in each figure.

Figure Number	Number of Squares

As part of a social studies class project on economics, Annette and Jeff are researching the benefits of membership in an online music club

1. Yearly membership with the online music club costs \$48. Members pay \$0.99 per song to download music. Nonmembers may download songs for \$1.29 each.
   a. Copy and complete the tables below to represent the yearly cost to download songs for members and nonmembers.

Members	
Number of Songs	**Total Cost**
0	
1	
2	
3	
4	

Nonmembers	
Number of Songs	**Total Cost**
0	
1	
2	
3	
4	

   b. Describe any patterns you notice in the tables.
   c. Represent the total cost of songs purchased for members and nonmembers as sequences. Tell whether each sequence has a common difference, and if so, identify it.
   d. Use the variable n to write expressions for the total cost of downloading n songs for members and for nonmembers.
   e. Use your expressions to determine the total cost of 8 songs for members and for nonmembers.

2. To determine whether becoming a member of the online music club is cost effective, Annette and Jeff must know at what point the costs for members and nonmembers are equal.
   a. Write an equation to represent the point at which the total cost of downloading songs as a club member is equal to the total cost of downloading songs as a nonmember. Then solve your equation and interpret your solution within the context of the problem.
   b. Assume you download 4 songs per week. Would it be beneficial for you to become a member of the online music club? Justify your response. (*Remember: There are 52 weeks in a year.*)

3. Members pay a \$48 membership fee each year. The literal equation $c = 48y + 0.99n$ represents the total cost c for a person who is a member of the music club for y years and who downloads n songs. Solve the equation for y.

4. The class is also working on creating family budgets. A sample monthly mortgage payment M can be represented by the equation $3250 - M = 2(M - 500) + 1500$. Jeff and Annette each solved the equation, but they disagree on the solution. Decide who is correct. For the correct solution, justify each step by writing a property or an explanation. For the incorrect solution, identify the error in the solution process.

Annette

$$3250 - M = 1500 + 2(M - 500)$$
$$3250 - M = 1500 + 2M - 1000$$
$$3250 - M = 2M + 1500 - 1000$$
$$3250 - M = 2M - 500$$
$$3250 - M + M = 2M + M - 500$$
$$3250 = 3M - 500$$
$$3250 + 500 = 3M - 500 + 500$$
$$\frac{3750}{3} = \frac{3M}{3}$$
$$1250 = M$$
$$M = \$1250$$

Jeff

$$3250 - M = 1500 + 2(M - 500)$$
$$3250 - M = 1500 + 2M - 1000$$
$$3250 - M = 1500 - 1000 + 2M$$
$$3250 - M = 500 + 2M$$
$$3250 - M + M = 500 + 2M + M$$
$$3250 = 500 + 3M$$
$$3250 - 500 = 500 - 500 + 3M$$
$$2750 = 3M$$
$$\frac{2750}{3} = \frac{3M}{3}$$
$$916.67 = M$$
$$M = \$916.67$$

Scoring Guide	**Exemplary**	**Proficient**	**Emerging**	**Incomplete**
	The solution demonstrates the following characteristics:			
Mathematics Knowledge and Thinking (Items 1a–e, 2a, 3, 4)	• Fluent use of patterns, sequences, and tables to write expressions and equations • Accuracy in solving a literal equation	• Adequate understanding of how to use patterns, sequences, and tables to write expressions and equations • Correct solution of a literal equation	• Partial understanding of how to use patterns, sequences, and tables to write expressions and equations • Partially solved literal equation	• Inaccurate or incomplete understanding of how to use patterns, sequences, and tables to write expressions and equations • No attempt to solve a literal equation
Problem Solving (Items 1e, 2a, 2b, 4)	• Appropriate and efficient strategy that results in a correct answer • Correct identification of an error in a solution process	• Strategy that may include unnecessary steps but results in a correct answer. • Correct identification of an error, but with an incorrect reason given	• Strategy that results in some incorrect answers • Correct identification of an error with no reason given	• No clear strategy when solving problems • No identification of an error in a solution process
Mathematical Modeling / Representations (Items 1a, 1d, 2a)	• Clear and accurate creation of a table to describe a real-world scenario • Effective understanding of how to write expressions and equations to represent a real-world scenario	• Little difficulty creating a table to describe a real-world scenario • Functional understanding of how to write expressions and equations to represent a real-world scenario	• Partially accurate table to describe a real-world scenario • Partial understanding of how to write expressions and equations to represent a real-world scenario	• Inaccurate or incomplete table to describe a real-world scenario • Little or no understanding of how to write expressions and equations to represent a real-world scenario
Reasoning and Communication (Items 1b, 2b, 4)	• Precise use of appropriate math terms and language to describe patterns and to justify each step in the solution of an equation • Clear and accurate conclusion drawn from an equation	• Adequate description of patterns and justification of each step in the solution of an equation • Reasonable conclusion drawn from an equation	• Confusing description of patterns and/or justification of the steps in the solution of an equation • Partially correct conclusion drawn from an equation	• Incomplete or inaccurate description of patterns and/or justification of the steps in the solution of an equation • Incomplete or inaccurate conclusion drawn from an equation

Solving Inequalities

Physical Fitness Zones

Lesson 3-1 Inequalities and Their Solutions

My Notes

Learning Targets:

- Understand what is meant by a solution of an inequality.
- Graph solutions of inequalities on a number line.

SUGGESTED LEARNING STRATEGIES: Levels of Questions, Think-Pair-Share, Interactive Word Wall, Construct an Argument, Quickwrite

Spartan Middle School students participate in Physical Education testing each semester. In order to pass, 12- and 13-year-old girls have to do at least 7 push-ups and 4 modified pull-ups. They also have to run one mile in 12 minutes or less.

You can use an inequality to express the passing marks in each test.

	Push-Ups, *p*	**Modified Pull-Ups, *m***	**One-Mile Run, *r***
Verbal	At least 7 push-ups	At least 4 pull-ups	12 minutes or less
Inequality	$p \geq 7$	$m \geq 4$	$r \leq 12$
Graph	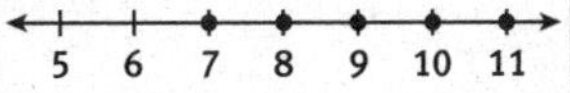	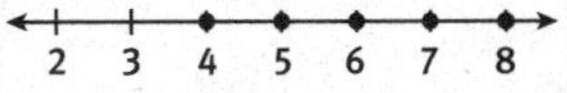	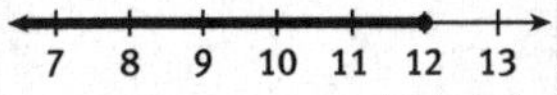

MATH TERMS

The **graph of an inequality** in one variable is all the points on a number line that make the inequality true.

1. Why do you think the graphs of push-ups and pull-ups are dotted but the graph of the mile run is a solid ray?

2. **Reason quantitatively.** Jamie ran one mile in 12 minutes 15 seconds, did 8 push-ups, and did 4 modified pull-ups. Did she pass the test? Explain.

3. Karen did 7 push-ups.
   a. Is this a passing number of push-ups? Which words in the verbal description indicate this? Explain.

   b. How is this represented in the inequality $p \geq 7$?

WRITING MATH

Other phrases that are equivalent to "at least" are "no less than" and "no fewer than." These phrases are also represented by the inequality symbol $\geq$.

The ***solution of an inequality*** in one variable is the set of numbers that make the inequality true. To verify a solution of an inequality, substitute the value into the inequality and simplify to see if the result is a true statement.

My Notes

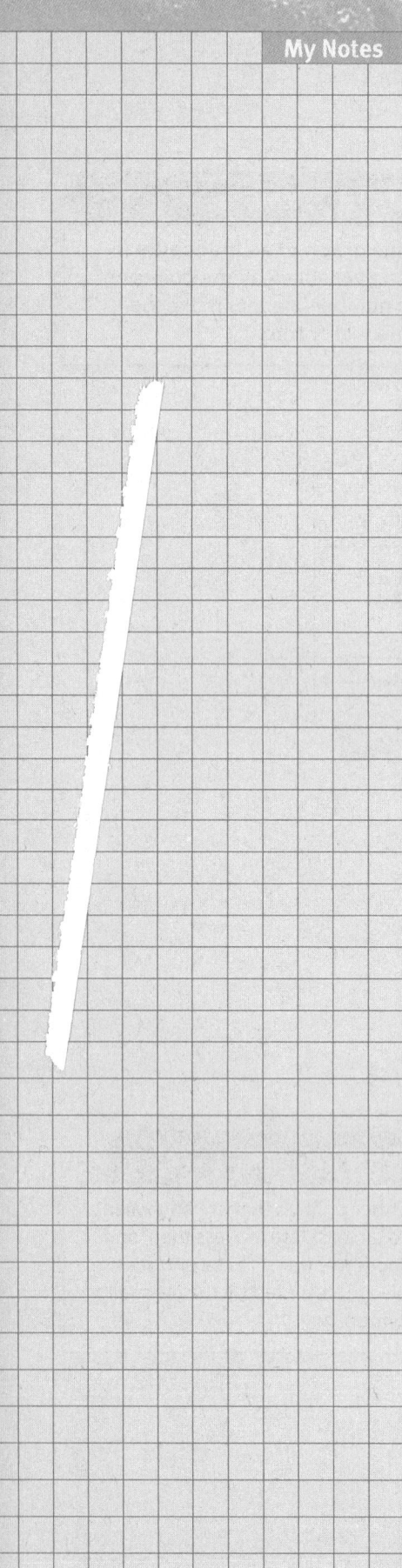

4. Use the table below to figure out which x-values are solutions to the equation and which ones are solutions to the inequality. Show your work in the rows of the table.

x-values	Solution to the equation? $2x + 3 = 5$	Solution to the inequality? $2x + 3 > 5$
–1		
0		
1		
2		
8.5		

5. How many solutions are there to the equation $2x + 3 = 5$? Explain.

6. Which numbers in the table are solutions to the inequality $2x + 3 > 5$? Are these the only solutions to the inequality? Explain.

7. Would 1 be a solution to the inequality $2x + 3 \geq 5$? Explain.

Here are the number line graphs of two different inequalities.

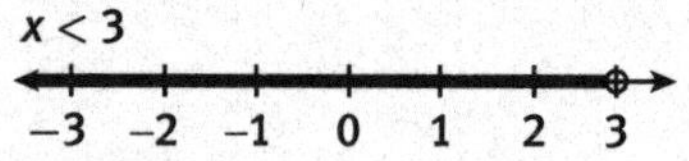

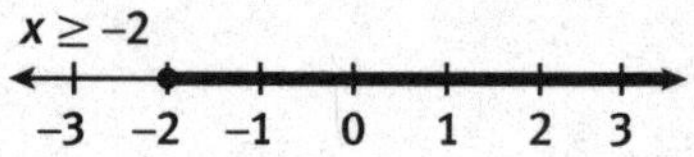

8. Use the graphic organizer to compare and contrast the two inequalities and graphs that are shown above.

Similarities	Differences

My Notes

9. Think about why the graphs are different.
   a. Why is one of the graphs showing a solid ray going to the left and the other graph showing a solid ray going to the right?

   b. Why does one graph have an open circle and the other graph a filled-in circle?

WRITING MATH

An open circle represents $<$ or $>$ inequalities, and a filled-in circle represents $\leq$ or $\geq$ inequalities.

Check Your Understanding

10. Write an inequality to represent each statement.
    a. x is less than 12.
    b. m is no greater than 35.
    c. Your height h must be at least 42 inches for you to ride a theme park ride.
    d. A child's age a can be at most 12 for the child to order from the children's menu.

11. How are the graphs of $x > -4$ and $x \geq -4$ alike and how are they different?

LESSON 3-1 PRACTICE

Graph each inequality on a number line.

12. $x < -2$

13. $x \geq 5$

14. $x < 4$

15. $x > -2$

16. $x \geq \frac{1}{2}$

17. Write a real-world statement that could be represented by the inequality $x \leq 6$.

18. **Attend to precision.** Consider the inequalities $x \leq -3$ and $x \geq -3$.
    a. Graph $x \leq -3$ and $x \geq -3$ on the same number line.
    b. Describe any overlap in the two graphs.
    c. Describe the combined graphs.

My Notes

Learning Targets:

- Write inequalities to represent real-world situations.
- Solve multi-step inequalities.

SUGGESTED LEARNING STRATEGIES: Create Representations, Guess and Check, Look for a Pattern, Think-Pair-Share, Identify a Subtask

1. **Make sense of problems.** Chloe and Charlie are taking a trip to the pet store to buy some things for their new puppy. They know that they need a bag of food that costs $7, and they also want to buy some new toys for the puppy. They find a bargain barrel containing toys that cost $2 each.
   a. Write an expression for the amount of money they will spend if the number of toys they buy is *t*.

   b. Chloe has $30 and Charlie has one-third of this amount with him. Use this information and the expression you wrote in Part (a) to write an inequality for finding the number of toys they can buy.

There are different methods for solving the inequality you wrote in the previous question. Chloe suggests that they guess and check to find the number of new toys that they could buy.

2. Use Chloe's suggestion to find the number of new puppy toys that Chloe and Charlie can buy with their combined money.

Charlie remembered that they could use algebra to solve inequalities. He imagined that the inequality symbol was an equal sign. Then he used equation-solving steps to solve the inequality.

3. Use Charlie's method to solve the inequality you wrote in Item 1b.

4. Did you get the same answer using Charlie's method as you did using Chloe's method? Explain.

My Notes

Check Your Understanding

5. How would you graph the solution to Charlie and Chloe's inequality?

6. Jaden solved an inequality as shown below. Describe and correct any errors in his work.

$$3x+5+6x>23$$
$$9x+5>23$$
$$9x>28$$
$$x>7\frac{1}{9}$$

Chloe liked the fact that Charlie's method for solving inequalities did not involve guess and check, so she asked him to show her the method for the inequality $-2x - 4 > 8$.

Charlie showed Chloe the work below to solve $-2x - 4 > 8$.

$$-2x-4>8$$
$$-2x-4+4>8+4$$
$$-2x>12$$
$$\frac{-2x}{-2}>\frac{12}{-2}$$
$$x>-6$$

When Chloe went back to check the solution by substituting a value for x back into the original inequality, she found that something was wrong.

7. Confirm or disprove Chloe's conclusion by substituting values for x into the original inequality.

Chloe tried the problem again but used a few different steps.

$$-2x-4>8$$
$$-2x+2x-4>8+2x$$
$$-4>8+2x$$
$$-4-8>8-8+2x$$
$$-12>2x$$
$$\frac{-12}{2}>\frac{2x}{2}$$
$$-6>x$$

Chloe concluded that $x<-6$.

8. Is Chloe's conclusion correct? Explain.

9. Explain what Chloe did to solve the inequality.

Charlie looked back at his work. He said that he could easily fix his work by simply switching the inequality sign.

10. **Critique the reasoning of others.** What do you think about Charlie's plan? Explain.

Although all of these methods worked, Charlie and Chloe wanted to know why they were working.

Here is an experiment to discover what went wrong with Charlie's first method. Look at what happens when you multiply or divide by a negative number.

Directions	Numbers	Inequality
Pick two different numbers.	2 and 4	$2<4$
Multiply both numbers by 3.	2(3) and 4(3)	$6<12$
Multiply both numbers by −3.	2(−3) and 4(−3)	$-6>-12$
Divide both numbers by 2.	$2\div2$ and $4\div2$	$1<2$
Divide both numbers by −2.	$2\div(-2)$ and $4\div(-2)$	$-1>-2$

11. Try this experiment again with two different numbers. Record your results in the *My Notes* section of this page. Compare your results to the rest of your class.

My Notes

12. Express regularity in repeated reasoning. What happens each time you multiply each side of an inequality by a negative number? What happens each time you divide each side of an inequality by a negative number?

13. How does this affect how you solve an inequality?

Example A

Solve and graph: $-5x + 8 \leq -2x + 23$

Step 1: Subtract 8 from both sides.

$$-5x + 8 - 8 \leq -2x + 23 - 8$$
$$-5x \leq -2x + 15$$

Step 2: Add $2x$ to both sides.

$$-5x + 2x \leq -2x + 2x + 15$$
$$-3x \leq 15$$

Step 3: Divide both sides by -3.
Remember to reverse the inequality sign.

$$\frac{-3x}{-3} \geq \frac{15}{-3}$$
$$x \geq -5$$

Solution: $x \geq -5$

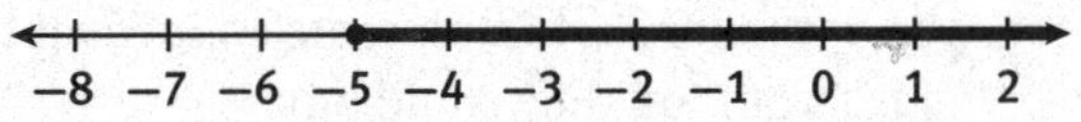

Try These A

Solve and graph each inequality.

a. $3 - 4x \leq 11$

b. $6 - 3(x + 2) > 15$

c. $2(x + 5) < 8(x - 3)$

MATH TIP

Substitute some sample answers back into the original inequality to check your work.

My Notes

Check Your Understanding

14. Write two different inequalities that have the solution graphed on the number line below.

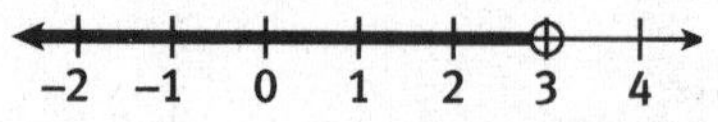

15. Explain why you reverse the inequality sign when you multiply or divide both sides of an inequality by a negative number.

MATH TIP

Remember to substitute some sample answers back into the original inequality to check your work.

LESSON 3-2 PRACTICE

Solve and graph each inequality.

16. $5 < 3x + 8$

17. $5 < -3x + 8$

18. $3x - 8 + 4x > 6$

19. $-5x + 2 \geq -8x$

20. $4 - 2(x + 1) < 18$

21. $-6x - 3 \leq -4x + 1$

22. $3(x + 7) \geq 2(2x + 8)$

23. $-2x - 3 + 8 < -3(3x + 5)$

24. **Model with mathematics.** Riley and Rhoda plan to buy several bags of dog food and a dog collar. Each bag of dog food costs \$7, and the dog collar costs \$5.

a. Use the information above to write an expression for the amount of money they will spend if they buy b bags of food.

b. Riley and Rhoda have \$30. Use this information and the expression you wrote for Part (a) to write an inequality for finding the number of bags of food they could buy.

c. Solve the inequality and graph the solutions. Check your answer in the original situation.

25. In Example A, $-5x + 8 \leq -2x + 23$ was solved by dividing each side of the inequality by -3 in the last step. Is there another way to solve this inequality so that you can avoid dividing by a negative number? Explain.

My Notes

Learning Targets:

- Graph compound inequalities.
- Solve compound inequalities.

SUGGESTED LEARNING STRATEGIES: Vocabulary Organizer, Look for a Pattern, Create Representations, Think-Pair-Share, Note Taking

Compound inequalities are two inequalities joined by the word *and* or by the word *or*. Inequalities joined by the word *and* are called ***conjunctions***. Inequalities joined by the word *or* are ***disjunctions***. You can represent compound inequalities using words, symbols, or graphs.

1. Complete the table. The first two rows have been done for you.

Verbal Description	Some Possible Solutions	Inequality	Graph
all numbers from 3 to 8, inclusive	3.5, 4, $4\frac{1}{3}$, 5, 6, 7.9, 8	$x \geq 3$ and $x \leq 8$	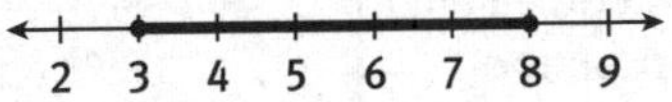
all numbers less than 5 or greater than 10	−2, 0, 3, 4, 4.8, $10\frac{3}{4}$, 11	$x < 5$ or $x > 10$	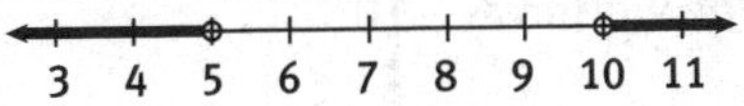
all numbers greater than −1 and less than or equal to 4			
all numbers less than or equal to 3 or greater than 6			

2. Use the graphic organizer below to compare and contrast the graphs for conjunctions and disjunctions.

Similarities	Differences

My Notes

Example A

Spartan Middle School distributes this chart to students each year to show what students must be able to do to pass the fitness test.

Age	Mile Run (min:sec)		Push-Ups		Modified Pull-Ups	
	Boys	Girls	Boys	Girls	Boys	Girls
12	8:00–10:30	9:00–12:00	10–20	7–15	7–20	4–13
13	7:30–10:00	9:00–12:00	12–25	7–15	8–22	4–13

Write and graph a compound inequality that describes the push-up range for 12-year-old boys.

Step 1: Choose a variable.
Let p represent the number of push-ups for 12-year-old boys.

Step 2: Determine the range and write an inequality.
The range is whole numbers p such that $p \geq 10$ and $p \leq 20$.

Solution: The compound inequality is $10 \leq p \leq 20$.

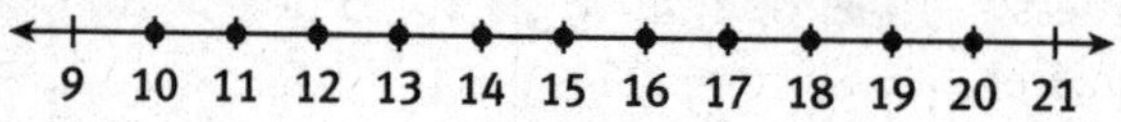

WRITING MATH

The inequality $p \geq 10$ is equivalent to the inequality $10 \leq p$. So the compound inequality "$p \geq 10$ and $p \leq 20$" can also be written as "$10 \leq p \leq 20$."

CONNECT TO AP

In upper-level mathematics classes, inequalities are expressed in interval notation. The interval notation for $x > 3$ is $(3, \infty)$.

Try These A

Write and graph a compound inequality for each range or score.

a. the push-up range for 13-year-old boys

b. the pull-up range for 13-year-old girls

c. the mile run range for 12-year-old girls

d. the mile run range for 13-year-old boys

e. a score outside the healthy fitness zone for girls' push-ups

3. Attend to precision. Why are individual points used in the graphs for Example A and some of the graphs in Try These A?

The solution of the conjunction will be *the solutions that are common to both parts.*

Example B

Solve and graph the conjunction: $3 < 3x - 6 < 8$

Step 1: Break the compound inequality into two parts.
$3 < 3x - 6$ and $3x - 6 < 8$

Step 2: Solve and graph $3 < 3x - 6$.

$$3 < 3x - 6$$
$$3 + 6 < 3x - 6 + 6$$
$$9 < 3x$$
$$3 < x \text{ or } x > 3$$

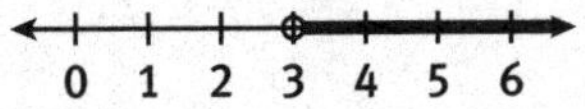

Step 3: Solve and graph $3x - 6 < 8$.

$$3x - 6 < 8$$
$$3x - 6 + 6 < 8 + 6$$
$$3x < 14$$
$$x < 4\frac{2}{3}$$

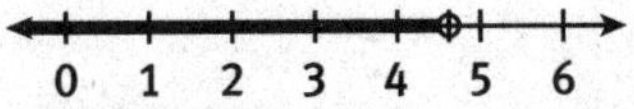

Step 4: Determine what is common to the solutions of each part. In the inequalities and graphs in Steps 2 and 3, the points between 3 and $4\frac{2}{3}$ are in common.

Solution: $3 < x < 4\frac{2}{3}$

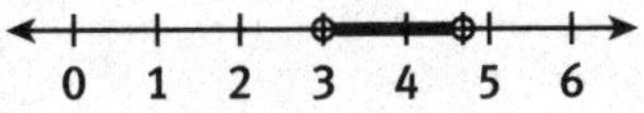

Try These B

Solve and graph each conjunction.

a. $-1 < 3x + 5 < 6$

b. $2 < \frac{x}{3} - 5 < 6$

c. $3 < 2(x + 2) - 7 \leq 13$

d. $-2 < 3(x + 6) < 18$

My Notes

MATH TIP

Remember to substitute some sample answers back into the original inequality to check your work.

My Notes

The solution of a disjunction will be *all the solutions from both its parts.*

Example C

Solve and graph the compound inequality: $2x - 3 < 7$ or $4x - 4 \geq 20$.

Step 1: Solve and graph $2x - 3 < 7$.

$2x - 3 + 3 < 7 + 3$

$2x < 10$

$x < 5$

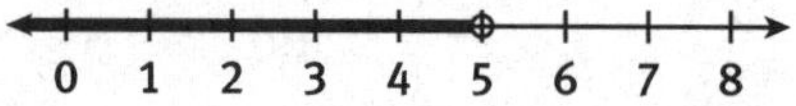

Step 2: Solve and graph $4x - 4 \geq 20$.

$4x - 4 \geq 20$

$4x - 4 + 4 \geq 20 + 4$

$4x \geq 24$

$x \geq 6$

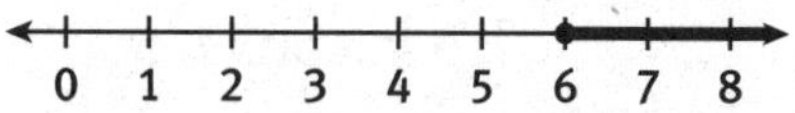

Step 3: Combine the solutions.

Solution: $x < 5$ or $x \geq 6$

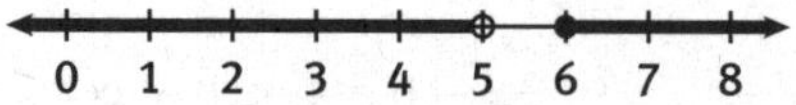

MATH TIP

Rememb to substitute some sample swers back into the origin nequality to check your wor

Try These C

Solve and graph each compound inequality.

a. $5x + 1 > 11$ or $x - 1 < -4$

b. $-5x > 20$ or $x - 2 \geq -7$

Check Your Understanding

4. The solutions of a conjunction are graphed below. What is the inequality?

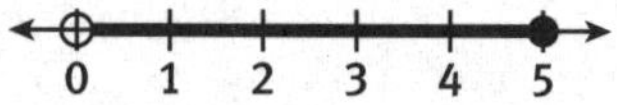

5. Describe the difference in the graph of the conjunction "$x > 2$ and $x < 10$" and the graph of the disjunction "$x > 2$ or $x > 10$."

LESSON 3-3 PRACTICE

6. Reason quantitatively. A Category 2 hurricane has wind speeds of at least 96 miles per hour and at most 110 miles per hour. Write the wind speed of a Category 2 hurricane as two inequalities joined by the word *or* or *and.*

Solve and graph each compound inequality on a number line.

7. $-2x + 3 < 8$ and $3(x + 4) - 11 < 10$

8. $-3x + 5 > -1$ or $2(x + 4) > 14$

9. Write a real-world statement that could be represented by the compound inequality $\$7.50 < p \leq \$18.50$.

ACTIVITY 3 PRACTICE

Write your answers on notebook paper. Show your work.

Lesson 3-1

1. Describe the similarities and differences in the solutions of $2x - 7 = 15$ and $2x - 7 \leq 15$.
2. For the equation $-3x + 2 = 8$ and the inequality $-3x + 2 > 8$, which x-values indicated below are solutions of the equation and which are solutions of the inequality?
   A. -3
   B. -2
   C. -1
   D. 0
3. Describe the graph of $x > -2$.
4. Describe the graph of $x \leq -2$.
5. Graph $x < 1\frac{1}{2}$ on a number line.
6. Graph $x \geq -3$ on a number line.

Lesson 3-2

7. Mayumi plans to buy pencils and a notebook at the school store. A pencil costs \$0.15, and a notebook costs \$1.59. Mayumi has \$5.00. Which inequality could she use to find the number of pencils she can buy?
   A. $5.00 < 0.15x + 1.59$
   B. $5.00 \geq 0.15x + 1.59$
   C. $5.00 < 0.15x - 1.59$
   D. $5.00 \geq 0.15x - 1.59$
8. Which values of a and b disprove the statement below?

   If $a > b$, then $a^2 > b^2$.

   A. $a = 2, b = 0$
   B. $a = 4, b = 1$
   C. $a = 3, b = -5$
   D. $a = \frac{3}{4}, b = \frac{1}{2}$

Solve the inequality and graph the solutions on a number line. Check your answers.

9. $5x - 4 > -4$
10. $8 > 6 + \frac{2}{5}x$
11. $5 - 3x \leq 8$
12. $3x - 4 \geq 6x + 11$
13. $x - 2 \geq -8x + 16$
14. $3x - 7 < 2(2x - 1)$
15. $2\left(\frac{1}{2}x - 4\right) < -(x - 5)$
16. $\frac{2x - 11}{3} \geq -x - 2$

Write an inequality that requires more than one step to solve and that has the given solution.

17. $x < -3$
18. $x > 1$
19. Roy is attending his cousin's graduation ceremony in another town. Roy has already driven 30 miles and the ceremony starts in 2 hours.
   a. Let r represent Roy's driving speed in miles per hour. Write an expression to show the total distance Roy will have driven in 2 hours.
   b. The graduation is 150 miles from Roy's home. Use this information and your expression from Part (a) to write an inequality showing the possible speeds Roy could drive to make it to the ceremony on time.
   c. Solve your inequality and graph the solutions. What does your solution mean in the context of the problem?

Lesson 3-3

Use the table for Items 20 and 21.

Average High and Low Monthly Temperatures		
	January	August
Austin, TX	41–62°F	75–97°F
Columbus, OH	20–36°F	63–83°F

20. Write a compound inequality for the range of temperatures for Austin, TX, in August.

21. Write a compound inequality for the range of temperatures for Columbus, OH, in January.

22. The sum of the lengths of any two sides of a triangle must be greater than the length of the third side.

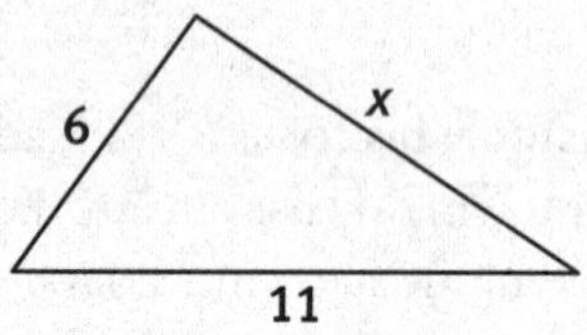

a. For the triangle shown, Anna said that any value of x greater than 5 is possible. Explain Anna's error.

b. Write a compound inequality that represents all possible values of x.

23. Find a value for n so that the compound inequality $-n < x < n$ has no solutions.

Solve each compound inequality and graph the solutions on a number line. Check your answers.

24. $-2 < 3x + 4 < 31$

25. $1 < 5x - 9 < 6$

26. $-2x + 7 > 1$ and $4x + 3 \geq -13$

27. $0 \leq \frac{6-x}{9}$ and $-2x \leq -10$

28. $5x - 2 < -7$ or $2x + 1 > 5$

29. $7x - 2 < -30$ or $4x + 5 > 13$

30. $\frac{1}{2} < \frac{2x-9}{2} \leq 3$

31. $3 \leq 2(x + 4) - 3 < 15$

32. $-2(x + 2) - 7 > 9$ or $3(x + 3) > -6$

MATHEMATICAL PRACTICES
Construct Viable Arguments and Critique the Reasoning of Others

33. The inequality $x + 5 < x + 4$ has no solutions. Explain why.

Absolute Value Equations and Inequalities

ACTIVITY 4

Student Distances

Lesson 4-1 Absolute Value Equations

My Notes

Learning Targets:

- Understand what is meant by a solution of an absolute value equation.
- Solve absolute value equations.

SUGGESTED LEARNING STRATEGIES: Paraphrasing, Create Representations, Think-Pair-Share, Note Taking, Identify a Subtask

Ms. Patel is preparing the school marching band for the homecoming show. She has the first row of band members stand in positions along a number line on the floor of the band room. The students' positions match the points on a number line as shown.

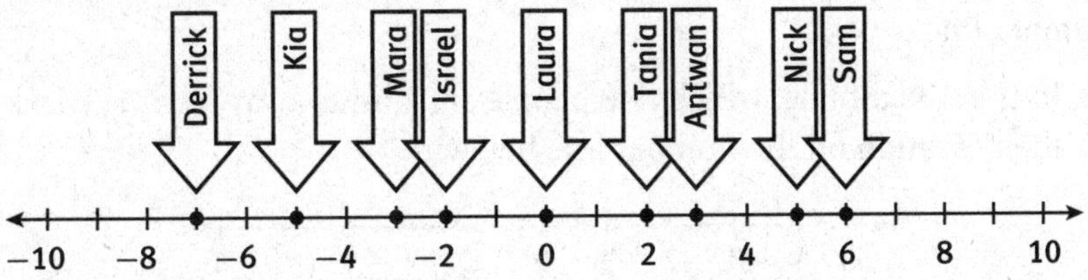

1. Use the number line to write each student's distance from 0 next to their name. For example, Tania is 2 units away from 0. Israel's distance from 0 is also 2 units even though he is at −2.

Derrick	Laura
Kia	Israel
Mara	Antwan
Tania	Nick
Sam	

The ***absolute value*** of a number is the distance from 0 to the number on a number line. Using ***absolute value notation,*** Mara's distance is $|-3|$ and Antwan's distance is $|3|$. Since Mara and Antwan are each 3 units from 0, $|-3| = 3$ and $|3| = 3$.

READING MATH

Read $|-3|$ as "the absolute value of negative three."

2. Attend to precision. Write each person's distance from 0 using absolute value notation.

Absolute value equations can represent distances on a number line.

MATH TERMS

An **absolute value equation** is an equation involving the absolute value of a variable expression.

3. The locations of the two students who are 5 units away from 0 are the solutions of the absolute value equation $|x| = 5$. Which two students represent the solutions to the equation $|x| = 5$?

My Notes

4. You can create a graph on a number line to represent the solutions of an absolute value equation. Graph the solutions of the equation $|x| = 5$ on the number line below. Then use the graph to help you explain why it makes sense that the equation $|x| = 5$ has two solutions.

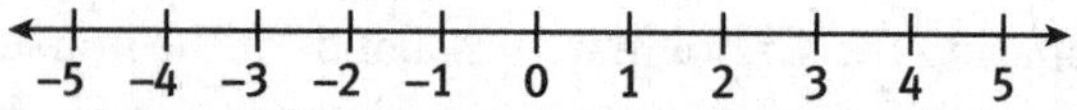

Absolute value equations can also represent distances between two points on a number line.

5. In the student line, which two people are 4 units away from 1? Mark their location on the number line below.

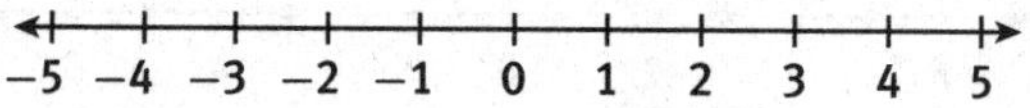

The equation $|x| = 4$ represents the numbers located 4 units away from 0. So the equation $|x| = 4$ can also be written as $|x - 0| = 4$, which shows the distance (4) away from the point 0. In Item 5, you were looking for the numbers located 4 units away from 1. So you can write the absolute value equation $|x - 1| = 4$ to represent that situation.

6. What are two possible values for $x - 1$ given that $|x - 1| = 4$? Explain.

7. Use the two values you found in Item 6 to write two equations showing what $x - 1$ could equal.

8. Solve each of the two equations that you wrote in Item 7.

The solutions in Item 8 represent the two points on the number line that are 4 units from 1.

9. How do the solutions relate to your answer for Item 5?

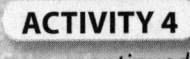

My Notes

10. Draw a number line to show the answer to each question. Then write an absolute value equation to represent the points described.

a. Which two points are 2 units away from 0?

b. Which two points are 5 units away from -2?

c. Which two points are 3 units away from 4?

11. Solve these absolute value equations.

a. $|x| = 10$

b. $|x| = -3$

c. $|x + 2| = 7$

d. $|2x - 1| = 5$

MATH TIP

If there are no values of x that make an equation true, the equation has **no solution**.

In the equation $|x - 3| = 7$, the 7 indicates that the distance between x and 3 is 7 units. There are two points that are 7 units from 3. These can be found by solving the equations $x - 3 = 7$ and $x - 3 = -7$. Rewriting an absolute value equation as two equations allows you to solve the absolute value equation using algebra.

Example A

Solve the equation $|x - 3| = 7$.

Step 1: Rewrite the equation as two equations that do not have absolute value symbols.

$x - 3 = 7$ $\qquad$ $x - 3 = -7$

Step 2: Solve $x - 3 = 7$.

$$x - 3 + 3 = 7 + 3$$
$$x = 10$$

Step 3: Solve $x - 3 = -7$.

$$x - 3 + 3 = -7 + 3$$
$$x = -4$$

Solution: 10, -4

My Notes

Try These A

Solve each absolute value equation.

a. $|x - 5| = 1$

b. $|x + 5| = 2$

c. $|2x + 3| = 11$

d. $|3x - 4| = 8$

You may need to first isolate the absolute value expression to solve an absolute value equation.

Example B

Solve the equation $6|x + 2| = 18$.

Step 1: Divide both sides of the equation by 6 so that the absolute value expression is alone on one side of the equation.

$$\frac{6|x+2|}{6} = \frac{18}{6}$$

$$|x + 2| = 3$$

Step 2: Rewrite the equation as two equations that do not have absolute value symbols.

$x + 2 = 3$ and $x + 2 = -3$

Step 3: Solve $x + 2 = 3$.

$x + 2 - 2 = 3 - 2$

$x = 1$

Step 4: Solve $x + 2 = -3$.

$x + 2 - 2 = -3 - 2$

$x = -5$

Solution: 1, −5

MATH TIP

Check your answers by substituting the solutions in the original equation.

$6|1 + 2| = 18$
$6|3| = 18$
$6(3) = 18$
$18 = 18$

$6|-5 + 2| = 18$
$6|-3| = 18$
$6(3) = 18$
$18 = 18$

Try These B

Solve each absolute value equation.

a. $3|x - 1| = 12$

b. $|x| - 14 = 6$

c. $|x + 4| + 5 = 8$

d. $3|x + 6| - 7 = 20$

My Notes

Check Your Understanding

12. Tell whether each statement is true or false. Explain your answers.
   a. For $x > 0$, $|x| = x$.
   b. For $x < 0$, $|x| = -x$.

13. Kate says that the opposite of $|-6|$ is 6. Is she correct? Explain.

LESSON 4-1 PRACTICE

Draw a number line to show the answer for each question. Then write an absolute value equation that has the numbers described as solutions.

14. Which two numbers are 3 units away from 0?

15. Which two numbers are 4 units away from -1?

16. Which two numbers are 3 units away from 3?

Solve each equation. Check your answers.

17. $|x - 5| = 8$

18. $|-2(x + 2)| = 1$

19. $|-(x - 5)| = 8.5$

20. $|3(x + 1)| = 15$

21. $2|x - 7| = -4$

22. $-2|x - 7| = -4$

23. **Make sense of problems.** Use the equations $|x - 3| = 7$ and $|x| - 3 = 7$ to answer the following questions.
   a. Describe the similarities and differences between the equations.
   b. Which of the following values are solutions of each equation: $-10, -4, 10$?
   c. Are the equations $|x - 3| = 7$ and $|x| - 3 = 7$ equivalent? Explain.
   d. Are the equations $|x| - 3 = 7$ and $|x| = 10$ equivalent? Explain.

My Notes

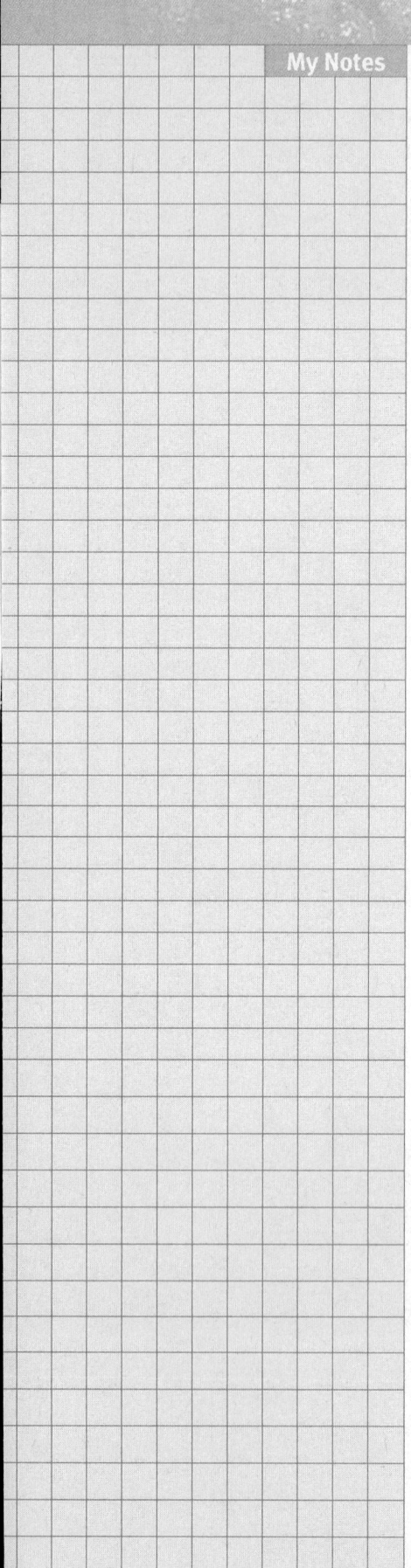

Learning Targets:

- Solve absolute value inequalities.
- Graph solutions of absolute value inequalities.

SUGGESTED LEARNING STRATEGIES: Role Play, Visualization, Create Representations, Guess and Check, Think-Pair-Share

Here is the marching band line-up once again.

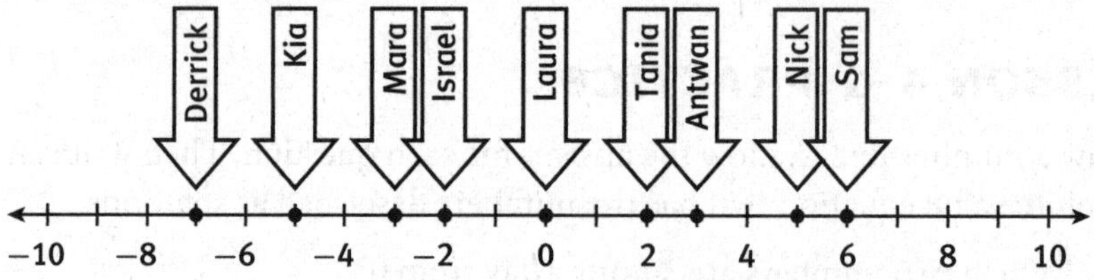

1. Which people in the line up are 3 or fewer units from 0?

2. Show the portion of the number line that includes numbers that are 3 or fewer units from 0.

The graph you created in Item 2 can be represented with an ***absolute value inequality***. The inequality $|x| \leq 3$ represents the numbers on a number line that are 3 or fewer units from 0.

3. Circle the numbers below that are solutions of $|x| \leq 3$. Explain why you chose those numbers.

$-1 \quad 3 \quad 0.5 \quad 4 \quad -3.1$

4. Reason abstractly. How many solutions does the inequality $|x| \leq 3$ have?

5. If you were to write a compound inequality for the graph of $|x| \leq 3$ that you sketched in Item 2, would it be a conjunction ("and" inequality) or a disjunction ("or" inequality)? Explain.

6. Write a compound inequality to represent the solutions to $|x| \leq 3$.

7. What numbers are more than 4 units away from 3 on a number line? Show the answer to this question on the number line.

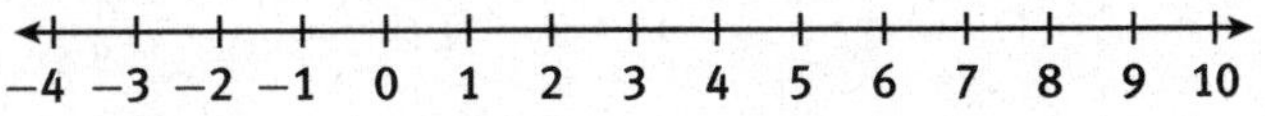

The absolute value inequality $|x - 3| > 4$ represents the situation in Item 7. A "greater than" symbol indicates that the distances are greater than 4.

8. Circle the numbers below that are solutions to the inequality $|x - 3| > 4$. Explain why you chose those numbers.

7.1   7   0   −2   6.9   100

9. **Construct viable arguments.** If you were to write a compound inequality for the graph of $|x - 3| > 4$ that you sketched in Item 7, would it be a conjunction or disjunction? Explain.

10. To solve $|x - 3| > 4$ for x, you need to write the absolute value inequality as a compound inequality.

a. Based on the graph from Item 7, the expression $x - 3$ is either greater than 4 or less than −4. Write this statement as a compound inequality.

b. Solve each of the inequalities you wrote in Part (a). Graph the solution.

My Notes

11. Make a graph that represents the answer to each question. Then write an absolute value inequality that has the solutions that are graphed. Finally, write each absolute value inequality as a compound inequality.

a. What numbers are less than 2 units from 0?

b. What numbers are 4 or more units away from 0?

c. What numbers are 4 or fewer units away from −2?

12. Describe the absolute value inequalities $|x| < 3$ and $|x| > 3$ as conjunctions or disjunctions and justify your choice in each case.

You can solve absolute value inequalities algebraically.

Example A

Solve the inequality $|2x| + 3 > 9$. Graph the solutions.

Step 1: Subtract 3 from both sides of the inequality so that the absolute value expression is alone on one side.

$|2x| + 3 - 3 > 9 - 3$

$|2x| > 6$

Step 2: Rewrite the equation as a compound inequality. Determine if the relationship is "and" or "or."

$2x > 6$ or $2x < -6$

Step 3: Solve the inequalities.

$2x > 6$ or $2x < -6$

$x > 3$ or $x < -3$

Solution: $x > 3$ or $x < -3$

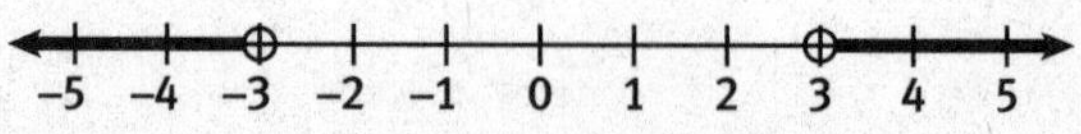

My Notes

Try These A

Solve each absolute value inequality and graph the solutions.

a. $|2x - 7| > 3$

b. $|3x + 8| < 5$

c. $|4x| - 3 \geq 5$

d. $2|x| + 7 \leq 11$

e. $-3|x - 9| \geq -21$

f. $-5|2x - 8| < -20$

MATH TIP

Remember that when multiplying or dividing each side of an inequality by a negative number, you must reverse the inequality symbol.

My Notes

Check Your Understanding

13. Describe the similarities and differences between solving an absolute value equation and solving an absolute value inequality.

LESSON 4-2 PRACTICE

14. Graph the following and then write an absolute value inequality that represents each question.

a. What numbers are 5 units or more away from -2 on a number line?

b. What numbers are 5 units or fewer from -2 on a number line?

Solve each inequality and graph the solutions.

15. $|x - 4| \leq 2$

16. $|x - 5| > 3$

17. $\left|\frac{2}{3}x + 5\right| > 4$

18. $\left|\frac{3x-5}{2}\right| \leq 7$

19. $|x| - 3 \leq 4$

20. $-3|x| < -12$

21. $6 \leq |2x - 9|$

22. $-5|x + 12| > -35$

23. **Critique the reasoning of others.** Isabelle was asked to write an absolute value inequality to represent the numbers that are less than 1 unit away from 7 on a number line. Isabelle wrote $|x - 1| < 7$. Explain and correct Isabelle's error.

ACTIVITY 4 PRACTICE

Write your answers on notebook paper.
Show your work.

Lesson 4-1

1. Use a number line to show the numbers that are 3 units from -1 on a number line. Then write an absolute value equation to describe the graph.

2. Explain why $|x| = -5$ does not have a solution.

3. Which graph shows the solutions of $|x - 11| = 7$?

A. 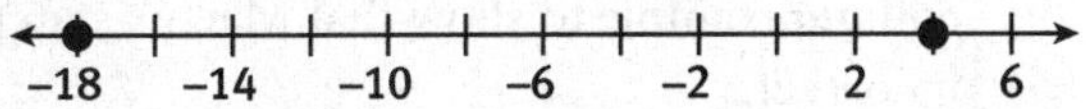

B. 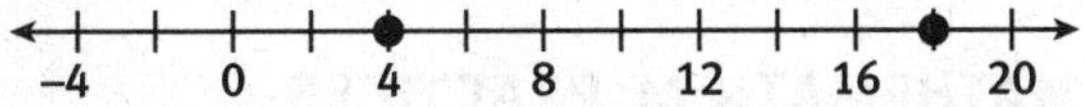

C. 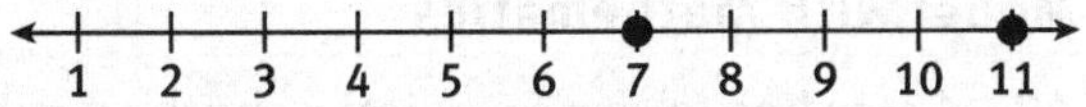

D. 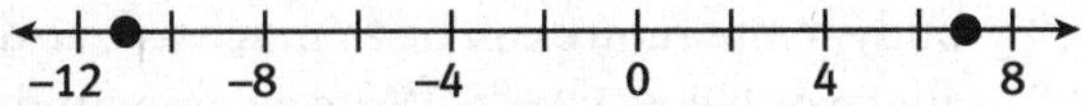

4. Suppose that x is negative and $|x| = n$. What conclusion can you draw?
A. x and n are opposites.
B. x and n are equal.
C. x is greater than n.
D. None of the above

For Items 5–17, solve each absolute value equation. Check your answers.

5. $|x| = 7$

6. $|x - 2| = -2$

7. $|x - (-2)| = 5$

8. $|3(x - 1)| = 15$

9. $\left|\frac{2}{5}x\right| = 4$

10. $|2(x - 3)| = 10$

11. $|3(x - 2)| = x$

12. $|4(x + 2)| + 9 = 15$

13. $|-2x + 3| = 7$

14. $|-3(x - 7)| = 21$

15. $-5|x - 2| = -20$

16. $-3|x + 5| + 7 = 4$

17. $\left|\frac{2x - 5}{7}\right| = 3$

Lesson 4-2

For Items 18–21, graph the solutions. Then write an absolute value inequality that represents each question.

18. What numbers are more than 3 units from -1 on a number line?

19. What numbers are less than 3 units from -1 on a number line?

20. What numbers are 5 or fewer units away from 3?

21. What numbers are 3 or more units away from 5?

For Items 22–25, graph the solutions of each absolute value inequality and write compound inequalities for the solutions.

22. $|x| > 3$

23. $|x| < 3$

24. $|x - 4| \geq 7$

25. $|x - 4| \leq 7$

26. Which describes the solutions of $|6x - 3| > 21$?
A. all numbers greater than 4
B. all numbers greater than -3 and greater than 4
C. all numbers between -3 and 4
D. all numbers less than -3 and greater than 4

27. Without solving, match each absolute value equation or inequality with its number of solutions. Justify your answers.

$\lvert x - 7\rvert < -2$	one solution
$\lvert x\rvert = 0$	no solutions
$\lvert x + 1\rvert > -5$	infinitely many solutions

28. The solutions to which absolute value inequality are shown in the graph below?

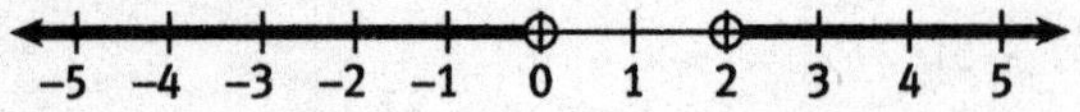

A. $|x + 1| < 1$
B. $|x + 1| > 1$
C. $|x - 1| < 1$
D. $|x - 1| > 1$

29. Create a graphic organizer that compares and contrasts the following equation and inequalities.

$|x - 3| = 6$

$|x - 3| \leq 6$

$|x| - 3 > 6$

For Items 30–40, solve each absolute value inequality and graph the solutions.

30. $|x - 2| > 3$

31. $|x - 5| < 2$

32. $|2x + 7| \geq 5$

33. $|3x + 2| \leq 11$

34. $\left|\frac{5x-3}{2}\right| < 6$

35. $|4(x - 1)| > 16$

36. $|x - 7| + 3 < 2$

37. $|x + 5| - 2 < 3$

38. $|2(x + 1)| - 7 \leq 1$

39. $\left|\frac{3x-1}{4}\right| \geq 5$

40. $-2|3x - 4| \leq -6$

41. Marty said that all absolute value inequalities that contain the symbol $<$ are conjunctions. Give a counterexample to show that Marty's statement is incorrect.

MATHEMATICAL PRACTICES
Model with Mathematics

42. According to some medical websites, normal body temperature can be as much as one degree above or below 98.6°F. Write a compound inequality that shows the range of normal body temperatures, *t*. Then write an absolute value inequality that shows the same information. Explain how you wrote the absolute value inequality, and include a number line graph in your explanation.

The table below shows ranges for the daily calorie needs of 15-year-old males according to the United States Department of Agriculture. Use the table for Items 1 and 2.

Daily Calories for 15-Year-Old Males	
Sedentary	no more than 2200 calories
Moderately Active	2400 to 2800 calories
Highly Active	2800 to 3200 calories

1. Write and graph an inequality for the daily number of calories that are recommended for a sedentary 15-year-old male.

2. Use the information for a moderately active 15-year-old male.
a. Draw a graph on a number line for the daily calorie requirements.
b. Write a compound inequality for the graph.

It is recommended that teenagers between 14 and 18 years old consume at least 46 grams of protein per day. The table shows the amounts of protein present in various foods. Use the table for Items 3 and 4.

Food	Amount of Protein
Milk	8 grams per cup
Chicken	7 grams per ounce
Beans	16 grams per cup
Yogurt	11 grams per cup

3. Darwin is 15 years old. So far today, he has consumed a total of 25 grams of protein. He plans to eat chicken at dinner.
a. Let c represent the number of ounces of chicken Darwin eats at dinner. Write an inequality to show how much chicken Darwin can eat and meet the minimum requirement for daily protein.
b. Solve your inequality from Part (a) and graph the solutions. How many ounces of chicken must Darwin eat?

4. Describe at least two other foods or combinations of foods from the table that Darwin could eat at dinner and meet the minimum requirement for daily protein. Justify your answers.

5. Darwin also keeps track of his target heart rate when he is exercising. The range for the number of heart beats per minute R for someone his age is $|R - 136| \leq 20$. Determine the solutions to the inequality and graph them on a number line.

Scoring Guide	Exemplary	Proficient	Emerging	Incomplete
	The solution demonstrates these characteristics:			
Math Knowledge and Thinking (Items 1, 2a, 2b, 3a, 3b, 5)	• Clear and accurate understanding of how to solve and graph inequalities, including compound and absolute value inequalities	• Largely correct understanding of how to solve and graph inequalities, including compound and absolute value inequalities	• Partial understanding of how to solve and graph inequalities, including compound and absolute value inequalities	• Inaccurate or incomplete understanding of how to solve and graph inequalities, including compound and absolute value inequalities
Problem Solving (Items 3b, 4, 5)	• Appropriate and efficient strategy that results in a correct answer	• Strategy that may include unnecessary steps but results in a correct answer	• Strategy that results in some incorrect answers	• No clear strategy when solving problems
Mathematical Modeling / Representations (Items 1, 2a, 2b, 3a, 5)	• Clear and accurate understanding of how to write and graph inequalities, including compound and absolute value inequalities, to represent real-world data or a real-world scenario	• Largely correct understanding of how to write and graph inequalities, including compound and absolute value inequalities, to represent real-world data or a real-world scenario	• Partial understanding of how to write and graph inequalities, including compound and absolute value inequalities, to represent real-world data or a real-world scenario	• Little or no understanding of how to write and graph inequalities, including compound and absolute value inequalities, to represent real-world data or a real-world scenario
Reasoning and Communication (Items 3b, 4)	• Clear and accurate conclusions drawn from an inequality and a table of data	• Reasonable conclusions drawn from an inequality and a table of data	• Partially correct conclusions drawn from an inequality and a table of data	• Incomplete or inaccurate conclusions drawn from an inequality and a table of data

Functions

2

Unit Overview

In this unit, you will build linear models and use them to study functions, domain, and range. Linear models are the foundation for studying slope as a rate of change, intercepts, and direct variation. You will learn to write linear equations given varied information and express these equations in different forms.

Key Terms

As you study this unit, add these and other terms to your math notebook. Include in your notes your prior knowledge of each word, as well as your experiences in using the word in different mathematical examples. If needed, ask for help in pronouncing new words and add information on pronunciation to your math notebook. It is important that you learn new terms and use them correctly in your class discussions and in your problem solutions.

Academic Vocabulary

- causation

Math Terms

- relation
- function
- vertical line test
- independent variable
- dependent variable
- continuous
- discrete
- y-intercept
- relative maximum
- relative minimum
- extrema
- x-intercept
- parent function
- absolute value function
- direct variation
- constant of variation
- indirect variation
- inverse function
- one-to-one
- arithmetic sequence
- explicit formula
- recursive formula
- slope-intercept form
- point-slope form
- standard form
- scatter plot
- trend line
- correlation
- line of best fit
- linear regression
- quadratic regression
- quadratic function
- exponential regression
- exponential function

ESSENTIAL QUESTIONS

- How can you show mathematical relationships?
- Why are linear functions useful in real-world settings?

EMBEDDED ASSESSMENTS

This unit has three embedded assessments, following Activities 8, 11, and 13. They will give you an opportunity to demonstrate what you have learned.

Getting Ready

Write your answers on notebook paper.
Show your work.

1. Copy and complete the table of values.

−1	−1
2	5
5	11
8	
11	23
	29

2. List the integers that make this statement true.

$-3 \le x < 4$

3. Evaluate for $a = 3$ and $b = -2$.

a. $2a - 5$ **b.** $3b + 4a$

4. Name the point for each ordered pair.

a. $(-3, 0)$ **b.** $(-1, 3)$ **c.** $(2, -2)$

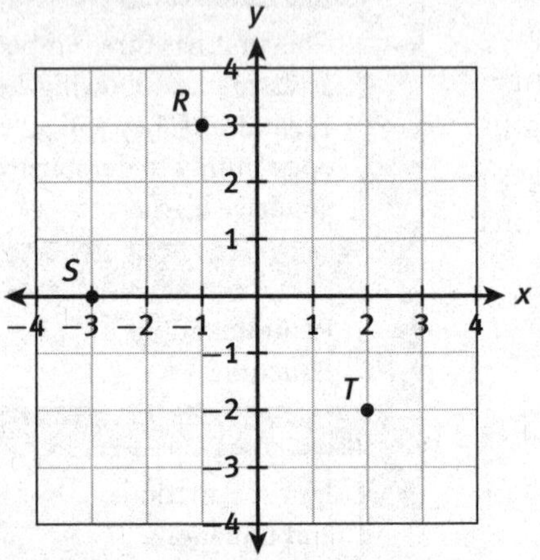

5. Explain how you would plot $(3, -4)$ on a coordinate plane.

6. Which of the following equations represents the data in the table?

x	1	3	5	7
y	2	8	14	20

A. $y = 2x - 1$ **B.** $y = 3x - 1$
C. $y = x + 1$ **D.** $y = 2x + 1$

7. If $2x + 6 = 2$, what is the value of x?

A. 4 **B.** 2 **C.** 0 **D.** −2

8. Which of the following are the coordinates of a point on this line?

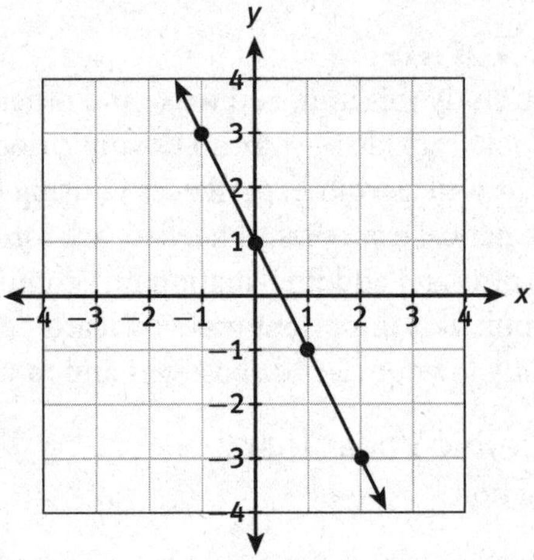

A. $(-1, 3)$ **B.** $(1, -3)$
C. $(-1, -3)$ **D.** $(1, 3)$

Functions and Function Notation

Vending Machines

Lesson 5-1 Relations and Functions

My Notes

Learning Targets:

- Represent relations and functions using tables, diagrams, and graphs.
- Identify relations that are functions.

SUGGESTED LEARNING STRATEGIES: Visualization, Create Representations, Think-Pair-Share, Interactive Word Wall, Paraphrasing

Use this machine to answer the questions below.

DVD Vending Machine

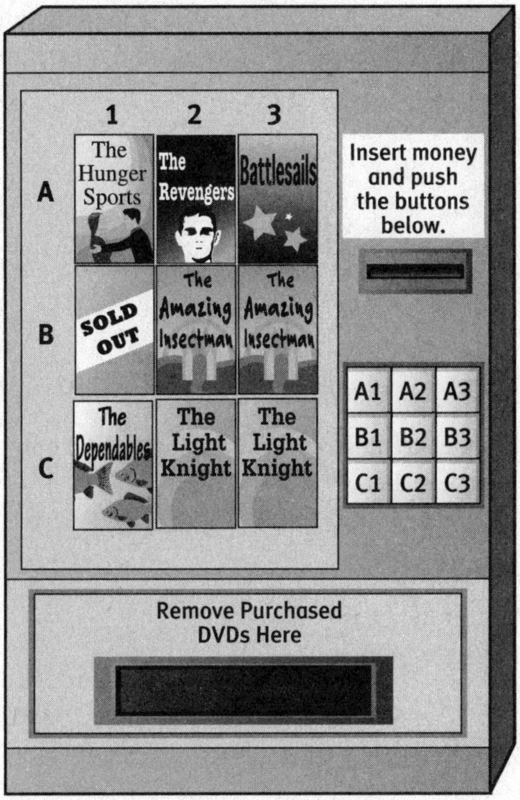

1. What DVD would you receive if you inserted your money and pressed:

a. A1?

b. C2?

c. B3?

2. Assuming the machine were filled properly, describe what would happen if you pressed the same button twice.

My Notes

Each time you press a button, an ***input***, you may receive a DVD, an ***output***.

3. In the DVD vending machine situation, does every input have an output? Explain your response.

MATH TERMS

A **mapping** is a visual-representation of a relation in which an arrow associates each input with its output.

4. Each combination of input and output can be expressed as a ***mapping*** written *input* → *output*. For example, B2 → The Amazing Insectman.

a. Write as mappings each of the possible combinations of buttons pushed and DVDs received in the vending machine.

b. Create a table to illustrate how the inputs and outputs of the vending machine are related.

MATH TERMS

An **ordered pair** shows the relationship between two elements, written in a specific order using parentheses notation and a comma separating the two values.

MATH TERMS

A **relation** is information that can be represented by a set of ordered pairs.

Mappings that relate values from one set of numbers to another set of numbers can be written as ***ordered pairs***. A ***relation*** is a set of ordered pairs.

Relations can have a variety of representations. Consider the relation {(1, 4),

My Notes

Relations can have a variety of representations. Consider the relation {(1, 4), (2, 3), (6, 5)}, shown here as a set of ordered pairs. This relation can also be represented in these ways.

Table

x	y
1	4
2	3
6	5

Mapping

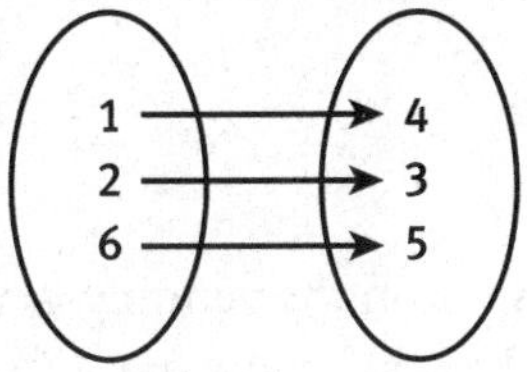

Graph

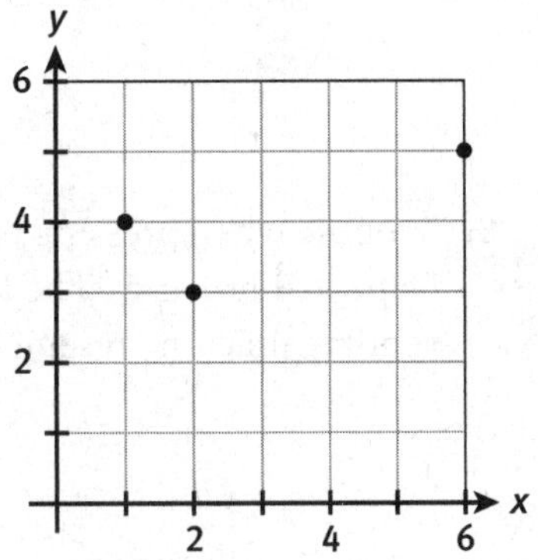

5. Write the following numerical mappings as ordered pairs.

Input		Output	Ordered Pairs
1	→	−2	(1, −2)
2	→	1	
3	→	4	
4	→	7	

Check Your Understanding

6. A vending machine at the Ocean, Road, and Air show creates souvenir coins. You select a letter and a number and the machine creates a souvenir coin with a particular vehicle imprinted on it. The graph shows the vending machine letter/number combinations for the different coins.

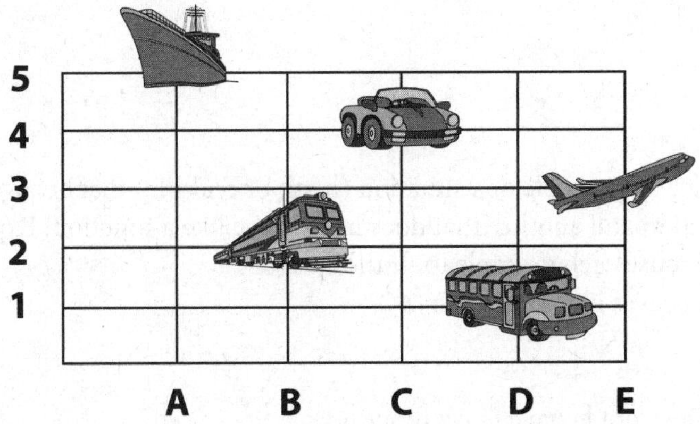

a. Make a table showing each coin's letter/number combination.
b. Write the letter/number combinations as a set of ordered pairs.
c. Write the letter/number combinations in a mapping diagram.

My Notes

A ***function*** is a relation in which each input is paired with exactly one output.

7. Compare and contrast the DVD Vending Machine with a function.

8. Suppose when pressing button C1 on the vending machine both "The Dependables" and "The Light Knight" come out. Describe how this vending machine resembles or does not resemble a function.

9. Imagine a machine where you input an age and the machine gives you the name of anyone who is that age. Compare and contrast this machine with a function. Explain by using examples and create a representation of the situation.

10. Create an example of a situation (math or real-life) that behaves like a function and another that does not behave like a function. Explain why you chose each example to fit the category.

a. Behaves like a function:

b. Does not behave like a function:

My Notes

11. Determine whether the ordered pairs and equations represent functions. Explain your answers.

a. {(5, 4), (6, 3), (7, 2)}

b. {(4, 5), (4, 3), (5, 2)}

c. {(5, 4), (6, 4), (7, 4)}

d. $y = 3x - 5$, where x represents input values and y represents output values

e. $y = -x + 4$, where x represents input values and y represents output values

12. Attend to precision. Using positive integers, write two relations as lists of ordered pairs below, one that is a function and one that is not a function.

Function:

Not a function:

Check Your Understanding

13. Does the mapping shown represent a function? Explain.

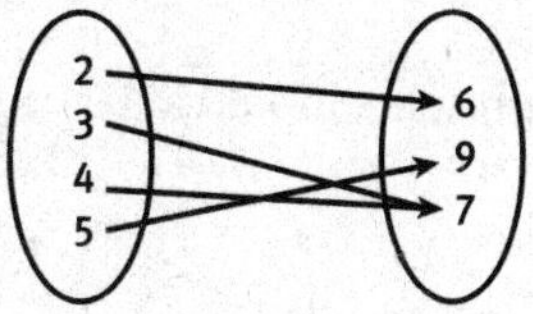

14. Does the graph shown represent a function? Explain.

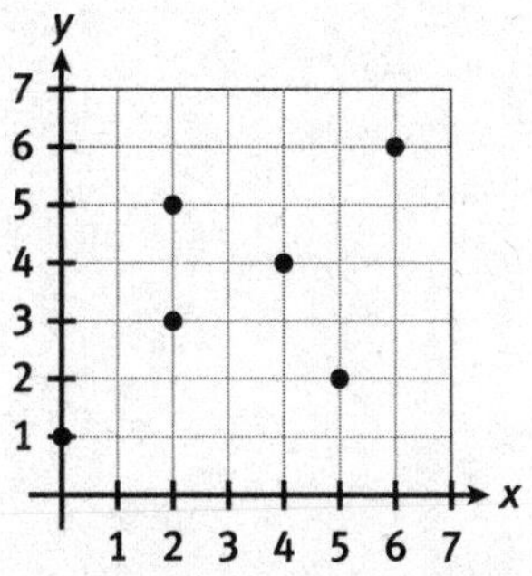

My Notes

LESSON 5-1 PRACTICE

For the Bingo card below, suppose that a combination of a column letter and a row number, such as B1, represents an input and the number at that location, such as 7, represents an output. Use this information for Items 15–17.

B	I	N	G	O
7	26	35	51	73
14	23	44	55	63
6	19	FREE	48	64
12	22	32	54	70
11	16	33	47	69

15. What output corresponds to I2?

16. What input corresponds to 54?

17. Does every input have a numerical output? Explain.

18. **Construct viable arguments.** Explain why each of the following is **not** a function.

a.

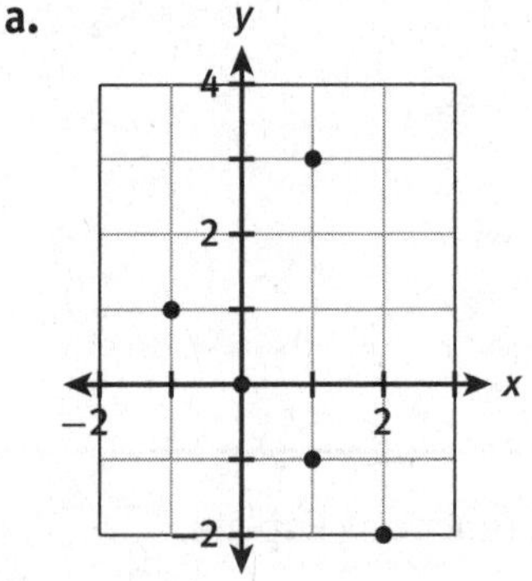

b.

x	y
12	−8
17	3
−4	9
17	−5

c. $y^2 = x$, where x represents input values and y represents output values.

My Notes

Learning Targets:

- Describe the domain and range of a function.
- Find input-output pairs for a function.

SUGGESTED LEARNING STRATEGIES: Quickwrite, Create Representations, Discussion Groups, Marking the Text, Sharing and Responding

The set of all inputs for a function is known as the ***domain*** of the function. The set of all outputs for a function is known as the ***range*** of the function.

1. Consider a vending machine where inserting 25 cents dispenses one pencil, inserting 50 cents dispenses 2 pencils, and so forth up to and including all 10 pencils in the vending machine.

a. Identify the domain in this situation.

b. Identify the range in this situation.

2. For each function below, identify the domain and range.

a.

input	output
7	6
3	−2
5	1

Domain:

Range:

b.

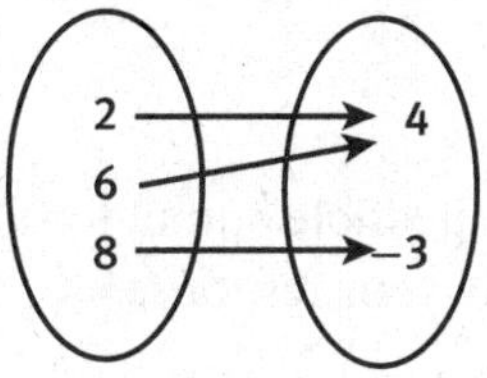

Domain:

Range:

c.

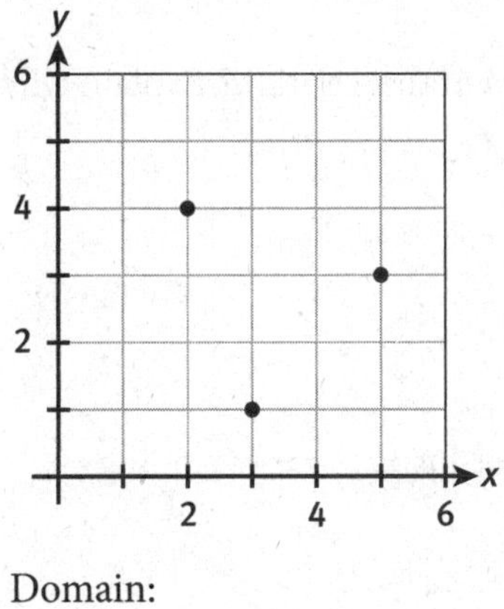

Domain:

Range:

d. {(−7, 0), (9, −3), (−6, 2.5)}

Domain:

Range:

WRITING MATH

The **domain** and **range** of a function can be written using set notation.

For example, for the function {(1, 2), (3, 4), (5, 6)}, the domain is {1, 3, 5} and the range is {2, 4, 6}.

My Notes

3. Consider a machine that exchanges quarters for dollar bills. Inserting one dollar bill returns four quarters and you may insert up to five one-dollar bills at a time.

a. Is 7 a possible input for the relation this change machine represents? Justify your response.

b. Could 3.5 be included in the domain of this relation? Explain why or why not.

c. **Reason abstractly.** What values are **not** in the domain? Justify your reasoning.

d. Is 8 a possible output for the relation this change machine represents? Justify your response.

e. Could 3 be included in the range of this relation? Explain why or why not.

f. What values are **not** in the range? Justify your reasoning.

4. Make sense of problems. Each of the functions that you have seen has a ***finite*** number of ordered pairs. There are functions that have an ***infinite*** number of ordered pairs. Describe any difficulties that may exist trying to represent a function with an infinite number of ordered pairs using the four representations of functions that have been described thus far.

My Notes

MATH TERMS

A **finite** set has a fixed countable number of elements. An **infinite** set has an unlimited number of elements.

5. Sometimes, machine diagrams are used to represent functions. In the function machine below, the inputs are labeled x and the outputs are labeled y. The function is represented by the expression $2x + 5$.

a. What is the output if the input is $x = 7$? $x = -2$? $x = \frac{1}{2}$?

b. Express regularity in repeated reasoning. Is there any limit to the number of input values that can be used with this expression? Explain.

Consider the function machine below.

6. Use the diagram to find the (input, output) ordered pairs for the following values.

a. $x = -5$ **b.** $x = \frac{3}{5}$ **c.** $x = -10$

My Notes

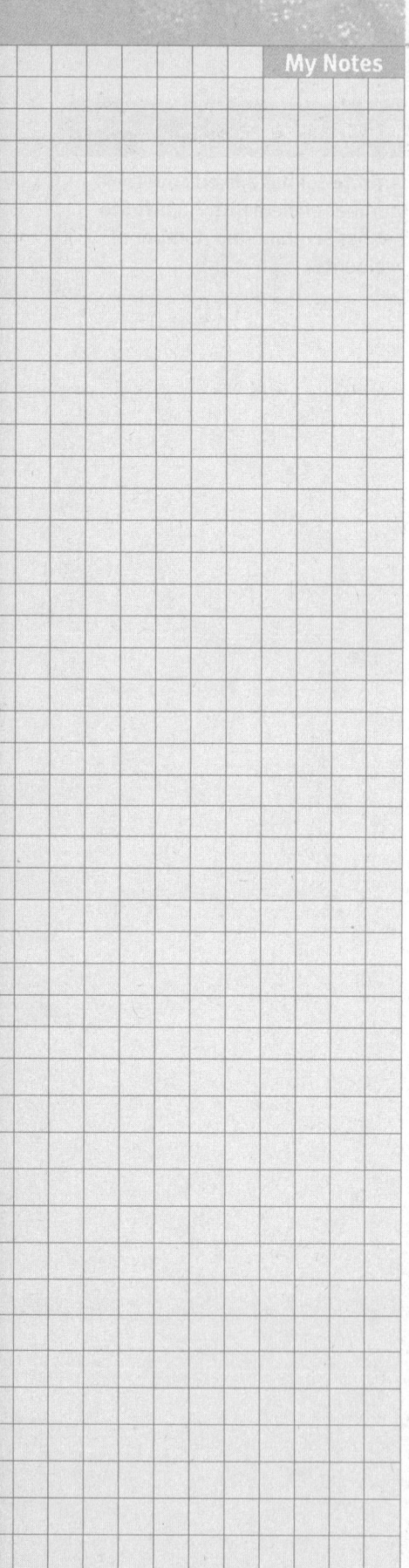

7. Make a function machine for the expression $10 - 5x$. Use it to find ordered pairs for $x = 3$, $x = -6$, $x = 0.25$, and $x = \frac{3}{4}$.

Creating a function machine can be time consuming and awkward. The function represented by the diagram in Item 5 can also be written algebraically as the equation $y = 2x + 5$.

8. For each function, find ordered pairs for $x = -2$, $x = 5$, $x = \frac{2}{3}$, and $x = 0.75$. Create tables of values.

a. $y = 9 - 4x$

b. $y = \frac{1}{x}$

Check Your Understanding

9. The set {(3, 5), (−1, 2), (2, 2), (0, −1)} represents a function. Identify the domain and range of the function.

10. Identify the domain and range for each function.

a.

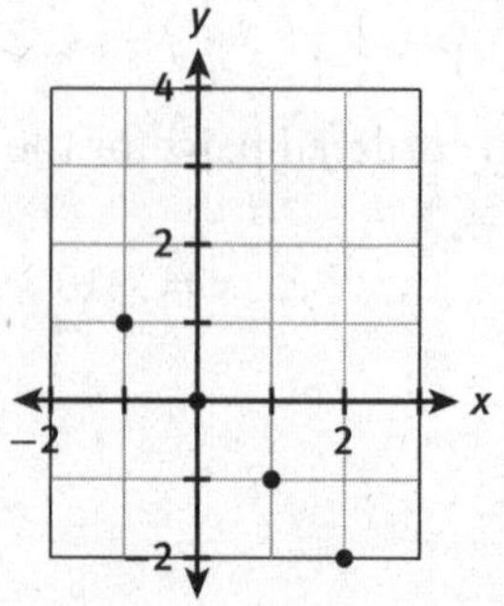

b.

x	y
12	−8
17	3
−4	9

My Notes

LESSON 5-2 PRACTICE

Identify the domain and range.

11.

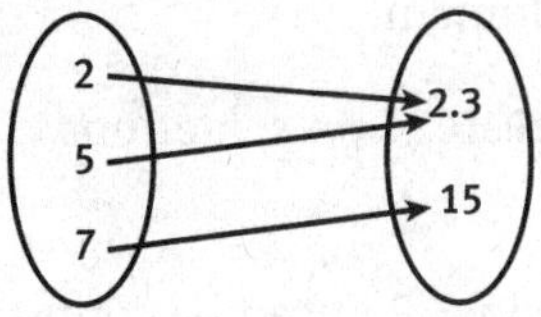

12.

x	y
1.5	4
−0.3	8
$\frac{1}{6}$	3

13. Model with mathematics. At an arcade, there is a machine that accepts game tokens and returns tickets that can be redeemed for prizes. Inserting 5 tokens returns 3 tickets and inserting 10 tokens returns 8 tickets. You must insert tokens in multiples of 5 or 10, and you have a total of 20 tokens.

a. Identify the domain in this situation.

b. Identify the range in this situation.

14. For the function machine shown, copy and complete the table of values.

x	y
−1	
0	
$\frac{1}{2}$	
1.2	

15. For each function below, find ordered pairs for $x = -1$, $x = 3$, $x = \frac{1}{2}$, and $x = 0.4$. Write your results as a set of ordered pairs.

a. $y = 4x$ **b.** $y = 2 - x^2$

My Notes

Learning Targets:

- Use and interpret function notation.
- Evaluate a function for specific values of the domain.

SUGGESTED LEARNING STRATEGIES: Create Representations, Discussion Groups

When referring to the functions in Item 8 in Lesson 5-2, it can be confusing to distinguish among them since each begins with "$y =$." Function notation can be used to help distinguish among different functions.

For instance, the function $y = 9 - 4x$ in Item 8a can be written:

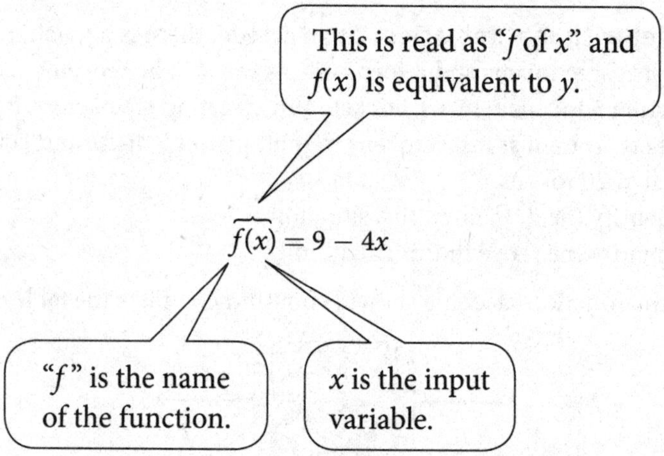
This is read as "*f* of *x*" and *f*(*x*) is equivalent to *y*.
$f(x) = 9 - 4x$
"*f*" is the name of the function.
x is the input variable.

MATH TIP

It is important to recognize that $f(x)$ **does not** mean f multiplied by x.

MATH TIP

Notice that $f(x) = y$. For a **domain** value x, the associated **range** value is $f(x)$.

1. To distinguish among different functions, it is possible to use different names. Use the name h to write the function from Item 8b using function notation.

Function notation is useful for evaluating functions for multiple input values. To evaluate $f(x) = 9 - 4x$ for $x = 2$, you substitute 2 for the variable x and write $f(2) = 9 - 4(2)$. Simplifying the expression yields $f(2) = 1$.

2. Use function notation to evaluate $f(x) = 9 - 4x$ for $x = 5$, $x = -3$, and $x = 0.5$.

My Notes

3. Use the values for x and $f(x)$ from Item 2. Display the values using each representation.

a. list of ordered pairs

b. table of values

c. mapping

d. graph

4. Given the function $f(x) = 9 - 4x$ as shown above, what value of x results in $f(x) = 1$?

5. Evaluate each function for $x = -5$ and $x = \frac{4}{3}$.

a. $f(x) = 2x - 7$

b. $g(x) = 6x - x^2$

c. $h(x) = \frac{2}{x^2}$

6. **Reason quantitatively.** Recall the money-changing machine from Item 3 in Lesson 5-2, in which customers can insert up to five one-dollar bills at a time and receive an equivalent amount of quarters. The function $f(x) = 4x$ represents this situation. What does x represent? What does $f(x)$ represent?

My Notes

A function whose domain is the set of positive consecutive integers forms a sequence. The terms of the sequence are the range values of the function. For the sequence 4, 7, 10, 13, …, $f(1) = 4$, $f(2) = 7$, $f(3) = 10$, and $f(4) = 13$.

7. Consider the sequence −4, −2, 0, 2, 4, 6, 8, ….

a. What is $f(3)$?

b. What is $f(7)$?

Check Your Understanding

8. Evaluate the functions for the domain values indicated.

a. $p(x) = 3x + 14$ for $x = -5, 0, 4$

b. $h(t) = t^2 - 5t$ for $t = -2, 0, 5, 7$

9. Consider the sequence −7, −3, 1, 5, 9, ….

a. What is $f(2)$?

b. What is $f(5)$?

LESSON 5-3 PRACTICE

Use the function $y = x^2 - 3x - 4$ for Items 10–12.

10. Write the function in function notation.

11. Evaluate the function for $x = -2$. Express your answer in function notation.

12. Make use of structure. For what value of x does $f(x) = -4$?

13. Consider the sequence $\frac{1}{2}, 1, \frac{3}{2}, 2, \frac{5}{2}, 3, \ldots$. What is $f(4)$?

ACTIVITY 5 PRACTICE

Write your answers on notebook paper.
Show your work.

Lesson 5-1

Use the Beverage Vending Machine to answer Items 1–6.

1. List all of the possible inputs.
2. List all of the possible outputs.
3. Which output results from an input of 2C?
   A. Juice
   B. Iced tea
   C. Latte
   D. Cocoa
4. Which number/letter combination would you input if you wanted the machine to output juice?
   A. 2A
   B. 1B
   C. 2B
   D. 1D
5. In a mapping of the relation shown by the vending machine, what drink would 1D map to?
6. In a table of the relation shown by the vending machine, what number/letter combination would correspond to cocoa?

For Items 7–9, two relations are given. One relation is a function and one is not. Identify each and explain.

7. $\{(5, -2), (-2, 5), (2, -5), (-5, 2)\}$

   $\{(5, -2), (-2, 5), (5, 2), (-5, 2)\}$

8.

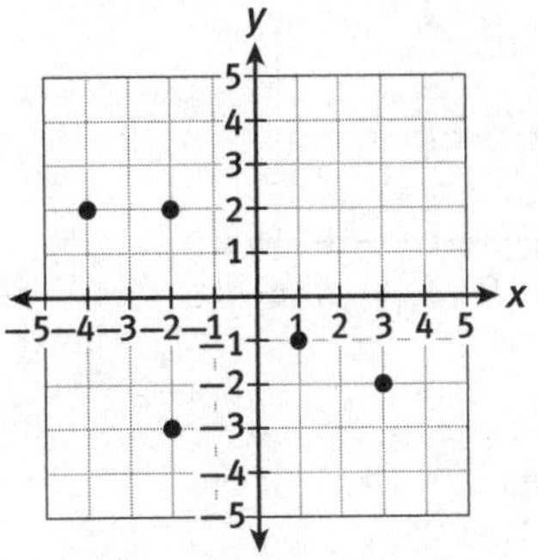

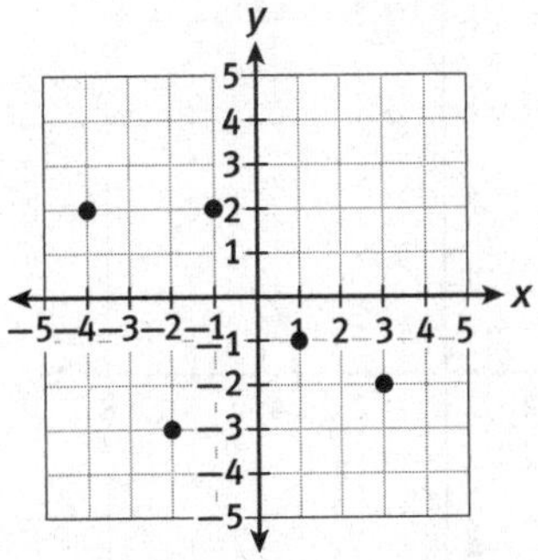

9.

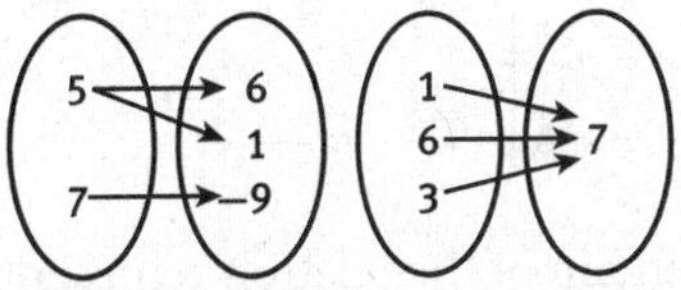

10. What value(s) of x in the relation below would create a set of ordered pairs that is not a function? Justify your answer.

$$\{(0, 5)\ (1, 5)\ (2, 6)\ (x, 7)\}$$

11. Does the graph shown represent a function? Explain.

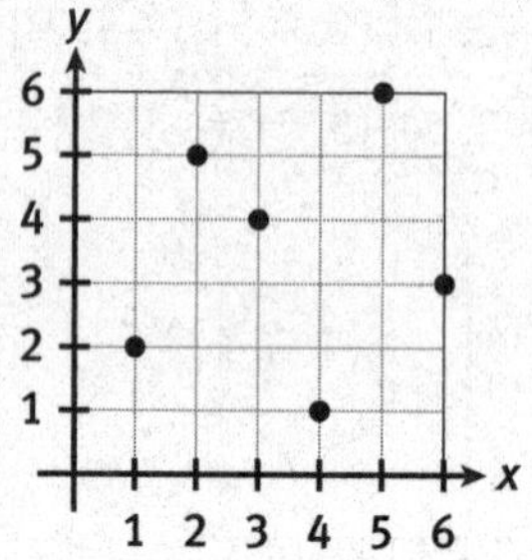

Lesson 5-2

Use the graph for Items 12–14.

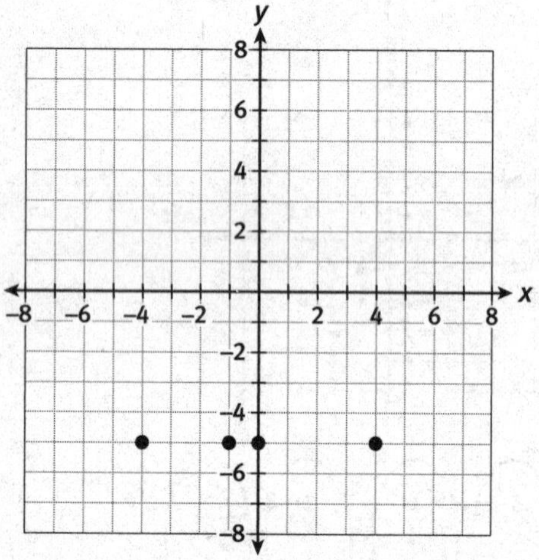

12. Identify the domain of the relation represented in the graph.

13. Identify the range of the relation represented in the graph.

14. Does the relation shown in the graph represent a function? Explain.

Lesson 5-3

Use the function machine for Items 15–17.

15. How would you write the function shown in the function machine in function notation?

16. What is the value of $f(-2)$?

17. What value(s) of x results in $f(x) = 8$?

18. Given the function $f(x) = -2x - 5$, determine the value of $f(-3)$.

The first seven numbers in the Fibonacci sequence are: 0, 1, 1, 2, 3, 5, 8. Use this information for Items 19 and 20.

19. What is $f(2)$?

20. What is $f(6)$?

MATHEMATICAL PRACTICES
Construct Viable Arguments and Critique the Reasoning of Others

21. Dora said that the mapping diagram below does not represent a function because each value in the domain is paired with the same value in the range. Explain the error in Dora's reasoning.

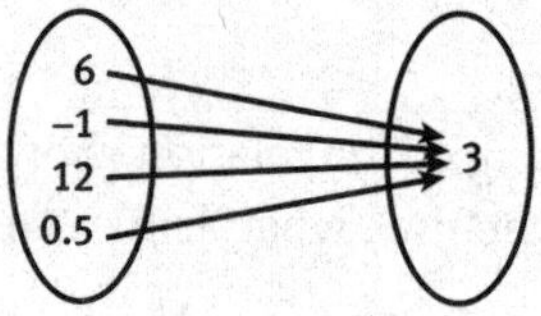

Interpreting Graphs of Functions

Shake, Rattle, and Roll

Lesson 6-1 Key Features of Graphs

My Notes

Learning Targets:

- Relate the domain and range of a function to its graph.
- Identify and interpret key features of graphs.

SUGGESTED LEARNING STRATEGIES: Marking the Text, Visualization, Interactive Word Wall, Discussion Groups

Roller coasters can be scary but fun to ride. Below is the graph of the heights reached by the cars of the Thunderball Roller Coaster over its first 1250 feet of track. The graph displays a function because each input value has one and only one output value. You can see this visually using the ***vertical line test***. Study this graph to determine the domain and range.

MATH TERMS

The **vertical line test** is a visual check to see if a graph represents a function. For a function, every vertical line drawn in the coordinate plane will intersect the graph in at most one point. This is equivalent to having each domain element associated with one and only one range element.

Thunderball Roller Coaster Graph

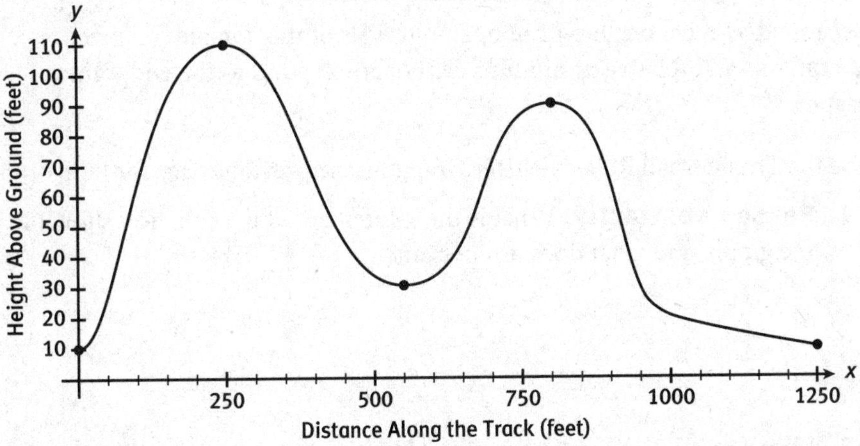

The domain gives all values of the ***independent variable***: in this case, the distance along the track in feet. The domain values are graphed along the horizontal or x-axis. The domain of the function above can be written in set notation as:

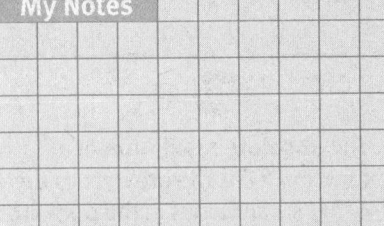

$$\{\text{all real values of } x: 0 \le x \le 1250\}$$

Read this notation as: *the set of all real values of x, between 0 and 1250, inclusive.*

MATH TERMS

An **independent variable** is the variable for which input values are substituted in a function.

A **dependent variable** is the variable whose value is determined by the input or value of the independent variable.

The range gives the values of the ***dependent variable***: in this case, the height of the roller coaster above the ground in feet. The range values are graphed on the vertical or y-axis. The range of the function above can be written in set notation as:

$$\{\text{all real values of } y: 10 \le y \le 110\}$$

Read this notation as: *the set of all real values of y, between 10 and 110, inclusive.*

My Notes

CONNECT TO AP

The **absolute maximum** of a function *f*(*x*) is the greatest value of *f*(*x*) for all values in the domain. The **absolute minimum** of a function *f*(*x*) is the least value of *f*(*x*) for all values in the domain. Unlike relative maximums and relative minimums, absolute maximums and absolute minimums may correspond to the endpoints of graphs.

MATH TIP

An open interval is an interval whose endpoints are not included. For example, $0 < x < 5$ is an open interval, but $0 \leq x \leq 5$ is not.

The graph above shows data that are ***continuous***. The points in the graph are connected, indicating that domain and range are sets of real numbers with no breaks in between. A graph of ***discrete*** data consists of individual points that are not connected by a line or curve.

Many other useful pieces of information about a function can be determined by looking at its graph.

- The ***y-intercept*** of a function is the point at which the graph of the function intersects the y-axis. The y-intercept is the point at which $x = 0$.
- A ***relative maximum*** of a function $f(x)$ is the greatest value of $f(x)$ for values in a limited open domain interval.
- A ***relative minimum*** of a function $f(x)$ is the least value of $f(x)$ for values in a limited open domain interval.

Because they must occur within open intervals of the domain, relative maximums and relative minimums cannot correspond to the endpoints of graphs.

Use the Thunderball Roller Coaster Graph on the previous page for Items 1–5.

1. **Reason abstractly.** What is the y-intercept of the function shown in the graph, and what does it represent?

2. Identify a relative maximum of the function represented by the graph.

3. Identify the absolute maximum of the function represented by the graph. Interpret its meaning in the context of the situation.

4. Identify a relative minimum of the function represented by the graph.

5. Identify the absolute minimum of the function represented by the graph. Interpret its meaning in the context of the situation.

My Notes

Suppose you got on a roller coaster called Cougar Mountain that immediately started climbing the track in a linear fashion, as shown in the graph.

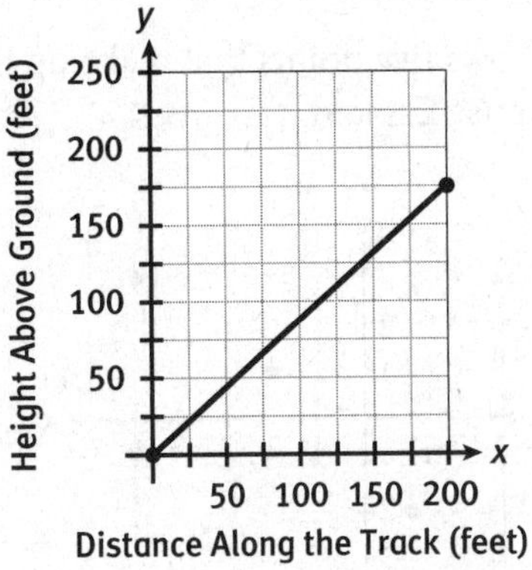

6. Identify the domain and range of the function.

7. Identify the y-intercept of the function.

8. Identify the absolute maximum and minimum of the function.

9. Does the function have any relative maximum or minimum values? Explain.

10. How are the ***extrema*** different on this linear graph versus the nonlinear graph for the Thunderball Roller Coaster?

MATH TERMS

Extrema refers to all maximum and minimum values.

My Notes

Check Your Understanding

11. The graph below shows five points that make up the function h. Is the function h continuous? Explain.

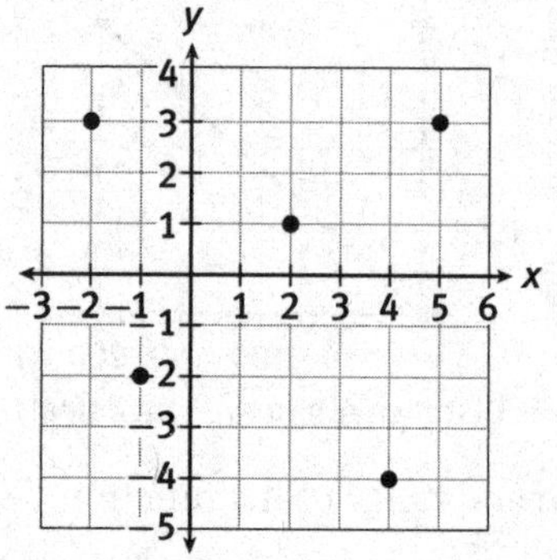

12. A function has three relative maximums: −2, 10.3, and 28. One of the relative maximums is also the absolute maximum. What is the absolute maximum?

Tell whether each statement is sometimes, always, or never true. Explain your answers.

13. A relative minimum is also an absolute minimum.

14. An absolute minimum is also a relative minimum.

Tom hiked along a circular trail known as the Juniper Loop. The graph shows his distance d from the starting point after t minutes.

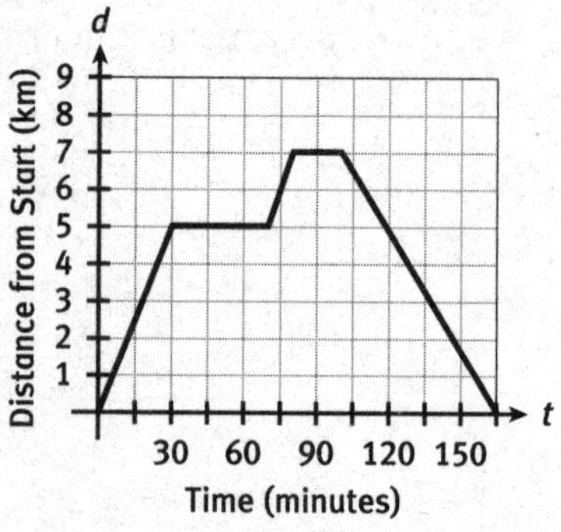

15. Identify the domain and range of the function shown in the graph.

16. Identify the absolute minimum of the function. What does it represent?

My Notes

17. In this function, the absolute minimum corresponds to two points on the graph. What are the two points? What do they represent in this context?

18. Identify the absolute maximum of the function. What does it represent?

19. What points on the graph correspond to the absolute maximum? What does this mean in the context of Tom's hike?

20. Identify any relative minimums for the function shown in the graph.

21. Identify any relative maximums for the function shown in the graph.

Check Your Understanding

22. What are the independent and dependent variables for the function representing Tom's hike?

23. Explain how to determine the maximum and minimum values of a function by examining its graph.

24. Is it possible for a function to have more than one absolute maximum or absolute minimum value? Explain.

My Notes

LESSON 6-1 PRACTICE

Model with mathematics. Use the graph below for Items 25–30.

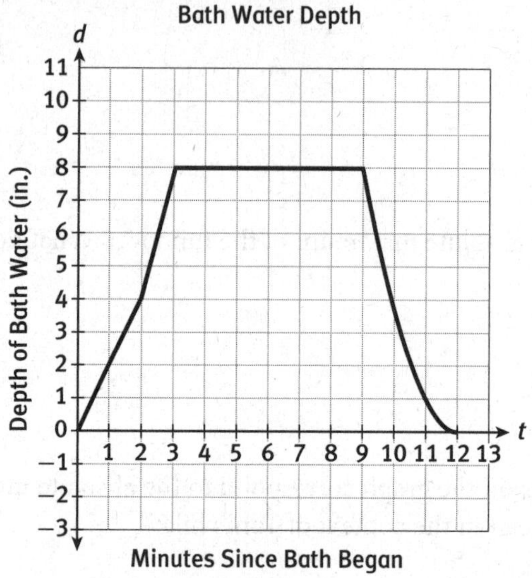

25. What are the independent and dependent variables? Explain.

26. Use set notation to write the domain and range of the function.

27. Is the function discrete or continuous? Explain.

28. What is the *y*-intercept? Interpret the meaning of the *y*-intercept in this context.

29. Identify any relative maximums or minimums of the function.

30. Identify the absolute maximum and absolute minimum values. Interpret their meanings in this context.

My Notes

Learning Targets:

- Relate the domain and range of a function to its graph and to its function rule.
- Identify and interpret key features of graphs.

SUGGESTED LEARNING STRATEGIES: Marking the Text, Levels of Questions, Think Aloud, Create Representations, Summarizing

Examine the graph of the function $f(x) = \frac{1}{(x-2)^2}$, graphed below.

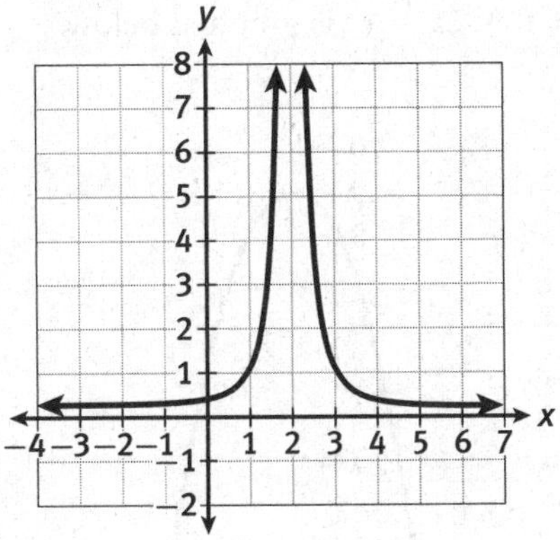

1. Describe how this graph is different from the graphs in Lesson 6-1.

Example A

Give the domain and range of the function $f(x) = \frac{1}{(x-2)^2}$.

Then find the y-intercept, the absolute maximum, and the absolute minimum.

To find the domain and range:

Step 1: Study the graph.
The sketch of this graph is a portion of the function represented by the equation $f(x) = \frac{1}{(x-2)^2}$.

Step 2: Look for values for which the domain causes the function to be undefined. Look how the graph behaves near $x = 2$.

Solution: The domain and range of $f(x) = \frac{1}{(x-2)^2}$ can be written:

Domain: {*all real values of* $x : x \neq 2$}

Range: {*all real values of* $y : y > 0$}

MATH TIP

Notice the result when $x = 2$ is substituted into $f(x)$.

$$f(2) = \frac{1}{(2-2)^2} = \frac{1}{0}$$

Division by zero is undefined in mathematics.

My Notes

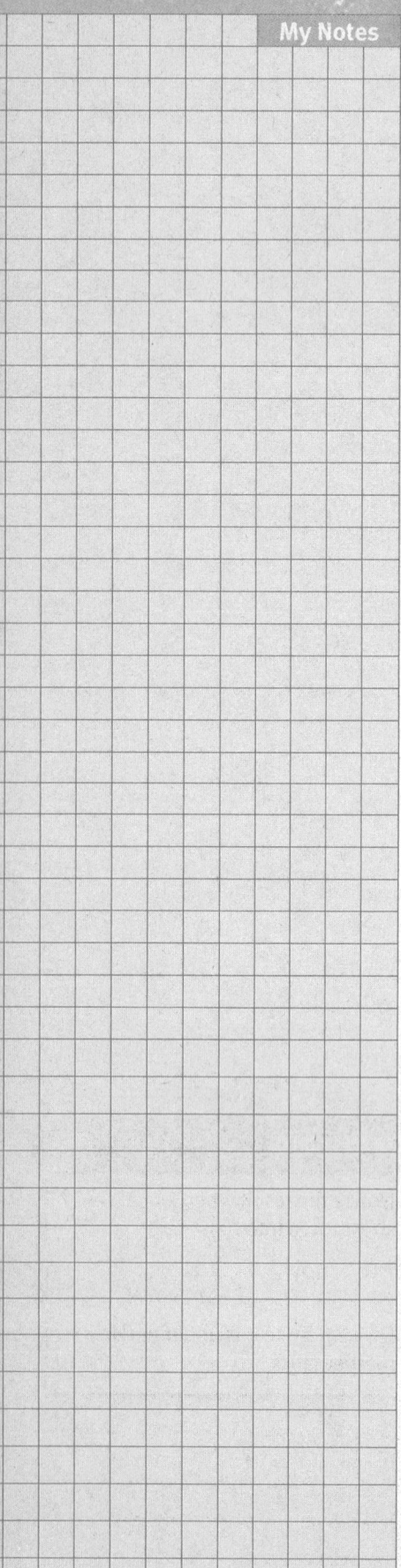

To determine the y-intercept and identify any maximums or minimums:

Study the graph. We can see that the function intersects the y-axis at (0, 0.25). The value of $f(x)$ keeps getting larger as x approaches 2 from both sides. The value of $f(x)$ approaches, but never reaches, 0 as x gets further from 2 on both sides.

Solution: The y-intercept is (0, 0.25). The function does not have an absolute maximum or minimum.

Try These A

The function $f(x) = 8 + 2x - x^2$ is graphed below.

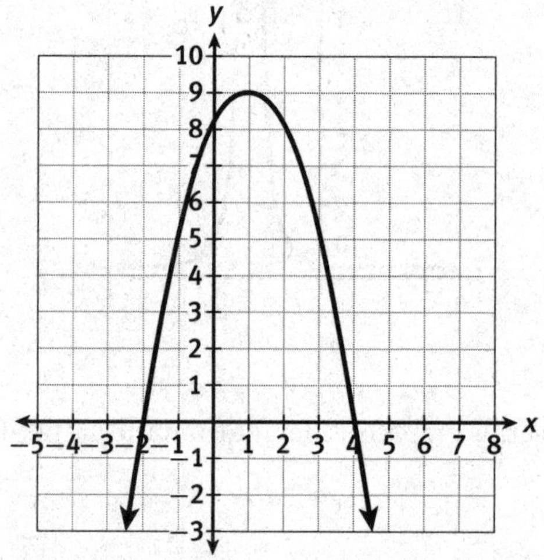

a. Identify the domain and range of the function.
Domain:

Range:

b. Identify the y-intercept.

c. Identify any relative or absolute minimums of the function.

d. Identify any relative or absolute maximums of the function.

2. The equation $y = 2x - 1$ is graphed below.

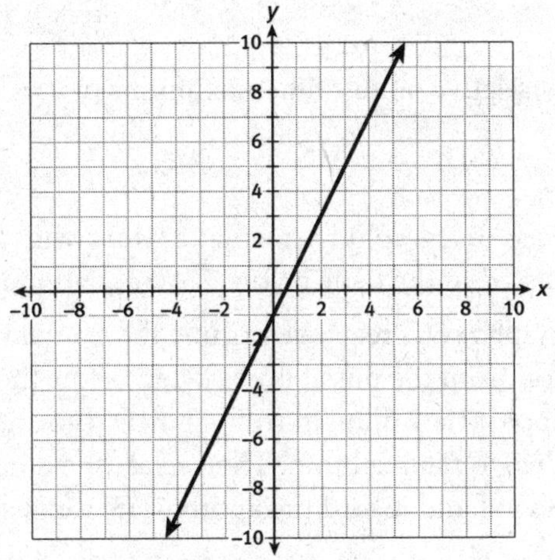

a. Identify the domain and range.
Domain:
Range:

b. What is the y-intercept of $y = 2x - 1$?

c. Identify any relative or absolute minimums of $y = 2x - 1$.

d. Identify any relative or absolute maximums of $y = 2x - 1$.

e. Construct viable arguments. Explain whether this equation represents a function and how you determined this.

3. The function $y = 2^x$ is graphed below.

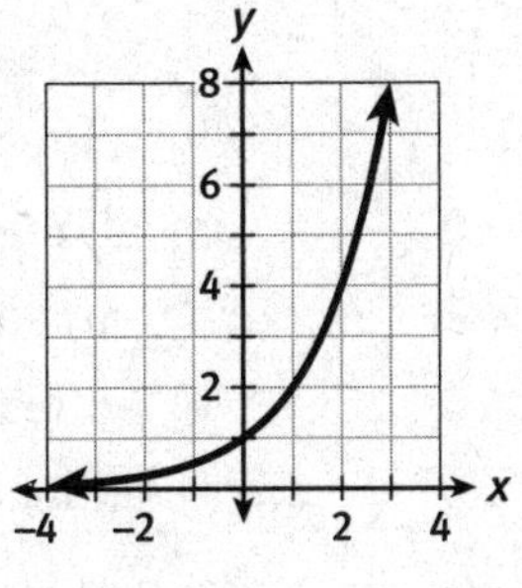

a. Identify the domain and range.
Domain:
Range:

b. What is the y-intercept of the function $y = 2^x$?

My Notes

c. Identify any relative or absolute minimums of $y = 2^x$.

d. Identify any relative or absolute maximums of $y = 2^x$.

MATH TIP

The domain is restricted to avoid situations where division by zero or taking the square root of a negative number would occur.

4. If you have access to a graphing calculator, work with a partner to graph the equations listed in the table below. Each equation is a function.

a. Using the graphs you create, determine the domain and range for each function from the possibilities listed below the chart.

b. Select the appropriate domain from choices 1–6 and record your answer in the Domain column. Then select the appropriate range from choices a–f and record the appropriate range in the Range column.

c. When the chart is complete, compare your answers with those from another group.

Function	Domain	Range
$y = -3x + 4$		
$y = x^2 - 6x + 5$		
$y = 9x - x^2$		
$y = \|x + 1\|$		
$y = 3 + \sqrt{x}$		
$y = \frac{4}{x}$		

Possible Domains

1) all real numbers
2) all real x, such that $x \neq -2$
3) all real x, such that $x \neq 0$
4) all real x, such that $x \neq 2$
5) all real x, such that $x \geq 0$
6) all real x, such that $x \leq 0$

Possible Ranges

a) all real numbers
b) all real y, such that $y \neq 0$
c) all real y, such that $y \geq -4$
d) all real y, such that $y \geq 0$
e) all real y, such that $y \leq 20.25$
f) all real y, such that $y \geq 3$

My Notes

Check Your Understanding

5. How can you determine from a function's graph whether the function has any maximum or minimum values?
6. How can you determine the domain of a function by examining its graph? By examining its function rule?
7. Give an example of a function that has a restricted domain. Justify your answer.

LESSON 6-2 PRACTICE

The function $f(x) = 2x^2 - 3$ is graphed below.

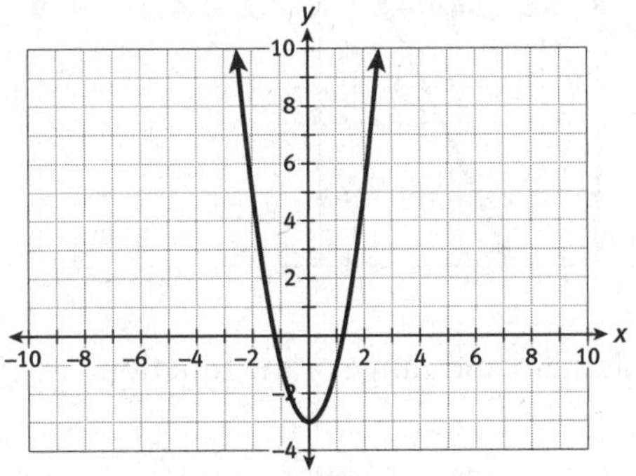

8. Give the domain, range, and y-intercept.
9. Identify any relative or absolute minimums.
10. Identify any relative or absolute maximums.
11. **Attend to precision.** Examine the graphs below. Explain why one function has an absolute minimum and an absolute maximum and the other function does not. Identify the absolute minimum and maximum values of the function for which they exist.

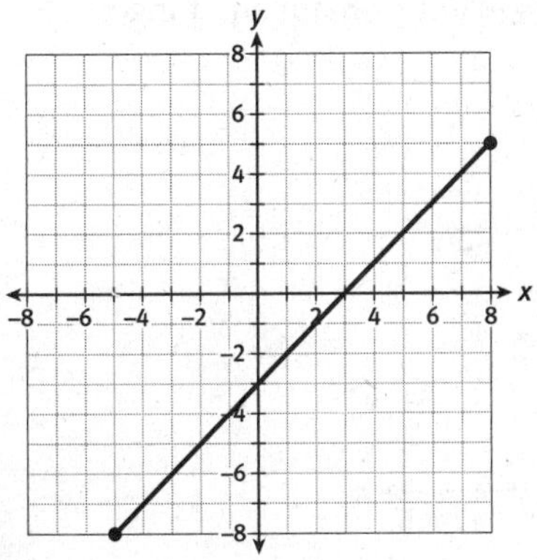

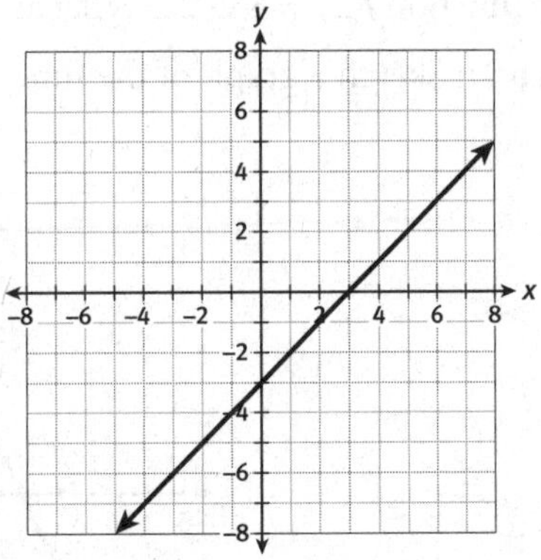

My Notes

Learning Targets:

- Identify and interpret key features of graphs.
- Determine the reasonable domain and range for a real-world situation.

SUGGESTED LEARNING STRATEGIES: Visualization, Discussion Groups, Look for a Pattern

The function $f(x) = 3 + 2x$ is graphed below.

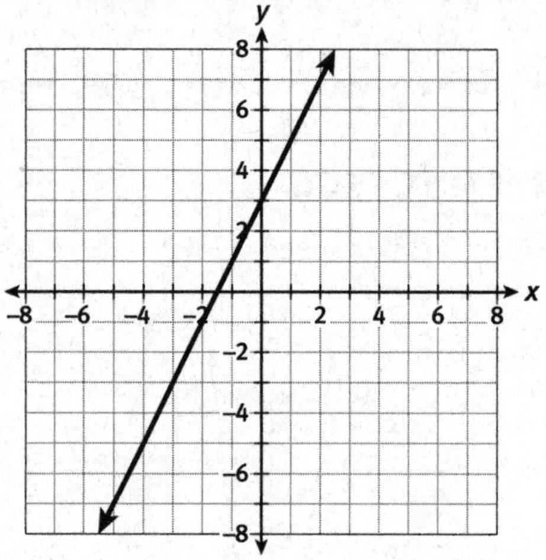

1. What are the domain and range of the function?
Domain:
Range:

In many real-world situations, not all values make sense for the domain and/or range. For example, distance cannot be negative; number of people cannot be a decimal or a fraction. In such situations, the values that make sense for the domain and range are called the *reasonable* domain and range.

MATH TIP

Graph a function by substituting several values for x and generating ordered pairs. You can organize the ordered pairs in a table. There are infinitely many other solutions because the graph has infinitely many points.

x	$f(x) = 3 + 2x$	(x, y)
0	3	(0, 3)
1	5	(1, 5)
2	7	(2, 7)

Example A

A taxi ride costs an initial rate of $3.00, which is charged as soon as you get in the cab, plus $2 for each mile traveled. The cost of traveling x miles is given by the function $f(x) = 3 + 2x$. What are the reasonable domain and range?

Step 1: Sketch a graph of the function.

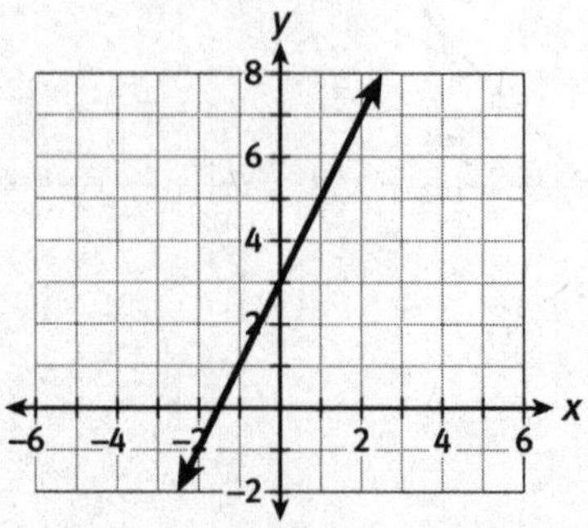

My Notes

Step 2: Determine the reasonable domain. Think about what the variable x represents. What values make sense?

The variable x represents the number of miles, so it does not make sense for x to be negative.

The reasonable domain is $\{x: x \geq 0\}$.

Step 3: Use the reasonable domain and the graph to determine the reasonable range.

From the graph, all y-values corresponding to the reasonable domain values are greater than or equal to 3. The reasonable range is $\{y: y \geq 3\}$.

Solution: The reasonable domain is $\{x: x \geq 0\}$. The reasonable range is $\{y: y \geq 3\}$.

Try These A

a. A banquet hall charges $15 per person plus a $100 setup fee. The cost for x people is given by the function $f(x) = 100 + 15x$. What are the reasonable domain and range?

b. Eight Ball Billiards charges $5 to rent a table plus $10 per hour of game play, rounded to the nearest whole hour. The cost of playing billiards for x hours is given by the function $f(x) = 5 + 10x$. What are the reasonable domain and range?

2. Reason quantitatively. Are the domain and range of $f(x) = 3 + 2x$ that you found in Item 1 the same as the reasonable domain and range of $f(x) = 3 + 2x$ found in Example A? Explain.

My Notes

3. The graph below represents a real-world situation.

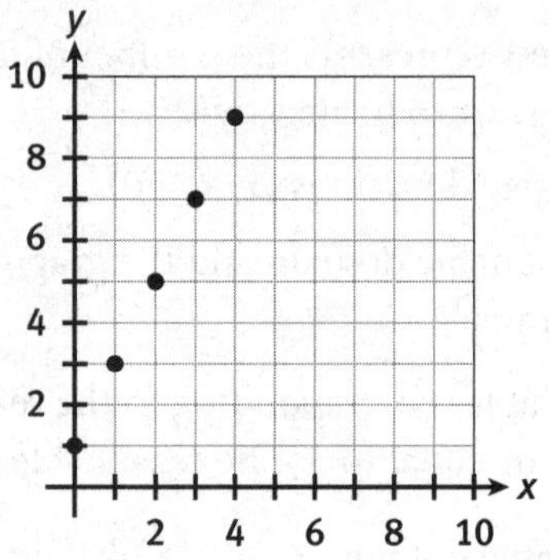

a. Identify the domain and range.

b. Describe a real-world situation that matches the graph. Your answers to Part (a) should be the reasonable domain and range for your situation.

c. Identify the independent and dependent variables in your real-world situation.

Check Your Understanding

4. For a function that models a real-world situation, the dependent variable y represents a person's height. What is a reasonable range? Explain.

5. A tour company charges \$25 to hire a tour director plus \$75 per tour member. The total cost for a group of x people is given by $f(x) = 25 + 75x$. What is the reasonable domain? Explain.

LESSON 6-3 PRACTICE

Talk the Talk Cellular charges a base rate of \$20 per month for unlimited texts plus \$0.15/minute of talk time. The monthly cost for x minutes is given by $f(x) = 20 + 0.15x$.

6. **Make sense of problems.** What is the independent variable and what is the dependent variable? Explain how you know.

7. What are the reasonable domain and range? Explain.

ACTIVITY 6 PRACTICE

Write your answers on notebook paper.
Show your work.

Lesson 6-1

Use the graph below for Items 1–5.

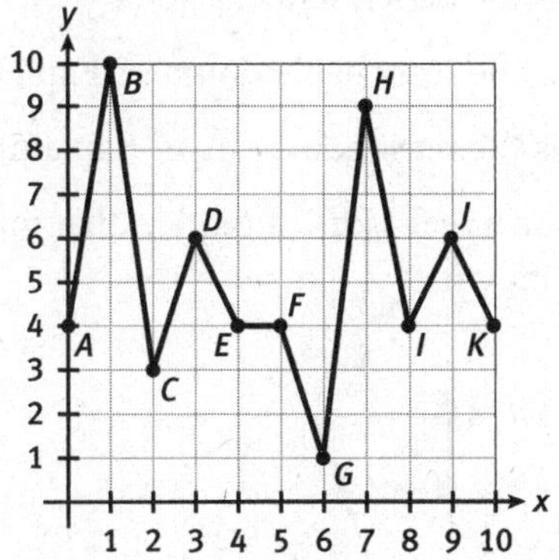

1. Which point corresponds to the absolute maximum of the function?
   A. *B*
   B. *D*
   C. *G*
   D. *H*
2. Which represents the range of the function shown in the graph?
   A. $\{0 \leq x \leq 10\}$
   B. $\{1 \leq x \leq 10\}$
   C. $\{0 \leq y \leq 10\}$
   D. $\{1 \leq y \leq 10\}$
3. Which point does **not** correspond to a relative minimum?
   A. *B*
   B. *C*
   C. *E*
   D. *I*
4. Is the function represented by the graph discrete or continuous? Explain.
5. What is the *y*-intercept of the function shown in the graph?
6. **a.** Give the domain and range for the function graphed below. Explain why this graph represents a function.

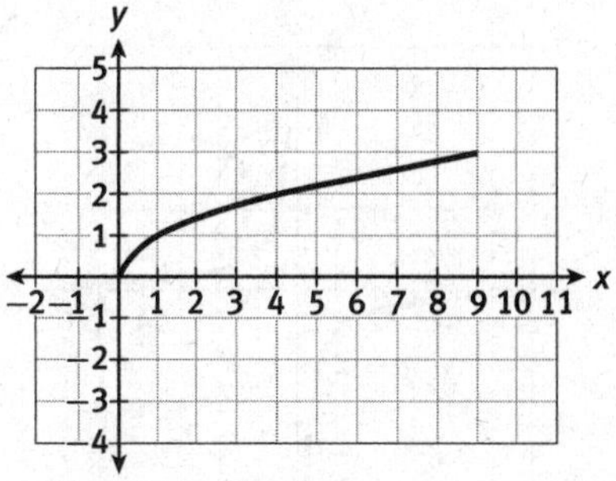

   b. What is the *y*-intercept of the function shown in the graph?
   c. Identify any extrema of the function shown in the graph.

Jeff walks at an average rate of 125 yards per minute. Mark's house is located 2000 yards from Jeff's house. The graph below shows how far Jeff still needs to walk to reach Mark's house. Use the graph for Items 7–10.

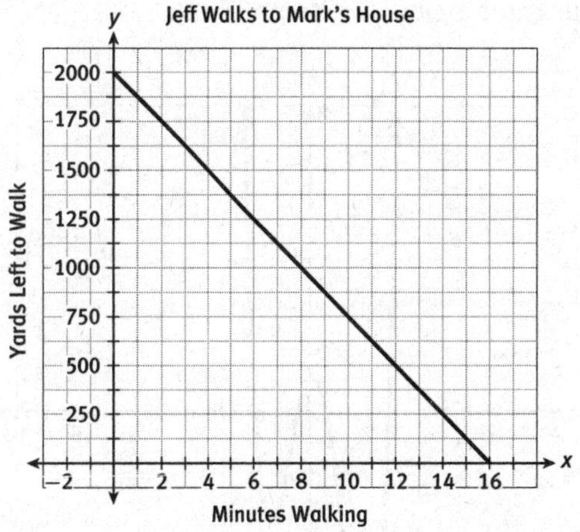

7. Identify the independent and dependent variables.
8. Identify the absolute minimum and absolute maximum values. What do these values represent?
9. Identify any relative maximums or minimums.
10. What is the *y*-intercept? What does it represent?

Lesson 6-2

Use the graph for Items 11–13.

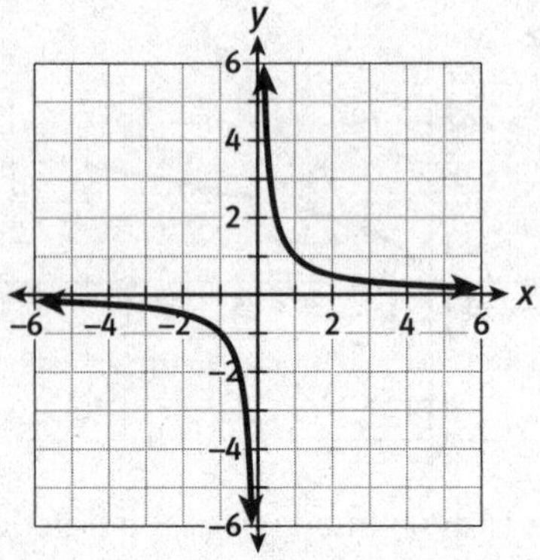

11. What is the domain of the function shown in the graph?

12. What is the range of the function shown in the graph?

13. What is the y-intercept of the function shown in the graph?

Use the graph below for Items 14–16.

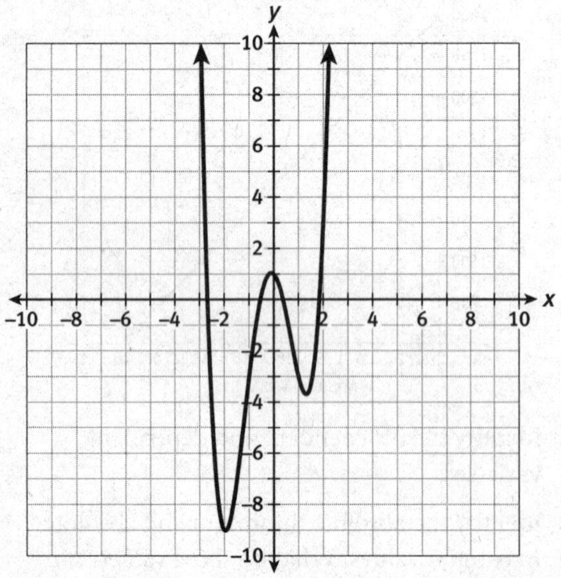

14. What is the y-intercept of the function shown in the graph?

15. Identify any relative maximums.

16. Identify any relative minimums.

Lesson 6-3

A fundraising organization will donate $250 plus half of the money it raises from a charity event. Use this information for Items 17–20.

17. What is the independent variable?

18. What is the dependent variable?

19. What is the reasonable domain? Explain.

20. What is the reasonable range? Explain.

21. Describe a real-world situation that matches the graph shown.

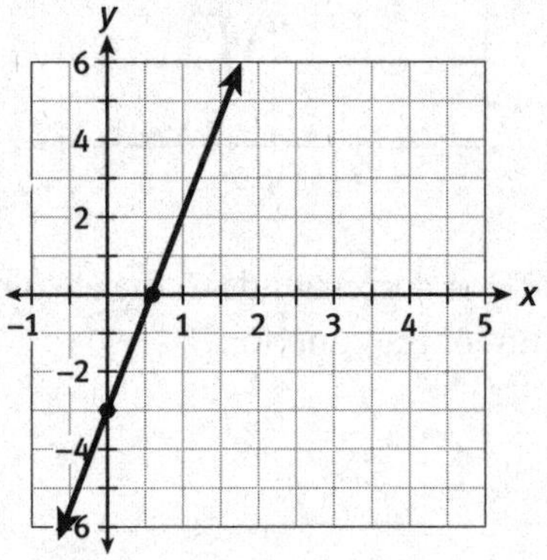

MATHEMATICAL PRACTICES
Look For and Make Use of Structure

22. The graph of a function is a horizontal line. What is true about the absolute maximum and absolute minimum values of this function? Explain.

Graphs of Functions

Experiment Experiences
Lesson 7-1 The Spring Experiment

My Notes

Learning Targets:

- Graph a function given a table.
- Write an equation for a function given a table or graph.

SUGGESTED LEARNING STRATEGIES: Discussion Groups, Look for a Pattern, Sharing and Responding, Think-Pair-Share, Create Representations, Construct an Argument

For the following experiment, you will need a paper cup, a rubber band, a paper clip, a measuring tape, and several washers.

A. Punch a small hole in the side of the paper cup, near the top rim.
B. Use the bent paper clip to attach the paper cup to the rubber band as shown in the diagram in the *My Notes* section.

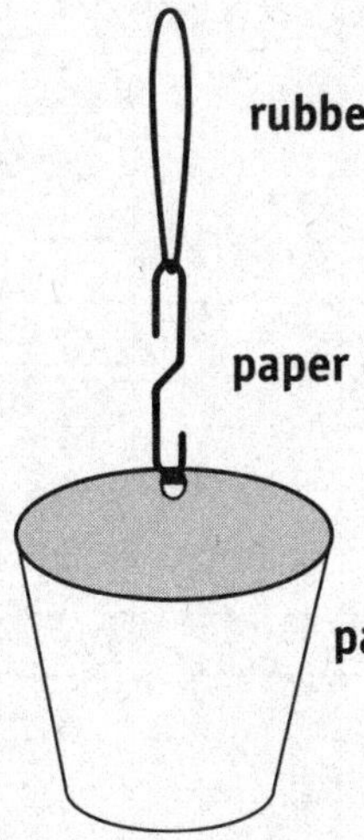

1. What is the length of the rubber band?

Drop washers one at a time into the cup. Each time you add a washer, measure the length of the rubber band. Subtract the original length you recorded in Item 1 to find the distance that the rubber band has stretched.

2. Make a table of your data.

Number of Washers x	Length of Stretch from Original Length y
1	
2	
3	
4	
5	

3. What patterns do you notice that might help you determine the relationship between the number of washers in the cup and the length of the rubber band stretch?

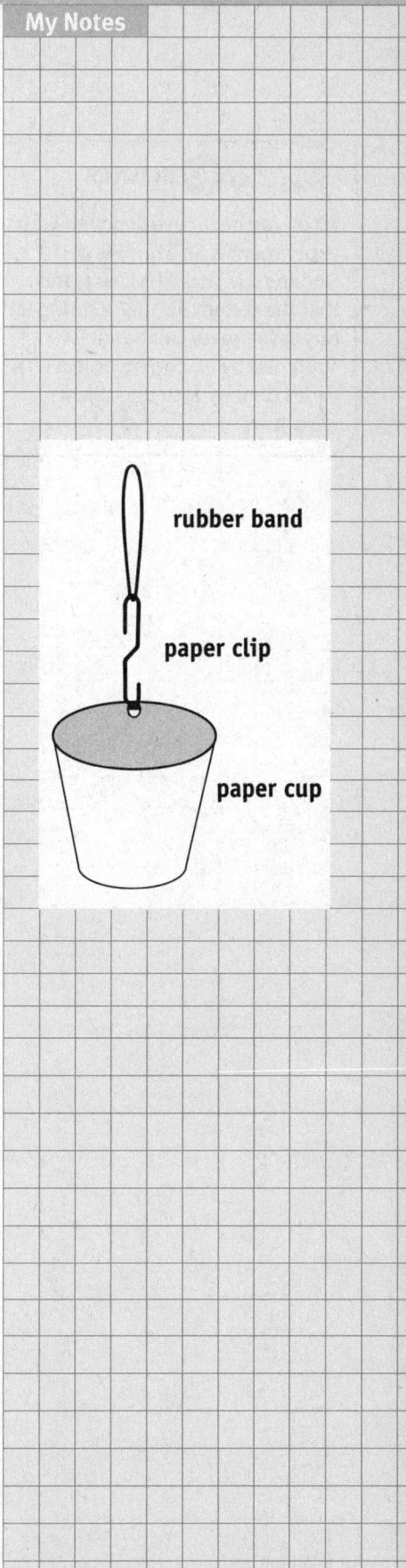

My Notes

CONNECT TO SCIENCE

What you have revealed with your experiment is an example of Hooke's Law. Hooke's Law states that the distance *d* that a spring (in this case the rubber band) is stretched by a hanging object varies directly with the object's weight *w*.

4. Use your table to make a graph. Be sure to label an appropriate scale and the units on the *y*-axis.

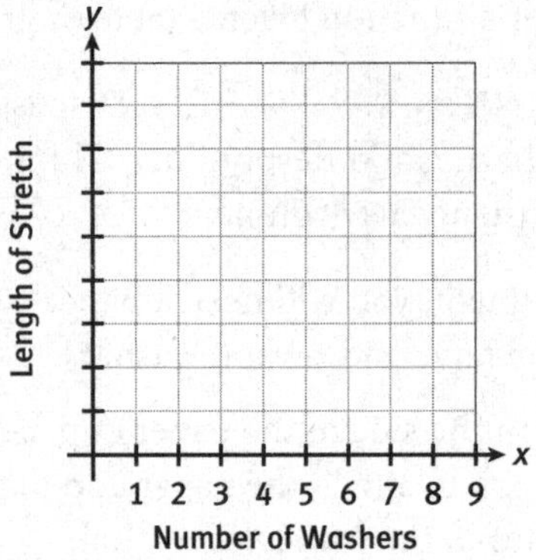

5. Describe your graph.

6. **Model with mathematics.** Use your graph and any patterns you described in Item 3 to write an equation that describes the relationship between the number of washers and the length of the stretch.

7. Use your graph or your equation to predict the length of the stretch for 8 washers and for 10 washers.

A group of students performed a similar experiment with a spring and various masses. The data they collected is shown in the table below.

Mass (g)	Spring Stretch (cm)
2	6
4	12
6	18
8	24
10	30
12	36

8. Make a graph of the data in the table.

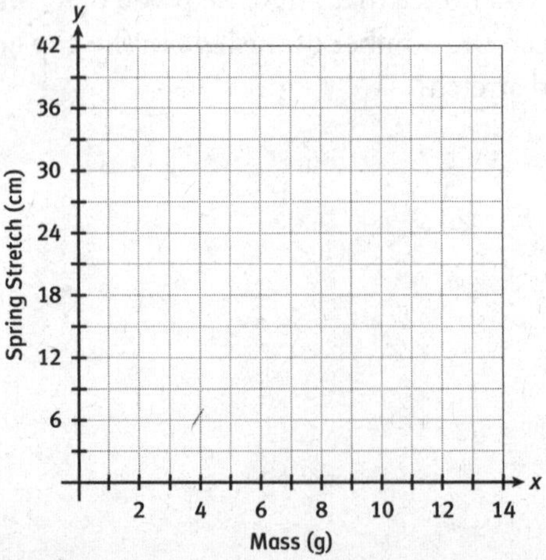

My Notes

9. **Reason quantitatively.** How much does the spring stretch for each additional gram of mass added? Explain how you found your answer.

10. **Reason abstractly.** Use the students' data to write an equation that gives the distance d that the spring will stretch in terms of the mass m. Explain your equation.

11. Use the equation or the graph to determine the length of the stretch for a mass of 1 gram. Graph the outcome on your graph.

12. Use the equation or the graph to determine the length of the stretch for a mass of 7 grams. Graph the outcome on your graph.

13. Use the equation or the graph to determine the length of the stretch for a mass of 13 grams. Graph the outcome on your graph.

14. **a.** What do you notice about the points you graphed in Items 11–13?

b. How could you represent the set of all possible masses and corresponding stretches?

15. What is the y-intercept of the graph? What does it represent?

16. What is the reasonable domain? Explain.

My Notes

Mr. Hardiff's class conducts an experiment with a spring and a set of weights. They record their data, but some of the information is missing.

Weight (oz)	Spring Stretch (in.)
5	12.5
8	20
10	25
12	
15	
16	

17. How much does the spring stretch for each additional ounce of weight?

18. Describe how to use your answer to Item 17 to write an equation for the data in the table.

19. Use your equation from Item 18 to complete the table.

Check Your Understanding

20. A 4.5-pound weight stretches a spring 18 inches and a 7.5-pound weight stretches the same spring 30 inches. How much does the spring stretch for each additional pound of weight? Explain how you found your answer.

LESSON 7-1 PRACTICE

Jeremy and his classmates conduct an experiment with a set of weights and a spring. They record their results in the table. Use the table to answer Items 21–24.

Student	Mass (lb)	Spring Stretch (in.)
Jeremy	5	7.5
Adele	8	12
Roberto	14	21
Shanice	21	36
Guillaume	28	42

21. Make a graph of the data.

22. **Critique the reasoning of others.** Which student made a mistake when taking their turn at the experiment? Explain how you know.

23. If the mistake in Item 22 were corrected, what would the correct data point be?

24. Write an equation to describe the students' data, using the corrected data point you identified in Item 23.

My Notes

Learning Target:

- Graph a function describing a real-world situation and identify and interpret key features of the graph.

SUGGESTED LEARNING STRATEGIES: Discussion Groups, Look for a Pattern, Construct an Argument, Think-Pair-Share, Summarizing, Sharing and Responding

1. The Empire State Building in New York City is 1454 feet tall. How long do you think it will take a penny dropped from the top of the Empire State Building to hit the ground?

In 1589, the mathematician and scientist Galileo conducted an experiment to answer a question much like the one in Item 1. Galileo dropped balls from the top of the Leaning Tower of Pisa in Italy and determined the time it took them to reach the ground. Galileo used several balls identical in shape but differing in mass. Because the balls all reached the ground in the same amount of time, he developed the theory that all objects fall at the same rate.

Galileo's findings can be represented with the equation $h(t) = 1600 - 16t^2$, where $h(t)$ represents the height in feet of an object t seconds after it has been dropped from a height of 1600 feet.

2. Make a table of values for Galileo's function $h(t) = 1600 - 16t^2$.

t (seconds)	$h(t)$ (feet)
0	
1	
2	
3	
4	
5	
6	
7	
8	
9	
10	

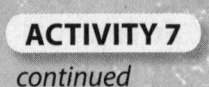

My Notes

3. **Construct viable arguments.** Why would negative domain values not be appropriate in this context?

4. Using your table of values, graph Galileo's function.

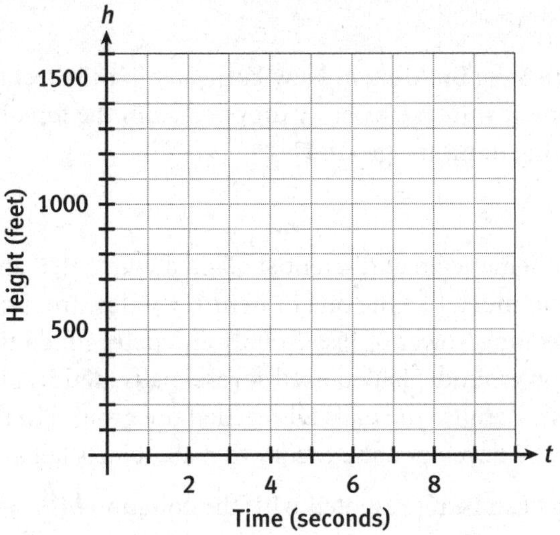

5. What is the reasonable domain of the function represented in your graph? What is the reasonable range?

6. What is the y-intercept?

7. What does the y-intercept represent?

8. What is the ***x*-intercept**? What does the x-intercept represent?

MATH TERMS

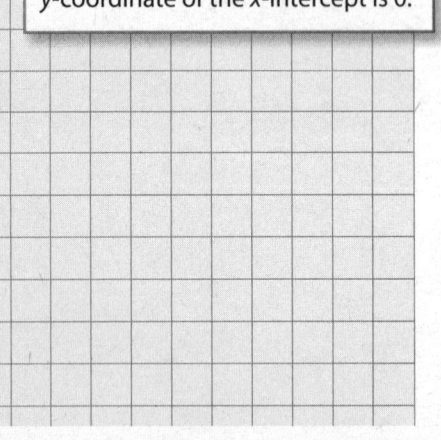

The ***x*-intercept** is the point where a graph crosses the x-axis. The y-coordinate of the x-intercept is 0.

9. Identify any extrema of the function shown in the graph. What do the extrema represent?

"Your homework assignment is to graph this function," your math teacher says. She then points to the following function on the board:

$$f(x) = x^2 - 2x$$

In this case, the function is not limited by a real-world situation. Therefore, it is important to use different types of domain values as you prepare to graph.

My Notes

10. Using various values for x, make a table of values for $f(x) = x^2 - 2x$.

x	$f(x)$

11. Using your table of values, graph the function.

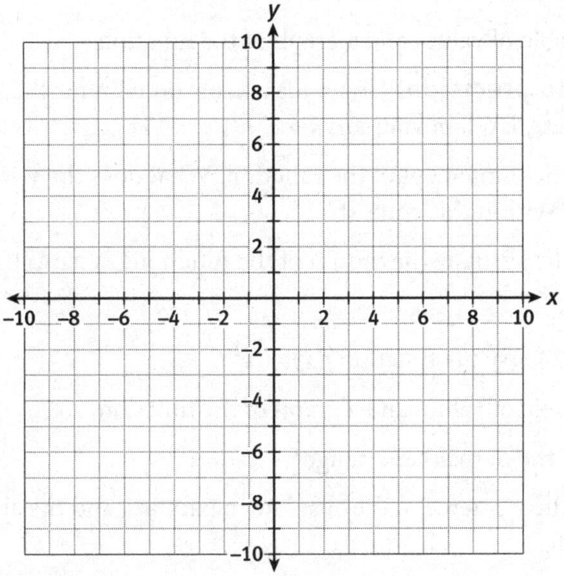

12. Describe the differences between the domain of $f(x) = x^2 - 2x$ and the domain of Galileo's function.

13. State the range of $f(x) = x^2 - 2x$.

14. Identify the y-intercept of $f(x) = x^2 - 2x$.

15. What is the absolute maximum of $f(x) = x^2 - 2x$? What is the absolute minimum?

My Notes

Check Your Understanding

16. Revisit your answer to Item 1 and revise it if necessary. About how long do you think it will take a penny dropped from the top of the Empire State Building to hit the ground? How can you use Galileo's equation to help you answer this question?

LESSON 7-2 PRACTICE

The area of a rectangle with a perimeter of 20 units is given by $f(w) = 10w - w^2$, where w is the width of the rectangle. Assume that w is a whole number. Use this function to answer Items 17–20.

17. Make a table of values and a graph of the function.

18. **Attend to precision.** Give a reasonable domain for the function in this context. Explain your answers.

19. Identify the y-intercept of the function. What does the y-intercept represent within this context?

20. What is the absolute maximum of the function? What is the absolute minimum?

For Items 21–23, use the function $f(x) = x^2 - 9$.

21. Make a table of values and a graph of the function.

22. What are the domain and range?

23. Identify the y-intercept, the absolute maximum, and the absolute minimum.

My Notes

Learning Targets:

- Given a verbal description of a function, make a table and a graph of the function.
- Graph a function and identify and interpret key features of the graph.

SUGGESTED LEARNING STRATEGIES: Discussion Groups, Look for a Pattern, Construct an Argument, Paraphrasing, Marking the Text, Think-Pair-Share

In the late nineteenth century, the scientist Marie Curie performed experiments that led to the discovery of radioactive substances.

A radioactive substance is a substance that gives off radiation as it decays. Scientists describe the rate at which a radioactive substance decays as its *half-life*. The half-life of a substance is the amount of time it takes for one-half of the substance to decay.

CONNECT TO SCIENCE

How much is half a life?
The half-life of a radioactive substance can be as little as 0.0018 seconds for Polonium-215 and as much as 4.5 billion years for Uranium-238.

1. Radium has a half-life of 1600 years. How much radium will be left from a 1000-gram sample after 1600 years?

2. How much radium will be left after another 1600 years?

3. Suppose a radioactive substance has a half-life of 1 second and you begin with a sample of 4 grams. Complete the table of values.

Time (seconds)	Amount Remaining (grams)
0	4
1	
2	
3	
4	
5	

My Notes

4. Graph the data from the table on the grid below.

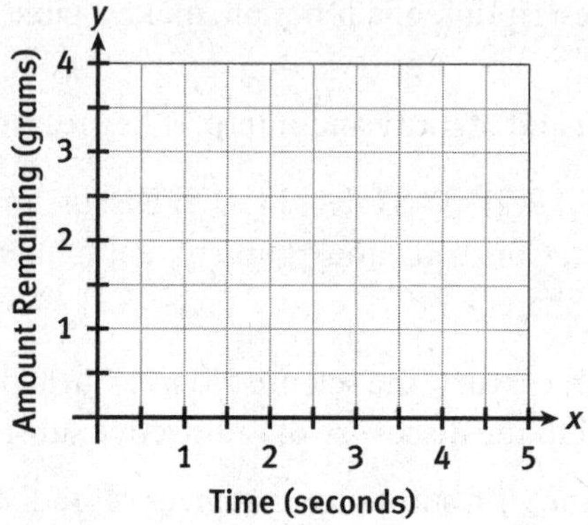

5. **Make use of structure.** Will the amount of the substance that remains ever reach 0? Explain.

6. What are the reasonable domain and range of the function represented in the graph? Explain.

7. What is the y-intercept and what does it represent?

8. Identify the absolute maximum and minimum of the function represented in the graph, and tell what they represent in the context.

My Notes

The function that describes the substance's decay is $f(x) = 4\left(\frac{1}{2}\right)^x$. The graph of this function when it does not model a real-world situation is shown below.

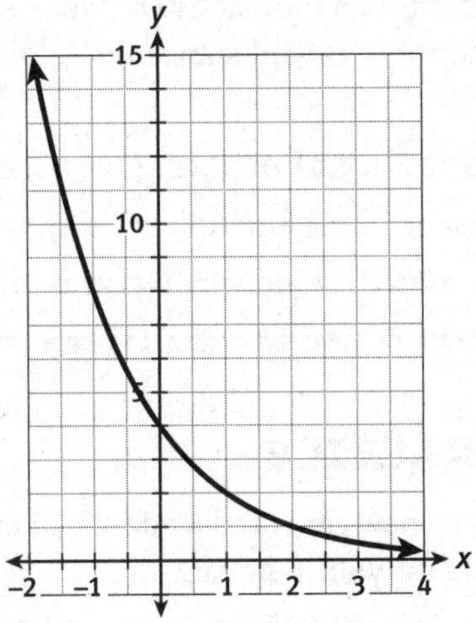

9. What are the domain and range of the function?

10. How is this graph different from your graph in Item 4?

11. How do the values of y change as the values of x increase?

12. How do the values of y change as the values of x decrease?

13. Identify the absolute maximum and absolute minimum of the function.

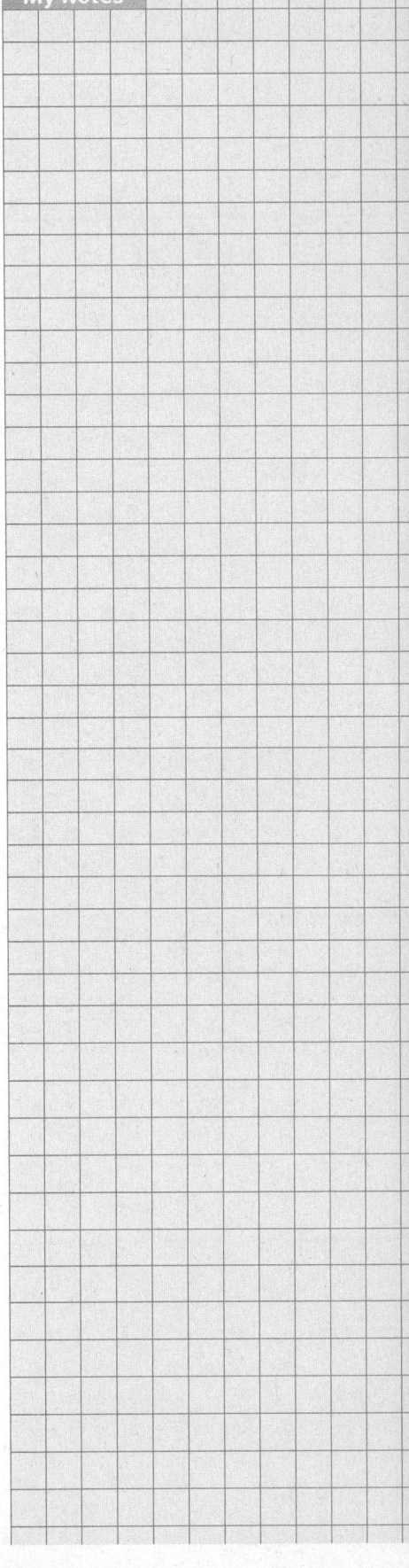

My Notes

Check Your Understanding

14. A scientist has *g* grams of a radioactive substance. Write an expression that shows the amount of the substance that remains after one half-life.

15. Critique the reasoning of others. Dylan looked at the function $f(x) = 4\left(\frac{1}{2}\right)^x$ and said, "This function is always greater than 0, so 0 is the absolute minimum." Explain why Dylan is incorrect.

LESSON 7-3 PRACTICE

Suppose the value of your new car is reduced by half every year that you own it. You paid $20,000 for your new car.

16. Describe how this situation is similar to the half-life of a radioactive substance.

17. Copy and complete the table below.

Time (years)	Value ($)
0	20,000
1	
2	
3	
4	
5	

18. Make sense of problems. For insurance purposes, a vehicle is considered scrap when its value falls below $500. After how many years will your new car be considered scrap?

ACTIVITY 7 PRACTICE

Write your answers on notebook paper. Show your work.

Lesson 7-1

A weight of 15 ounces stretches a spring 10 inches. A weight of 24 ounces stretches the same spring 16 inches. Use this information to answer Items 1–4.

1. How many inches does the spring stretch per ounce of additional weight?
   A. $\frac{2}{3}$ inch
   B. $\frac{3}{2}$ inches
   C. 25 inches
   D. 150 inches
2. Write an equation to describe the relationship between the distance d that the spring stretches and the weight w that is attached to it.
3. How much will the spring stretch for a weight of 9 ounces?
4. The spring is stretched 14 inches. How many ounces is the weight that is attached to it?

A spring stretches 2.5 inches for each ounce of weight. Use this information for Items 5–7.

5. Determine a function that represents this situation.
6. If you were to graph the function represented by this situation, what would be the reasonable domain? Explain.
7. Which of the following data points would **not** lie on the graph representing this function?
   A. (0, 0)
   B. (1, 2.5)
   C. (2.5, 1)
   D. (10, 25)

Lesson 7-2

Suppose that the height of an object after x seconds is given by $f(x) = 100 - 4x^2$, as shown in the graph below.

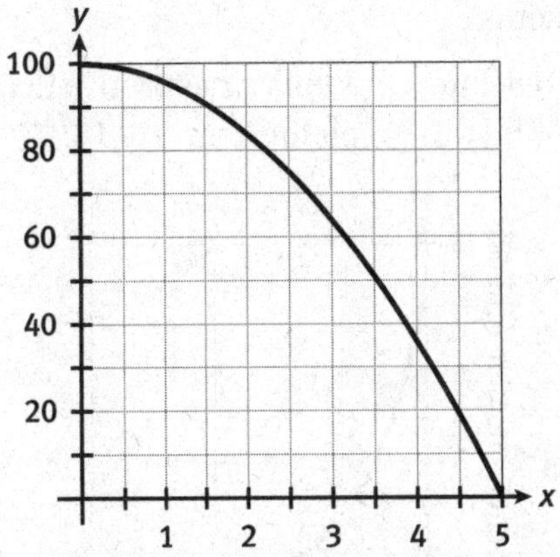

Use the function or the graph for Items 8–14.

8. What is the reasonable domain of the function?
9. What is the reasonable range of the function?
10. Identify the y-intercept of the function.
11. What does the y-intercept represent?
12. Identify the x-intercept of the function.
13. What does the x-intercept represent?
14. Loni says that because of the negative sign in front of $4x^2$, the reasonable domain for this function is only negative values. Is her reasoning correct? Explain.

Lesson 7-3

15. The half-life of a radioactive substance is 1 hour. If you begin with 100 ounces of the substance, how many hours does it take for 12.25 ounces to remain?

The graph below represents a radioactive decay situation. Use this graph for Items 16–18.

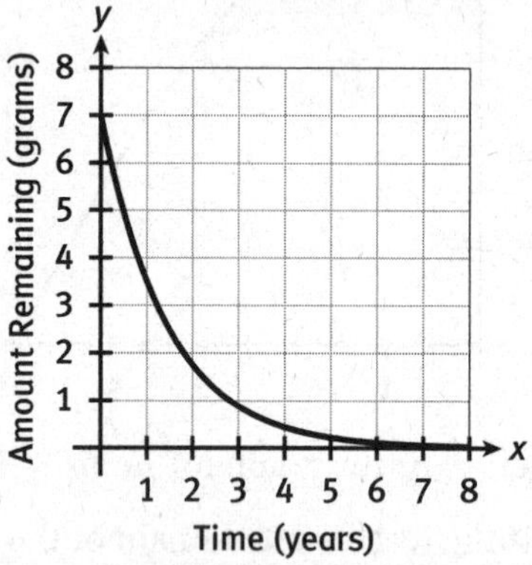

16. What is the original amount of the radioactive substance? Explain how you know.

17. What are the reasonable domain and range?

18. Identify the absolute maximum and absolute minimum values of the function. What do these values represent?

Barry has a piece of paper whose area is 150 square inches. He cuts the paper in half and discards one of the pieces. He repeats this procedure several times. Use this information for Items 19–24.

19. Copy and complete the table below to show the area of the remaining piece of paper after x cuts.

Number of Cuts, x	Area of Remaining Piece, y
0	150
1	
2	
3	
4	

20. Describe how this situation is similar to the half-life of a radioactive substance.

21. If you were to graph the points from the table, would you connect the points? Explain.

22. Describe how the reasonable domain in this situation is different from the reasonable domain in a radioactive decay situation.

23. Identify the y-intercept. What does it represent?

24. Identify the absolute maximum value. What does it represent?

MATHEMATICAL PRACTICES
Construct Viable Arguments and Critique the Reasoning of Others

25. Maude receives $100 for her birthday. “I am going to spend half of my birthday money each day until none is left,” she decides. Is it reasonable for her to believe that she will eventually spend all of the money? Justify your answer.

Transformations of Functions

Transformers

Lesson 8-1 Exploring $f(x) + k$

My Notes

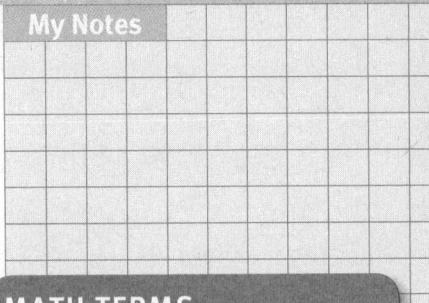

Learning Targets:

- Identify the effect on the graph of replacing $f(x)$ by $f(x) + k$.
- Identify the transformation used to produce one graph from another.

SUGGESTED LEARNING STRATEGIES: Look for a Pattern, Interactive Word Wall, Think-Pair-Share, Create Representations, Discussion Groups

The equation and the graph of $y = x$ or $f(x) = x$ are referred to as the linear ***parent function***. The graph of $f(x) = x$ is shown below.

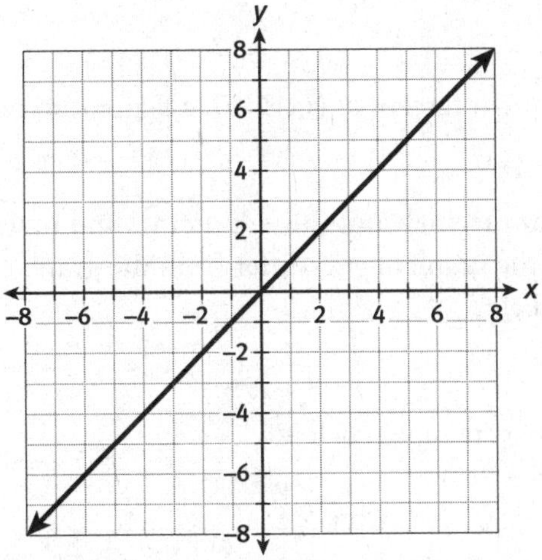

MATH TERMS

A **parent function** is the most basic function of a particular category or type.

1. Complete the table for $g(x) = x + 5$.

x	$f(x) = x$	$g(x) = x + 5$
−3	−3	2
−2	−2	
−1	−1	
0	0	
1	1	
2	2	
3	3	

2. **Make use of structure.** How do the y-values for $g(x)$ compare to the y-values for $f(x)$? Make a conjecture about the graph of $g(x)$. As you share your ideas with your group, be sure to use mathematical terms and academic vocabulary precisely. Make notes to help you remember the meaning of new words and how they are used to describe mathematical concepts.

My Notes

3. Test your conjecture by using a graphing calculator to graph $g(x) = x + 5$. Graph this on the grid in Item 1.

a. What is the y-intercept of the parent function?

b. What is the y-intercept of $g(x)$?

c. What is the x-intercept of the parent function? What is the zero of the function $f(x)$?

MATH TIP

The x-coordinate of the x-intercept is called a zero of the function. You will learn more about zeros of functions when you study quadratic functions later in this course.

d. What is the x-intercept of $g(x)$? What is the zero of the function?

e. Revisit your original conjecture in Item 2 and revise it if necessary. How does the graph of $g(x)$ differ from the graph of the parent function, $f(x) = x$?

The graph of $f(x) = x^3$ is shown below.

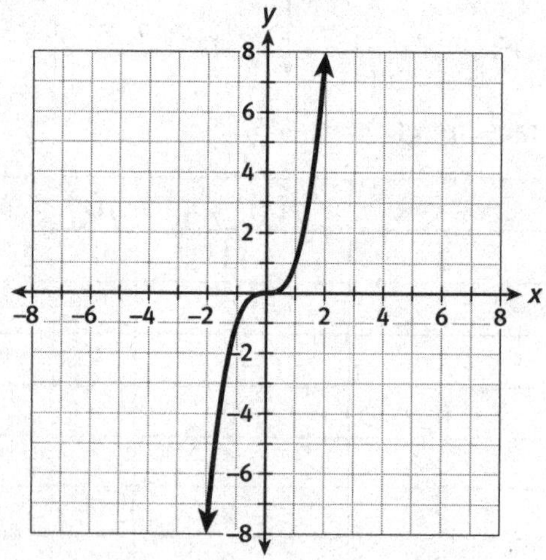

4. Make a conjecture about the graph of $g(x) = x^3 - 4$.

5. Graph both $f(x)$ and $g(x)$ on a graphing calculator. Sketch the graph of $g(x)$ on the grid above. Label a few points on each graph.

6. Revisit your original conjecture in Item 4 about the graph of $g(x)$ and revise it if necessary. How does the graph of $g(x)$ differ from the graph of $f(x)$?

7. Express regularity in repeated reasoning. How does the value of k in the equation $g(x) = f(x) + k$ change the graph of $f(x)$?

My Notes

A change in the position, size, or shape of a graph is a ***transformation***. The changes to the graphs in Items 1–6 are examples of a transformation called a ***vertical translation***.

MATH TERMS

A **vertical translation** of a graph shifts the graph up or down. A vertical translation preserves the shape of the graph.

8. In the figure, the graphs of $g(x)$ and $h(x)$ are vertical translations of the graph of $f(x) = 2^x$.

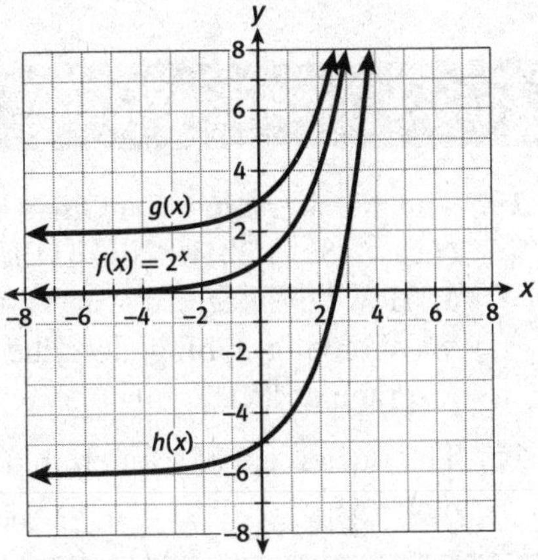

a. Write the equation for $g(x)$.

b. Write the equation for $h(x)$.

Check Your Understanding

9. Without graphing, describe the transformation from the graph of $f(x) = x^2$ to the graph of $g(x) = x^2 + 7$.

10. Suppose $f(x) = x - 2$. Describe the transformation from the graph of $f(x)$ to the graph of $g(x) = x + 3$. Use a graphing calculator to check your answer.

Ray's Gym charges an initial sign-up fee of \$25.00 and a monthly fee of \$15.00.

11. Reason abstractly. Write a function that describes the gym's total membership fee for x months.

12. Graph the function you wrote in Item 11 on the grid below. Label several points on the graph.

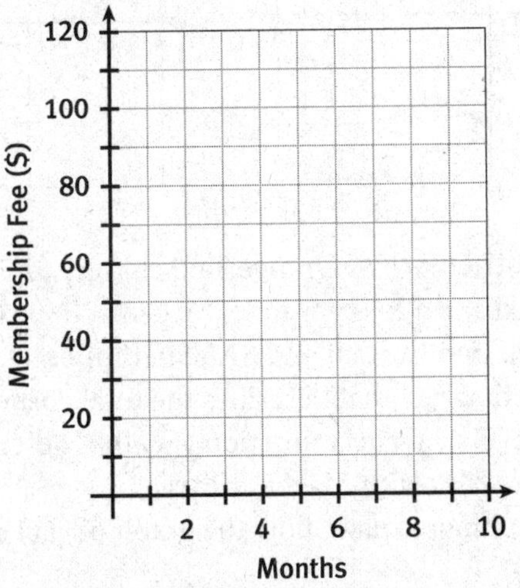

13. Identify the y-intercept. What does the y-intercept represent?

My Notes

14. How would the function change if the initial sign-up fee were increased by $5.00? How would the graph change?

Check Your Understanding

15. The membership fee at Gina's Gym is given by the function $g(x) = 15x + 20$, where x is the number of months.
 a. How do the fees at Gina's Gym compare to those at Ray's Gym?
 b. Without graphing, describe how the graph of $g(x)$ compares to the graph of $f(x)$.

16. The y-intercept of a function $f(x)$ is $(0, b)$. What is the y-intercept of $f(x) + k$?

LESSON 8-1 PRACTICE

Identify the transformation from the graph of $f(x) = x^2$ to the graph of $g(x)$. Then graph $f(x)$ and $g(x)$ on the same coordinate plane.

17. $g(x) = x^2 - 7$

18. $g(x) = x^2 + 10$

Write the equation of the function described by each of the following transformations of the graph of $f(x) = x^3$.

19. Translated up 9 units

20. Translated down 5 units

Each graph shows a vertical translation of the graph of $f(x) = x$. Write an equation to describe each graph.

21.

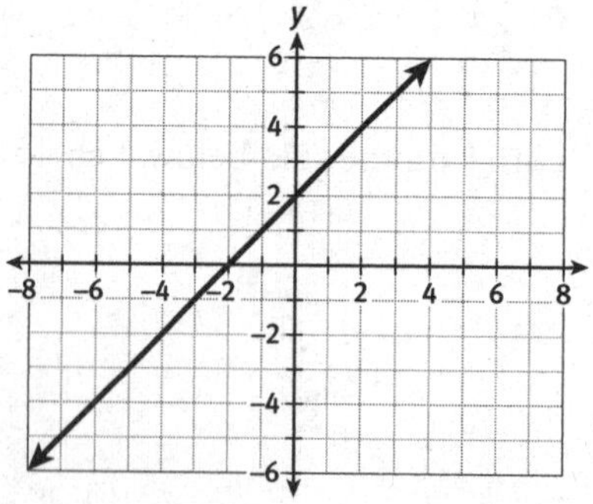

22.

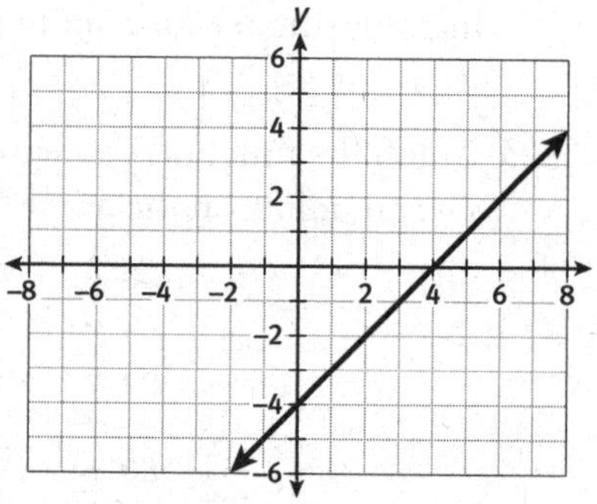

23. Model with mathematics. Orange Taxi charges $2.75 as soon as you step into the taxi and $2.50 per mile. Magenta Taxi charges $3.25 as soon as you step into the taxi and $2.50 per mile.
 a. Write a function $f(x)$ that describes the total cost of a ride of x miles with Orange Taxi. Write a function $g(x)$ that describes the total cost of a ride of x miles with Magenta Taxi.
 b. Without graphing, explain how the graph of $g(x)$ compares to the graph of $f(x)$.
 c. Check your answer to Part (b) by graphing the functions.

My Notes

Learning Targets:

- Identify the effect on the graph of replacing $f(x)$ by $f(x + k)$.
- Identify the transformation used to produce one graph from another.

SUGGESTED LEARNING STRATEGIES: Predict and Confirm, Look for a Pattern, Create Representations, Think-Pair-Share, Discussion Groups

The function $f(x) = |x|$ is graphed below.

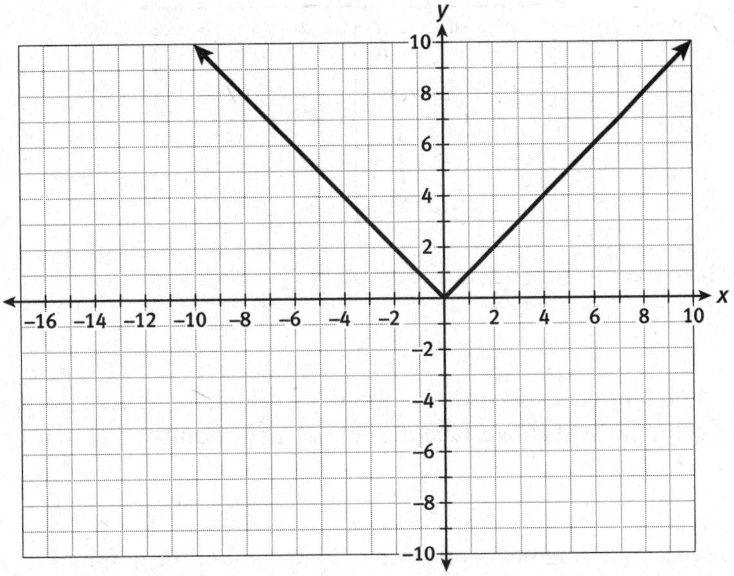

1. Write a new function, $g(x)$, by replacing x with $x + 7$.

2. Graph both $f(x) = |x|$ and $g(x)$ on a graphing calculator. Sketch the graph of $g(x)$ on the grid above, labeling at least a few points on each graph.

3. What is the x-intercept of $f(x) = |x|$?

4. What is the x-intercept of $g(x)$?

5. Describe the transformation from the graph of $f(x) = |x|$ to the graph of $g(x)$.

Note that the function $g(x)$ can be written as $f(x + 7)$. This means that x is replaced with $x + 7$ in the function $f(x)$.

MATH TERMS

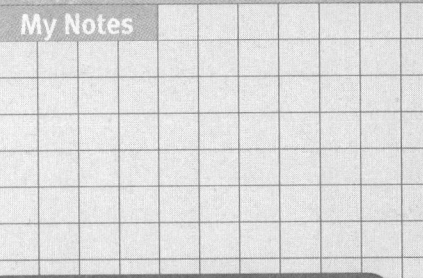

An **absolute value function** is written as $f(x) = |x|$ and is defined by

$$f(x) = \begin{cases} -x \text{ if } x < 0 \\ x \text{ if } x \geq 0 \end{cases}$$

CONNECT TO AP

The vertex of an absolute value function is an example of a **cusp** in a graph. A graph has a cusp at a point where there is an abrupt change in direction.

My Notes

The graph of $f(x) = x^3$ is shown below.

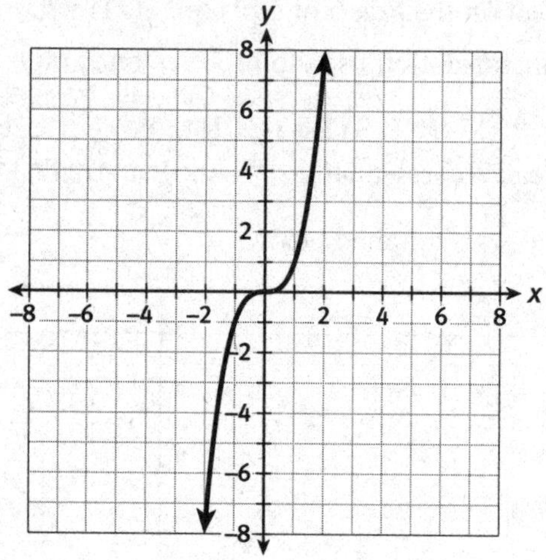

6. Make a conjecture about the graph of $g(x) = (x - 3)^3$.

7. Graph both $f(x)$ and $g(x)$ on a graphing calculator. Sketch the graph of $g(x)$ on the grid above, labeling at least a few points on each graph.

8. Revisit your original conjecture in Item 6 about the graph of $g(x)$ and revise it if necessary. How does the graph of $g(x)$ differ from the graph of $f(x)$?

9. How does the value of k in the equation $g(x) = f(x + k)$ change the graph of the function $f(x)$?

The changes to the graphs in Items 1–8 are examples of a transformation called a ***horizontal translation***.

MATH TERMS

A **horizontal translation** of a graph shifts the graph left or right. Like a vertical translation, a horizontal translation preserves the shape of the graph.

10. The figure shows the graph of the function $f(x) = 2^x$.
   a. Without using a graphing calculator, sketch the graph of $g(x) = f(x + 8) = 2^{x+8}$ on the grid.

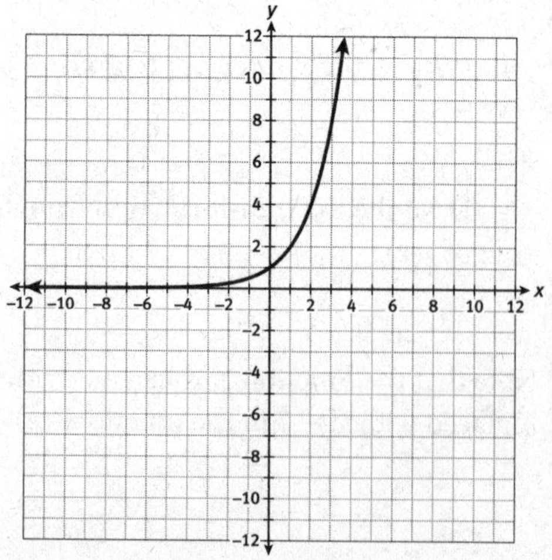

   b. Use a graphing calculator to check your graph in Part (a). Revise your graph if necessary.

My Notes

Check Your Understanding

11. Without graphing, describe the transformation from the graph of $f(x) = x^2$ to the graph of $g(x)$.

a. $g(x) = (x + 4)^2$ **b.** $g(x) = f(x - 7)$

c. $g(x) = (x - 2)^2 + 5$ **d.** $g(x) = (x + 9)^2 - 1$

12. The function $f(x) = x^2$ and another function, $g(x)$, are graphed below. Write the equation for $g(x)$. Explain how you found your answer.

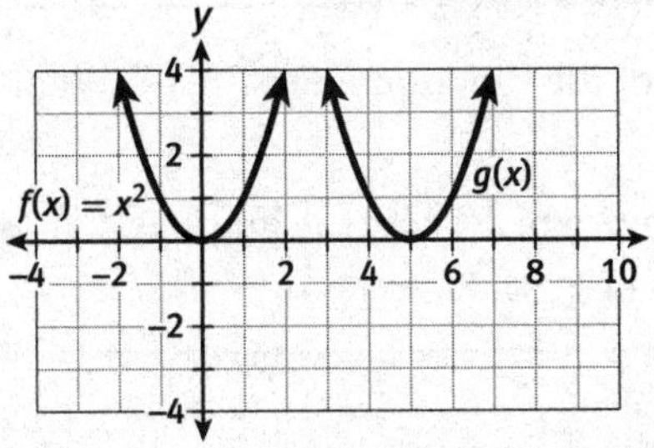

13. Make sense of problems. Julio went to a theme park in July. He paid \$15 to enter the park and \$3.00 for each ride. He went on x rides.

a. Write a function that describes the total cost of Julio's trip to the theme park.

b. Julio went back to the theme park in September. The entrance fee was the same and each ride still cost \$3.00. However, this time Julio went on 5 more rides. Use your function from Part (a) to describe Julio's second trip.

c. How does the equation for Julio's second trip to the park change the graph of the first trip?

d. What kind of transformation describes the change from the first graph to the second graph?

e. Julio went to the park again in October and went on 8 fewer rides than he did in July. Use your function from Part (a) to describe Julio's third trip. How does this change the initial graph?

My Notes

f. Julio goes to the park again in November. Now it is the off-season and the entrance fee is $10 less than it was in July. He goes on the same number of rides as he did in July. Write a function to describe Julio's fourth trip. How does the graph of the initial trip change with this new situation?

Check Your Understanding

14. The x-intercept of the function $f(x)$ is $(a, 0)$. What is the x-intercept of the function $f(x + k)$?

15. Without graphing, explain how the graph of $y = (x - 4)^3$ is related to the graph of $y = (x + 4)^3$.

LESSON 8-2 PRACTICE

Identify the transformation from the graph of $f(x) = x^2$ to the graph of $g(x)$. Then graph $f(x)$ and $g(x)$ on the same coordinate plane.

16. $g(x) = (x - 1)^2$

17. $g(x) = (x + 3)^2$

Write the equation of the function described by each of the following transformations of the graph of $f(x) = x^3$.

18. Translated 7 units to the left

19. Translated 8 units to the right

20. Each graph shows a horizontal translation of the graph of $f(x) = x$. Write an equation to describe each graph.

a.

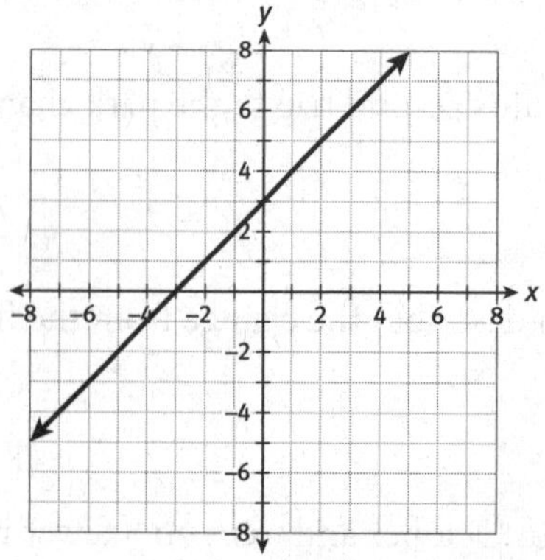

b.

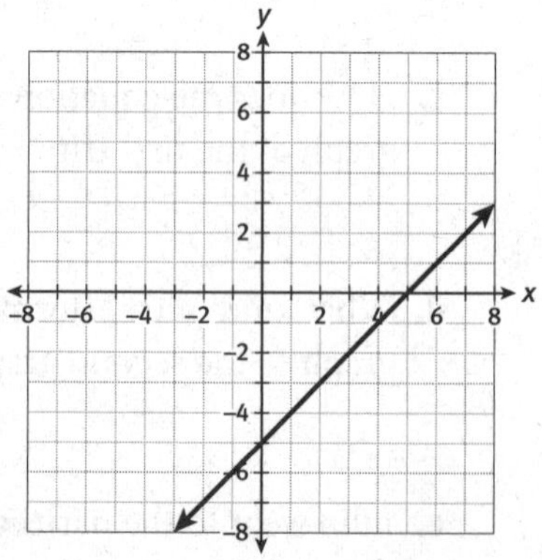

c. Critique the reasoning of others. Molly said that the graphs above are also vertical translations of the graph of $f(x) = x$. Is Molly correct? Explain.

21. How does the graph of $h(x) = |x - 4|$ compare with the graph of $f(x) = |x|$?

ACTIVITY 8 PRACTICE

Write your answers on notebook paper. Show your work.

Lesson 8-1

In Items 1–4, identify the transformation from the graph of $f(x) = x^3$ to the graph of $g(x)$.

1. $g(x) = x^3 + 11$
2. $g(x) = x^3 - 4$
3. $g(x) = x^3 + 0.1$
4. $g(x) = -2 + x^3$
5. The graph of $f(x) = x^2$ is translated 9 units down to create the graph of $g(x)$. Which of the following is the equation for $g(x)$?
   A. $g(x) = x^2 + 9$
   B. $g(x) = x^2 - 9$
   C. $g(x) = (x + 9)^2$
   D. $g(x) = (x - 9)^2$

In Items 6 and 7, each graph shows a vertical translation of the graph of $f(x) = x$. Write an equation to describe the graph. Identify the zeros of each function.

6.

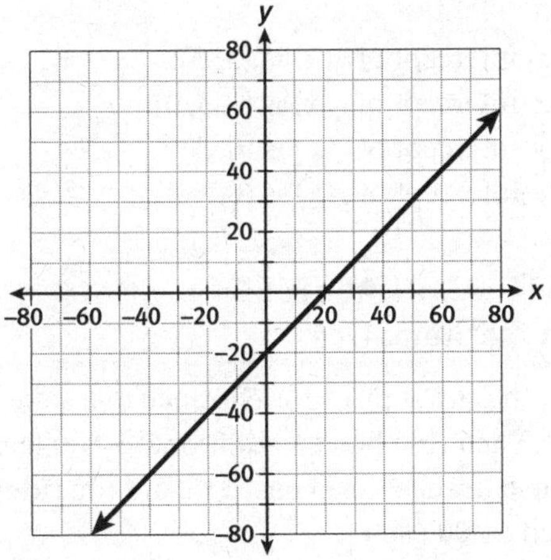

7.

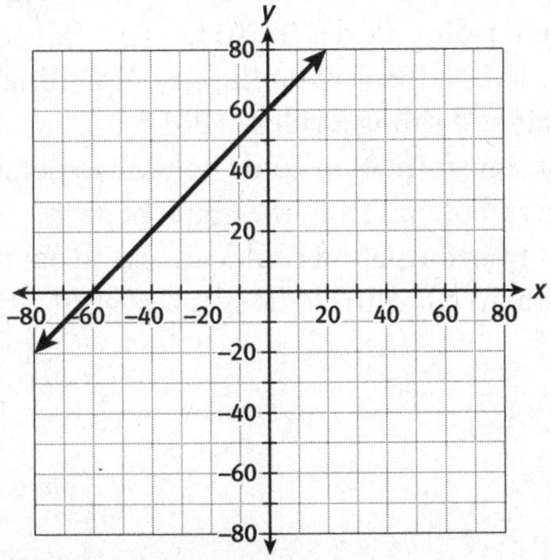

For Items 8 and 9, determine the equation of the function described by each of the following transformations of the graph of $f(x) = 3^x$.

8. Translated 15 units down
9. Translated 2.1 units up
10. An air conditioner costs \$450 plus \$40 per month to operate.
    a. Write a function that describes the total cost of buying and operating the air conditioner for x months.
    b. Use your calculator to graph the function.
    c. What is the y-intercept? What does it represent?
    d. How would the function change if the price of the air conditioner were reduced to \$425? How would the graph change?

Given that $g(x) = f(x) + k$, with $k \neq 0$, determine whether each statement is always, sometimes, or never true.

11. The graph of $g(x)$ is a vertical translation of the graph of $f(x)$.
12. The graphs of $f(x)$ and $g(x)$ are both lines.
13. The graph of $f(x)$ has the same y-intercept as the graph of $g(x)$.
14. Caitlin drew the graph of $f(x) = x^2$. Then she translated the graph 6 units up to get the graph of $g(x)$. Next, she translated the graph of $g(x)$ 8 units down to get the graph of $h(x)$. Which of these is an equation for $h(x)$?
    A. $h(x) = x^2 + 14$
    B. $h(x) = x^2 + 2$
    C. $h(x) = x^2 - 2$
    D. $h(x) = x^2 - 14$

Lesson 8-2

In Items 15–18, identify the transformation from the graph of $f(x) = 2^x$ to the graph of $g(x)$.

15. $g(x) = 2^x - 3$

16. $g(x) = 2^{(x-3)}$

17. $g(x) = 2^x + 4$

18. $g(x) = 2^{(x+4)}$

19. The graph of which function is a translation of the graph of $f(x) = x^2$ five units to the right?

A. $g(x) = x^2 - 5$
B. $g(x) = (x + 5)^2$
C. $g(x) = (x - 5)^2$
D. $g(x) = x^2 + 5$

Write the equation of the function described by each of the following transformations of the graph of $f(x) = x^3$.

20. Translated 7 units up

21. Translated 4 units down

22. Translated 2 units right

23. Translated 5 units down

24. Translated 3 units left

25. The figure shows the graph of $f(x) = x^4$ and the graph of $g(x)$. Write an equation for the graph of $g(x)$.

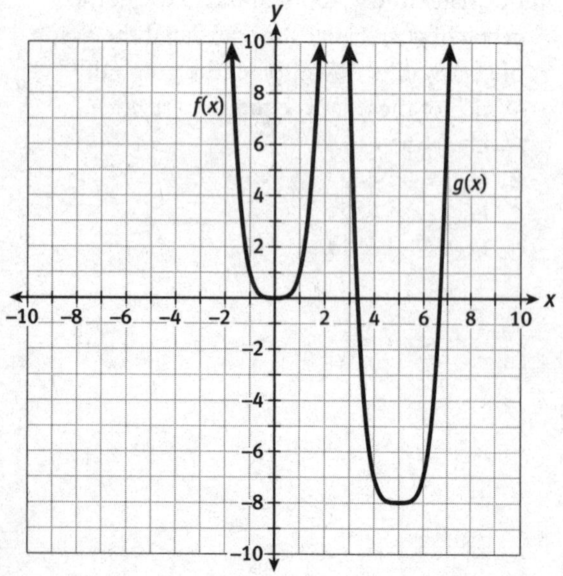

Without graphing, describe the transformation from the graph of $f(x) = x^2$ to the graph of $g(x)$.

26. $g(x) = (x - 7)^2 + 1$

27. $g(x) = f(x + 4)$

28. $g(x) = (x + 9)^2 - 0.2$

29. $g(x) = f(x - 2) - 3$

30. The graph of $f(x)$ is shown below. Which of the following is a true statement about the graph of $g(x) = f(x + 3)$?

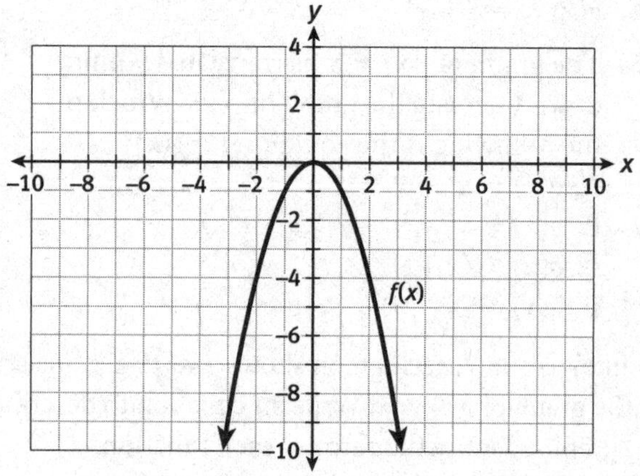

A. The x-intercept of $g(x)$ is (3, 0).
B. The x-intercept of $g(x)$ is (−3, 0).
C. The y-intercept of $g(x)$ is (0, 3).
D. The y-intercept of $g(x)$ is (0, −3).

MATHEMATICAL PRACTICES
Model with Mathematics

31. In 2011, the ticket price for entrance to a state fair was \$12. Each ride had an additional \$4.00 fee. In 2012, the entrance ticket cost \$15 and the rides remained \$4.00 each.

a. Write a function $f(x)$ for the cost of visiting the fair and riding x rides in 2011.
b. Write a function $g(x)$ for the cost of visiting the fair and riding x rides in 2012.
c. What transformation could you use to obtain the graph of $g(x)$ from the graph of $f(x)$?
d. What transformation could you use to obtain the graph of $f(x)$ from the graph of $g(x)$?

While on vacation, Jorge and Jackie traveled to Bryce Canyon National Park in Utah. They were impressed by the differing elevations at the viewpoints along the road. The graph describes the elevations for several viewpoints in terms of the time since they entered the park.

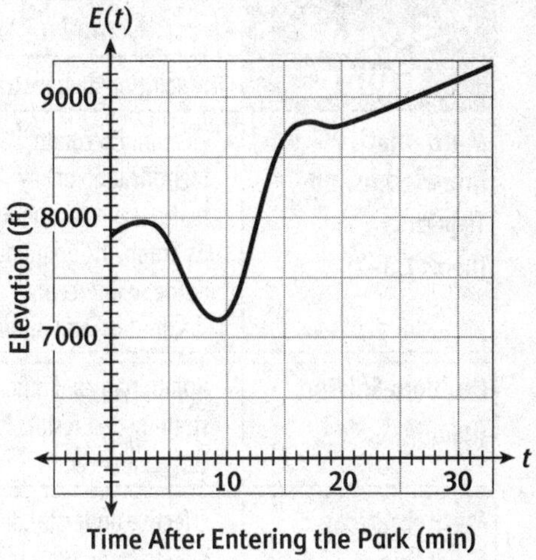

1. The graph represents a function $E(t)$. Describe why the graph represents a function. Identify the domain and range of the function.

2. Is this discrete or continuous data? Explain.

3. What is the y-intercept? Interpret the meaning of the y-intercept in the context of the problem.

4. Identify a relative maximum of the function represented by the graph.

5. What is the absolute maximum of the function represented by the graph? What does it represent?

6. Identify a relative minimum of the function represented by the graph.

7. What is the absolute minimum of the function represented by the graph? What does it represent?

While at Bryce Canyon National Park, Jorge and Jackie hiked at an average speed of about 2 miles per hour.

8. Copy and complete the table below to show the distance hiked by a person whose constant speed is 2 miles per hour.

Time (hours)	Distance (miles)
0	0
1	2
2	
3	
4	
5	

9. Write a function $f(x)$ to describe the data in the table. What are the reasonable domain and range?

10. Create a graph of the function.

11. How long will it take this person to hike 5 miles? Justify your answer.

12. On the same coordinate grid that you used in Item 9, create a graph of another function by translating the graph 5 units up.

13. Write a function to describe the graph you created in Item 12. Explain how you determined your answer.

Representations of Functions

BRYCE CANYON HIKING

Scoring Guide	Exemplary	Proficient	Emerging	Incomplete
	The solution demonstrates the following characteristics:			
Mathematics Knowledge and Thinking (Items 1, 3–7)	• Clear and accurate identification of key features of the function and its graph, including domain, range, y-intercept, maximums, and minimums	• Correct identification of most of the key features of the function and its graph, including domain, range, y-intercept, maximums, and minimums	• Partially correct identification of some of the key features of the function and its graph, including domain, range, y-intercept, maximums, and minimums	• Inaccurate or incomplete identification of key features of the function and its graph, including domain, range, y-intercept, maximums, and minimums
Problem Solving (Item 11)	• Appropriate and efficient strategy that results in a correct answer	• Strategy that may include unnecessary steps but results in a correct answer	• Strategy that results in some incorrect answers	• No clear strategy when solving problems
Mathematical Modeling / Representations (Items 8–10, 12, 13)	• Effective understanding of how to complete a table of real-world data, and how to write, graph, and interpret the associated function • Fluency in translating a graph and writing the associated function	• Largely correct understanding of how to complete a table of real-world data, and how to write, graph, and interpret the associated function • Little difficulty translating a graph and writing the associated function	• Partial understanding of how to complete a table of real-world data, and how to write, graph, and interpret the associated function • Some difficulty translating a graph and writing the associated function	• Inaccurate or incomplete understanding of how to complete a table of real-world data, and how to write, graph, and interpret the associated function • Significant difficulty translating a graph and writing the associated function
Reasoning and Communication (Items 1–3, 5, 7, 13)	• Precise use of appropriate math terms and language to describe key features of a graph and to explain how a function rule was determined from a translated graph • Clear and accurate interpretations of the graph of a function	• Adequate description of key features of a graph • Reasonable interpretations of the graph of a function • Adequate explanation of how a function rule was determined from a translated graph	• Confusing description of key features of a graph • Partially correct interpretations of the graph of a function • Confusing explanation of how a function was determined from a translated graph	• Incomplete or inaccurate description of key features of a graph • Incomplete or inaccurate interpretation of the graph of a function • Incomplete or inaccurate explanation of how a function was determined from a translated graph

Learning Targets:

- Determine the slope of a line from a graph.
- Develop and use the formula for slope.

SUGGESTED LEARNING STRATEGIES: Close Reading, Summarizing, Sharing and Responding, Discussion Groups, Construct an Argument, Identify a Subtask

Margo's grandparents are moving in with her family. The family needs to make it easier for her grandparents to get in and out of the house. Margo has researched the specifications for building stairs and wheelchair ramps. She found the government website that gives the Americans with Disabilities Act (ADA) accessibility guidelines for wheelchair ramps and discovered the following diagram:

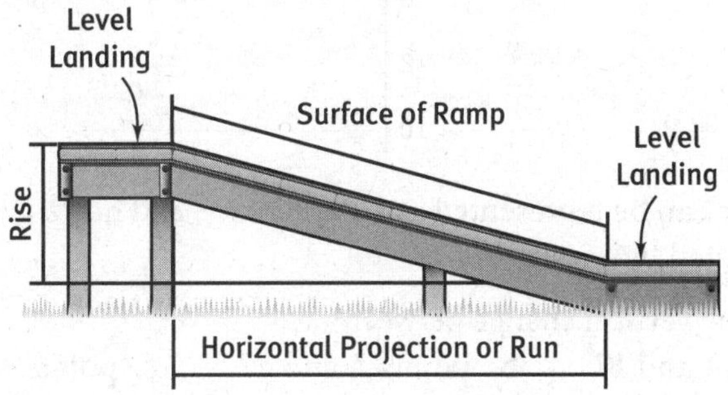

Then, Margo decided to look for the requirements for building stairs and found the following diagram:

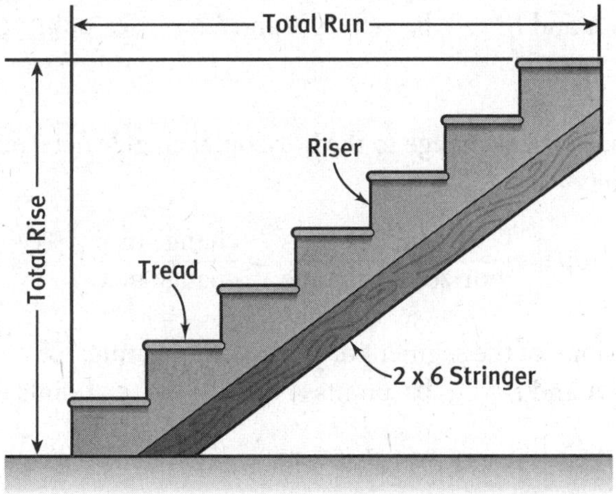

Review with your group the background information that is given as you solve the following items.

1. What do you think is meant by the terms *rise* and *run* in this context?

My Notes

CONNECT TO SOCIAL SCIENCE

The table gives information from the ADA website about the slope of wheelchair ramps.

Slope	Maximum Rise		Maximum Run	
	in.	mm	ft	m
$\frac{1}{16} < m \le \frac{1}{12}$	30	760	30	9
$\frac{1}{20} \le m < \frac{1}{16}$	30	760	40	12

My Notes

Consider the line in the graph below:

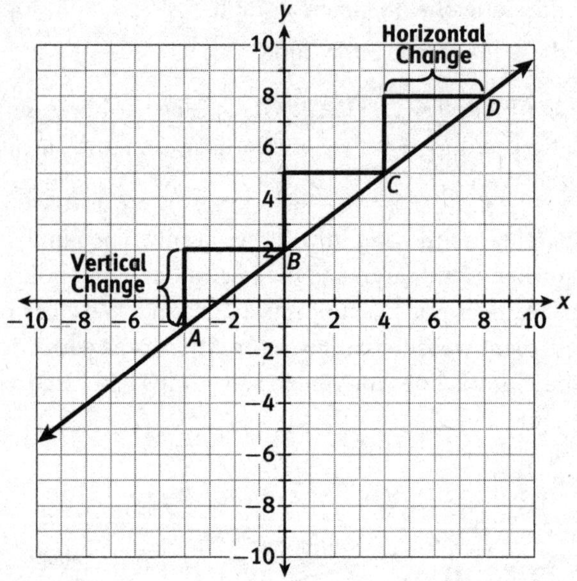

Vertical change can be represented as a *change in y*, and horizontal change can be represented by a *change in x*.

2. What is the vertical change between:
a. points *A* and *B*? **b.** points *A* and *C*? **c.** points *C* and *D*?

3. What is the horizontal change between:
a. points *A* and *B*? **b.** points *A* and *C*? **c.** points *C* and *D*?

The ratio of the vertical change to the horizontal change determines the ***slope*** of the line.

$$\text{slope} = \frac{\text{vertical change}}{\text{horizontal change}} = \frac{\text{change in } y}{\text{change in } x} = \frac{\Delta y}{\Delta x}$$

MATH TERMS

Slope is a measure of the amount of decline or incline of a line. The variable *m* is often used to represent slope.

4. Find the slope of the segment of the line connecting:
a. points *A* and *B* **b.** points *A* and *C* **c.** points *C* and *D*

5. What do you notice about the slope of the line in Items 4a, 4b, and 4c?

6. What does your answer to Item 5 indicate about points on a line?

My Notes

7. Slope is sometimes referred to as $\frac{\text{rise}}{\text{run}}$. Explain how the ratio $\frac{\text{rise}}{\text{run}}$ relates to the ratios for finding slope mentioned above.

8. **Reason quantitatively.** Would the slope change if you counted the run (horizontal change) before you counted the rise (vertical change)? Explain your reasoning.

WRITING MATH

In mathematics the Greek letter Δ (delta) represents a change or difference between mathematical values.

9. Determine the slope of the line graphed below.

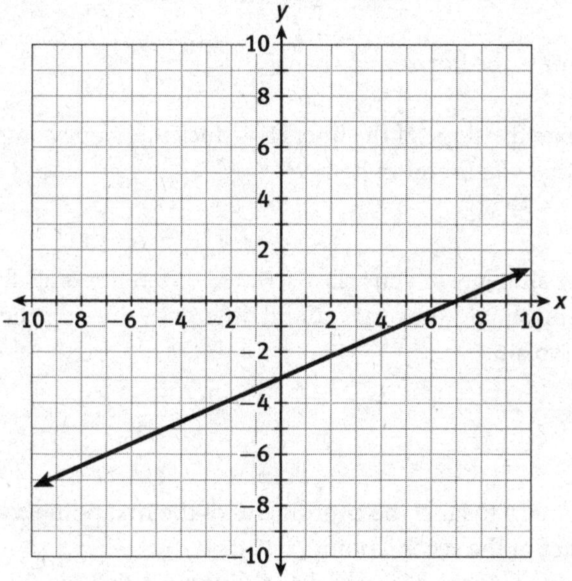

MATH TIP

Select two points on the line and use them to compute the slope.

My Notes

Although the slope of a line can be calculated by looking at a graph and counting the vertical and horizontal change, it can also be calculated numerically.

10. Recall that the slope of a line is the ratio $\frac{\text{change in } y}{\text{change in } x}$.

a. Identify two points on the graph above and record the coordinates of the two points that you selected.

	x-coordinate	*y*-coordinate
1st point		
2nd point		

b. Which coordinates relate to the vertical change on the graph?

c. Which coordinates relate to the horizontal change on a graph?

d. Determine the vertical change.

e. Determine the horizontal change.

f. Calculate the slope of the line. How does this slope compare to the slope that you found in Item 9?

g. If other students in your class selected different points for this problem, should they have found different values for the slope of this line? Explain.

11. It is customary to label the coordinates of the first point (x_1, y_1) and the coordinates of the second point (x_2, y_2).

a. Write an expression to calculate the vertical change, Δy, of the line through these two points.

b. Write an expression to calculate the horizontal change, Δx, of the line through these two points.

c. Write an expression to calculate the slope of the line through these two points.

My Notes

Check Your Understanding

12. Use the slope formula to determine the slope of a line that passes through the points (4, 9) and (−8, −6).

13. Use the slope formula to determine the slope of the line that passes through the points (−5, −3) and (9, −10).

14. Explain how to find the slope of a line from a graph.

15. Explain how to find the slope of a line when given two points on the line.

LESSON 9-1 PRACTICE

16. Find Δx and Δy for the points (7, −2) and (9, −7).

17. **Critique the reasoning of others.** Connor determines the slope between (−2, 4) and (3, −3) by calculating $\frac{4-(-3)}{-2-3}$. April determines the slope by calculating $\frac{3-(-2)}{-3-4}$. Explain whose reasoning is correct.

18. When given a table of ordered pairs, you can find the slope by choosing any two ordered pairs from the table. Determine the slope represented in the table below.

x	5	7	9	11
y	5	3	1	−1

19. Determine the slope of the given line.

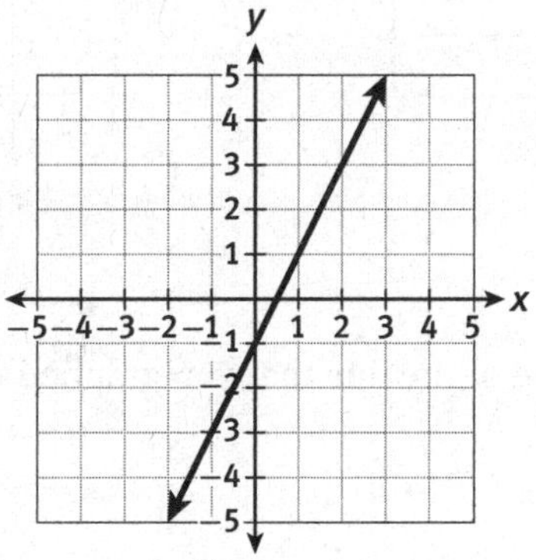

My Notes

Learning Targets:

- Calculate and interpret the rate of change for a function.
- Understand the connection between rate of change and slope.

SUGGESTED LEARNING STRATEGIES: Discussion Groups, Create Representations, Look for a Pattern, Think-Pair-Share

The ***rate of change*** for a function is the ratio of the change in y, the dependent variable, to the change in x, the independent variable.

1. Margo went to the lumberyard to buy supplies to build the wheelchair ramp. She knows that she will need several pieces of wood. Each piece of wood costs $3.

   a. **Model with mathematics.** Write a function $f(x)$ for the total cost of the wood pieces if Margo buys x pieces of wood.

   b. Make an input/output table of ordered pairs and then graph the function.

Pieces of Wood, x	Total Cost, $f(x)$

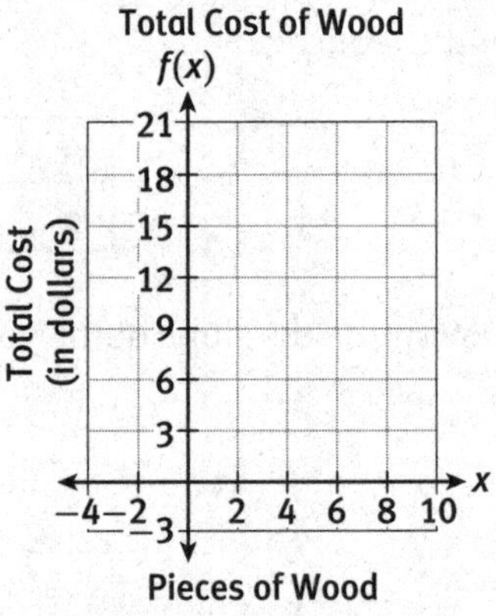

   c. What is the slope of the line that you graphed?

   d. By how much does the cost increase for each additional piece of wood purchased?

My Notes

e. How does the slope of this line relate to the situation with the pieces of wood?

f. Is there a relationship between the slope of the line and the equation of the line? If so, describe that relationship.

2. Margo is going to work with a local carpenter during the summer. Each week she will earn $10.00 plus $2.00 per hour.

a. Write a function $f(x)$ for Margo's total earnings if she works x hours in one week.

b. Make an input/output table of ordered pairs and then graph the function. Label your axes.

Hours, x	Earnings, $f(x)$ (dollars)

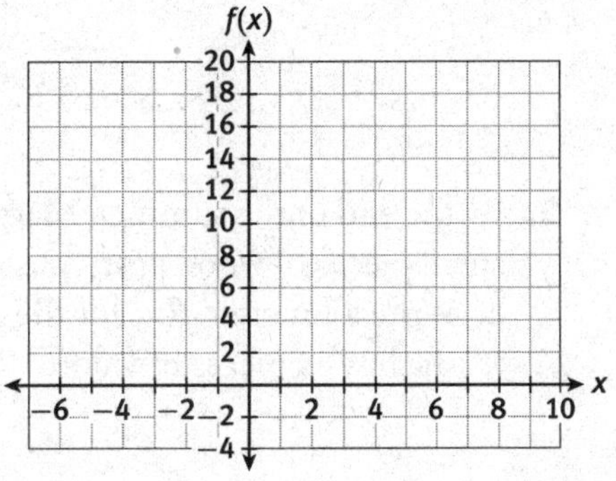

My Notes

c. How much will Margo's earnings change if she works 6 hours instead of 2? If she works 4 hours instead of 3? How much do Margo's earnings change for each additional hour worked?

d. Does the function have a constant rate of change? If so, what is it?

e. What is the slope of the line that you graphed?

f. Describe the meaning of the slope within the context of Margo's job.

g. Describe the relationship between the slope of the line, the rate of change, and the equation of the line.

h. How much will Margo earn if she works for 8 hours in one week?

3. By the end of the summer, Margo has saved $375. Recall that each of the small pieces of wood costs $3.

a. Write a function $f(x)$ for the amount of money that Margo still has if she buys x pieces of wood.

My Notes

b. Make an input/output table of ordered pairs and then graph the function.

Pieces of Wood, x	Money Remaining, $f(x)$ (dollars)

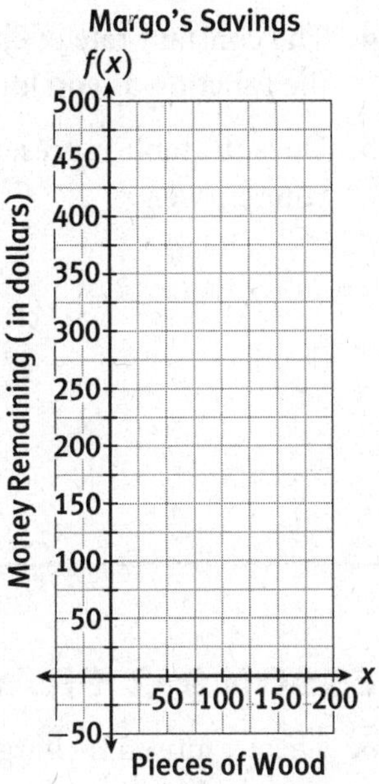

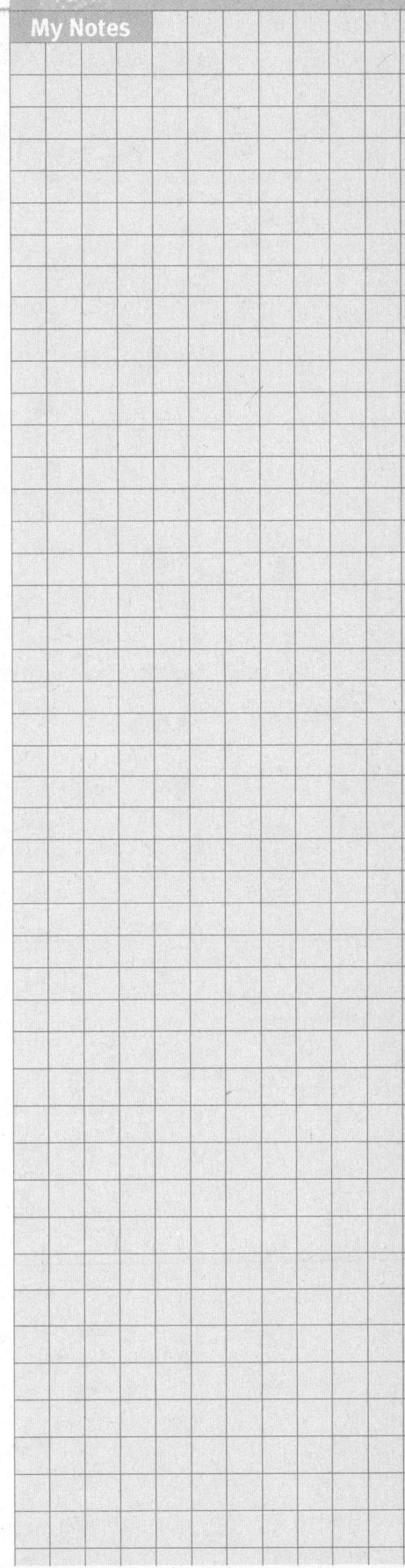

c. How much will the amount Margo has saved change if she buys 100 instead of 25 pieces of wood? If she buys 50 instead of 0 pieces of wood? For each additional piece of wood? Explain.

d. Does the function have a constant rate of change? If so, what is it?

e. What is the slope of the line that you graphed?

f. How are the rate of change of the function and the slope related?

g. Describe the meaning of the slope within the context of Margo's savings.

h. How does this slope differ from the other slopes that you have seen in this activity?

My Notes

Check Your Understanding

4. The constant rate of change of a function is −5. Describe the graph of the function as you look at it from left to right.

5. Does the table represent data with a constant rate of change? Justify your answer.

x	*y*
2	−5
4	5
7	20
11	40

LESSON 9-2 PRACTICE

6. The art museum charges an initial membership fee of $50.00. For each visit the museum charges $15.00.

a. Write a function $f(x)$ for the total amount charged for x trips to the museum.

b. Make a table of ordered pairs and then graph the function.

c. What is the rate of change? What is the slope of the line?

d. How does the slope of this line relate to the number of museum visits?

7. Critique the reasoning of others. Simone claims that the slope of the line through (−2, 7) and (3, 0) is the same as the slope of the line through (2, 1) and (12, −13). Prove or disprove Simone's claim.

My Notes

Learning Targets:

- Show that a linear function has a constant rate of change.
- Understand when the slope of a line is positive, negative, zero, or undefined.
- Identify functions that do not have a constant rate of change and understand that these functions are not linear.

SUGGESTED LEARNING STRATEGIES: Look for a Pattern, Think-Pair-Share, Construct an Argument, Sharing and Responding, Summarizing

You have seen that for a linear function, the rate of change is constant and equal to the slope of the line. This is because linear functions increase or decrease by equal differences over equal intervals. Look at the graph below.

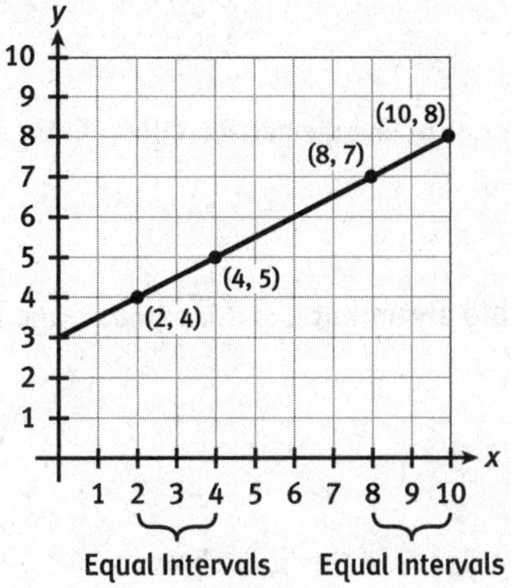

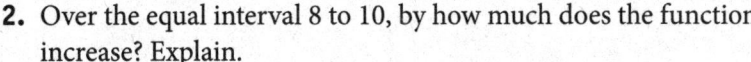

1. Over the interval 2 to 4, by how much does the function increase? Explain.

2. Over the equal interval 8 to 10, by how much does the function increase? Explain.

"Equal differences over equal intervals" is an equivalent way of referring to constant slope. "Differences" refers to Δy, and "intervals" refers to Δx. "Equal differences over equal intervals" means $\frac{\Delta y}{\Delta x}$, which represents the slope, will always be the same.

My Notes

3. The table below represents a function.

x	y
−8	62
−6	34
−1	−1
1	−1
5	23
7	47

a. Determine the rate of change between the points (−8, 62) and (−6, 34).

b. Determine the rate of change between the points (−1, −1) and (1, −1).

c. Construct viable arguments. Is this a linear function? Justify your answer.

4. a. Determine the slopes of the lines shown.

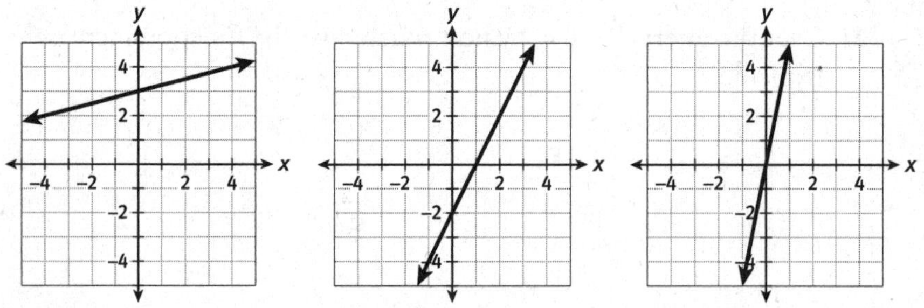

b. Express regularity in repeated reasoning. Describe the slope of any line that rises as you view it from left to right.

My Notes

5. a. Determine the slopes of the lines shown.

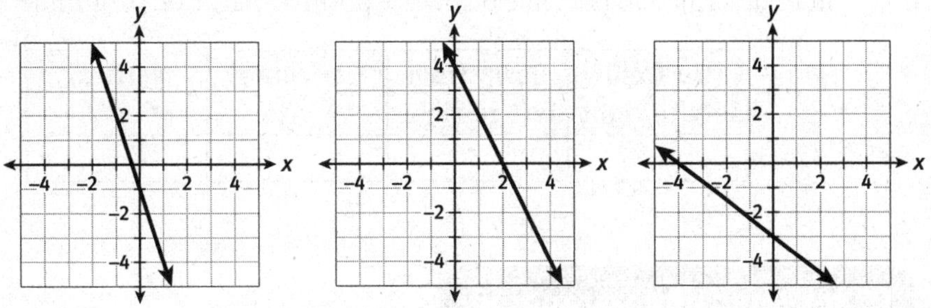

b. Express regularity in repeated reasoning. Describe the slope of any line that falls as you view it from left to right.

6. a. Determine the slopes of the lines below.

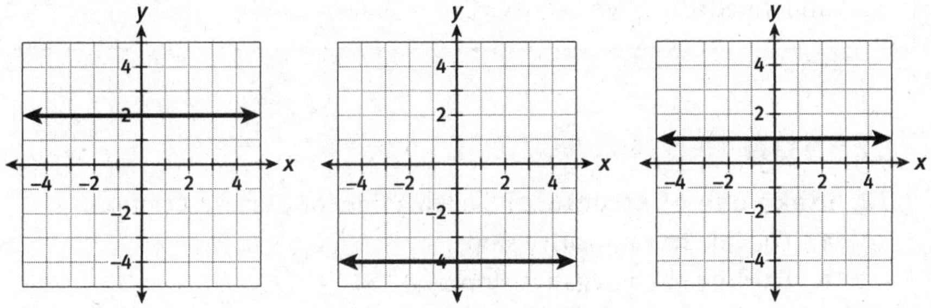

b. What is the slope of a horizontal line?

7. a. Determine the slopes of the lines shown.

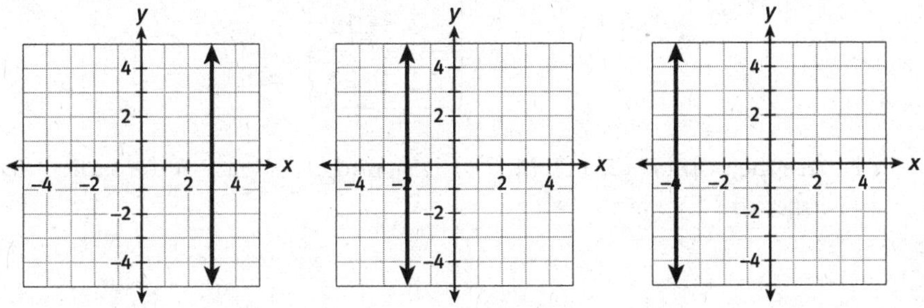

b. What is the slope of a vertical line?

My Notes

8. Summarize your findings in Items 4–7. Tell whether the slopes of the lines described in the table below are positive, negative, 0, or undefined.

Up from left to right	Down from left to right	Horizontal	Vertical

Check Your Understanding

9. Suppose you are given several points on the graph of a function. Without graphing, how could you determine whether the function is linear?
10. How can you tell from a graph if the slope of a line is positive or negative?
11. Describe a line having an undefined slope. Why is the slope undefined?

LESSON 9-3 PRACTICE

12. **Make use of structure.** Sketch a line for each description.
    a. The line has a positive slope.
    b. The line has a negative slope.
    c. The line has a slope of 0.
13. Does the table represent a linear function? Justify your answer.

x	y
1	−1
4	9
7	19
11	29

14. Are the points (12, 11), (2, 7), (5, 9), and (1, 5) part of the same linear function? Explain.

ACTIVITY 9 PRACTICE

Write your answers on notebook paper. Show your work.

Lesson 9-1

1. Find Δx and Δy for each of the following pairs of points.
   a. (2, 6), (−6, −8)
   b. (0, 9), (4, −8)
   c. (−3, −3), (7, 10)

For Items 2 and 3, use the table to calculate the slope.

2.

x	y
−5	−1
0	2
5	5
10	8

3.

x	y
−4	20
−3	14
0	−4
2	−16

4. Two points on a line are (−10, 1) and (5, −5). If the y-coordinate of another point on the line is −3, what is the x-coordinate?

For Items 5–7, determine the slope of the line that passes through each pair of points.

5. (−4, 11) and (1, −9)
6. (−10, −3) and (−5, 1)
7. (−2, −7) and (−8, −4)
8. Are the three points (2, 3), (5, 6), and (0, −2) on the same line? Explain.
9. Which of the following pairs of points lies on a line with a slope of $-\frac{3}{5}$?
   A. (4, 0), (−2, 10)
   B. (4, 2), (10, 4)
   C. (−4, −10), (0, −2)
   D. (10, −2), (0, 4)

For Item 10, determine the slope of the line that is graphed.

10.

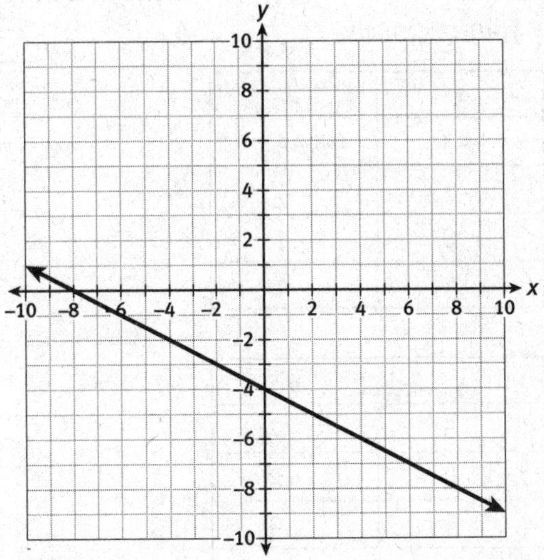

Lesson 9-2

11. Juan earns \$7 per hour plus \$20 per week making picture frames.
    a. Write a function $g(x)$ for Juan's total earnings if he works x hours in one week.
    b. Without graphing the function, determine the slope.
    c. Describe the meaning of the slope within the context of Juan's job.
12. The graph shows the height of an airplane as it descends to land.

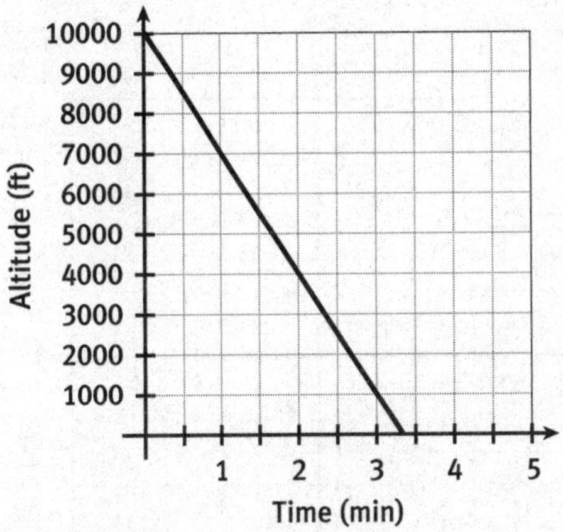

    a. Does the function have a constant rate of change? If so, what is it?
    b. What is the slope of the line?
    c. How are the rate of change and the slope of the line related?
    d. Describe the meaning of the slope within the context of the situation.

Lesson 9-3

For Items 13–15, tell whether the function is linear. Justify your response.

13.

x	y
−3	44
−1	4
0	−1
1	4

14.

x	y
−5	−7
0	−8
5	−9
10	−10

15.

x	y
4	−30
6	−46
8	−62
9	−70

16. One point on the line described by $y = -2x + 3$ is shown below. Use your knowledge of slope to give the coordinates of three more points on the line.

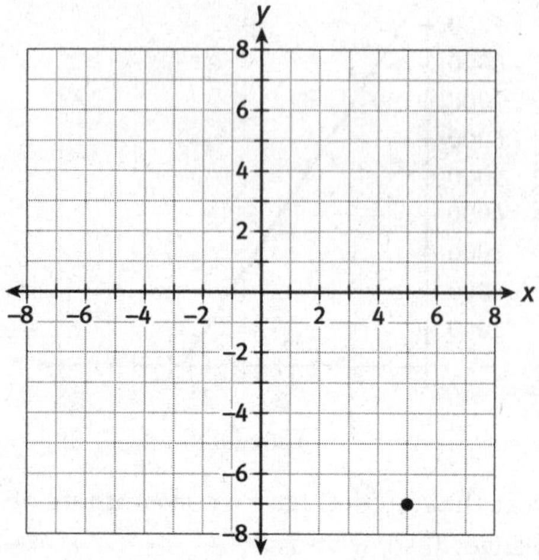

17. Which of the following is **not** a linear function?

A. (4, −6), (7, −12), (8 −14), (10, −18), (2, −2)
B. (−2, −6), (1, 0), (4, −30), (0, 2), (7, −96)
C. (−4, 9), (0, 7), (2, 6), (6, 4), (8, 3)
D. (2, 18), (6, 50), (−3, −22), (0, 2), (3, 26)

For Items 18 and 19, identify the slope of the line in each graph as positive, negative, 0, or undefined.

18.

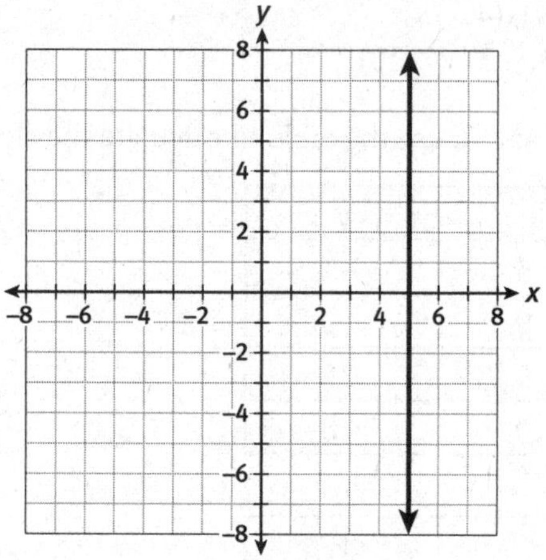

19.

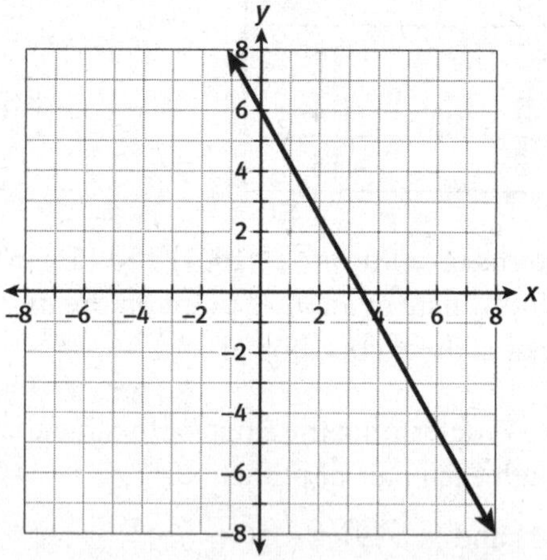

20. The slope of a line is 0. It passes through the point (−3, 4). Identify two other points on the line. Justify your answers.

MATHEMATICAL PRACTICES
Look For and Make Use of Structure

21. Describe three different ways to determine the slope of a line and the similarities and differences between the methods.

Linear Models
Stacking Boxes
Lesson 10-1 Direct Variation

My Notes

Learning Targets:
- Write and graph direct variation.
- Identify the constant of variation.

SUGGESTED LEARNING STRATEGIES: Create Representations, Interactive Word Wall, Marking the Text, Sharing and Responding, Discussion Groups

You work for a packaging and shipping company. As part of your job there, you are part of a package design team deciding how to stack boxes for packaging and shipping. Each box is 10 cm high.

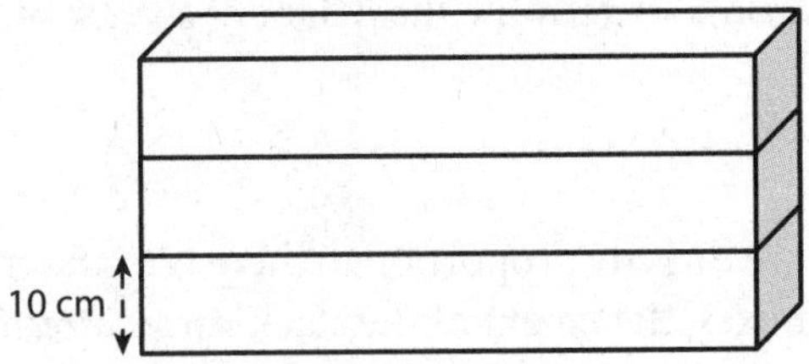

1. Complete the table and make a graph of the data points (number of boxes, height of the stack).

Number of Boxes	Height of the Stack (cm)
0	0
1	10
2	20
3	
4	
5	
6	
7	

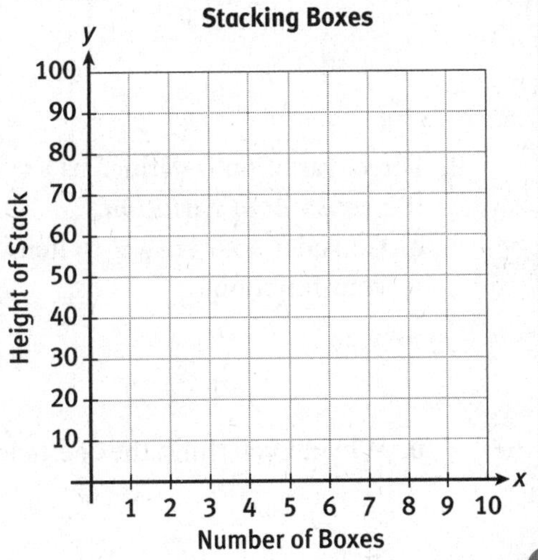

2. Write a function to represent the data in the table and graph above.

WRITING MATH

Either y or $f(x)$ can be used to represent the output of a function.

3. What is a reasonable and realistic domain for the function? Explain.

4. What is a reasonable and realistic range for the function? Explain.

My Notes

5. What do $f(x)$, or y, and x represent in your equation from Item 2?

6. Describe any patterns that you notice in the table and graph representing your function.

MATH TERMS

A **direct proportion** is a relationship in which the ratio of one quantity to another remains constant.

7. The number of boxes is ***directly proportional*** to the height of the stack. Use a proportion to determine the height of a stack of 12 boxes.

When two values are directly proportional, there is a ***direct variation***. In terms of stacking boxes, the height of the stack *varies directly* as the number of boxes.

8. Using variables x and y to represent the two values, you can say that y varies directly as x. Use your answer to Item 6 to explain this statement.

9. Direct variation is defined as $y = kx$, where $k \neq 0$ and the coefficient k is the ***constant of variation.***

a. Consider your answer to Item 2. What is the constant of variation in your function?

b. Why do you think the coefficient is called the constant of variation?

c. **Reason quantitatively.** Explain why the value of k cannot be equal to 0.

d. Write an equation for finding the constant of variation by solving the equation $y = kx$ for k.

My Notes

10. a. Interpret the meaning of the point (0, 0) in your table and graph.

b. True or False? Explain your answer. "The graphs of all direct variations are lines that pass through the point (0, 0)."

c. Identify the slope and *y*-intercept in the graph of the stacking boxes.

d. Describe the relationship between the constant of variation and the slope.

Direct variation can be used to answer questions about stacking and shipping your boxes.

11. The height *y* of a different stack of boxes varies directly as the number of boxes *x*. For this type of box, 25 boxes are 500 cm high.

a. Find the value of *k*. Explain how you found your answer.

b. Write a direct variation equation that relates *y*, the height of the stack, to *x*, the number of boxes in the stack.

c. How high is a stack of 20 boxes? Explain how you would use your direct variation equation to find the height of the stack.

My Notes

12. At the packaging and shipping company, you get paid each week. One week you earned \$48 for 8 hours of work. Another week you earned \$30 for 5 hours of work.

a. Write a direct variation equation that relates your wages to the number of hours you worked each week. Explain the meaning of each variable and identify the constant of variation.

b. How much would you earn if you worked 3.5 hours in one week?

Check Your Understanding

13. Tell whether the tables, graphs, and equations below represent direct variations. Justify your answers.

a.

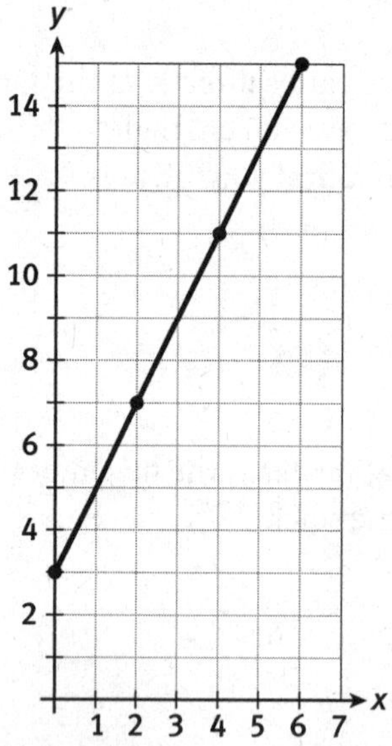

b.

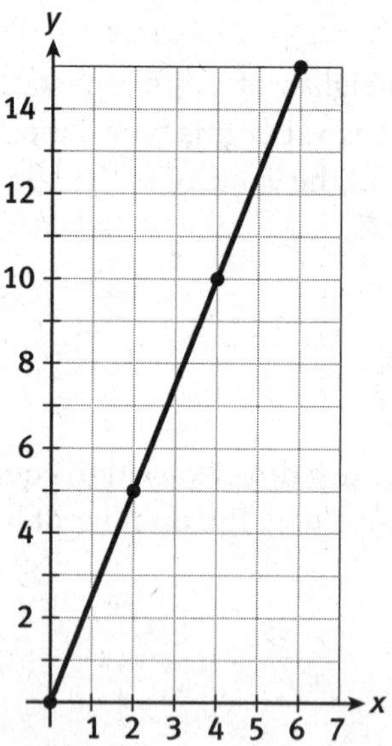

c.

x	y
2	12
4	24
6	36

d.

x	y
2	8
4	12
6	16

e. $y = 20x$

f. $y = 3x + 2$

My Notes

LESSON 10-1 PRACTICE

14. In the equation $y = 15x$, what is the constant of variation?

15. In the equation $y = 8x$, what is the constant of variation?

16. The value of y varies directly with x and the constant of variation is 7. What is the value of x when $y = 63$?

17. The value of y varies directly with x and the constant of variation is 12. What is the value of y when $x = 5$?

18. Model with mathematics. The height of a stack of boxes varies directly with the number of boxes. A stack of 12 boxes is 15 feet high. How tall is a stack of 16 boxes?

19. Jan's pay is in direct variation to the hours she works. Jan earns $54 for 12 hours of work. How much will she earn for 18 hours work?

My Notes

Learning Targets:

- Write and graph indirect variations.
- Distinguish between direct and indirect variation.

SUGGESTED LEARNING STRATEGIES: Create Representations, Marking the Text, Sharing and Responding, Think-Pair-Share, Discussion Groups

When packaging a different product, your team at the packaging and shipping company determines that all boxes for this product will have a volume of 400 cubic inches and a height of 10 inches. The lengths and the widths will vary.

MATH TIP

The volume of a rectangular prism is found by multiplying length, width, and height: $V = lwh$.

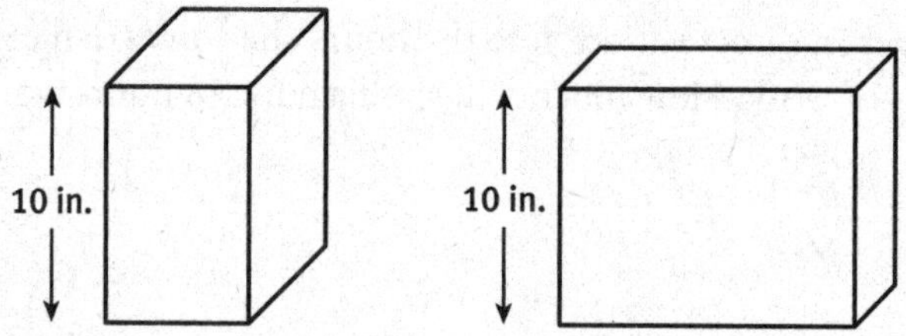

1. To explore the relationship between length and width, complete the table and make a graph of the points.

Width (x)	Length (y)
1	40
2	20
4	10
5	
8	
10	
20	

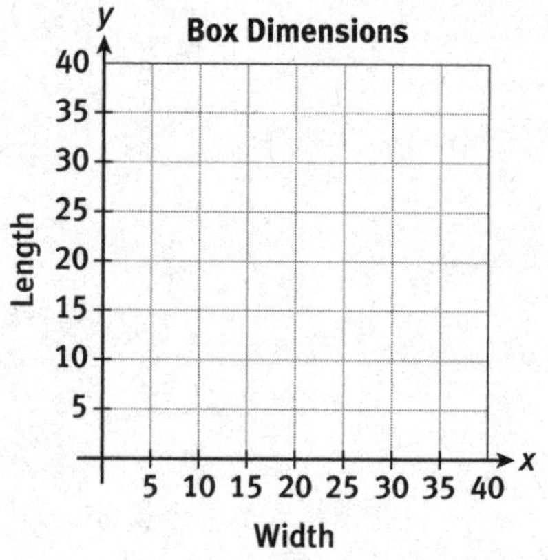

2. How are the lengths and widths in Item 1 related? Write an equation that shows this relationship.

3. Use the equation you wrote in Item 2 to write a function to represent the data in the table and graph above.

4. Describe any patterns that you notice in the table and graph representing your function.

My Notes

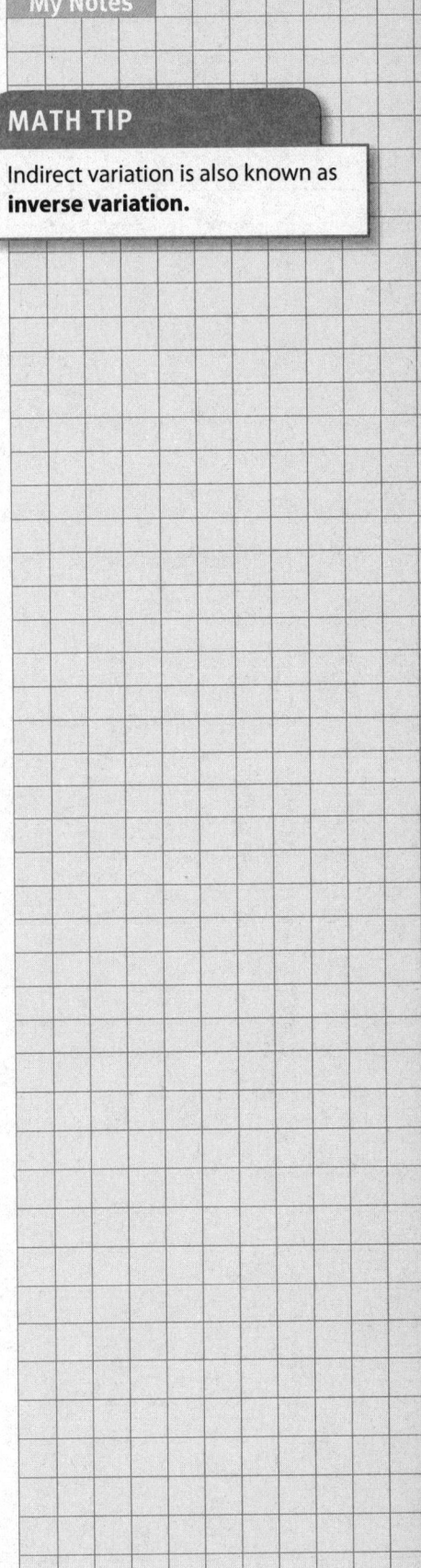

In terms of box dimensions, the length of the box varies indirectly as the width of the box. Therefore, this function is called an ***indirect variation***.

MATH TIP

Indirect variation is also known as **inverse variation.**

5. Recall that direct variation is defined as $y = kx$, where $k \neq 0$ and the coefficient k is the constant of variation.

a. How would you define indirect variation in terms of y, k, and x?

b. Are there any limitations on these variables as there are on k in direct variation? Explain.

c. Write an equation for finding the constant of variation by solving for k in your answer to Part (a).

6. Reason abstractly. Compare and contrast the equations of direct and indirect variation.

7. Compare and contrast the graphs of direct and indirect variation.

8. Use your function in Item 3 to determine the following measurements for your company.

a. Find the length of a box whose width is 80 inches.

b. Find the length of a box whose width is 0.4 inches.

My Notes

9. The time, y, needed to load the boxes on a truck for shipping varies indirectly as the number of people, x, working. If 10 people work, the job is completed in 20 hours.

   a. Explain how to find the constant of variation. Then find it.

   b. Write an indirect variation equation that relates the time to load the boxes to the number of people working.

   c. How long does it take 8 people to finish loading the boxes? Use your equation to answer this question.

   d. On the grid below, make a graph to show the time needed for 2, 4, 5, 8, 10, and 25 people to load the boxes on the truck.

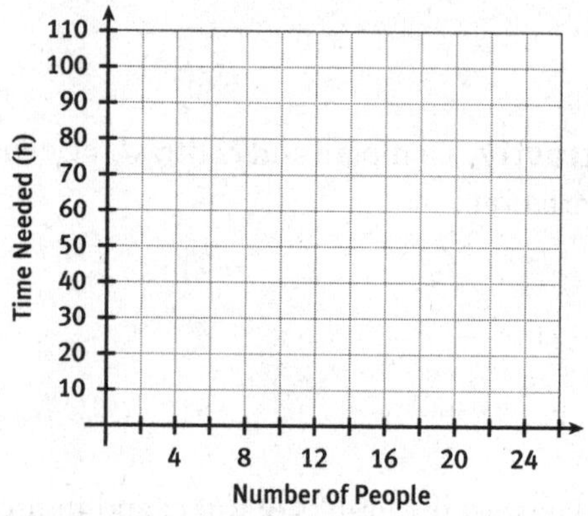

10. The cost for the company to ship the boxes varies indirectly with the number of boxes being shipped. If 25 boxes are shipped at once, it will cost \$10 per box. If 50 boxes are shipped at once, the cost will be \$5 per box.

   a. Write an indirect variation equation that relates the cost per box to the number of boxes being shipped.

   b. How much would it cost to ship only 10 boxes?

11. Is an indirect variation function a linear function? Explain.

My Notes

Check Your Understanding

12. Identify the following graphs as direct variation, indirect variation, neither, or both.

a.

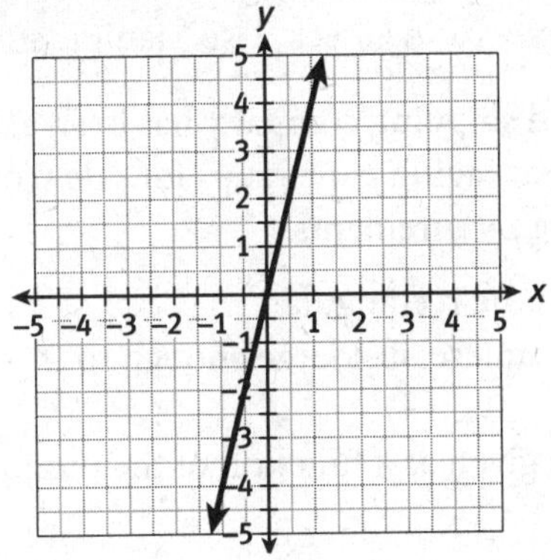

b.

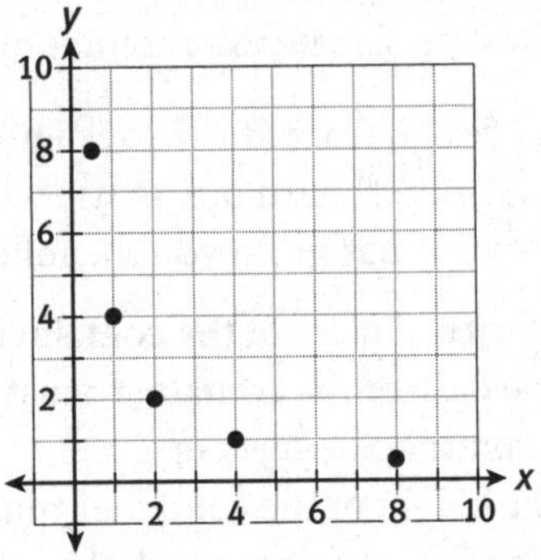

c.

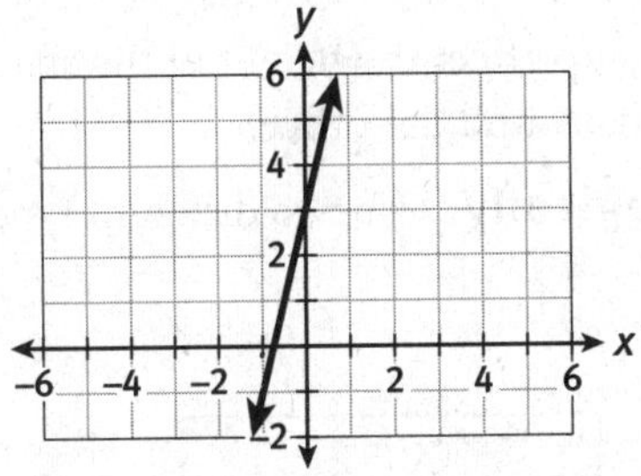

13. Which equations are examples of indirect variation? Justify your answers.

A. $y = 2x$ **B.** $y = \frac{x}{2}$

C. $y = \frac{2}{x}$ **D.** $xy = 2$

14. In the equation $y = \frac{80}{x}$, what is the constant of variation?

LESSON 10-2 PRACTICE

15. Graph each function. Identify whether the function is an indirect variation.

a.

x	−4	−2	−1	1	2	4
y	$-\frac{3}{4}$	$-\frac{3}{2}$	−3	3	$\frac{3}{2}$	$\frac{3}{4}$

b.

x	−4	−2	−1	1	2	4
y	11	5	3	−4	−7	−13

16. Make sense of problems. For Parts (a) and (b) below, y varies indirectly as x.

a. If $y = 6$ when $x = 24$, find y when $x = 16$.

b. If $y = 8$ when $x = 20$, find the value of k.

My Notes

CONNECT TO GEOMETRY

The carton will be a right rectangular prism. A **rectangular prism** is a closed, three-dimensional figure with three pairs of opposite parallel faces that are congruent rectangles.

Learning Targets:

- Write, graph, and analyze a linear model for a real-world situation.
- Interpret aspects of a model in terms of the real-world situation.

SUGGESTED LEARNING STRATEGIES: Marking the Text, Discussion Groups, Create Representations, Guess and Check, Use Manipulatives

Your design team at the packaging and shipping company has been asked to design a cardboard box to use when packaging paper cups for sale. Your supervisor has given you the following requirements.

- All lateral faces of the container must be rectangular.
- The base of the container must be a square, just large enough to accommodate one cup.
- The height of the container must be given as a function of the number of cups the container will hold.
- All measurements must be in centimeters.

To help discover which features of the cup affect the height of the stack, collect data on two types of cups found around the office.

1. **Use appropriate tools strategically.** Use two different types of cups to complete the tables below.

CUP 1	
Number of Cups	Height of Stack
1	
2	
3	
4	
5	
6	

CUP 2	
Number of Cups	Height of Stack
1	
2	
3	
4	
5	
6	

2. **Express regularity in repeated reasoning.** What patterns do you notice that might help you figure out the relationship between the height of the stack and the number of cups in that stack?

My Notes

Use your data for Cup 1 to complete Items 3–13.

3. Make a graph of the data you collected.

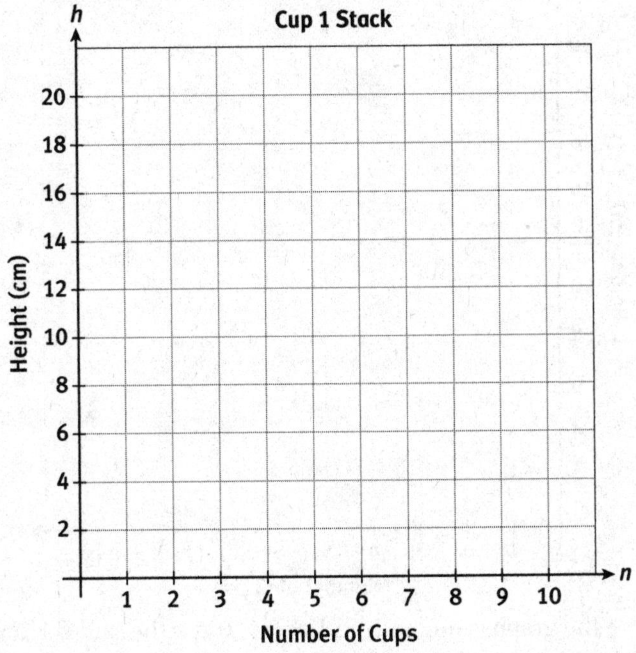

4. Predict, without measuring, the height of a stack of 16 cups. Explain how you arrived at your prediction.

5. Predict, without measuring, the height of a stack of 50 cups. Explain how you arrived at your prediction.

6. Write an equation that gives the height of a stack of cups, h, in terms of n, the number of cups in the stack.

7. Use your equation from Item 6 to find h when $n = 16$ and when $n = 50$. Do your answers to this question agree with your predictions in Items 4 and 5?

My Notes

8. Sketch the graph of your equation from Item 6.

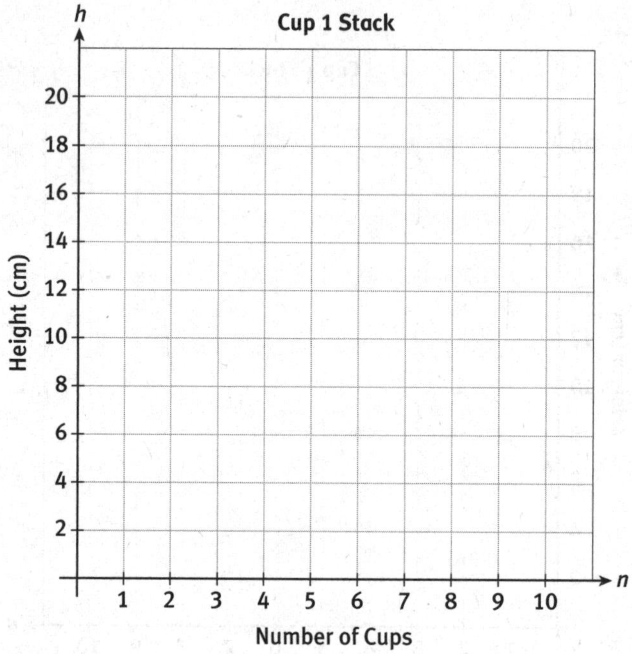

9. How are the graphs you made in Items 3 and 8 the same? How are they different?

10. Do the graphs in Items 3 and 8 represent direct variation, indirect variation, or neither? Explain.

11. Remember that you are designing a container with a square base. What dimension(s), other than the height of the stack, do you need to design your cup container? Use Cup 1 to find this/these dimension(s).

12. Find the dimensions of a container that will hold a stack of 25 cups.

My Notes

13. Your team has been asked to communicate its findings to your supervisor. Write a report to her that summarizes your findings about the cup container design. Include the following information in your report.

- The equation your team discovered to find the height of the stack of Cup 1 style cups
- A description of how your team discovered the equation and the minimum number of cups needed to find it
- An explanation of how the numbers in the equation relate to the physical features of the cup
- An equation that could be used to find the height of the stack of Cup 2 style cups

MATH TIP

When writing your answer to Item 13, you can use a RAFT.

- Role—team leader
- Audience—your boss
- Format—a letter
- Topic—stacks of cups

Check Your Understanding

14. A group of students performed the cup activity described in this lesson. For their Cup 1, they found the equation $h = 0.25n + 8.5$, where h is the height in inches of a stack of cups and n is the number of cups.

a. What would be the height of 25 cups? Of 50 cups?

b. Graph this equation. Describe your graph.

LESSON 10-3 PRACTICE

15. Reason quantitatively. A group of students performed the cup activity in this lesson using plastic drinking cups. Their data is shown below.

CUP 1	
Number of Cups	Height of Stack
1	14.5 cm
2	16 cm
3	17.5 cm
4	19 cm
5	20.5 cm

CUP 2	
Number of Cups	Height of Stack
1	10.5 cm
2	11.75 cm
3	13 cm
4	14.25 cm
5	15.5 cm

For each cup, write and graph an equation. Describe your graphs.

16. A consultant earns a flat fee of \$75 plus \$50 per hour for a contracted job. The table shows the consultant's earnings for the first four hours she works.

Hours	0	1	2	3	4
Earnings	\$75	\$125	\$175	\$225	\$275

The consultant has a 36-hour contract. How much will she earn?

My Notes

Learning Targets:

- Write the inverse function for a linear function.
- Determine the domain and range of an inverse function.

SUGGESTED LEARNING STRATEGIES: Visualization, Create Representations, Think-Pair-Share, Discussion Groups, Construct an Argument

After reading your report, your supervisor was able to determine the equation for the height of the stack for the specific cup that the company will manufacture. The company will use the function $S(n) = 0.5n + 12.5$.

1. What do S, n, and $S(n)$ represent?

2. What do the numbers 0.5 and the 12.5 in the function S tell you about the physical features of the cup?

3. Evaluate $S(1)$ to find the height of a single cup.

4. How tall is a stack of 35 cups? Show your work using function notation.

5. If you add 2 cups to a stack, by how much does the height of the stack increase?

6. If you add 20 cups to a stack, by how much does the height of the stack increase?

7. **Critique the reasoning of others.** A member of one of the teams stated: "If you double the number of cups in a stack, then the height of the stack is also doubled." Is this statement correct? Explain.

My Notes

8. If you were to graph the function $S(n) = 0.5n + 12.5$, you would see that the points lie on a line.

a. What is the slope of this line?

b. Interpret the slope of the line as a rate of change that relates a change in height to a change in the number of cups.

Check Your Understanding

Use this table for Items 9 and 10.

x	1	2	3	4	5
y	1	5	9	13	17

9. Write an equation for y in terms of x.

10. Explain how the numbers in your equation relate to the numbers in the table.

11. Evaluate the function you wrote in Item 9 for each of the following values of x.

a. $x = 8$ **b.** $x = 12$ **c.** $x = 15$ **d.** $x = 0$

12. a. The supervisor wanted to increase the height of a container by 5 cm. How many more cups would fit in the container?

b. If the supervisor wanted to increase the height of a container by 6.4 cm, how many more cups would fit in the container?

c. How many cups fit in a container that is 36 cm tall?

d. How many cups fit in a container that is 50 cm tall?

My Notes

13. The function $S(n) = 0.5n + 12.5$ describes the height S in terms of the number of cups n.

a. Solve this equation for n to describe the number of cups n in terms of the height S.

b. How many cups fit in a carton that is 85 cm tall? Compare your method of answering this question to your method used in Items 12c and 12d.

c. What is the slope of the line represented by your equation in Part (a)? Interpret it as a rate of change and compare it to the rate of change found in Item 8b.

READING MATH

$f^{-1}(x)$ is read as "f inverse of x". It does **not** mean "f to the negative one power."

An ***inverse function*** is a function that interchanges the independent and dependent variables of another function. In Item 13, you found the inverse function for $S(n)$. In general, the inverse function for $f(x)$ is $f^{-1}(x)$.

Example A

Use the table below to fill in the steps to find the inverse function for $f(x) = 2x + 3$.

Write the function, replacing $f(x)$ with y.	
Switch x and y.	
Solve for y in terms of x.	
Replace y with $f^{-1}(x)$.	

Try These A

Determine the inverses of each of the following of functions.

a. $f(x) = -4x - 5$ **b.** $f(x) = \frac{2}{3}x + 2$ **c.** $f(x) = -\frac{1}{2}x + 4$

Only those functions that are ***one-to-one*** functions have an inverse function. Functions that are not one-to-one must have their domain restricted for an inverse function to exist.

14. Is $S(n) = 0.5n + 12.5$ a one-to-one function? Explain.

15. Do the following graphs of functions show one-to-one functions? Justify your answers.

a.

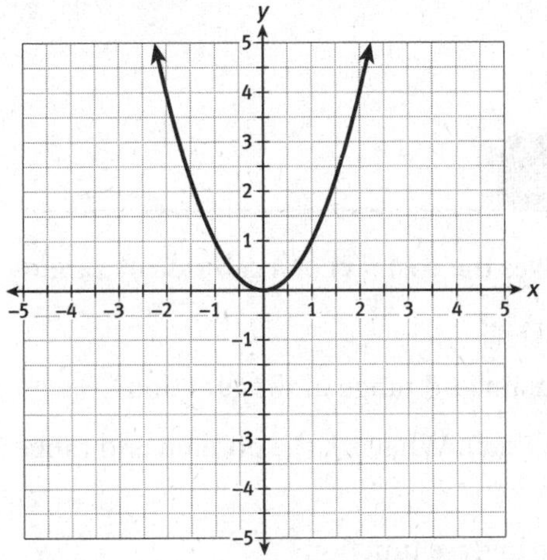

b.

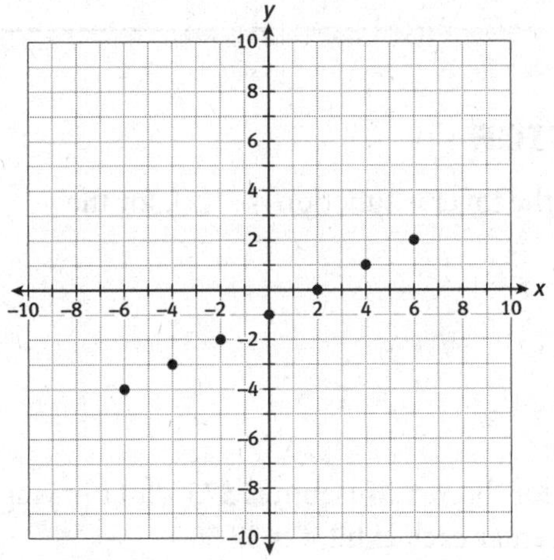

A visual test for a one-to-one function is the horizontal line test. If you can draw a horizontal line that intersects the graph of a function in more than one place, that function is not one-to-one.

16. Construct viable arguments. Are linear functions one-to-one functions? Justify your response.

My Notes

MATH TERMS

For a function to be **one-to-one** means that no two values of *x* are paired with the same value of *y*.

My Notes

17. A function is defined by the ordered pairs $\{(-3, -1), (-1, 0), (1, 1), (3, 2), (5, 3)\}$. What are the domain and range of the function?

Because inputs and outputs are switched when writing the inverse of a function, the domain of a function is the range of its inverse function, and the range of a function is the domain of its inverse function.

18. What are the domain and range of the inverse function for the function in Item 17?

Check Your Understanding

The function $f(x) = 2.5x + 3.5$ gives the cost $f(x)$ of a cab ride of x miles.

19. What is the cost of a 6-mile ride?

20. What are the reasonable domain and range of the function?

21. Write the inverse function, $f^{-1}(x)$. What are the domain and range of $f^{-1}(x)$?

22. What does x represent in the inverse function?

23. A cab ride costs $46. Show how to use the inverse function to find the distance of the cab ride in miles.

LESSON 10-4 PRACTICE

Make use of structure. Find the inverse function, $f^{-1}(x)$, for the functions in Items 24–26.

24. $f(x) = 3x - 5$

25. $f(x) = -2x + 10$

26. $f(x) = \frac{7x}{3} - \frac{1}{6}$

27. The yearly membership fee for the Art Museum is $75. After paying the membership fee, the cost to enter each exhibit is $7.50.

a. Write a function for the total cost of a member for one year of attending the art museum.

b. What is the total cost for a member who sees 12 exhibits?

c. What are the domain and range for the function?

d. What is $f^{-1}(x)$? What are the domain and range for $f^{-1}(x)$?

e. What does x represent in $f^{-1}(x)$?

f. How many exhibits can a member see in a year for a total of $210, including the membership fee?

ACTIVITY 10 PRACTICE

Write your answers on notebook paper.
Show your work.

Lesson 10-1

1. The value of y varies directly as x and $y = 125$ when $x = 25$. What is the value of y when $x = 2$?

2. Which is the graph of a direct variation?

A.

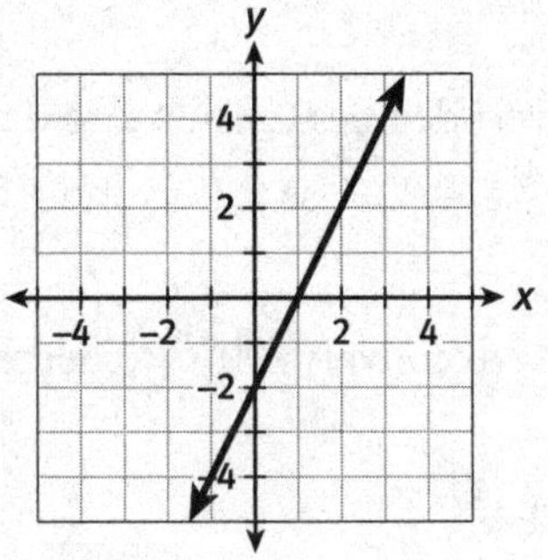

B.

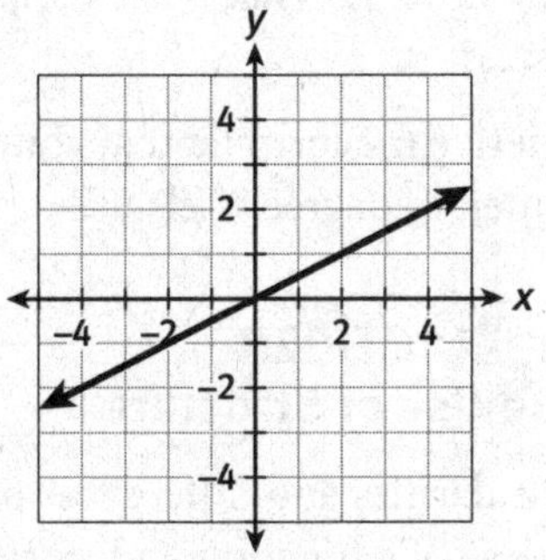

C.

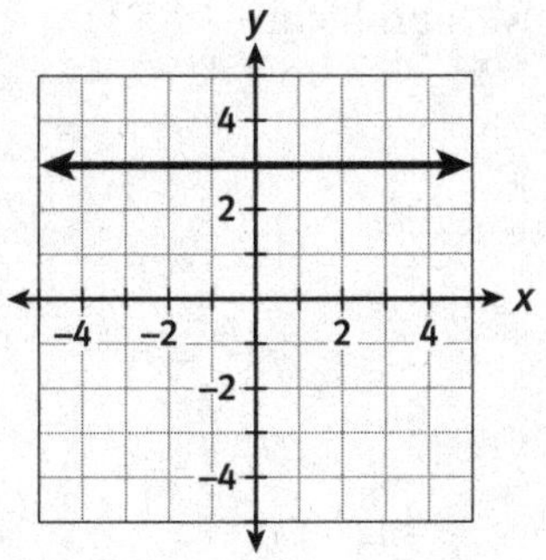

D.

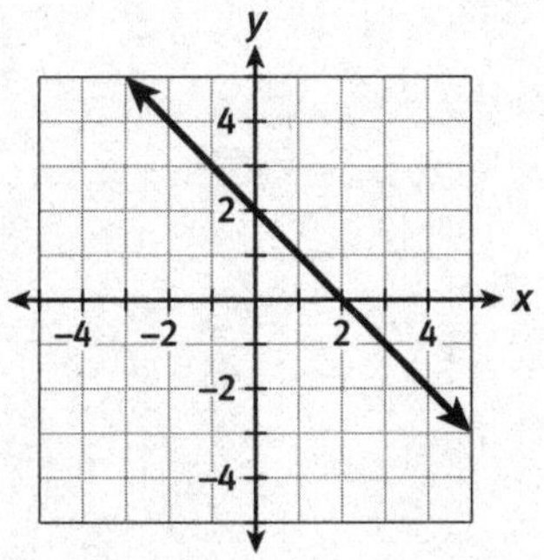

3. Which equation **does not** represent a direct variation?
   A. $y = \frac{x}{3}$
   B. $y = \frac{2}{5}x$
   C. $y = \frac{3}{x}$
   D. $y = \frac{5x}{2}$

4. The value of y varies directly as x and $y = 9$ when $x = 6$. What is the value of y when $x = 15$?

5. The tailor determines that the cost of material varies directly with the amount of material. The cost is \$42 for 14 yards of material. What is the cost for 70 yards of material?

Lesson 10-2

6. The value of y varies indirectly as x and $y = 4$ when $x = 20$. What is the value of y when $x = 40$?
   A. $y = 2$
   B. $y = 8$
   C. $y = 50$
   D. $y = 80$

7. The temperature varies indirectly as the distance from the city. The temperature equals 3°C when the distance from the city is 40 miles. What is the temperature when the distance is 20 miles from the city?

8. The amount of gas left in the gas tank of a car varies indirectly to the number of miles driven. There are 9 gallons of gas left after 24 miles. How much gas is left after the car is driven 120 miles?

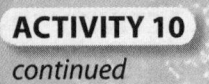

Lesson 10-3

The Pete's Pets chain of pet stores is growing. The table below shows the number of stores in business each month. Use the table for Items 9–12.

Month	Stores
January	0
February	3
March	6
April	9
May	12
June	15

9. According to the table, how many new stores open per month?

10. How many stores will be in business by December?

11. Are the Pete's Pets data an example of indirect variation, direct variation, or neither? Explain your reasoning.

12. What is the slope of this function? Interpret the meaning of the slope.

Jeremy collected the following data on stacking chairs. Use the data for Items 13 and 14.

Number of Chairs	Height (in.)
1	30
2	33
3	36
4	39
5	42
6	45

13. Write a linear function that models the data.

14. Chairs cannot be stacked higher than 5 feet. What is the maximum number of chairs Jeremy can stack? Justify your answer.

Lesson 10-4

Write the inverse function for each of the following.

15. $f(x) = -8x + 4$

16. $f(x) = \frac{1}{4}x - 3$

17. $f(x) = 8x - 15$

18. $f(x) = x + 1$

19. $f(x) = -x + 1$

The formula to convert degrees Celsius C to degrees Fahrenheit F is $F = \frac{9}{5}C + 32$. Use this formula for Items 20–23.

20. Use the formula to convert 100°C to degrees Fahrenheit.

21. What is the slope?

22. The temperature is 50°F. What is the temperature in degrees Celsius?

23. Solve for C to derive the formula that converts degrees Fahrenheit to degrees Celsius.

MATHEMATICAL PRACTICES
Look for and Make Use of Structure

24. Describe the similarities and differences between finding the inverse of a function and working backward to solve a problem.

Arithmetic Sequences

Picky Patterns

Lesson 11-1 Identifying Arithmetic Sequences

ACTIVITY 11

My Notes

Learning Targets:

- Identify sequences that are arithmetic sequences.
- Use the common difference to determine a specified term of an arithmetic sequence.

SUGGESTED LEARNING STRATEGIES: Look for a Pattern, Create Representations, Discussion Groups, Marking the Text, Use Manipulatives

1. Use toothpicks to make the following models.

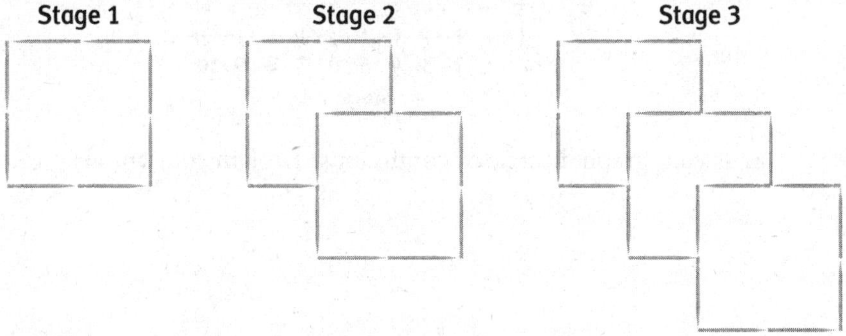

2. Continue to Stage 4 and Stage 5. Draw your models below.

Stage 4 Stage 5

3. Complete the table for the number of toothpicks used for each stage of the models up through Stage 5.

Stage	Number of Toothpicks
1	8
2	
3	
4	
5	

My Notes

4. Model with mathematics. Use the following grid to make a graph of the data in the table.

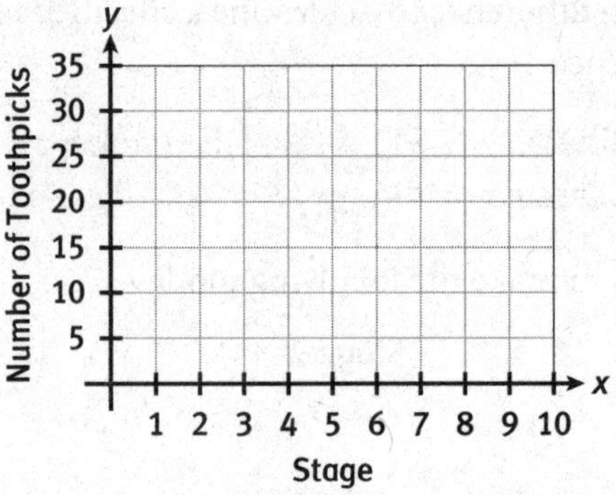

a. Is your graph discrete or continuous? Explain your answer.

b. Is your graph the graph of a linear function? Explain your answer.

READING MATH

Read a_n as "*a* sub *n*."

An ordered list of numbers is called a ***sequence***. The numbers in a sequence are ***terms***. To refer to the *n*th term in a sequence, you can use either function notation, $f(n)$, or the indexed variable a_n.

The toothpick data form a sequence. The numbers of toothpicks at each stage are the terms of the sequence.

5. What are the first four terms of the toothpick sequence?

$a_1 =$ $a_2 =$ $a_3 =$ $a_4 =$

6. In a sequence, what is the distinction between a term and a term number?

Check Your Understanding

7. For the sequence 7, −5, −3, 1, 1, . . . , what is a_4?

8. For the sequence 1, 5, 9, 13, 17, . . . , what is a_5?

READING MATH

A common difference may also be called a **constant difference**.

An ***arithmetic sequence*** is a sequence in which the difference between terms is constant. The difference between consecutive terms in an arithmetic sequence is called the ***common difference***.

9. Explain why the toothpick sequence is an arithmetic sequence.

My Notes

10. What is the rate of change for the toothpick data?

11. Look back at the graph in Item 4.

a. Determine the slope between any two points on the graph.

b. Describe the connections between the slope, the rate of change, and the common difference.

Check Your Understanding

Tell whether each sequence is an arithmetic sequence. For each arithmetic sequence, find the common difference.

12. 9, 16, 23, 30, 37, . . .

13. $-24, -20, -14, -10, -4, 0, \ldots$

14. $-2.8, -2.2, -1.6, -1.0, \ldots$

15. 3, 5, 8, 12, 17, . . .

16. $\frac{1}{4}, \frac{1}{2}, \frac{3}{4}, 1, \ldots$

17. Reason abstractly. Can the common difference in an arithmetic sequence be negative? If so, give an example. If not, explain why not.

LESSON 11-1 PRACTICE

Tell whether each sequence is arithmetic. If the sequence is arithmetic, identify the common difference and find the indicated term.

18. $-9, -4, 1, 6, 11, \ldots; a_7 = ?$

19. $2, 4, 7, 11, 16, 22, \ldots; a_9 = ?$

20. $-7, -1, 5, 11, 17, \ldots; a_6 = ?$

21. $1.2, 1.9, 2.6, 3.3, 4.0, \ldots; a_8 = ?$

22. $3, \frac{5}{2}, \frac{3}{2}, -\frac{3}{2}, \ldots; a_7 = ?$

23. Write an arithmetic sequence in which the last digit of each term is 4. What is the common difference for your sequence?

24. Critique the reasoning of others. Jim said that the terms in an arithmetic sequence must always increase, because you must add the common difference to each term to get the next term. Is Jim correct? Justify your reasoning.

My Notes

Learning Targets:

- Develop an explicit formula for the *n*th term of an arithmetic sequence.
- Use an explicit formula to find any term of an arithmetic sequence.
- Write a formula for an arithmetic sequence given two terms or a graph.

SUGGESTED LEARNING STRATEGIES: Look for a Pattern, Create Representations, Interactive Word Wall, Predict and Confirm, Think-Pair-Share

1. Rewrite the terms of the toothpick sequence and identify the common difference.

2. Find the next three terms in the sequence without building toothpick models. Explain how you found your answers.

3. Why might it be difficult to find the 100th term of the toothpick sequence using repeated addition of the common difference?

MATH TERMS

An **explicit formula** for an arithmetic sequence describes any term in the sequence using the first term and the common difference.

An ***explicit formula*** for a sequence allows you to compute any term in a sequence without computing all of the terms before it.

4. Develop an explicit formula for the toothpick sequence using the first term and the common difference.

The first term of the sequence is $a_1 = 8$.

The second term is $a_2 = 8 + 6$.

The third term is $a_3 = 8 + 6 + 6$, or $a_3 = 8 + 2(6)$.

a. Write an expression for the fourth term using the value of a_1 and the common difference.

$a_4 =$

b. **Express regularity in repeated reasoning.** Use the patterns you have observed to determine the 15th term. Justify your reasoning.

$a_{15} =$

c. Write an expression that can be used to find the *n*th term of the toothpick sequence.

$a_n =$

My Notes

The formula you wrote in Item 4c is the explicit formula for the toothpick sequence.

For any arithmetic sequence, a_1 refers to the first term and d refers to the common difference.

5. Write an explicit formula for finding the nth term of any arithmetic sequence.

Example A

Write the explicit formula for the arithmetic sequence 3, −3, −9, −15, −21, ... Then use the formula to find the value of a_{10}.

Step 1: Find the common difference.

3 → −3 → −9 → −15 → −21

−6 −6 −6 −6

The common difference is −6.

Step 2: Write the explicit formula and simplify.

$a_n = a_1 + (n - 1)d$

$a_n = 3 + (n - 1)(-6) = 9 - 6n$

Step 3: Use the formula to find a_{10} by substituting for n.

$a_{10} = 9 - 6(10) = -51$

Solution: The explicit formula is $a_n = 9 - 6n$ and $a_{10} = -51$.

Try These A

For the following arithmetic sequences, find the explicit formula and the value of the indicated term.

a. 2, 6, 10, 14, 18, . . . ; a_{21}

b. −0.6, −1.0, −1.4, −1.8, −2.2, . . . ; a_{15}

c. $\frac{1}{3}, 1, \frac{5}{3}, \frac{7}{3}, \ldots; a_{37}$

An arithmetic sequence can be graphed on a coordinate plane. In the ordered pairs the term numbers (1, 2, 3, . . .) are the x-values and the terms of the sequence are the y-values.

6. Look back at the sequence in Example A. Make a prediction about its graph.

My Notes

7. On the grid below, create a graph of the arithmetic sequence in Example A. Revise your prediction in Item 6 if necessary.

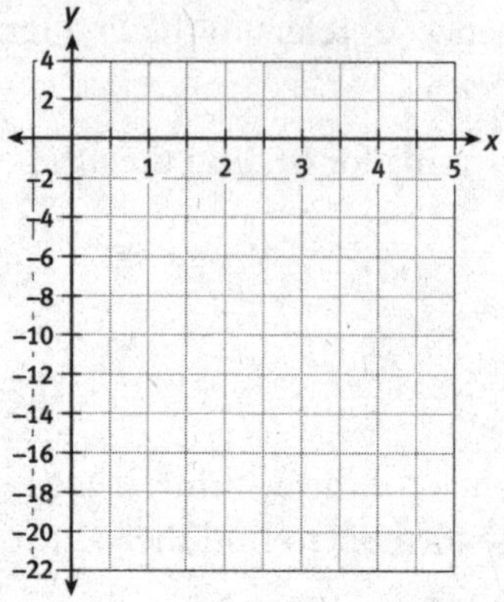

8. Determine the slope between any two points on your graph in Item 7. How does the slope compare to the common difference of the sequence?

9. The first three terms of the arithmetic sequence 2, 5, 8, ... are graphed below. Determine the common difference. Then graph the next three terms.

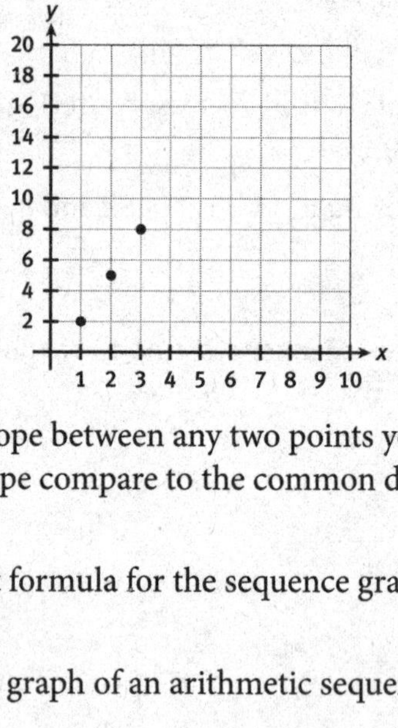

10. Determine the slope between any two points you graphed in Item 9. How does the slope compare to the common difference of the sequence?

11. Write the explicit formula for the sequence graphed in Item 9.

12. If you are given a graph of an arithmetic sequence, how do you find the explicit formula?

The 11th term of an arithmetic sequence is 59 and the 14th term is 74.

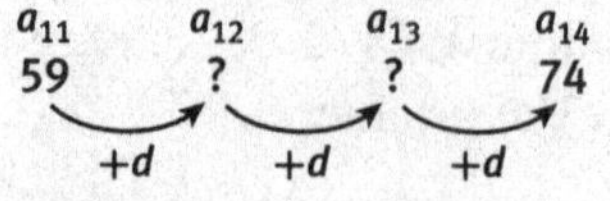

13. Reason quantitatively. How could you determine the value of *d*? What is the value of *d*?

My Notes

14. How could you determine the value of a_1? What is this value?

15. Write the explicit formula for the nth term of the sequence.

16. Determine a_{30}, the 30th term of the sequence.

Check Your Understanding

17. The explicit formula for the nth term of an arithmetic sequence is $a_n = a_1 + (n - 1)d$. What does each variable in the explicit formula represent?

18. What is a_{50} of the arithmetic sequence 45, 40, 35, 30, . . . ?

19. The 8th term of an arithmetic sequence is 12.5 and the 13th term is 20. What is a_{25}?

20. **Construct viable arguments.** Could an arithmetic sequence also be a direct variation? Justify your answer.

21. Why is the graph of an arithmetic sequence made up of discrete points?

LESSON 11-2 PRACTICE

For the following arithmetic sequences, find the explicit formula and the value of the term indicated.

22. 2, 11, 20, 29, . . . ; a_{30}

23. 0.5, 0.75, 1, . . . ; a_{18}

24. $\frac{1}{6}, 0, -\frac{1}{6}, -\frac{1}{3}, \ldots; a_{42}$

25. The 3rd term of an arithmetic sequence is -1 and the 7th term is -13. Find the explicit formula for this sequence. What is a_{22}?

26. **Make use of structure.** Write the explicit formula for each arithmetic sequence graphed below. Then find the 25th term.

a.

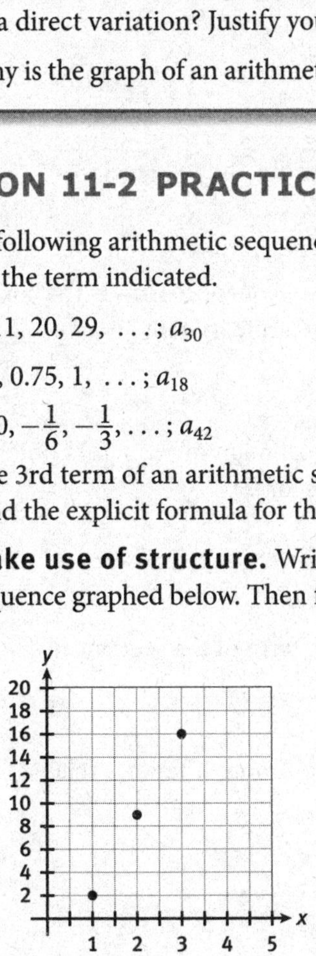

b.

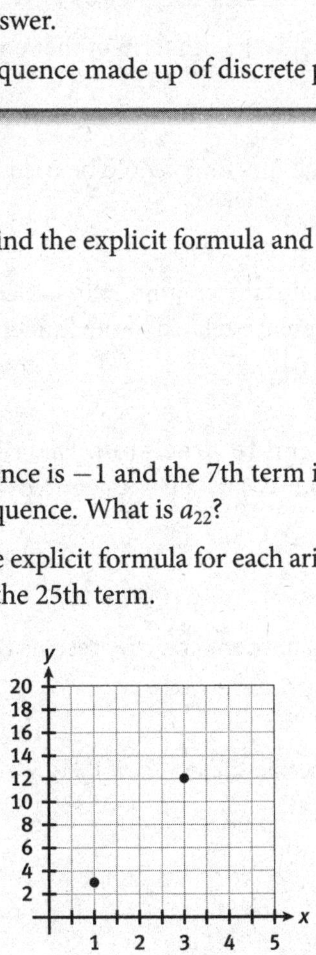

c.

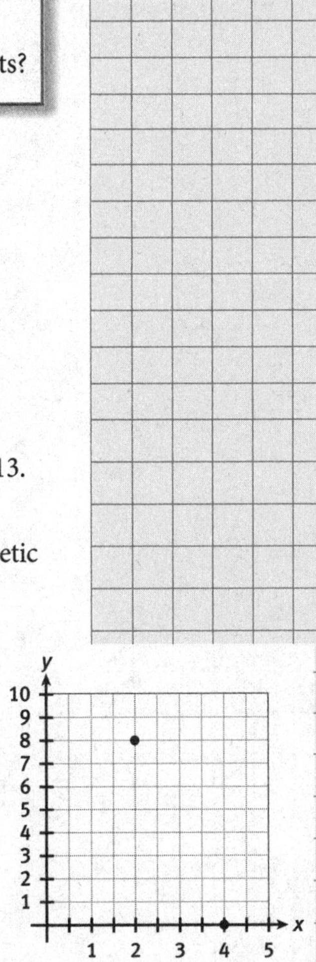

My Notes

Learning Targets:

- Use function notation to write a general formula for the *n*th term of an arithmetic sequence.
- Find any term of an arithmetic sequence written as a function.

SUGGESTED LEARNING STRATEGIES: Look for a Pattern, Create Representations, Discussion Groups, Sharing and Responding, Group Presentation

An arithmetic sequence is a special case of a linear function. The terms of the sequence are the functional values $f(1), f(2), f(3), \ldots, f(n)$ for some *n*.

1. Fill in the next three terms of the arithmetic sequence.

$a_1 = 7 = f(1)$

$a_2 = 10 = f(2)$

$a_3 = 13 = f(3)$

$a_4 =$ __________ $= f(4)$

$a_5 =$ __________ $= f(5)$

$a_6 =$ __________ $= f(6)$

2. What is the *n*th term of the sequence?

3. What function *f* could be used to describe the sequence?

4. What is the common difference of the sequence? How is the common difference related to the function you wrote in Item 3?

5. **Attend to precision.** Describe the domain of *f* using set notation. (*Hint*: What values are used as inputs for *f*?)

6. What ordered pair represents the *n*th term of the sequence?

7. Describe the graph of *f*. How is the common difference related to the graph?

My Notes

Check Your Understanding

For Items 8–11, use the arithmetic sequence –5, 1, 7, 13, ...

8. What is $f(1)$?

9. What is $f(4)$?

10. Write a function to describe the sequence.

11. Use your function to find $f(14)$.

LESSON 11-3 PRACTICE

Write a function to describe each arithmetic sequence.

12. 10, 14, 18, 22, ...

13. 8.5, 10.3, 12.1, 13.9, ...

14. $\frac{1}{3}, \frac{7}{12}, \frac{5}{6}, \frac{13}{12}, \ldots$

15. −7, −4.5, −2, 0.5, ...

16.

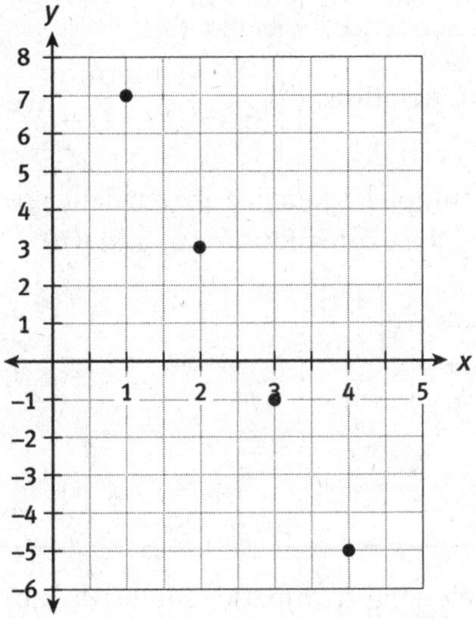

The 1st term of an arithmetic sequence is 5, and the common difference is 1.5. Use this information for Items 17–19.

17. What is $f(3)$?

18. Write a function to describe this arithmetic sequence.

19. Determine the 25th term of the sequence. Use function notation in your answer.

20. **Make sense of problems.** The 3rd term of an arithmetic sequence is −2, and the 8th term is −32. Write a function to describe this sequence.

My Notes

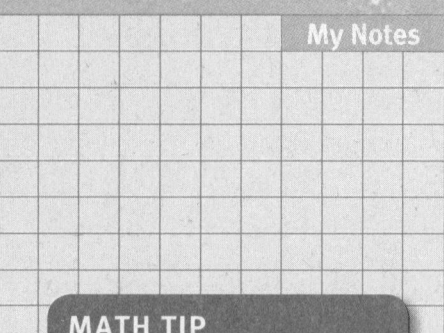

Learning Targets:

- Write a recursive formula for a given arithmetic sequence.
- Use a recursive formula to find the terms of an arithmetic sequence.

SUGGESTED LEARNING STRATEGIES: Look for a Pattern, Create Representations, Close Reading, Marking the Text, Discussion Groups

In a sequence, the term before $f(n) = a_n$ is $f(n - 1) = a_{n-1}$.

The first four terms of the toothpick sequence can be written as

$a_1 = 8$	$f(1) = 8$
$a_2 = 14 = a_1 + 6$	$f(2) = 14 = f(1) + 6$
$a_3 = 20 = a_2 + 6$	$f(3) = 20 = f(2) + 6$
$a_4 = 26 = a_3 + 6$	$f(4) = 26 = f(3) + 6$

1. For any value of n, how can you find the value of $f(n - 1) = a_n - 1$?

MATH TIP

In a sequence, $f(n - 1) = a_{n-1}$ refers to the term before $f(n) = a_n$. Item 1 is asking "For any value of n, how can you find the term before $f(n) = a_n$?"

A ***recursive formula*** can be used to represent an arithmetic sequence. Recursion is the process of choosing a starting term and repeatedly applying the same process to each term to arrive at the following term.

A recursive formula for an arithmetic sequence looks like this:

$$\begin{cases} a_1 = \text{ 1st term} \\ a_n = a_{n-1} + d \end{cases}, \text{ or in function notation: } \begin{cases} f(1) = \text{ 1st term} \\ f(n) = f(n-1) + d \end{cases}$$

2. The recursive formulas for the toothpick sequence are partially given below. Complete them by writing the expressions for a_n and $f(n)$.

$$\begin{cases} a_1 = 8 \\ a_n = \end{cases} \quad \text{and} \quad \begin{cases} f(1) = 8 \\ f(n) = \end{cases}$$

Check Your Understanding

Write the recursive formula for the following arithmetic sequences. Include the recursive formula in function notation.

3. 2, 4, 6, 8, . . .

4. −2, −5, −8, −11, . . .

5. $-3, -\frac{3}{2}, 0, \frac{3}{2}, \ldots$

6. Suppose that $a_{n-1} = -4$.

a. Find the value of a_n for the arithmetic sequence with the recursive formula $\begin{cases} a_1 = 6 \\ a_n = a_{n-1} + (-5) \end{cases}$.

b. What term did you find? (In other words, what is n equal to?)

My Notes

7. An arithmetic sequence has the recursive formula below.

$$\begin{cases} f(1) = \frac{1}{2} \\ f(n) = f(n-1) + 2 \end{cases}$$

a. Determine the first five terms of the sequence.

b. Write the explicit formula for the sequence using function notation.

8. An arithmetic sequence has the explicit formula $a_n = 3n - 8$.

a. What are the values of a_1 and a_2?

b. How can you use the values of a_1 and a_2 to find d? What is d?

c. Use your answers to Parts (a) and (b) to write the recursive formula for the sequence.

In the 12th century, Leonardo of Pisa, also known as Fibonacci, first described a sequence known as the ***Fibonacci sequence.*** The sequence can be described by the recursive formula below.

$$\begin{cases} a_1 = 1 \\ a_2 = 1 \\ a_n = a_{n-1} + a_{n-2}, \text{ for } n > 2 \end{cases}$$

Notice that the first two terms of the sequence are 1 and that the expression describing a_n applies to those terms after the 2nd term.

9. Use the recursive formula to determine the first 10 terms of the Fibonacci sequence.

10. Is the Fibonacci sequence an arithmetic sequence? Justify your response.

My Notes

Check Your Understanding

11. **Attend to precision.** Compare and contrast the explicit and recursive formulas for an arithmetic sequence.

12. Explain how to find any term of the Fibonacci sequence.

LESSON 11-4 PRACTICE

Write the recursive formula for each arithmetic sequence. Include the recursive formula in function notation.

13. 1, 6, 11, 16, . . .

14. 1, 4, 7, 10, 13, . . .

15. $a_n = 11 - 3n$

16. $a_n = \frac{1}{4} + \frac{3}{20}n$

17.

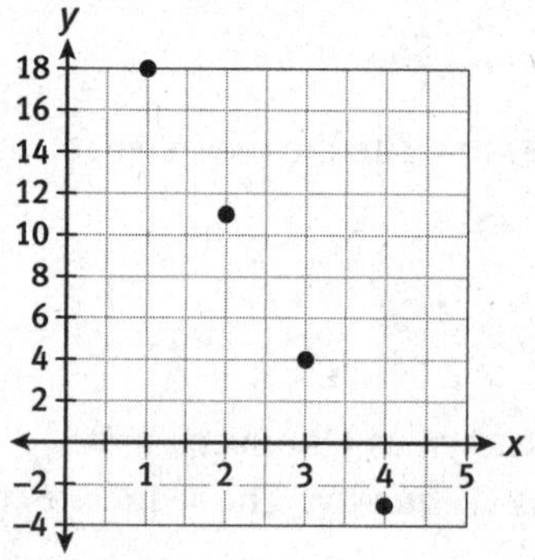

18. Given $f(n-1) = 1.2$, use the recursive formula below to find $f(n)$.

$$\begin{cases} f(1) = -0.6 \\ f(n) = f(n-1) + 0.3 \end{cases}$$

19. **Reason quantitatively.** Describe how each sequence is similar to the Fibonacci sequence. Then find the next two terms.

a. 4, 4, 8, 12, 20, 32, . . .

b. 2, 2, 4, 6, 10, 16, . . .

ACTIVITY 11 PRACTICE

Write your answers on notebook paper. Show your work.

Lesson 11-1

For Items 1–3, refer to the toothpick pattern shown below.

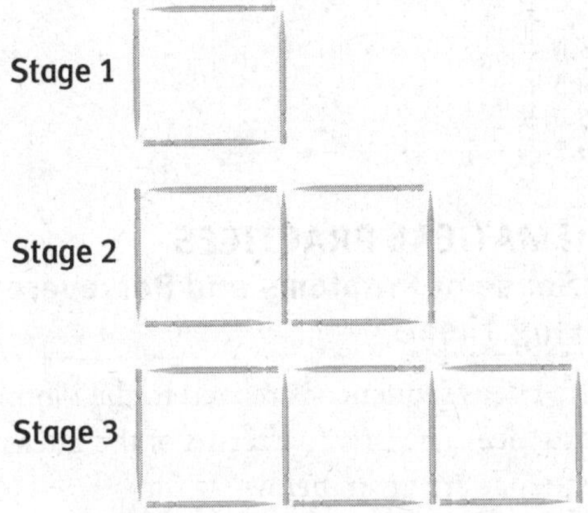

1. Copy and complete the table below.

Stage	Number of Toothpicks
1	
2	
3	
4	
5	

2. Write the number of toothpicks as a sequence.

3. Is the sequence you wrote in Item 2 an arithmetic sequence? If so, determine the common difference. If not, explain why not.

4. Which of the following is **not** an arithmetic sequence?

A. $\frac{1}{2}, 1, \frac{3}{2}, 2 \ldots$
B. 11, 14, 17, 20, …
C. 2, 4, 8, 16, …
D. 5, 2, −1, −4, …

For Items 5 and 6, find the common difference for each arithmetic sequence.

5. −2.3, −1.1, 0.1, 1.3, …

6. −8, −13, −18, −23, …

7. What are the next three terms in the arithmetic sequence −6, −10, −14, … ?

8. Write an arithmetic sequence in which some of the terms are whole numbers and the common difference is $\frac{3}{4}$.

Lesson 11-2

For Items 9 and 10, determine the explicit formula for each arithmetic sequence.

9. 3, −3, −9, −15, …

10. 1, 6, 11, 16, …

11. What is the 20th term of the arithmetic sequence: 7, 4, 1, … ?

12. The 9th and 10th terms of an arithmetic sequence are −24 and −30, respectively. What is the 30th term?

13. The 15th and 21st terms of an arithmetic sequence are −67 and −97, respectively. What is the 30th term?

14. The 9th and 14th terms of an arithmetic sequence are 23 and 33, respectively. What is the 1st term?

15. For the sequence graphed below, write the explicit formula. Then find the 57th term of the sequence.

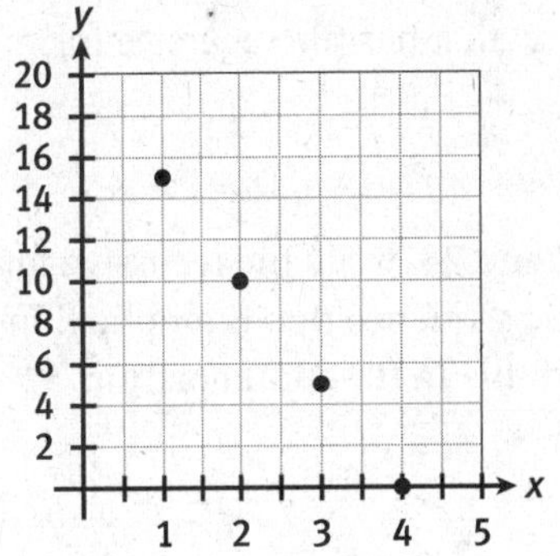

Lesson 11-3

16. Write a function to describe the arithmetic sequence graphed below. Then find the 5th term of the sequence. Use function notation in your answer.

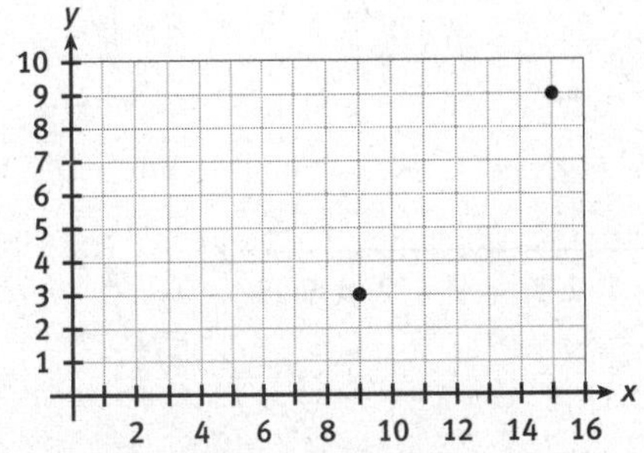

17. For an arithmetic sequence, $f(1) = \frac{4}{5}$ and the common difference is -1. What is $f(20)$?

A. $18\frac{1}{5}$

B. $-\frac{1}{5}$

C. $-20\frac{1}{5}$

D. $-18\frac{1}{5}$

18. An arithmetic sequence is described by the function $f(n) = -3n + 7$. Determine the first five terms of this sequence.

Lesson 11-4

19. What are the first five terms in the arithmetic sequence with the recursive formula below?

$$\begin{cases} a_1 = 5 \\ a_n = a_{n-1} + 4 \end{cases}$$

20. What is the recursive formula for the arithmetic sequence described by the function below?

$$f(n) = \frac{1}{2}n - 2$$

21. What is the explicit formula for the arithmetic sequence that has the recursive formula below?

$$\begin{cases} f(1) = -2.5 \\ f(n) = f(n-1) - 4 \end{cases}$$

For Items 22 and 23, write the recursive formula for the arithmetic sequence that is graphed. Include the recursive formula in function notation.

22.

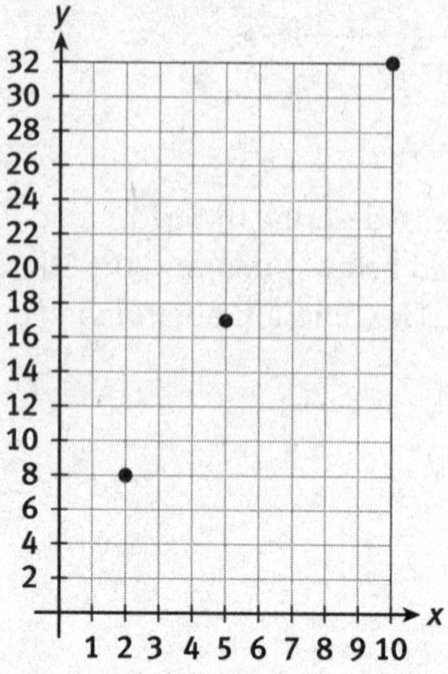

23.

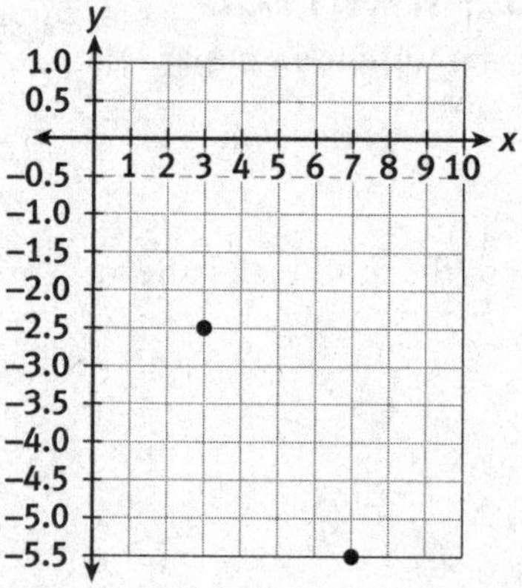

MATHEMATICAL PRACTICES
Make Sense of Problems and Persevere in Solving Them

24. The Lucas sequence is related to the Fibonacci sequence. The first five terms of the Lucas sequence are given below.

$a_1 = 1$

$a_2 = 1 + 2 = 3$

$a_3 = 1 + 3 = 4$

$a_4 = 2 + 5 = 7$

$a_5 = 3 + 8 = 11$

a. Beginning with a_2, what do you observe about the first addends in each sum? What do you observe about the second addends?

b. What are the next two terms of the Lucas sequence? Explain how you determined your answer.

TEXT MESSAGE PLANS

Pedro is planning to add a text messaging feature to his cell phone plan. He has gathered information about the two different plans offered by his wireless phone company.

Plan A: $4.00 per month plus 4 cents for each message
Plan B: 5 cents per message

1. Use the mathematics you have been studying in this unit to provide Pedro with the following information for each plan.
   a. Plan A
   - a table of data
   - a graph of the data
   - the linear function that fits this plan
   - the domain and range of the function

   b. Plan B
   - a table of data
   - a graph of the data
   - the linear function that fits this plan
   - the domain and range of the function
2. If Pedro sends 360 messages on average each month, which plan would you recommend that he choose? Support your recommendation using mathematical evidence.
3. If Pedro knows that his average usage is going to increase to 500 text messages per month, should he change to a different plan? Explain and justify your reasoning.
4. Explain whether either of the plans represents a direct variation.
5. Pedro's friend Chenetta is considering another text messaging plan that advertises the following: "A one-time joining fee of $3.00 and $0.08 per message."
   a. Write an explicit formula for the text messaging plan.
   b. Chenetta knows that she sends and receives about 1800 text messages per month. Use an example and other mathematical evidence to let Chenetta know if you think this plan would be a good deal for her.

Scoring Guide	Exemplary	Proficient	Emerging	Incomplete
	The solution demonstrates the following characteristics:			
Mathematics Knowledge and Thinking (Items 1, 4, 5a)	• Clear and accurate understanding of linear models, including direct variation • Effective understanding of arithmetic sequences	• Largely correct understanding of linear models, including direct variation • Adequate understanding of arithmetic sequences	• Partial understanding of linear models, including direct variation • Some difficulty with arithmetic sequences	• Inaccurate or incomplete understanding of linear models, including direct variation • Little or no understanding of arithmetic sequences
Problem Solving (Items 2, 3, 5b)	• Appropriate and efficient strategy that results in a correct answer	• Strategy that may include unnecessary steps but results in a correct answer	• Strategy that results in some incorrect answers	• No clear strategy when solving problems
Mathematical Modeling / Representations (Items 1, 5a)	• Clear and accurate tables of real-world data, graphs of the data, and linear functions to model the data, including reasonable domain and range • Fluency in writing an explicit formula to model a real-world scenario	• Correct tables of real-world data, graphs of the data, and linear functions to model the data, including reasonable domain and range • Little difficulty writing an explicit formula to model a real-world scenario	• Partially correct tables of real-world data, graphs of the data, and linear functions to model the data, including reasonable domain and range • Some difficulty writing an explicit formula to model a real-world scenario	• Inaccurate or incomplete tables of real-world data, graphs of the data, and linear functions to model the data, including reasonable domain and range • Significant difficulty writing an explicit formula to model a real-world scenario
Reasoning and Communication (Items 2–4, 5b)	• Precise use of appropriate math terms and language to make and justify a recommendation • Clear and accurate explanation of whether one of the plans represents a direct variation	• Appropriate recommendations with adequate justifications • Largely correct explanation of whether one of the plans represents a direct variation	• Misleading or confusing recommendations and/or justifications • Partially correct explanation of whether one of the plans represents a direct variation	• Incomplete or inaccurate recommendations and/or justifications • Incomplete or inaccurate explanation of whether one of the plans represents a direct variation

Forms of Linear Functions

Under Pressure

Lesson 12-1 Slope-Intercept Form

My Notes

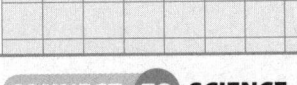

Learning Targets:

- Write the equation of a line in slope-intercept form.
- Use slope-intercept form to solve problems.

SUGGESTED LEARNING STRATEGIES: Create Representations, Think-Pair-Share, Marking the Text, Discussion Groups

When a diver descends in a lake or ocean, pressure is produced by the weight of the water on the diver. As a diver swims deeper into the water, the pressure on the diver's body increases at a rate of about 1 *atmosphere of pressure* per 10 meters of depth. The table and graph below represent the total pressure, y, on a diver given the depth, x, under water in meters.

x	y
0	1
1	1.1
2	1.2
3	1.3
4	1.4
5	1.5
6	1.6

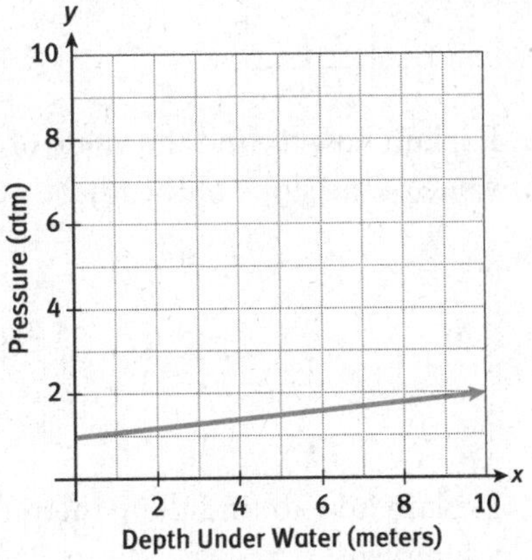

CONNECT TO SCIENCE

Pressure is force per unit area. *Atmospheric pressure* is defined using the unit atmosphere. 1 atm is 14.6956 pounds per square inch.

1. Write an equation describing the relationship between the pressure exerted on a diver and the diver's depth under water.

2. What is the slope of the line? What are the units of the slope?

3. What is the y-intercept? Explain its meaning in this context.

MATH TERMS

A **linear equation** is an equation that can be written in standard form $Ax + By = C$ where A, B, and C are constants and A and B cannot both be zero.

Slope-Intercept Form **of a** ***Linear Equation***

$$y = mx + b$$

where m is the slope of the line and $(0, b)$ is the y-intercept.

MATH TIP

Linear equations can be written in several forms.

4. Identify the slope and y-intercept of the line described by the equation $y = -2x + 9$.

My Notes

5. Create a table of values for the equation $y = -2x + 9$. Then plot the points and graph the line.

x	y
−2	
0	
2	
4	
5	

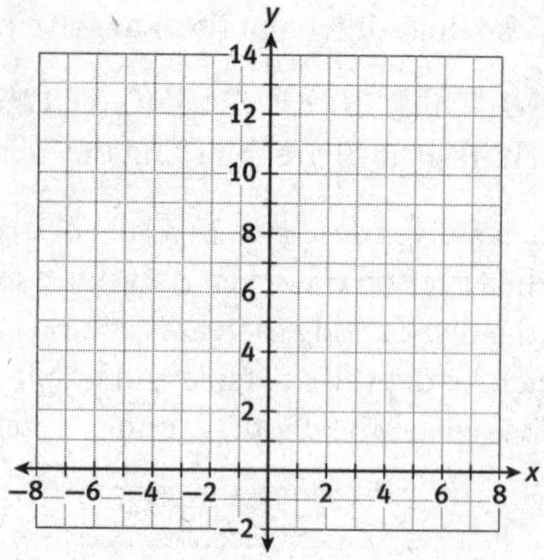

6. Explain how to find the value of the slope from the table. What is the value of the slope of the line?

7. Explain how to find the y-intercept from the table. What is the y-intercept?

8. Explain how to find the value of the slope from the graph. What is the value of the slope?

9. Explain how to find the y-intercept from the graph. What is the y-intercept?

My Notes

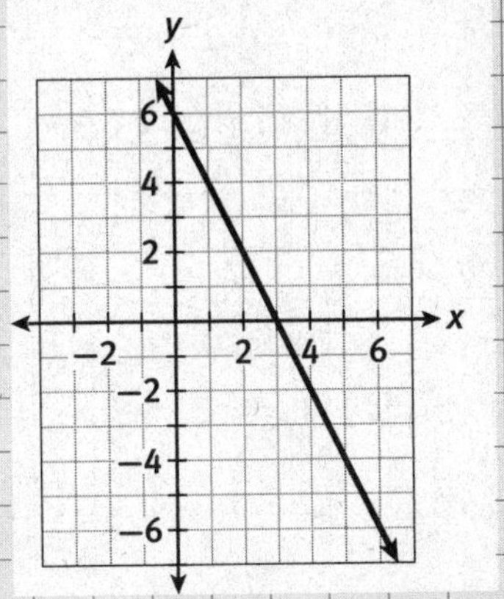

Check Your Understanding

10. What are the slope and y-intercept of the line described by the equation $y = -\frac{4}{5}x - 10$?

11. Write the equation in slope-intercept form of the line that is represented by the data in the table.

x	−2	−1	0	1	2	3
y	9	7	5	3	1	−1

12. Write the equation, in slope-intercept form, of the line with a slope of 4 and a y-intercept of (0, 5).

13. Write an equation of the line graphed in the *My Notes* section of this page.

Monica gets on an elevator in a skyscraper. The elevator starts to move at a rate of −20 ft/s. After 6 seconds on the elevator, Monica is 350 feet from the ground floor of the building.

14. The rate of the elevator is negative. What does this mean in the situation? What value in the slope-intercept form of an equation does this rate represent?

15. **a.** How many feet was Monica above the ground when she got on the elevator? Show how you determined your answer.

b. What value in the slope-intercept form does your answer to Part (a) represent?

16. **Model with mathematics.** Write an equation in slope-intercept form for the motion of the elevator since it started to move. What do x and y represent?

a. What does the y-intercept represent?

b. Use the equation you wrote to determine, at this rate, how long it will take after Monica enters the elevator for her to exit the elevator on the ground floor. Explain how you found your answer.

My Notes

Check Your Understanding

17. Write the equation $3x - 2y = 16$ in slope-intercept form. Explain your steps.

18. A flowering plant stands 6.5 inches tall when it is placed under a growing light. Its growth is 0.25 inches per day. Today the plant is 11.25 inches tall.

a. Write an equation in slope-intercept form for the height of the plant since it was placed under the growing light.

b. In your equation, what do x and y represent?

c. Use the equation to determine how many days ago the plant was placed under the light.

LESSON 12-1 PRACTICE

19. What are the slope, m, and y-intercept, $(0, b)$, of the line described by the equation $3x + 6y = 12$?

20. Write an equation in slope-intercept form for the line that has a slope of $\frac{2}{3}$ and y-intercept of $(0, -5)$.

21. Write an equation in slope-intercept form for the line that passes through the points $(6, -3)$ and $(0, 2)$.

22. Matt sells used books on the Internet. He has a weekly fee he has to pay for his website. He has graphed his possible weekly earnings, as shown.

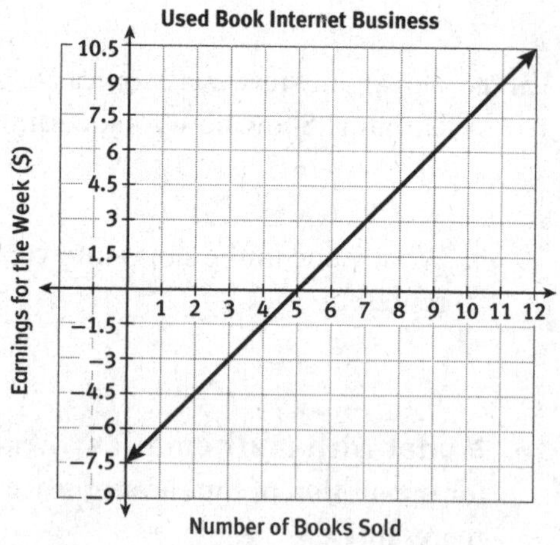

a. What is the weekly fee that Matt pays for his website? How do you know?

b. How much does Matt make for each book sold? How do you know?

c. Write the equation in slope-intercept form for the line in Matt's graph.

d. How many books does Matt have to sell to make $30 for the week? Explain.

23. Make use of structure. Without graphing, describe the graph of each equation below. Tell whether the line is ascending or descending from left to right and where the line crosses the y-axis.

a. $y = 3x$ **b.** $y = 5x + 2$ **c.** $y = -2x - 5$ **d.** $y = -6x + 4$

My Notes

Learning Targets:

- Write the equation of a line in point-slope form.
- Use point-slope form to solve problems.

SUGGESTED LEARNING STRATEGIES: Create Representations, Marking the Text, Note Taking, Think-Pair-Share, Critique Reasoning, Sharing and Responding

Another form of the equation of a line is the point-slope form. The point-slope form of the equation is found by solving the slope formula $m = \frac{y - y_1}{x - x_1}$ for $y - y_1$, by multiplying both sides by $x - x_1$. You may use this form when you know a point on the line and the slope.

Point-Slope Form of a Linear Equation

$$y - y_1 = m(x - x_1)$$

where m is the slope of the line and (x_1, y_1) is a point on the line.

CONNECT TO AP

In calculus, the point-slope form of a line is used to write the equation of the line tangent to a curve at a given point.

Example A

Write an equation of the line with a slope of $\frac{1}{2}$ that passes through the point (2, 5). Graph the line.

Step 1: Substitute the given values into point-slope form.

$$y - y_1 = m(x - x_1)$$
$$y - 5 = \frac{1}{2}(x - 2)$$

Step 2: Graph $y - 5 = \frac{1}{2}(x - 2)$. Plot the point (2, 5) and use the slope to find another point.

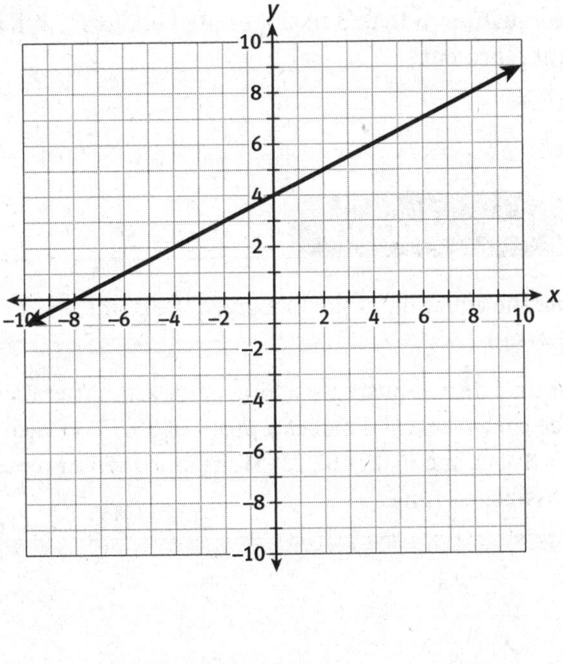

MATH TIP

If you needed to express the solution to Example A in slope-intercept form, you could apply the Distributive Property and combine like terms.

My Notes

Try These A

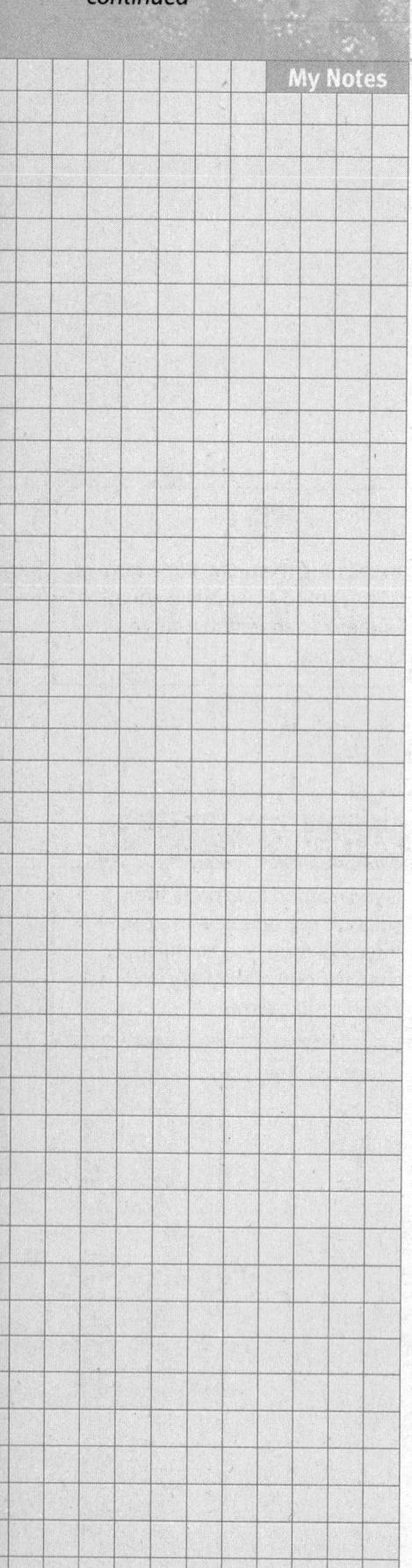

Find an equation of the line given a point and the slope.

a. $(-2, 7), m = \frac{2}{3}$ **b.** $(6, -1), m = -\frac{5}{4}$

Determine the slope and a point on the line for each equation.

c. $y + 4 = \frac{3}{8}(x - 3)$ **d.** $y - 6 = -\frac{5}{2}(x + 3)$

The town of San Simon charges its residents for trash pickup and water usage on the same bill. Each month the city charges a flat fee for trash pickup and a fee of $0.25 per gallon for water used. In January, one resident used 44 gallons of water, and received a bill for $16.

1. If x is the number of gallons of water used during a month, and y represents the bill amount in dollars, write a point (x_1, y_1).

2. What does $0.25 per gallon represent?

3. Reason abstractly. Use point-slope form to write an equation that represents the bill cost y in terms of the number of gallons of water x used in a month.

4. Write the equation in Item 3 in slope-intercept form. What does the y-intercept represent?

Check Your Understanding

5. Determine the equation of the line given the point (86, 125) and the slope $m = -18$.

6. Violet has an Internet business selling paint sets. After an initial website fee each week, she makes a profit of $0.75 on each set she sells. If she sells 8 sets, she makes $2.25. Write an equation representing her weekly possible earnings.

My Notes

7. **Critique the reasoning of others.** Jamilla and Ryan were asked to write the equation of the line through the points (6, 4) and (3, 5). Both Jamilla and Ryan determined that the slope was $-\frac{1}{3}$. Jamilla wrote the equation of the line as $y-4=-\frac{1}{3}(x-6)$. Ryan wrote the equation of the line as $y-5=-\frac{1}{3}(x-3)$.

   a. Rewrite each student's equation in slope-intercept form and compare the results.

   b. Whose equation was correct? Justify your response.

8. Find the equation in point-slope form of the line shown in the graph.

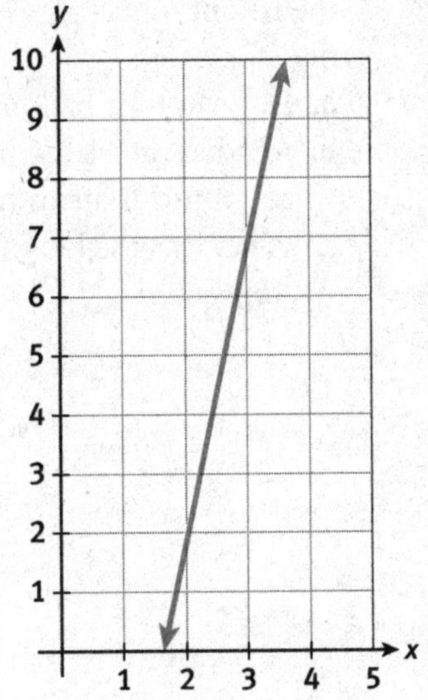

9. Write the equation of the line in slope-intercept form.

Check Your Understanding

10. Explain the process you would use to write an equation of a line in point-slope form when given two points on the line.

11. Describe the similarities and differences between point-slope form and slope-intercept form.

My Notes

LESSON 12-2 PRACTICE

12. Write an equation of the line with a slope of 0.25 that passes through the point $(-1, -8)$.

13. Find the slope and a point on the line for the lines with the following equations.

a. $y - 9 = -\frac{3}{4}(x - 4)$

b. $y = 3 - \frac{2}{3}(x + 4)$

14. Write the equation of the line through the points $(-3, 3)$ and $(7, 5)$ in slope-intercept form. What is the y-intercept?

15. Jay pays a flat fee each month for basic cable service. He also pays \$3.50 for each movie he orders during the month. Last month, he ordered 5 movies and his total bill came to \$54.

a. Write an equation in point-slope form that represents the total bill, y, in terms of the number of movies, x.

b. Write the equation in slope-intercept form.

c. What is the monthly fee for basic cable service? How do you know?

d. Next month, Jay plans to order 7 movies. What will be his total bill for the month?

e. This month, Jay's total bill is \$78.50. How many movies did he order this month?

16. Attend to precision. The equation $y - 160 = 40(x - 1)$ represents the height in feet, y, of a hot-air balloon x minutes after the pilot started her stopwatch.

a. Is the hot-air balloon rising or descending? Justify your answer.

b. At what rate is the hot-air balloon rising or descending? Be sure to use appropriate units.

c. What was the height of the balloon when the pilot started her stopwatch?

My Notes

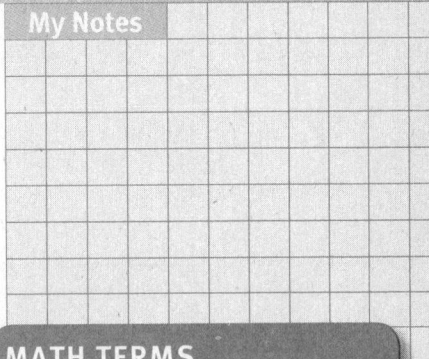

Learning Targets:

- Write the equation of a line in standard form.
- Use the standard form of a linear equation to solve problems.

SUGGESTED LEARNING STRATEGIES: Create Representations, Note Taking, Discussion Groups, Think-Pair-Share, Identify a Subtask

A **linear equation** can be written in the form $Ax + By = C$ where A, B, and C are constants and A and B are not both zero.

MATH TERMS

The **greatest common factor** of two or more integers is the greatest integer that is a divisor of all the integers.

Standard Form **of a Linear Equation**

$$Ax + By = C$$

where $A \geq 0$, A and B are not both zero, and A, B, and C are integers whose ***greatest common factor*** is 1.

1. **Reason abstractly.** You can use the coefficients of this form of an equation to find the x-intercept, y-intercept, and slope.
   a. Determine the x-intercept.

   b. Determine the y-intercept.

   c. Write $Ax + By = C$ in slope-intercept form to find the slope.

The definition of standard form states that both A and B are not 0. However, one of A or B may be equal to 0.

2. Write the standard form if $A = 0$.

   a. Suppose $A = 0$, $B = -1$, and $C = 3$. Write the equation of the line in standard form.

   b. Graph the line on the grid in the *My Notes* section. Describe the graph. What is the slope?

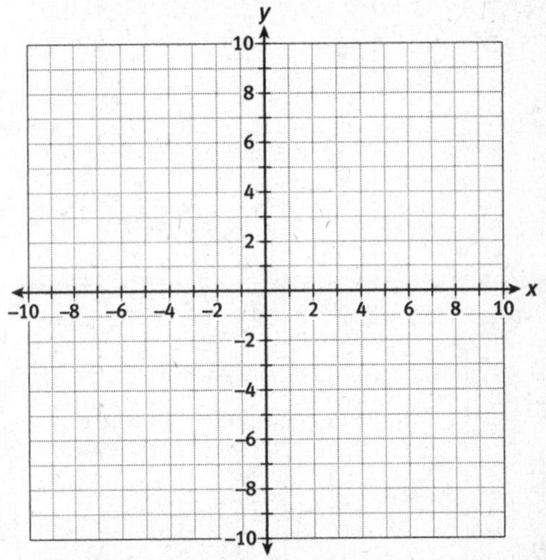

3. Write the standard form if $B = 0$.

   a. Suppose $A = 1$, $B = 0$, and $C = -6$. Write the equation of the line in standard form.

   b. Graph the equation on the grid in the *My Notes* section. Describe the graph. What is the slope?

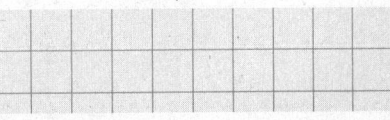

My Notes

4. Write $3x + 2y = 8$ in slope-intercept form.

5. Write the equation $y - 7 = 2(x + 1)$ in standard form.

Check Your Understanding

6. Write the equation $2x + 3y = 18$ in slope-intercept form.

7. Write the equation $y = -\frac{6}{5}x - 4$ in standard form.

8. Describe the graph of any line whose equation, when written in standard form, has $A = 0$.

9. Susheila is making a large batch of granola to sell at a school fundraiser. She needs to buy walnuts and almonds to make the granola. Walnuts cost $3 per pound and almonds cost $2 per pound. She has $30 to spend on these ingredients.

a. Write an equation that represents the different amounts of walnuts, x, and almonds, y, that Susheila can buy.

b. Graph the x- and y-intercepts on the coordinate plane below. Use these to help you graph the line.

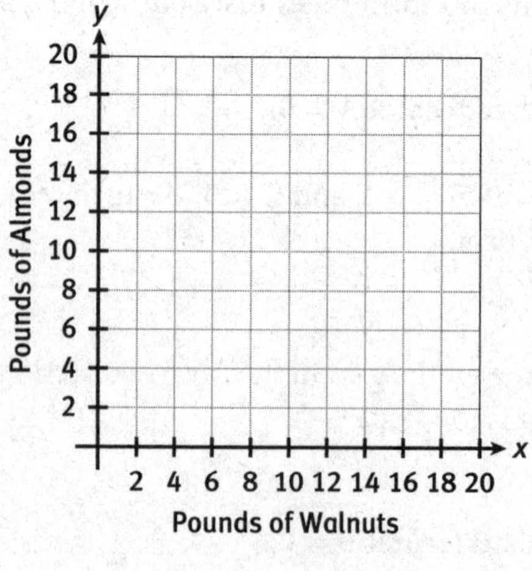

c. If Susheila buys 4 pounds of walnuts, how many pounds of almonds can she buy?

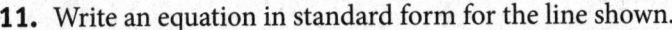

10. Refer to the graph you made in Item 9b. What is the x-intercept? What does it represent?

11. Write an equation in standard form for the line shown.

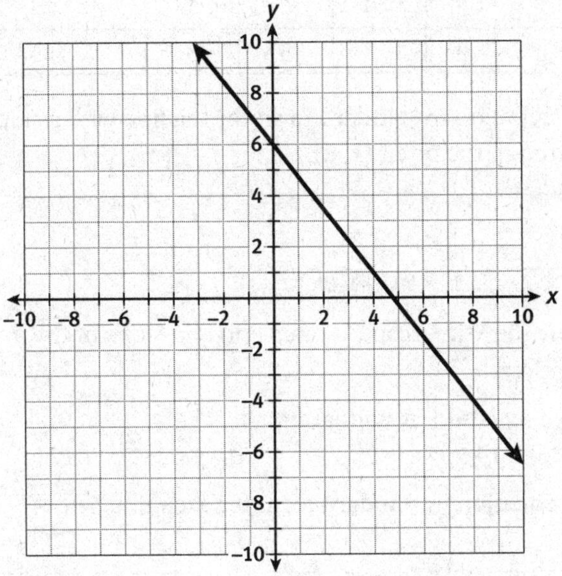

12. Make use of structure. The equation $2x - 5y = 20$, the table below, and the graph below represent three different linear functions.

x	y
−3	1
−2	4
−1	7
0	10
1	13
2	16
3	19

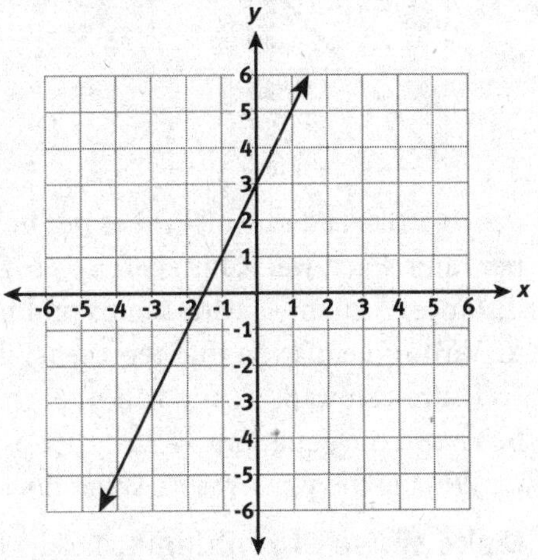

Which function represents the line with the greatest slope? Explain your reasoning.

My Notes

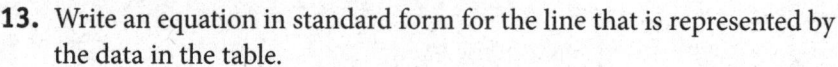

Check Your Understanding

13. Write an equation in standard form for the line that is represented by the data in the table.

x	−2	−1	0	1	2	3
y	9	7	5	3	1	−1

14. Write an equation in standard form for the line with a slope of 7 that passes through the point (1, 2).

LESSON 12-3 PRACTICE

15. Determine the *x*-intercept, *y*-intercept, and slope of the line described by $-3x + 7y = -21$.

16. Write each equation in standard form.

a. $8x = 26 + 14y$ **b.** $y = -\frac{6}{7}x + 12$

17. Write an equation in standard form for each line below.

a.

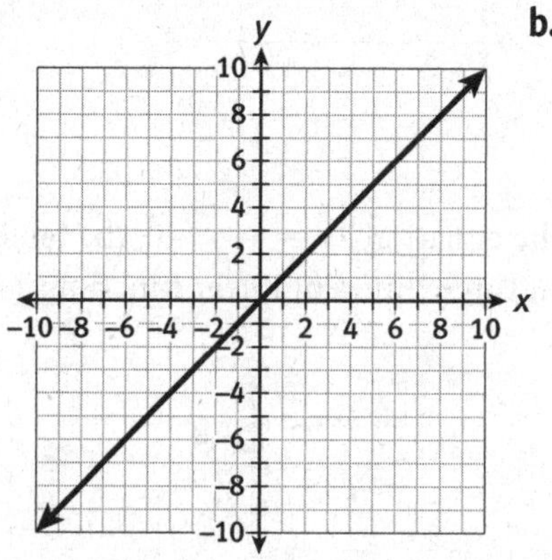

b.

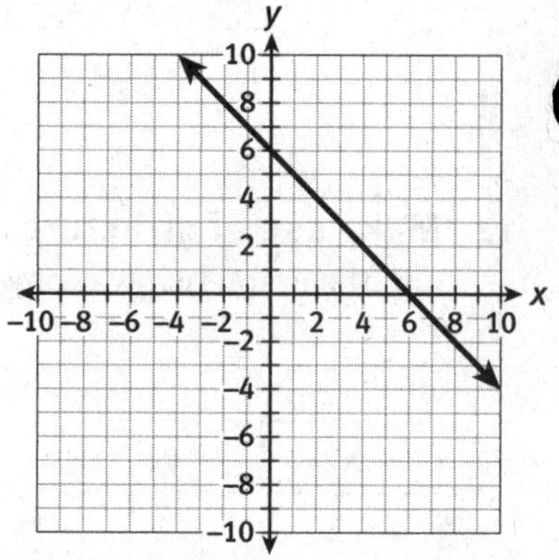

18. Pedro walks at a rate of 4 miles per hour and runs at a rate of 8 miles per hour. Each week, his exercise program requires him to cover a total distance of 20 miles with some combination of walking and/or running.

a. Write an equation that represents the different amounts of time Pedro can walk, *x*, and run, *y*, each week.

b. Graph the equation.

c. What is the *y*-intercept? What does this tell you?

19. **Make sense of problems.** Keisha bought a discount pass at a movie theater. It entitles her to a special discounted admission price for every movie she sees. Keisha wrote an equation that gives the total cost *y* of seeing *x* movies. In standard form, the equation is $7x - 2y = -31$.

a. What was the cost of the pass?

b. What is the discounted admission price for each movie?

My Notes

Learning Targets:

- Describe the relationship among the slopes of parallel lines and perpendicular lines.
- Write an equation of a line that contains a given point and is parallel or perpendicular to a given line.

SUGGESTED LEARNING STRATEGIES: Think-Pair-Share, Predict and Confirm, Create Representations, Look for a Pattern, Discussion Groups

Parallel lines and perpendicular lines are pairs of lines that have special relationships.
Parallel lines in a plane are equidistant from each other at all points.

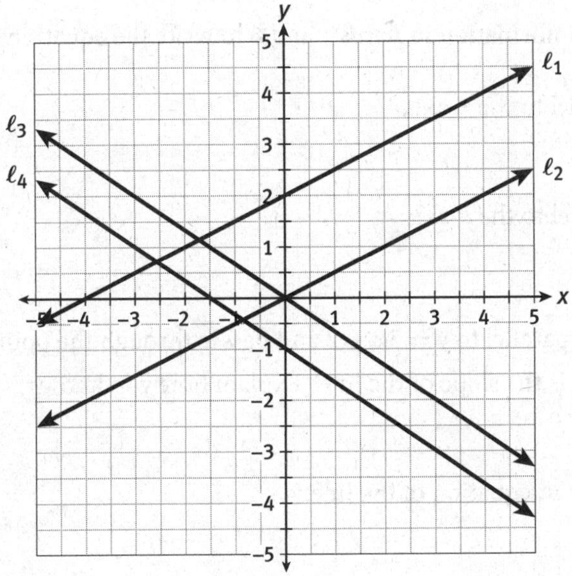

1. Consider lines l_1, l_2, l_3, and l_4 on the graph above. Determine the slope of each line.

2. Reason quantitatively. In the graph above, l_1 is parallel to l_2 and l_3 is parallel to l_4. Write a conjecture about the slopes of parallel lines.

3. Determine the slope of a line that is parallel to the line whose equation is $y = -3x + 4$.

4. Write the equation of a line that is parallel to the line $y = \frac{3}{4}x - 1$ and has a y-intercept of (0, 5).

My Notes

5. Horizontal lines are described by equations of the form $y = number$. For example, the equation of the x-axis is $y = 0$, because all points on the x-axis have y-coordinate 0. Explain why any two horizontal lines are parallel.

6. Vertical lines are described by equations of the form $x = number$. For example, the equation of the y-axis is $x = 0$, because all points on the y-axis have x-coordinate 0. Do you think that any two vertical lines are parallel? Explain why or why not.

7. Use the information in Items 5 and 6 to write the equation of a line that is

a. parallel to the x-axis.

b. parallel to the y-axis.

8. A line is parallel to $y = 3x + 2$ and passes through the point (1, 4).

a. What is the slope of the line? Explain how you know.

b. Write an equation of the line.

MATH TIP

Perpendicular lines intersect to form right angles.

9. Graph and label each line described below on the grid in the *My Notes* section. Which lines appear to be perpendicular?

- l_5 has slope $-\frac{4}{3}$ and contains the point (0, 2).
- l_6 has slope $-\frac{3}{4}$ and contains the point (0, 0).
- l_7 has slope $\frac{3}{4}$ and contains the point (−2, −1).

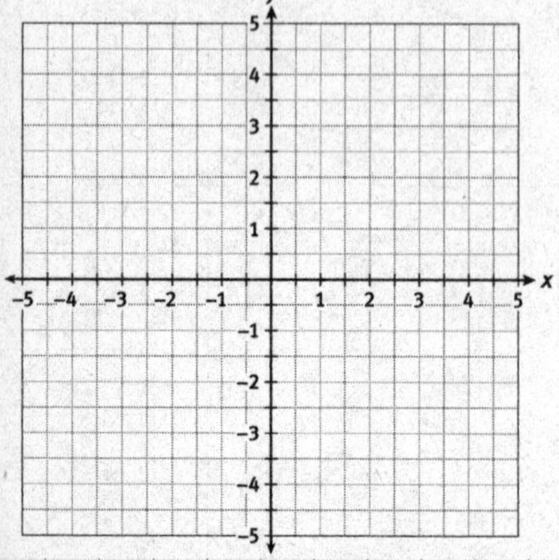

10. Write a conjecture about the slopes of perpendicular lines.

My Notes

11. Use your prediction from Item 10 to write the equations of two lines that are perpendicular. On the grid in the *My Notes* section on the previous page, graph both lines and confirm that they are perpendicular.

12. In the coordinate plane, what is true about a line that is perpendicular to a horizontal line?

13. Line l_1 contains the points (0, −1) and (3, 1). It is perpendicular to line l_2 that contains the point (−1, 2).

a. What is the slope of each line? Explain how you know.

b. Write the equation of each line.

Check Your Understanding

14. Determine whether the lines with the given slopes are parallel, perpendicular, or neither.

a. $m_1 = -4, m_2 = \frac{1}{4}$ **b.** $m_1 = -3, m_2 = 3$

c. $m_1 = \frac{10}{12}, m_2 = -1\frac{1}{5}$ **d.** $m_1 = \frac{1}{2}, m_2 = \frac{1}{2}$

15. The equation of line l_1 is $y = \frac{1}{3}x - 2$.

a. Write the equation of a line parallel to l_1. Explain.

b. Write the equation of a line perpendicular to l_1. Explain.

16. Write the equation of a line that is parallel to the line $3x + 4y = 4$ and contains the point (8, 1).

17. Write an equation of a line that is perpendicular to the line $y = 5x + 1$ and contains the point (−10, 2).

My Notes

LESSON 12-4 PRACTICE

18. Determine whether the lines with the given slopes are parallel, perpendicular, or neither.

a. $m_1 = 5, m_2 = \frac{1}{5}$

b. $m_1 = -6, m_2 = \frac{1}{6}$

c. $m_1 = -\frac{2}{3}, m_2 = -\frac{2}{3}$

19. The slopes of three lines are given below.

$m_1 = -\frac{1}{2}$ $\qquad m_2 = 3$ $\qquad m_3 = 0$

a. Determine the slope of a line that is parallel to a line with each given slope.

b. Determine the slope of a line that is perpendicular to a line with each given slope.

20. Determine the slope of any line that is parallel to the line described by $y = -\frac{1}{2}x + 5$.

21. Write the equation of a line that is parallel to the line described by $x - 4y = 8$. Explain how you know the lines are parallel.

22. Determine the slope of any line that is perpendicular to the line described by $y = \frac{3}{4}x - 9$.

23. Write an equation of the line that is perpendicular to the line $2x + 5y = -15$ and contains the point $(-8, 3)$.

24. Determine the equation of a line perpendicular to the x-axis that passes through the point $(4, -1)$.

25. **Construct viable arguments.** A line a passes through points with coordinates $(-3, 5)$ and $(0, 0)$ and a line b passes through points with coordinates $(3, 5)$ and $(0, 0)$. Are lines a and b parallel, perpendicular, or neither? Explain your answer.

ACTIVITY 12 PRACTICE

Write your answers on notebook paper. Show your work.

Lesson 12-1

1. Write the equation of a line in slope-intercept form that has a slope of -8 and a y-intercept of $(0, 3)$.
2. Write the equation of a line in slope-intercept form that passes through the point $(0, -7)$ and has a slope of $\frac{3}{4}$.
3. Find the slope and the y-intercept of the line whose equation is $-5x + 3y - 8 = 0$.
4. Which of the following is the slope-intercept form of the equation of the line in the graph?

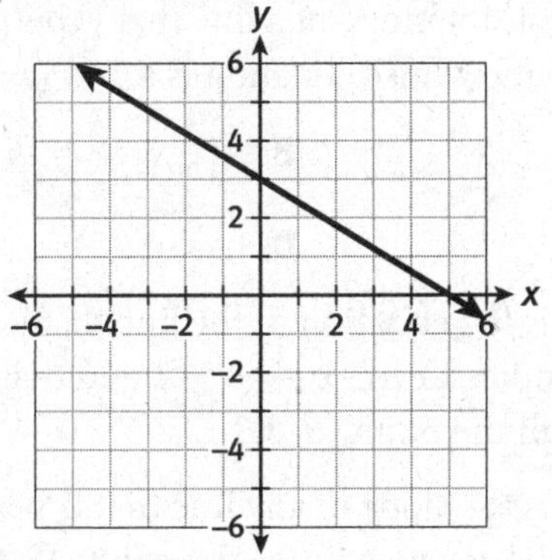

A. $y = -\frac{5}{3}x + 3$

B. $y = -\frac{3}{5}x + 5$

C. $y = -\frac{3}{5}x + 3$

D. $y = -\frac{5}{3}x + 5$

After paying an initial fee each week, Mike can sell packs of baseball cards in a sports shop. He displays his possible earnings for one week on the following graph. Use the graph for Items 5–9.

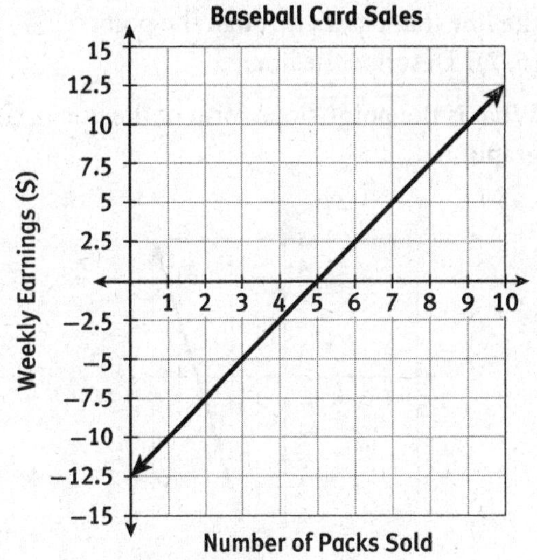

5. What is the initial fee Mike pays each week?
6. How many packs does Mike have to sell to break even?
7. What is the price of one pack of cards?
8. What is the equation in slope-intercept form for the line shown in graph?
9. How many packs of cards must Mike sell to make \$40? Explain.

Lesson 12-2

10. What is the equation in point-slope form of the line that passes through $(-9, 12)$ with a slope of $\frac{5}{6}$?
11. What is the equation in slope-intercept form of the line that has a slope of 0.25 and passes through the point $(6, -8)$?
12. What is the equation in point-slope form of the line that passes through the points $(2, -3)$ and $(-5, 8)$?

13. Write an equation in slope-intercept form of the line that passes through the points (4, 2) and (1, −7).

14. What is the equation in slope-intercept form of the line that passes through the points (2, 7) and (6, 7)? Describe the line.

15. What is the point-slope form of the line in the graph?

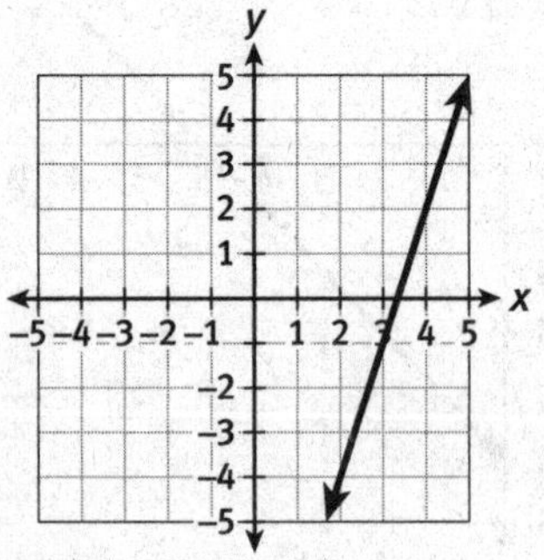

Lesson 12-3

16. Write the equation of the line in the graph from Item 15 in standard form.

17. David is ordering tea from an online store. Black tea costs \$0.80 per ounce and green tea costs \$1.20 per ounce. He plans to spend a total of \$12 on the two types of tea.

a. Write an equation that represents the different amounts of black tea, x, and green tea, y, that David can buy.

b. Graph the equation.

c. What is the x-intercept? What does it represent?

d. Suppose David decides to buy 10 ounces of black tea. How many ounces of green tea will he buy?

18. Is the equation $6x - 15y = -12$ in standard form? Why or why not?

19. Which is a true statement about the line $x - 4y = 8$?

A. The x-intercept of the line is (2, 0).

B. The y-intercept of the line is (0, 2).

C. The slope of the line is $\frac{1}{4}$.

D. The line passes through the origin.

20. Write the equation of a line in standard form that has an x-intercept of (3, 0) and a y-intercept of (0, 5).

Lesson 12-4

21. What is the slope of a line parallel to a line whose equation is $3x + 5y = 12$?

22. What is the slope of a line perpendicular to a line whose equation is $-4x - 2y + 18 = 0$?

23. Which is the slope of a line that is perpendicular to the line whose equation is $5x - 3y = -10$?

A. $\frac{3}{5}$ **B.** $-\frac{3}{5}$

C. $\frac{5}{3}$ **D.** $-\frac{5}{3}$

24. What is the equation of the line that is perpendicular to $2x + 4y = 1$ and that passes through the point (6, 8)?

25. What is the slope of any line that is perpendicular to the line that contains the points (8, 8) and (12, 12)?

MATHEMATICAL PRACTICES
Construct Viable Arguments and Critique the Reasoning of Others

26. Aidan stated that for any value of b, the line $y = 2x + b$ is parallel to the line that passes through (2, 5) and (−1, −1). Do you agree with Aidan? Explain why or why not.

Equations From Data

Pass the Book

Lesson 13-1 Scatter Plots and Trend Lines

My Notes

Learning Targets:

- Use collected data to make a scatter plot.
- Determine the equation of a trend line.

SUGGESTED LEARNING STRATEGIES: Predict and Confirm, Sharing and Responding, Create Representations, Look for a Pattern, Interactive Word Wall

How fast can you and your classmates pass a textbook from one person to the next until the book has been relayed through each person in class?

1. Suppose your entire class lined up in a row. Estimate the length of time you think it would take to pass a book from the first student in the row to the last. Assume that the book starts on a table and the last person must place the book on another table at the end of the row.

   Estimated time to pass the book: ______________

2. As a class, experiment with the actual time it takes to pass the book using small groups of students in your class. Use the table below to record the times.

Number of students passing the book	3	6	9	11	13	15
Time to pass the book (nearest tenth of a second)						

3. **Reason quantitatively.** Based on the data you recorded in the table above, would you revise your estimated time from Item 1? Explain the reasoning behind your answer.

My Notes

MATH TERMS

A **scatter plot** displays the relationship between two sets of numerical data. It can reveal trends in data.

4. Graph the data in your table from Item 2 as a ***scatter plot*** on the coordinate grid.

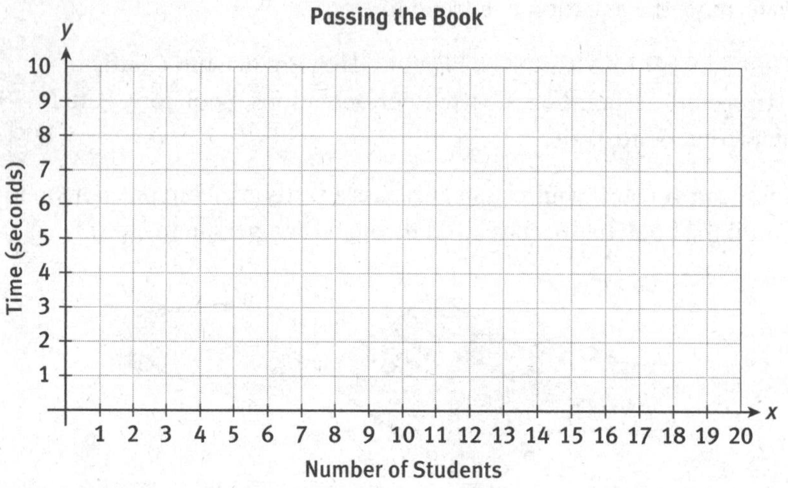

5. Are the data that you collected linear data?

a. Explain your answer using the scatter plot.

b. Explain your answer using the table of data.

6. Describe how the time to pass the book changes as the number of students increases.

7. Work as a group to predict the number of seconds it will take to pass the book through the whole class.

a. Place a ***trend line*** on the scatter plot in Item 4 in a position that your group feels best models the data. Then, mark two points on the line.

b. In the spaces provided below, enter the coordinates of the two points identified in Part (a).

Point 1: (________, ________) Point 2: (________, ________)

c. Why does your group think that this line gives the best position for modeling the scatter plot data?

MATH TERMS

A **trend line** is a line drawn on a scatter plot to show the **correlation**, or association, between two sets of data.

My Notes

8. Use the coordinate pairs you recorded in Item 7b to write the equation for your trend line (or linear model) of the scatter plot.

9. Explain what the variables in the equation of your linear model represent.

10. **Reason abstractly.** Interpret the meaning of the slope in your linear model.

11. Use your model to predict how long it would take to pass the book through all the students in your class.

Predicted time to pass the book: ________________

12. Using all of the students in your class, find the actual time it takes to pass the book.

Actual time to pass the book: ________________

13. How do your estimate from Item 1 and your predicted time from Item 11 compare to the actual time that it took to pass the book through the entire class?

14. **Attend to precision.** Suppose that another class took 1 minute and 47 seconds to pass the book through all of the students in the class. Use your linear model to estimate the number of students in the class.

My Notes

Check Your Understanding

The table shows the number of days absent and the grades for several students in Ms. Reynoso's Algebra 1 class.

Days Absent	0	3	6	1	2	2	4
Grade (percent)	98	88	69	89	90	86	77

15. Create a scatter plot of the data using Days Absent as the independent variable.

16. Are the data linear? Explain using the scatter plot and the table of data.

17. Based on the data, how do grades change as the number of days absent increases?

18. Draw a trend line on your scatter plot. Identify two points on the trend line and write an equation for the line containing those two points.

19. What is the meaning of the x and y variables in the equation you wrote?

20. Interpret the meaning of the slope and the y-intercept of your trend line.

21. Use your equation to predict the grade of a student who is absent for 5 days.

LESSON 13-1 PRACTICE

Model with mathematics. The scatter plot shows the day of the month and total rainfall for January.

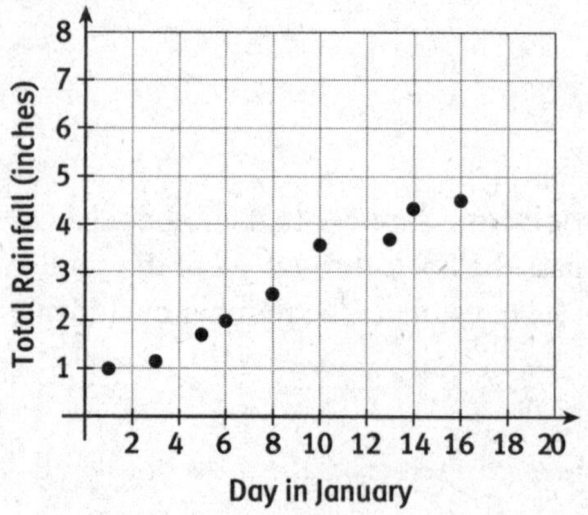

22. Copy the scatter plot and draw a trend line on the scatter plot. Identify two points on the trend line and write a linear equation to model the data containing those two points.

23. Explain the meaning of x and y in your equation.

24. Interpret the meaning of the slope and the y-intercept of your trend line.

My Notes

Learning Targets:

- Use a linear model to make predictions.
- Use technology to perform a linear regression.

SUGGESTED LEARNING STRATEGIES: Marking the Text, Interactive Word Wall, Look for a Pattern, Think-Pair-Share, Quickwrite

There is a *correlation* between two variables if they share some kind of relationship.

1. Is there a correlation between the variables of your linear model in Item 4 in Lesson 13-1? Explain.

MATH TERMS

A scatter plot will show a **positive correlation** if *y* tends to increase as *x* increases. Other data may have a **negative correlation**, where *y* tends to decrease as *x* increases, or **no correlation.** A correlation is sometimes called an association.

Examples of data with two variables that illustrate a ***positive correlation***, a ***negative correlation***, and ***no correlation*** are shown below. The more closely the data resemble a line, the stronger the linear correlation.

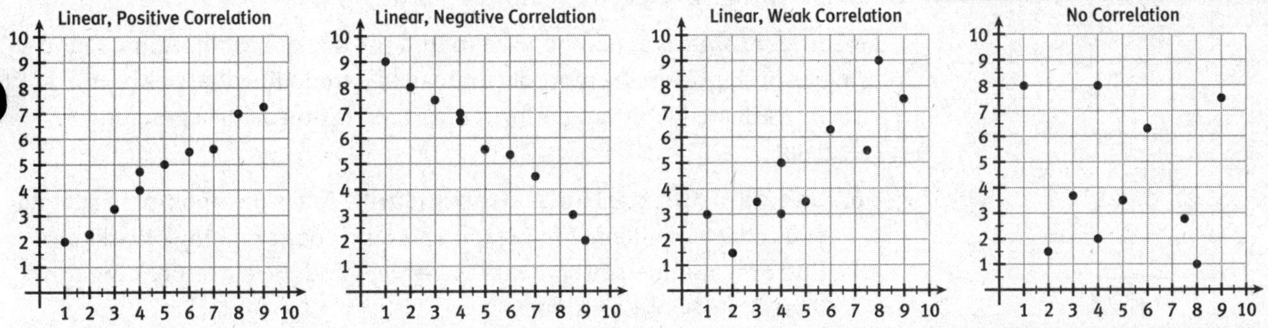

2. Look back at your linear model in Item 4 in Lesson 13-1. Does your linear model represent a positive correlation, a negative correlation, or no correlation? Explain.

There is ***causation*** between two variables if a change in one variable causes the other variable to change. For example, doing more exercise causes a greater number of calories to be burned.

3. Does there seem to be causation between the variables of your linear model in Item 4 in Lesson 13-1? Explain.

ACADEMIC VOCABULARY

The idea of ***causation*** is important in physics. For example, a cause can be represented by a force acting on an object.

My Notes

Correlation does not imply causation. Just because there is a correlation between two variables does not mean that there is causation between them; there may be other factors affecting the situation.

Check Your Understanding

4. Consider the following two variables: your shoe size each year since you were born and the average price of a movie ticket each year since you were born.
   a. Is there a correlation between the variables? Explain.
   b. Is there causation between the variables? Explain.

5. Give an example of two variables for which there is both correlation and causation.

A scatter plot and a ***line of best fit***, the most accurate trend line, can be created using a graphing calculator, a spreadsheet program, or other Computer Algebra Systems (CAS).

Linear regression is a method used to find the line of best fit. A line found using linear regression is more accurate than a trend line that has been visually estimated. You can perform linear regression using a graphing calculator.

6. Use appropriate tools strategically. Enter the book-passing data you collected in Item 2 in Lesson 13-1 into your graphing calculator. Enter the numbers of students as x-values and the corresponding times to pass the book as y-values.

   a. To find the equation of the line of best fit, use the linear regression feature of your calculator.

   The calculator should return values for a and b. Write these values below.

   $a =$

   $b =$

   b. The value of a is the slope of the line of best fit, and $(0, b)$ is the y-intercept. Round a and b to the nearest hundredth and write the equation of the line of best fit in the form $y = ax + b$. Describe how this equation is different from or similar to your equation in Item 8 in Lesson 13-1.

My Notes

Check Your Understanding

7. Enter the following data into your graphing calculator. Make sure that any previous data have been cleared.

   (6, 1), (9, 0), (12, −3), (3, 3), (0, 5), (−3, 7), (−5, 9), (−7, 13)

   a. Find the equation of the line of best fit. Round values to the nearest hundredth.

   b. Use the equation of the line of best fit to predict the value of y when $x = 20$.

8. Compare using a graphing calculator to using paper and pencil when plotting data and finding a trend line.

LESSON 13-2 PRACTICE

The owner of a café kept records on the daily high temperature and the number of hot apple ciders sold on that day. Some of the owner's data are shown below.

Daily High Temperature (°F)	32	75	80	48	15
Number of Hot Apple Ciders Sold	51	22	12	40	70

9. Create a scatter plot of the data.
10. Is there a correlation between the variables? If so, what type?
11. Determine the equation of the line of best fit. Round values to the nearest hundredth.
12. What is the slope? What does the slope represent?
13. Identify the y-intercept. What does the y-intercept represent?
14. **Model with mathematics.** Use your model to predict the number of hot apple ciders the café would sell on a day when the high temperature is 92°F. Explain.

My Notes

Learning Targets:

- Use technology to perform quadratic and exponential regressions, and then make predictions.
- Compare and contrast linear, quadratic, and exponential regressions.

SUGGESTED LEARNING STRATEGIES: Look for a Pattern, Create Representations, Quickwrite, Think-Pair-Share, Discussion Groups

Online shopping has experienced tremendous growth since the year 2000. One way to measure the growth is to track the average number of daily hits at the websites of online stores. The tables show the average number of daily hits for three different online stores in various years since the year 2000.

Nile River Retail					
Years Since 2000	0	2	4	6	8
Daily Hits (thousands)	52.1	56.2	60.0	64.1	68.0

eBuy					
Years Since 2000	0	2	4	6	8
Daily Hits (thousands)	1.0	4.9	17.2	37.2	64.9

Spendco					
Years Since 2000	0	2	4	6	8
Daily Hits (thousands)	2.0	6.6	20.9	68.1	220.4

1. **Make sense of problems.** Compare and contrast the growth of the three online stores based on the data in the tables.

My Notes

2. Plot the data for the three online stores on the graphs below.

Nile River Retail

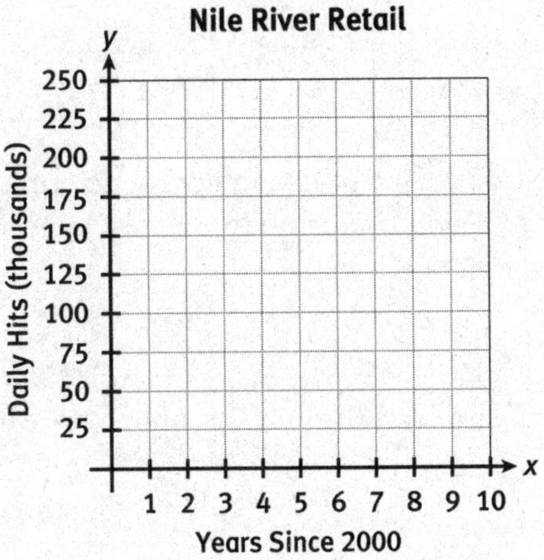

eBuy

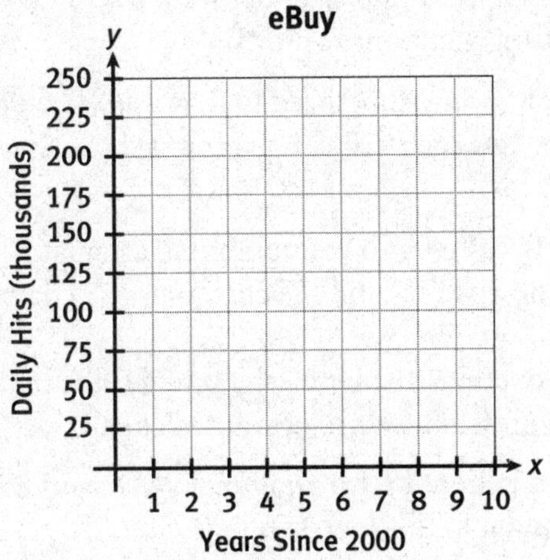

Spendco

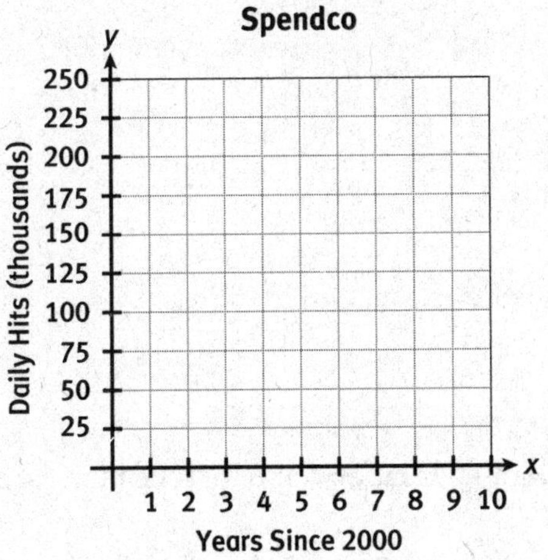

My Notes

3. Which online store's growth could best be modeled by a linear function? Explain.

4. For the online store you identified in Item 3, determine the equation of the line of best fit. Round values to the nearest hundredth.

5. Predict the number of daily hits for this online store in 2015.

When a line does not appear to be a good fit for a set of data, you may want to model the data using a nonlinear model.

Quadratic regression is a method used to find a ***quadratic function*** that models a set of data. You can perform quadratic regression using a graphing calculator.

MATH TERMS

A **quadratic function** is a nonlinear function that can be written in the form $y = ax^2 + bx + c$, where $a \neq 0$. You will study quadratic functions in more detail later in this book.

6. Enter the data for eBuy into your graphing calculator. Enter the years since 2000 as the x-values and the corresponding daily hits in thousands as the y-values.
   a. To find the quadratic equation that models the data, use the quadratic regression feature of your calculator.

   The calculator should return values for a, b, and c. Write these values below, rounding to the nearest hundredth.

   $a =$

   $b =$

   $c =$

   b. Write the quadratic equation in the form $y = ax^2 + bx + c$.

   c. Use the quadratic equation to predict the number of daily hits for eBuy in 2015.

My Notes

When a set of data shows very rapid growth or decay, an exponential model may be the best choice for modeling the data.

Exponential regression is a method used to find an ***exponential function*** that models a set of data. You can perform exponential regression using a graphing calculator.

MATH TERMS

An **exponential function** is a nonlinear function that can be written in the form $y = ab^x$. You will study exponential functions in more detail later in this book.

7. Enter the data for Spendco into your graphing calculator. Enter the years since 2000 as the x-values and the corresponding daily hits in thousands as the y-values.

a. To find the exponential equation that models the data, use the exponential regression feature of your calculator.

The calculator should return values for a and b. Write these values below, rounding to the nearest hundredth.

$a =$

$b =$

b. Write the exponential equation in the form $y = ab^x$.

c. Use the exponential equation to predict the number of daily hits for Spendco in 2015.

8. Construct viable arguments. Based on your predictions for the number of daily hits for each online store in 2015, which type of function has the fastest growth: linear, quadratic, or exponential? Explain.

My Notes

Check Your Understanding

9. How are quadratic regression and exponential regression similar to and different from linear regression?

10. Do you think the exponential model would be appropriate for predicting the number of daily hits for Spendco in any future year? Explain your reasoning.

LESSON 13-3 PRACTICE

The population of Williston, North Dakota, has grown rapidly over the past decade due to an oil boom. The table gives the population of the town in 2007, 2009, and 2011.

Years Since 2000	7	9	11
Population (thousands)	12.4	13.0	16.0

11. Use your calculator to find the equation of the line of best fit for the data.

12. **Reason quantitatively.** What is the slope of the line? What does it tell you about the population growth of the town?

13. Use your calculator to find a quadratic equation that models the growth of the town.

14. Use your quadratic equation to predict the population of Williston in 2020.

15. Use your calculator to find an exponential equation that models the growth of the town.

16. Use your exponential equation to predict the population of Williston in 2020.

17. According to the exponential model, in what year will the town have a population greater than 40,000 for the first time? (*Hint*: Use the table feature of your calculator.) What assumptions do you make when you use the exponential model to answer this question?

ACTIVITY 13 PRACTICE

Write your answers on notebook paper. Show your work.

Lesson 13-1

The scatter plot shows the relationship between the day of the month and a frozen yogurt stand's daily profit during the month of the July.

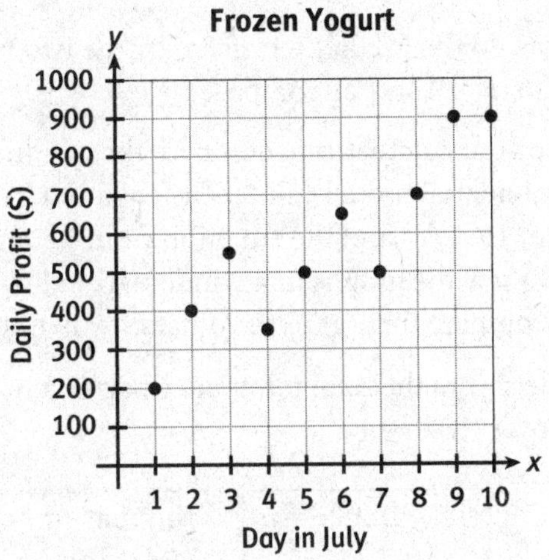

1. Are the data linear? Explain.
2. Draw a trend line on the scatter plot and name two points that your trend line passes through.
3. Write the equation of the trend line you drew in Item 2.
4. What do the variables in your equation represent?
5. What is the slope of the trend line? What does this tell you?
6. Use your trend line to predict the yogurt stand's daily profit on July 20.
7. The owner of a competing frozen yogurt stand finds that her daily profit each day in July is exactly $100 more than that of the stand in the scatter plot. Write the equation of a trend line for the competing stand.
8. The manager of a local history museum experiments with different prices for admission to the museum. For each price, the manager notes the number of visitors who enter the museum on that day. The table shows the data.

Price	$2.75	$3.50	$4.25	$5.75
Number of Daily Visitors	112	88	66	63

Which is a true statement about the data?

A. A trend line on the scatter plot has a positive slope.
B. The *y*-intercept of the trend line is above the *x*-axis.
C. The trend line predicts at least 70 visitors when the admission price is $6.25.
D. The trend line fits the data perfectly because the data is linear.

Lesson 13-2

Use your calculator to perform a linear regression for the following data. Use your linear regression for Items 9–12.

$(-6, -3), (-8, -4), (-2, 1), (1, 4), (3, 6), (5, 8), (7, 13)$

9. What is the equation of the line of best fit?
10. What is the value of the slope? What does this tell you about the relationship between x and y?
11. According to your model, what is the value of y when $x = -19$?
12. According to your model, for what value of x is $y = 100$?

13. Look at the scatter plot on the previous page showing the daily profits of a frozen yogurt stand. What type of correlation, if any, does the scatter plot show?

14. Which of the following pairs of variables are likely to show a negative correlation?
A. the length of a shoe; the size of the shoe
B. the number of miles on a car's odometer; the age of the car
C. the weight of a watermelon; the price of the watermelon
D. the number of minutes you have waited for a bus; the number of minutes remaining until the bus arrives

15. At several times during the school year, Emilio collected data on the height of a plant in the classroom and the total number of quizzes he had taken so far in his science class. The data are shown below.

Height of Plant (cm)	16	19	22	26
Total Number of Quizzes	4	6	7	9

a. Is there a correlation between the variables? Explain.
b. Is there causation between the variables? Explain.

Lesson 13-3

The table shows the number of employees at a software company in various years.

Years Since 2000	4	6	8	10
Number of Employees	32	40	75	124

16. Make a scatter plot of the data.

17. Do you think a linear equation would be a good model for the data? Justify your answer.

18. Use your calculator to find a quadratic equation that models the growth of the company.

19. Use the quadratic model to predict the number of employees in the year 2015.

20. Use your calculator to find an exponential equation that models the growth of the company.

21. Use the exponential model to predict the number of employees in the year 2015.

22. How do the predictions given by the two models in Items 19 and 21 compare?

23. Use your calculator to compare the quadratic and exponential models. Enter the equation from Item 19 as Y_1 and the equation from Item 21 as Y_2. View the graphs in a window that allows you to compare their growth. What do you notice?

The table shows the total number of bacteria in a sample over five hours.

Hour	Number of Bacteria
1	12
2	144
3	1728
4	20,736
5	248,832

24. Use your calculator to find an exponential equation that models the bacteria data.

25. If this trend continues, how many bacteria will be growing in the sample after 9 hours?

MATHEMATICAL PRACTICES
Look for and Make Use of Structure

26. Is it possible to tell from the equation of a line of best fit whether there is a positive or negative correlation between two variables? If so, explain how. If not, explain why not.

Linear Models and Slope as Rate of Change

A 10K RUN

CONNECT TO METRIC MEASUREMENT

A "10K Run" means that the length of the course for the foot race is 10 kilometers, or 10,000 meters.

$10\text{K} = 10\text{ km} \times \frac{1000\text{ m}}{1\text{ km}} = 10{,}000\text{ m}$

Jim was serving as a finish-line judge for the Striders 10K Run. He was interested in finding out how three of his friends were doing out on the course. He was able to get the following data from racing officials.

Runner: J. Matuba					
Time (min)	4	5	7	12	20
Distance (m)	1090	1380	2040	3640	6300

Runner: E. Rodriguez					
Time (min)	1	6	10	18	25
Distance (m)	500	2000	3280	5510	7700

Runner: T. Donovan					
Time (min)	2	4	9	15	20
Distance (m)	620	1250	2900	4690	6250

Answer Items 1–3 below, based on the information Jim received about his three running friends. Use x as the number of minutes elapsed since the race began and y as the number of meters completed.

1. Make a scatter plot showing the data for each runner.
2. Perform a linear regression to find the equation of the line of best fit for each runner. Round values in the equations to the nearest tenth.
3. Explain the order in which the runners will finish the race based on the models you formed using the data.

Answer the following questions for the linear models you formed. Explain your answers.

4. What is the standard form of the linear model for Matuba?
5. What is the domain of the linear model for Rodriguez?
6. What is the slope of the linear model for Donovan? What is its significance in the context of the problem situation?

Linear Models and Slope as Rate of Change

A 10K RUN

Scoring Guide	Exemplary	Proficient	Emerging	Incomplete
	The solution demonstrates the following characteristics:			
Mathematics Knowledge and Thinking (Items 1, 2, 4–6)	• Clear and accurate understanding of scatter plots, linear regression, standard form of a linear model, domain, and slope	• Largely correct understanding of scatter plots, linear regression, standard form of a linear model, domain, and slope	• Partial understanding of scatter plots, linear regression, standard form of a linear model, domain, and slope	• Inaccurate or incomplete understanding of scatter plots, linear regression, standard form of a linear model, domain, and slope
Problem Solving (Item 3)	• Appropriate and efficient strategy that results in a correct answer	• Strategy that may include unnecessary steps but results in a correct answer	• Strategy that results in a partially incorrect answer	• No clear strategy when solving problems
Mathematical Modeling / Representations (Items 1, 2, 4, 6)	• Clear and accurate scatter plot • Fluency in fitting a linear model to real-world data, including how to interpret and draw accurate conclusions from the model	• Largely correct scatter plot • Adequate understanding of how to fit a linear model to real-world data, including how to interpret and draw accurate conclusions from the model	• Partially correct scatter plot • Partial understanding of how to fit a linear model to real-world data, including how to interpret and draw accurate conclusions from the model	• Inaccurate or incomplete scatter plot • Little or no understanding of how to fit a linear model to real-world data, including how to interpret and draw accurate conclusions from the model
Reasoning and Communication (Items 3–6)	• Precise use of appropriate math terms and language to explain the order in which the runners will finish, including justification based on the model • Clear and accurate descriptions of how to find the standard form, identify a reasonable domain, and identify and interpret the slope of a linear model	• Adequate explanation and justification of the order in which the runners will finish • Largely correct description of how to find the standard form, identify a reasonable domain, and identify and interpret the slope of a linear model	• Misleading or confusing explanation and justification of the order in which the runners will finish • Partially correct description of how to find the standard form, identify a reasonable domain, and identify and interpret the slope of a linear model	• Incomplete or inaccurate explanation and justification of the order in which the runners will finish • Incorrect or incomplete description of how to find the standard form, identify a reasonable domain, and identify and interpret the slope of a linear model

Extensions of Linear Concepts

3

Unit Overview

In this unit, you will extend your study of linear concepts to the study of piecewise-defined functions and systems of linear equations and inequalities. You will learn to solve systems of equations and inequalities in a variety of ways.

Key Terms

As you study this unit, add these and other terms to your math notebook. Include in your notes your prior knowledge of each word, as well as your experiences in using the word in different mathematical examples. If needed, ask for help in pronouncing new words and add information on pronunciation to your math notebook. It is important that you learn new terms and use them correctly in your class discussions and in your problem solutions.

Math Terms

- piecewise-defined function
- linear inequality
- solutions of a linear inequality
- boundary line
- half-plane
- closed half-plane
- open half-plane
- system of linear equations
- substitution method
- elimination method
- parallel
- coincident
- slope-intercept form
- independent
- dependent
- inconsistent
- consistent
- system of linear inequalities
- solution region

ESSENTIAL QUESTIONS

? Why would you use multiple representations of linear equations and inequalities?

? How are systems of linear equations and systems of linear inequalities useful in analyzing real-world situations?

EMBEDDED ASSESSMENTS

These assessments, following Activities 16 and 18, will give you an opportunity to demonstrate what you have learned about piecewise-defined functions, inequalities, and systems of equations and inequalities.

Getting Ready

Write your answers on notebook paper.
Show your work.

1. Which of the following tables of values represents linear data?

A.

x	y
2	3
4	6
6	6
8	3

B.

x	y
1	1
2	4
3	7
4	10

C.

x	y
2	3
3	4
4	3
6	4

D.

x	y
1	3
2	6
3	12
4	24

2. Give an algebraic representation of these data.

x	1	4	7	10
y	1	7	13	19

3. You open a savings account with a deposit of \$100. In each month after your initial deposit, you add \$25 to your account. Provide a table to display the total amount of money in the account after 1, 2, 3, and 4 months. Give a graphical and an algebraic representation that would allow you to determine the amount of money A in dollars you have deposited after m months.

4. Graph $2x + 3y = 4$.

5. Describe the graph of $y = 3$.

6. Which ordered pair is a solution of $y > x + 5$?

A. (2, 8) **B.** (−5, 0)
C. (1, 6) **D.** (0, −5)

7. Compare and contrast the graphs of these two compound statements:
$-1 < x \leq 3$
$x < -1$ or $x \geq 3$

8. Which of the following represents a function with a constant rate of change?

A.

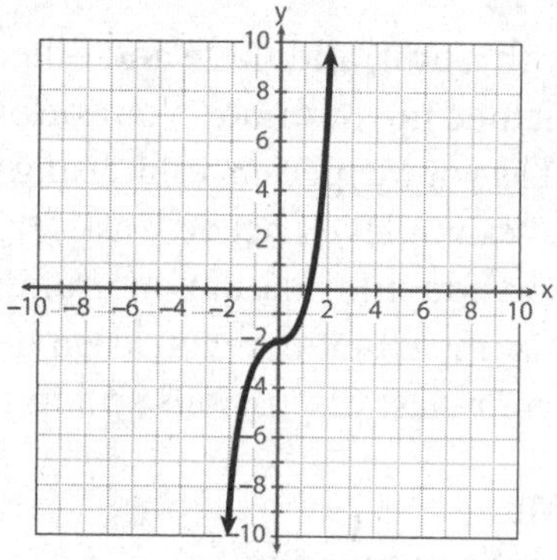

B.

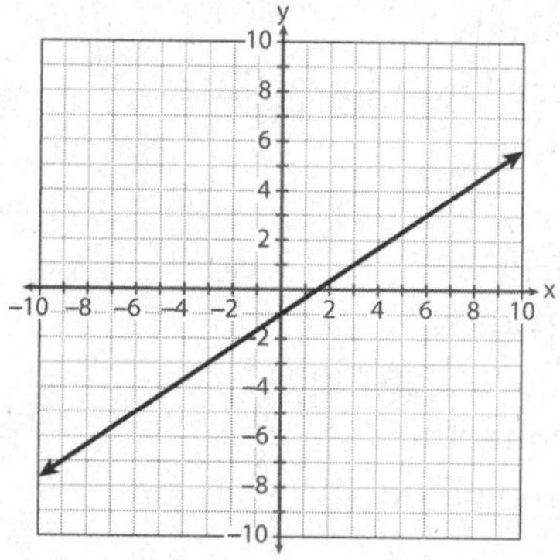

C. $y = \frac{4}{x}$

D.

x	1	4	16	64
y	5	10	15	20

Piecewise-Defined Linear Functions

Breakfast for Bowser

Lesson 14-1 Function Notation and Rate of Change

My Notes

Learning Targets:

- Use function notation and interpret statements that use function notation in terms of a context.
- Calculate the rate of change of a linear function presented in multiple representations.

SUGGESTED LEARNING STRATEGIES: Summarizing, Look for a Pattern, Think-Pair-Share, Create Representations, Construct an Argument

Miriam has accepted a job at a veterinarian's office. Her first assignment is to feed the dogs that are housed there. The bag of dog food was already torn open, and she found only part of the label that described the amount of dog food to feed each dog.

Barko Dog Food Feeding Chart

Weight of Dog (pounds)	Daily Amount of Barko (ounces)
3	3
12	12
20	20

Each dog has an information card that gives the dog's name and weight. Using this information, Miriam fed each dog the amount of food she thought was appropriate.

1. How many ounces of dog food per day do you think Miriam gave each dog? Complete the table below.

Dog	Dog's Weight (pounds)	Amount of Dog Food (ounces)
Muffy	4	
Trixie	14	
Rags	6	
Hercules	68	

2. a. Based on the dog's weight and the partial label at the top of the page, write a verbal rule that Miriam could use to determine the ounces of dog food per day to feed each dog.

b. Reason abstractly: Could this verbal rule be represented with function notation? Why or why not?

My Notes

DISCUSSION GROUP TIPS

As you discuss ideas for the function of the scenario, make notes and listen to what your group members have to contribute. Ask and answer questions to clearly aid comprehension and to ensure understanding of all group members' ideas.

3. **Model with mathematics.** Let w be a dog's weight in pounds and let $A(w)$ be the amount of food in ounces that the dog should be fed each day. Use the table in Item 1 and function notation to write a function that expresses $A(w)$ in terms of w.

After several hours, all of the dogs had finished their food except for Hercules. The vet asked, "How much food did you give Hercules?"

After hearing Miriam's answer, the vet told Miriam that she had overfed Hercules and suggested that Miriam check the complete feeding chart posted on the office wall.

Barko Dog Food Feeding Chart

Weight of Dog (pounds)	Daily Amount of Barko (ounces)
3	3
12	12
20	20
50	35
100	60
over 100 pounds	60 ounces plus 1 ounce for each additional 10 pounds of weight

4. Miriam had assumed that each dog should be fed as many ounces of dog food as the dog weighed in pounds. Explain why the data in the new chart indicate that her original assumption was not correct.

Adult dogs that weigh less than 20 pounds are classified as *small* dogs. Dogs that weigh 20 to 100 pounds are classified as *mid-size* dogs. Finally, dogs that weigh more than 100 pounds are classified as *large* dogs. Miriam had not known that the formulas for feeding small, mid-size, and large dogs are all different.

My Notes

5. The vet tells Miriam that the data for feeding mid-size dogs are linear. Verify the vet's statement using the feeding chart.

6. Determine a rate of change that describes the number of additional ounces of food a mid-size dog should be fed for each pound of weight greater than 20. Explain how you found your answer.

7. Using the rate of change you found in Item 6 and the feeding chart, use function notation to write a function that expresses $A(w)$, the ounces of dog food, in terms of w, the dog's weight in pounds, for mid-size dogs.

8. How many ounces was Hercules overfed? Justify your response.

Check Your Understanding

Use the table for Items 9 and 10.

Grapes (pounds)	Cost (dollars)
0.5	\$1.20
1.5	\$3.60
2.0	\$4.80

9. Determine the rate of change of the data. Express the rate as cost per pound.

10. Do the data in the table represent a linear relationship? Explain.

11. Tom has read the first 40 pages of a book. He plans to read another 12 pages each day. Use function notation to write a function that expresses the number of pages read p after d days.

My Notes

LESSON 14-1 PRACTICE

12. Todd had five gallons of gasoline in his motorbike. After driving 100 miles, he has three gallons left. What is Todd's rate of change in miles per gallon?

13. This table shows how the balance in John's savings account has changed over the course of a year.

Month	Balance (dollars)
1	$ 400
3	$ 600
7	$1000
10	$1300
12	$1500

How much did John save per month during the year?

14. Linda purchased a house for $144,000. Thinking of possibly refinancing after 11 years, she had her house appraised and found that it is now worth $245,000. Find the rate of change of the value of the house in dollars per year.

15. Make sense of problems. The cost in dollars of producing x vehicles for a company is given by $C(x) = 1200x + 5500$. Interpret the rate of change of this linear function.

16. A 500-liter tank full of oil is being drained at the constant rate of 20 liters per minute. Use function notation to write a linear function expressing the number of liters in the tank V after t minutes.

My Notes

Learning Targets:

- Write linear equations in two variables given a table of values, a graph, or a verbal description.
- Determine the domain and range of a linear function, determine their reasonableness, and represent them using inequalities.

SUGGESTED LEARNING STRATEGIES: Identify a Subtask, Think-Pair-Share, Create Representations, Construct an Argument, Interactive Word Wall

1. Rewrite the function that you wrote in Item 7 in Lesson 14-1.

2. a. Remember that this function is true only for mid-size dogs. Describe the appropriate input values for the function.

b. Use your description of the appropriate input values to express the domain of the function using set notation.

domain = { *w*: ____________________}

Miriam knows that she also has several large dogs to feed. When she looks at the chart, she reads the instruction "60 ounces plus 1 ounce for each additional 10 pounds of weight."

3. How much additional dog food should large dogs be fed for each pound of weight greater than 100 pounds?

4. Critique the reasoning of others. A 140-pound Great Dane has arrived for a short stay at the kennel. Miriam says that the dog should be fed 64 ounces of food per day. Her friend Chase says that the dog requires 100 ounces of food per day. Who is correct? Justify your response.

My Notes

5. Write a function that expresses the amount of food, $A(w)$, in ounces that a large dog should be fed, as a function of w, the weight of the dog in pounds. Determine the domain and range of the function.

Miriam realizes that there are three different algebraic feeding rules to follow because dogs are different sizes. She organizes her feeding rules in a list so that she can quickly refer to them whenever she has to decide how much food to feed a dog.

6. Complete Miriam's list by writing the appropriate function. Indicate the domain by writing the appropriate inequality symbols.

$A(w) =$ ____________, when 0 ____________ w ____________ 20

$A(w) =$ ____________, when 20 ____________ w ____________ 100

$A(w) =$ ____________, when w ____________ 100

7. The vet has a German Shepard named Max, and Miriam knows that the vet feeds Max 63 ounces of food each day. Miriam also knows that the vet feeds her cocker spaniel Min 32 ounces of food each day. If the two dogs are being correctly fed, what is each dog's weight? Explain your reasoning.

DISCUSSION GROUP TIPS

As you listen to the group discussion, take notes to aid comprehension and to help you describe your own ideas to others in your group. Ask questions to clarify ideas and to gain further understanding of key concepts.

My Notes

Miriam decides to make a table that lists the weight of the dogs she will be feeding, in the order that she will feed them. A portion of Miriam's table is shown below.

8. Complete the table using the rules you wrote in Item 6.

Weight of Dog (pounds)	Amount of Dog Food (ounces)
21	
70	
16	
120	
15	
56	
27	
107	
7	
25	

The functions in Item 6 can be written as ***piecewise-defined functions.*** Every possible weight w has exactly one feeding amount A assigned to it, but the rule for determining that feeding amount changes for dogs of different sizes. The different feeding rules, along with their domains, are considered to be the pieces of a single piecewise-defined function.

MATH TERMS

A **piecewise-defined function** is a function that is defined differently for different disjoint intervals in its domain.

9. Miriam decides to write a piecewise-defined function to represent the functions she wrote in Item 6.

a. Explain how you know that each equation in Item 6 represents a function.

b. Complete the piecewise-defined function for the feeding rules.

$$A(w) = \begin{cases} w, \text{\_\_\_\_\_\_\_\_\_\_\_\_\_\_\_\_} \\ \text{\_\_\_\_\_\_\_\_\_\_\_\_}, \text{ when } 20 \leq w \leq 100 \\ 0.1w + 50, \text{ when } \text{\_\_\_\_\_\_\_\_\_\_\_\_} \end{cases}$$

c. Explain why the equation from part b is also a function. Use the domain of each rule to help justify your answer.

My Notes

Check Your Understanding

A wholesale grocery store has the following sale on mixed nuts:

SALE!
Mixed nuts
Less than 4 lbs: \$2/lb
4–10 lbs: \$6 + \$0.50/lb

10. Write a function to represent the cost $C(x)$ of buying less than four pounds of mixed nuts. Identify the domain and range.

11. Write a function to represent the cost $C(x)$ of buying 4 to 10 pounds of mixed nuts. Identify the domain and range.

12. Make a table to show the cost of 4, 5, and 8 pounds of mixed nuts.

13. Use your answers to Items 10 and 11 to write a piecewise-defined function to represent the cost $C(x)$ of x pounds of mixed nuts.

LESSON 14-2 PRACTICE

14. Speed Cell Wireless offers a plan of \$40 for the first 400 minutes, and an additional \$0.50 for every minute over 400. Let t represent the total talk time in minutes. Write a piecewise-defined function to represent the cost $C(t)$.

15. If Pam works more than 40 hours per week, her hourly wage for every hour over 40 is 1.5 times her normal hourly wage of \$7. Write a piecewise-defined function that gives Pam's weekly pay $P(h)$ in terms of the number of hours h that she works.

16. A man walks for 45 minutes at a rate of 3 mi/h, then jogs for 75 minutes at a rate of 5 mi/h, then rests for 30 minutes, and finally walks for 90 minutes at a rate of 3 mi/h. Write a piecewise-defined function expressing the distance $D(t)$ he traveled as a function of time t.

17. **Model with mathematics.** A state income tax law reads as follows:

Annual Income (dollars)	Tax
$<$ \$2000	\$0 tax
\$2000 – \$6000	2% of income over \$2000
$>$ \$6000	\$80 plus 5% of income over \$6000

Write a piecewise-defined function to represent the income tax law.

My Notes

Learning Targets:

- Evaluate a function at specific inputs within the function's domain.
- Graph piecewise-defined functions.

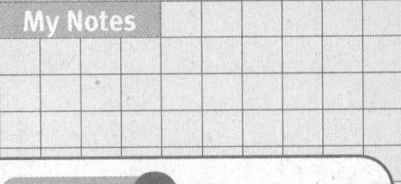

SUGGESTED LEARNING STRATEGIES: Identify a Subtask, Think-Pair-Share, Construct an Argument, Create Representations, Discussion Groups

CONNECT TO AP

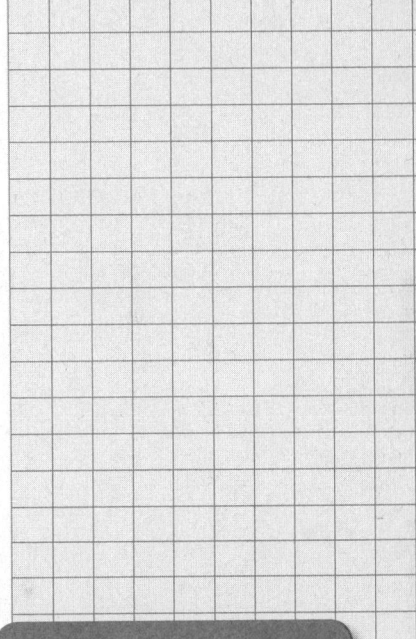

This is a connection to the concept of continuity in AP Calculus.

1. Rewrite the piecewise-defined feeding function that you wrote in Item 9 in Lesson 14-2.

2. Miriam must feed a dog that weighs 57 pounds.

a. Which piece of the feeding function should Miriam use? Explain your answer.

b. Use your answer to part a to evaluate $A(57)$.

3. Make use of structure. When graphing a piecewise-defined function, it is necessary to graph each piece of the function only for its appropriate interval of the domain. Graph the feeding function on the axes below. When finished, your graph should consist of three line segments.

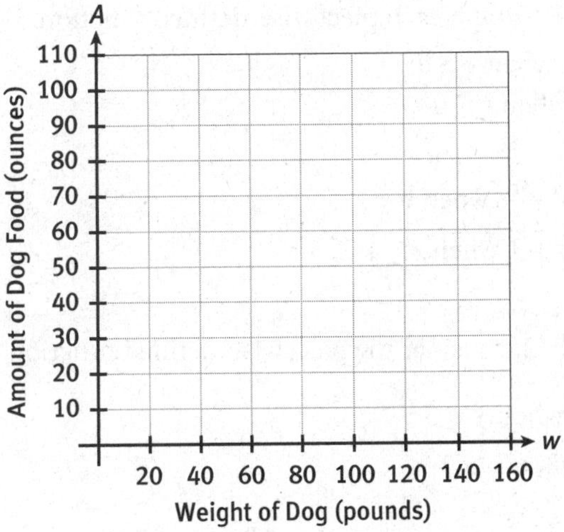

TECHNOLOGY TIP

To graph a piecewise-defined function in your calculator, enter the function into $Y =$, in dot mode, using parentheses to indicate the domain intervals. For example

$$f(x)=\begin{cases}-x^2-1, & \text{when } x<1\\ x+2, & \text{when } x\geq 1\end{cases}$$

would be entered as $Y_1=(-x^2-1)(x<1)+(x+2)(x\geq 1)$.

My Notes

4. Why is the graph of a piecewise-defined function more useful than three separate graphs?

5. How can you conclude that the graph represents a function?

Check Your Understanding

Use the piecewise-defined function for Items 6 and 7.

$$f(x) = \begin{cases} x+1, \text{ when } -3 < x \le 1 \\ 2x, \text{ when } 1 < x \le 4 \\ x-1, \text{ when } 4 < x < 10 \end{cases}$$

6. Find $f(1)$ and $f(4)$. Explain which piece of the function you used to find each value.

7. Sketch a graph of the function.

LESSON 14-3 PRACTICE

8. **Model with mathematics.** Graph the piecewise-defined function that you wrote in Item 15 in Lesson 14-2 to describe Pam's pay.

For Items 9 and 10, graph each piecewise-defined function.

9. $f(x) = \begin{cases} 2x, \text{ when } x > 0 \\ x, \text{ when } x \le 0 \end{cases}$

10. $f(x) = \begin{cases} \frac{1}{2}x + \frac{3}{2}, \text{ when } x < 1 \\ -x + 3, \text{ when } x \ge 1 \end{cases}$

For Items 11 and 12, consider the piecewise-defined function

$$f(x) = \begin{cases} \frac{1}{2}x - 10, \text{ when } x \le 3 \\ -x - 1, \text{ when } x > 3 \end{cases}.$$

11. Find $f(-4)$.

12. Find $f(6)$.

My Notes

Learning Target:

- Compare the properties of two functions each represented in a different way.

SUGGESTED LEARNING STRATEGIES: Think-Pair-Share, Discussion Groups, Construct an Argument, Create Representations

When Miriam looks at the graph of the feeding function, she notices that for small dogs, each increase of 1 pound in weight causes an increase of 1 ounce of food. She concludes that the rate of change of the feeding rule for small dogs is 1 ounce per pound.

1. **Express regularity in repeated reasoning.** What is the rate of change of the feeding rule for large dogs?

2. Miriam compares the rates of change of the feeding rules for each type of dog. She concludes that the feeding rule for small dogs has the greatest rate of change.
   a. How does the graph in Item 3 in Lesson 14-3 support Miriam's conclusion?

   b. What parts of the algebraic feeding rules in Item 6 in Lesson 14-2 support Miriam's conclusion?

3. Dogs in one size category have a greater rate of change in feeding than other dogs. Explain why you think this might be true.

My Notes

The graph below shows how many ounces of water a dog should drink daily, based on the dog's weight.

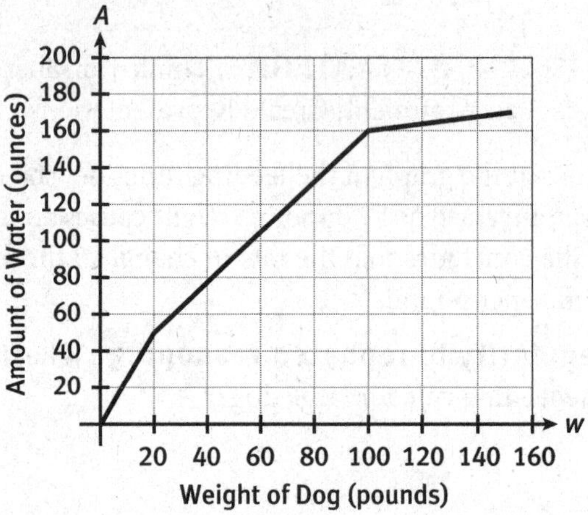

4. How much water should a 20-pound dog consume each day?

5. How much water should a 100-pound dog consume each day?

6. Compare the water function with the feeding function. Describe the similarities and differences.

7. Write a piecewise-defined function for the graph.

My Notes

Check Your Understanding

8. Write a piecewise-defined function for the graph, including the domain for each part.

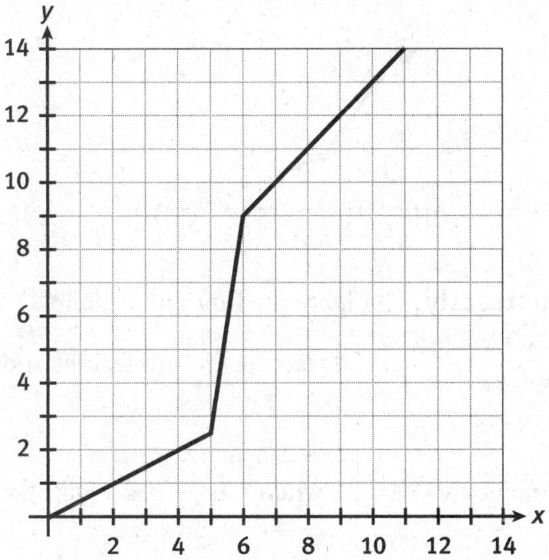

Use the piecewise-defined function below for Items 9 and 10.

$$f(x)=\begin{cases} x-1, & \text{when } 0<x\leq 5 \\ 1.5x, & \text{when } 5<x\leq 11 \end{cases}$$

9. Compare the function with the function from Item 8. Describe the similarities and differences.

10. Is the value $x = 5.5$ in the domain of the function you wrote in Item 8? If so, what is the value of the function when $x = 5.5$? Justify your response.

My Notes

LESSON 14-4 PRACTICE

11. Write a piecewise-defined function for the graph shown.

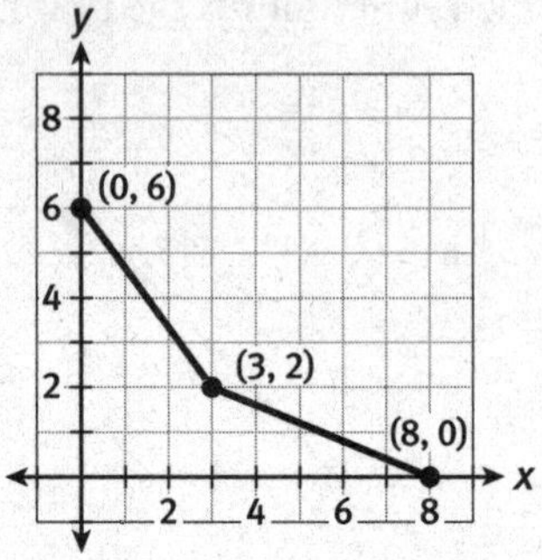

12. Reason abstractly. Compare the function in Item 11 to the function $f(x)=\begin{cases}2x+1, \text{when } x<1\\ -x+4, \text{when } x\geq 1\end{cases}$. Describe the similarities and differences.

13. Graph the function $f(x)=\begin{cases}2x+8, \text{when } x<-2\\ 4, \text{when } -2\leq x<2\\ -2x+8, \text{when } x\geq 2\end{cases}$. State the domain and range.

14. How does the function in Item 13 compare to the function shown in the graph below?

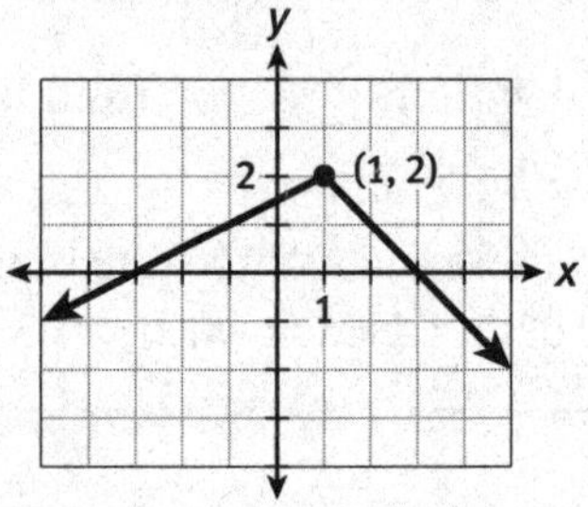

ACTIVITY 14 PRACTICE

Write your answers on notebook paper.
Show your work.

Lesson 14-1

Use the tables for Items 1–5.

x	*y*
0	2
1	3
2	4
3	5
4	6
5	7

x	*y*
6	−5
7	−6
8	−7
9	−8
10	−9
11	−10

1. Find the rate of change for each table.

2. At what *x*-value does the rate of change switch patterns?

3. Find a function that represents the first table of data, and write the domain for the function.

4. Find a function that represents the second table of data, and write the domain for the function.

Lesson 14-2

5. Write the piecewise-defined function that represents the data in the tables.

6. A store is having a special sale on designer soaps. For every three bars of soap purchased, one is given for free. The bars of soap cost \$2 each, and there is a limit of eight bars of soap per customer (including free ones).

a. Write a piecewise-defined function $C(b)$ that gives the total cost C for b bars of soap.

b. Write the domain for this function in terms of the context.

7. A museum has the following prices for admission:

Children under 12: free
Kids age 12–17: \$3
Adults age 18–64: \$8
Senior citizens 65 and over: \$4

Write a piecewise-defined function that gives the cost $C(n)$ for a museum visitor who is n years old. What is the range of the function?

Lesson 14-3

8. A piecewise-defined function is shown.

$$h(x)=\begin{cases}x+3, \text{ when } x<3\\ bx, \text{ when } x\geq 3\end{cases}$$

What value of b is needed so that $h(3)=6$?

A. 1 **B.** 2
C. 3 **D.** 6

9. Sketch a graph of the function.

$$f(x)=\begin{cases}x, & \text{when } x<2\\ -x+4, & \text{when } x\geq 2\end{cases}$$

10. Ashley participated in a triathlon.

- She swam for 10 minutes at a rate of 40 meters/min.
- Then she biked for 40 minutes at a rate of 400 meters/min.
- Finally, she ran for 25 minutes at a rate of 200 meters/min.

a. Write a piecewise-defined function expressing the distance $d(t)$ in meters that Ashley traveled as a function of time t in minutes.

b. Graph your function from part a.

Lesson 14-4

11. Refer back to Item 10. Wanda competed in the same triathlon as Ashley. A graph of the distance that Wanda covered as a function of time is shown below.

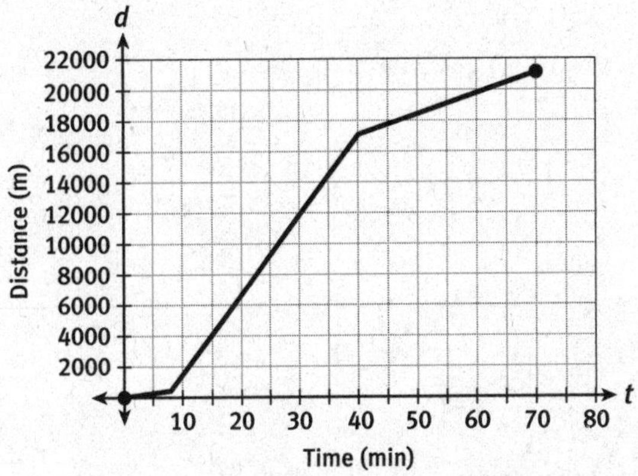

a. Tell who completed each leg of the triathlon (swimming, biking, and running) faster, Ashley or Wanda. Justify your answers.

b. Who completed the triathlon first, Ashley or Wanda? Justify your answer.

Use the graph for Items 12–15.

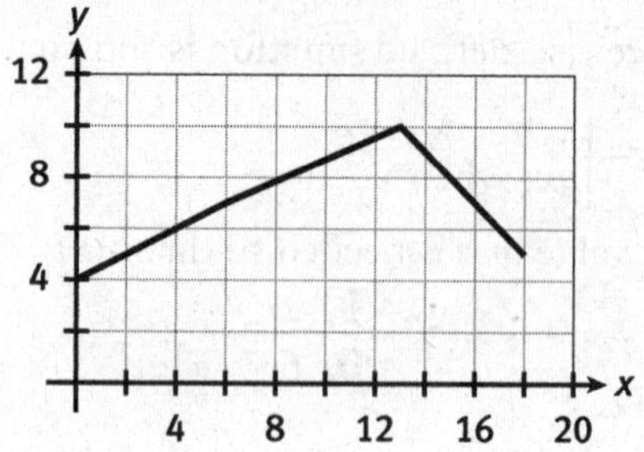

12. What is the domain of the piecewise-defined function?

A. $0 \leq x \leq 18$ **B.** $0 \leq x < 10$
C. $4 \leq x \leq 10$ **D.** $x \neq 6$

13. What is the value of the function for $x = 6$?

A. 5 **B.** 6
C. 7 **D.** 10

14. Which of the following defines the function for the domain $0 \leq x < 6$?

A. $y = 0.5x$ **B.** $y = 0.5x + 4$
C. $y = 2$ **D.** $y = -x + 2$

15. What is the range of the piecewise-defined function?

A. $0 \leq y \leq 18$ **B.** $0 \leq y \leq 10$
C. $4 \leq y \leq 10$ **D.** $y \neq 7$

A function gives Tanya's distance y (in miles) from home x minutes after she leaves her friend's house.

Use the graph of the function for Items 16–18.

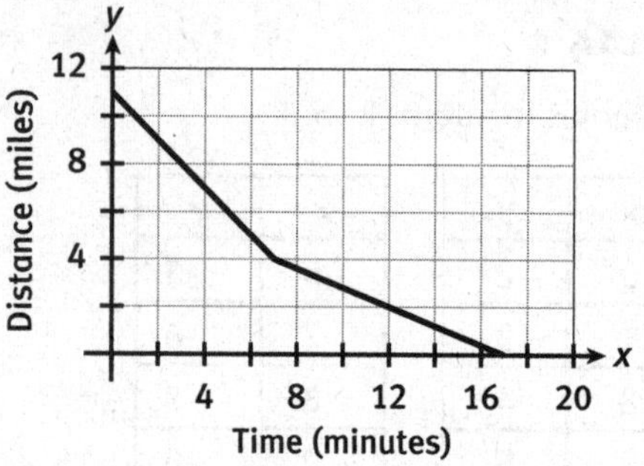

16. What are the domain and range of the function?

A. $0 \leq x \leq 11, 0 \leq y \leq 17$
B. $0 \leq x \leq 17, 0 \leq y \leq 11$
C. $0 \leq x \leq 7, 0 \leq y \leq 4$
D. $0 \leq x \leq 4, 0 \leq y \leq 7$

17. Which statement is true?

A. After 7 minutes, Tanya's average speed increased.
B. After 7 minutes, Tanya's average speed decreased.
C. After 4 minutes, Tanya's average speed increased.
D. After 4 minutes, Tanya's average speed decreased.

18. How far does Tanya live from her friend's house?

A. 4 miles **B.** 7 miles
C. 11 miles **D.** 17 miles

MATHEMATICAL PRACTICES
Attend to Precision

19. Explain why a restricted domain and a restricted range are sometimes appropriate for a piecewise-defined function. Describe instances when the domain and range need not be restricted.

Comparing Equations

A Tale of a Trucker

Lesson 15-1 Writing Equations from Graphs and Tables

Learning Targets:

- Write a linear equation given a graph or a table.
- Analyze key features of a function given its graph.

SUGGESTED LEARNING STRATEGIES: Summarizing, Visualization, Look for a Pattern, Create Representations, Discussion Groups, RAFT

Travis Smith and his brother, Roy, are co-owners of a trucking company. One of their regular weekly jobs is to transport fruit grown in Pecos, Texas, to a Dallas, Texas, distributing plant. Travis knows that his customers are concerned about the speed with which the brothers can deliver the produce, because fruit will spoil after a certain length of time.

1. Travis wants to address his customers' concerns with facts and figures. He knows that a typical trip between Pecos and Dallas takes 7.5 hours. He makes the following graph.

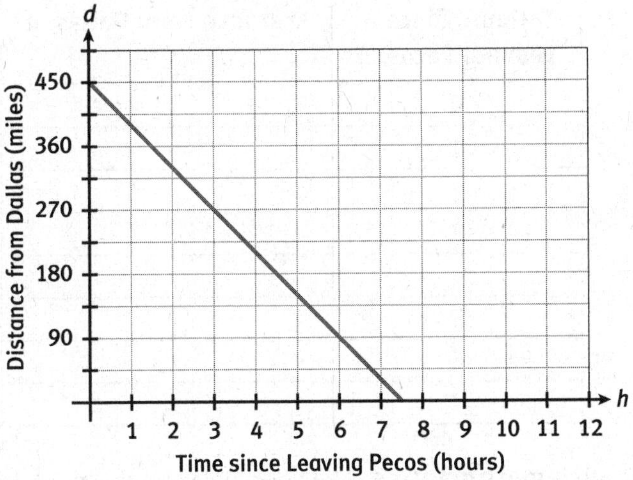

a. What information does the graph provide?

b. Locate the point with coordinates of (3, 270) on the graph. Label it point A. Describe the information these coordinates provide.

My Notes

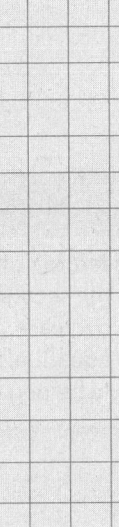

CONNECT TO BUSINESS

Companies that ship fruit from distribution plants to stores around the country use refrigerated trucks to keep the fruit fresher longer. For example, the optimal temperature range for shipping cantaloupes is 36–41°F. What would the temperature range look like on a number line graph?

My Notes

c. According to the graph, how many hours will it take Travis to reach Dallas from Pecos? Explain how you determined your answer.

d. Interstate 20 is the direct route between Pecos and Dallas. Based upon his graph, at what average speed does Travis expect to travel? Explain how you determined your answer.

2. Complete the table to show Travis's distance from Dallas at each hour.

Hours Since Leaving Pecos, *h*	Distance from Dallas, *d*
0	
1	
2	
3	
4	
5	
6	
7	

3. Model with mathematics. Use your table and the graph to write an equation that expresses Travis's distance d from Dallas as a function of the number of hours h since he left Pecos.

WRITING MATH

The graph falls from left to right, and the numbers in the second column of the table decrease. When you write the equation, the coefficient of h should be negative.

4. The graph of an equation in two variables is the set of all solutions of the equation plotted in the coordinate plane.

a. The graph appears to pass through the point (6, 90). Use your equation to verify this.

My Notes

b. Generate four more ordered-pair solutions to your equation by substituting 1, 5, 8, and 10 for h. Does each point appear on the graph showing Travis's distance from Dallas? Explain.

c. Which of the solution points you generated in part b make sense in this context?

d. Explain why only some solution points make sense in this context.

e. Use what you know about the solution points and the graph to state a reasonable domain and range for the function in the given context.

f. Use the graph to identify the zero of the function.

Check Your Understanding

The graph shows the amount of water remaining in a tank after a leak occurs. Use the graph for Items 5–7.

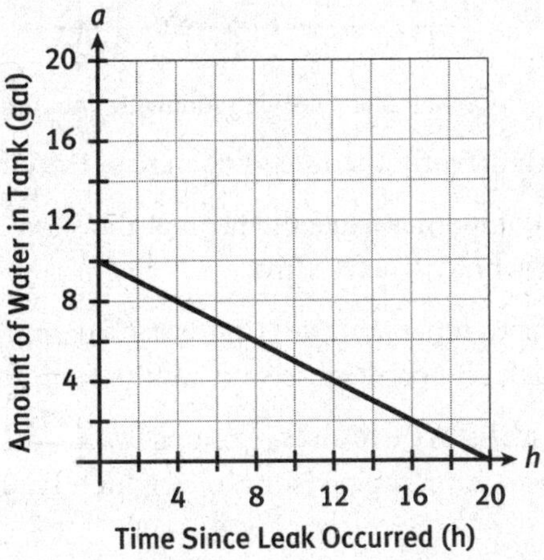

5. Make a table to show the amount of water in the tank at five different times. Use the graph to identify the zero of the function.

6. **Make use of structure.** Use your table and the graph to write an equation that expresses the amount of water in the tank a as a function of the number of hours h since the leak occurred.

7. State a reasonable domain and range for the function in the given context.

My Notes

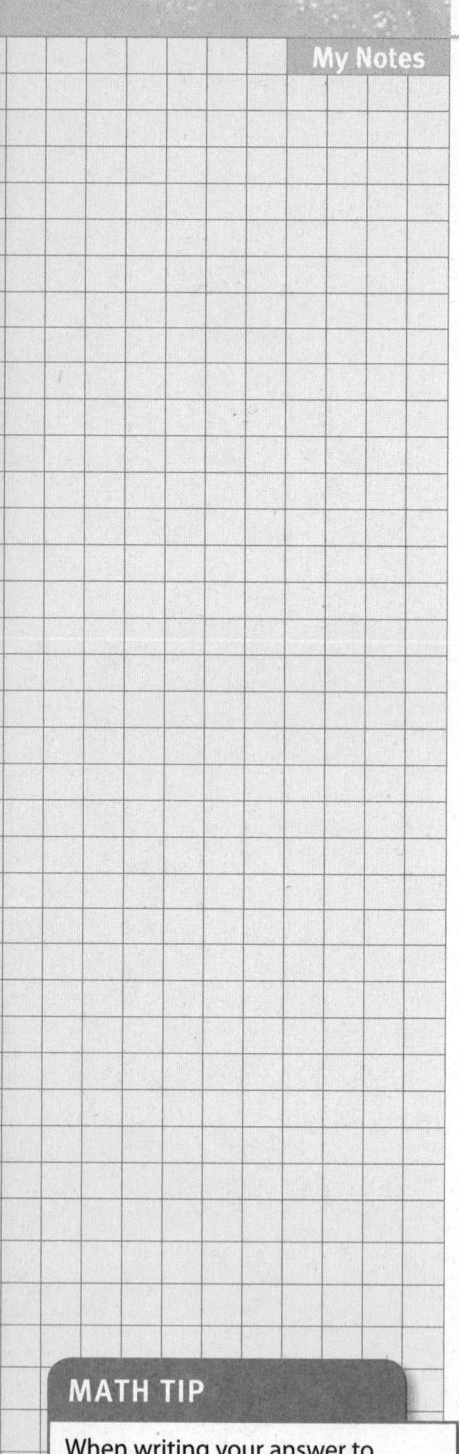

LESSON 15-1 PRACTICE

A farm in Plainville, Texas, will ship fruit to the distribution plant in Dallas. The farm is 350 miles from the plant.

8. The table shows part of a trip that Travis and Roy made from the farm to the distribution plant.

Distance from Dallas After Leaving Plainville

Time h (hours)	Distance d (miles)
2	230
2.5	200
3	170
3.5	140

Determine and interpret the rate of change of the data in the table.

9. A competing shipping company guarantees the fastest delivery times. The graph shows a recent trip from the farm to the distribution plant.

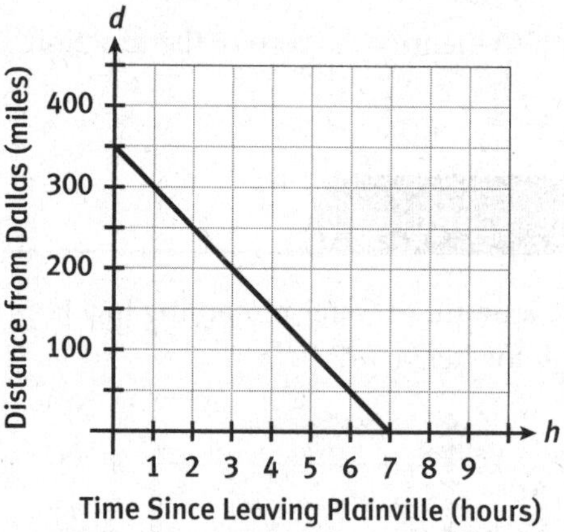

Determine and interpret the rate of change of the data in the graph.

10. Suppose both companies leave the farm at the same time and head to Dallas. Predict who will arrive first.

11. Write equations to represent the trip for each company. Determine how long it takes each company to drive from the farm to the plant in Dallas.

12. **Construct viable arguments.** Suppose you work at the farm. Write a memo to your supervisor about which trucking company to use and why.

MATH TIP

When writing your answer to Item 12, you can use a RAFT.

- Role—farm worker
- Audience—your supervisor
- Format—a memo
- Topic—which trucking company to use and why

My Notes

Learning Targets:

- Graph and analyze functions on the same coordinate plane.
- Write inequalities to represent real-world situations.

SUGGESTED LEARNING STRATEGIES: Think-Pair-Share, Create Representations, Sharing and Responding, Discussion Groups

Travis knows that his first graph represents a model for an ideal travel scenario. In reality, he assumes that his average speed will be lower because he will need to stop to refuel. He also decides that he must account for concerns about road construction and traffic.

1. Experience tells Travis that his average speed will decrease to 45 mi/h. Travis's ideal travel scenario is shown on the grid below. On these same axes, draw the graph of his distance from Dallas, based upon his assumption that he will maintain a 45 mi/h average speed throughout his trip.

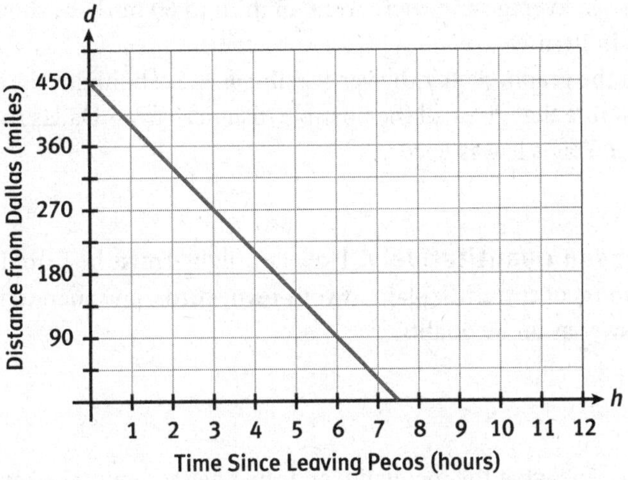

2. Write an equation for the line you drew in Item 1. Interpret the meaning of the constant and the coefficient of h in terms of the context.

My Notes

3. Travis can average anywhere from 45 mi/h to 60 mi/h, as shown on the two previous graphs.
   a. On the graph in Item 1, sketch the vertical line $h = 3$. Highlight the segment of that line that gives all the possible distances from Dallas three hours after Travis leaves Pecos.

   b. How far might Travis be from Dallas after 3 hours of travel? Explain in your own words and then write your answer as an inequality.

The distances you found in Item 3b are the possible distances after three hours of travel. The line segment can be described as the set of ordered pairs (h, d) where $h = 3$ and the value of d is between the values indicated in the inequality you wrote in Item 3b.

4. Travis can average anywhere from 45 mi/h to 60 mi/h, as shown on the graph in Item 1.
   a. On the graph, sketch the vertical line $h = 5$. Highlight the segment of that line that gives all the possible distances from Dallas five hours after Travis leaves Pecos.

   b. **Reason quantitatively.** How far might Travis be from Dallas after 5 hours of travel? Explain in your own words, and then write your answer as an inequality.

   c. Describe what the inequality and the line segment tell Travis about his trip.

WRITING MATH

If an inequality includes the greatest and least numbers, write the inequality using $\leq$ or $\geq$. If the greatest and least numbers are not included, write the inequality using $<$ or $>$.

5. Use the graph in Item 1.
   a. Draw the line segment for $h = 7.5$. How far might Travis be from Dallas after 7.5 hours of travel? Write your answer as an inequality.

   b. Describe what the inequality and the line segment tell Travis about his trip.

My Notes

6. Use the graph in Item 1.

a. Draw the line segment for $h = 9$. How far might Travis be from Dallas after 9 hours of travel? Write your answer as an inequality.

b. Describe what the inequality and the line segment tell Travis about his trip.

7. There is a region that could be filled by similar vertical line segments for all values of h that Travis could be on his trip. Shade this region on the graph in Item 1.

8. Travis likes to stop for a break after driving half the distance to Dallas.

a. On the graph in Item 1, sketch the horizontal line that represents half the distance to Dallas. What is the equation of this line? Highlight the segment of the line that gives all possible times that he may stop.

b. How much time might remain in the trip when Travis stops for a break? Explain how you know.

9. After Travis drives for 4 hours and 20 minutes at 45 mi/h, he wonders how much farther he has to drive.

a. Would you prefer to use the graph or the equation to determine the answer? Explain your choice.

b. How much farther does Travis have to drive? Justify your response.

My Notes

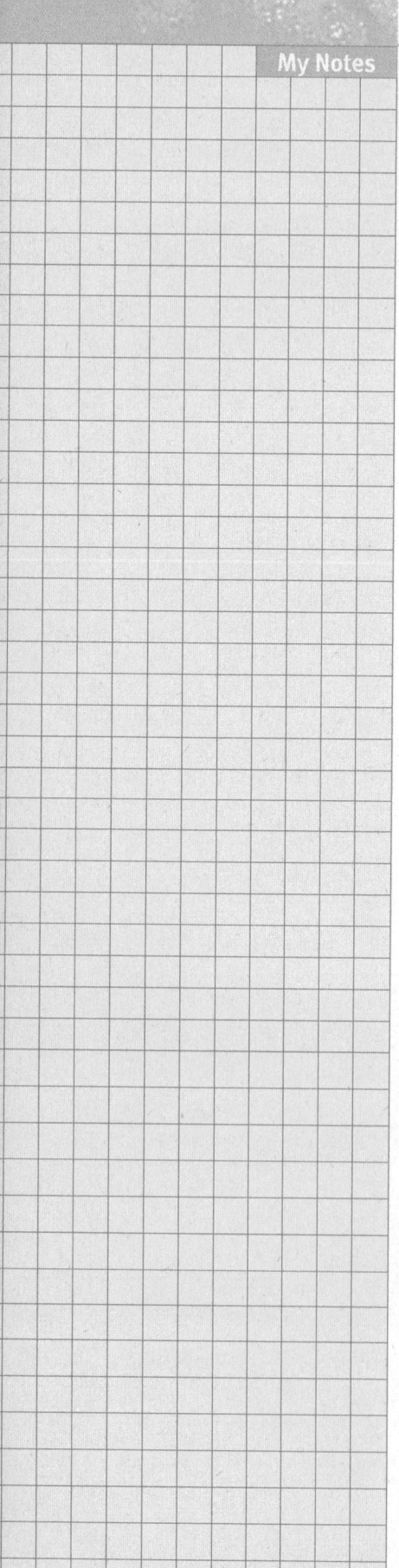

Check Your Understanding

10. Explain how you can use the graph in Item 1 to determine how far Travis might be from Dallas after 6 hours of travel.

11. Assuming Travis can average anywhere from 45 mi/h to 60 mi/h, is it possible for him to be 300 miles from Dallas after 2 hours of travel? Justify your answer.

12. Suppose Travis maintains an average speed of 50 mi/h throughout his trip. Describe how the graph of his distance from Dallas would compare to the graph in Item 1.

LESSON 15-2 PRACTICE

Travis's sister, Amy, lives in Midland, which is 100 miles from Pecos. After visiting his sister, Travis plans to drive his truck to Dallas. The graph shows his planned trip.

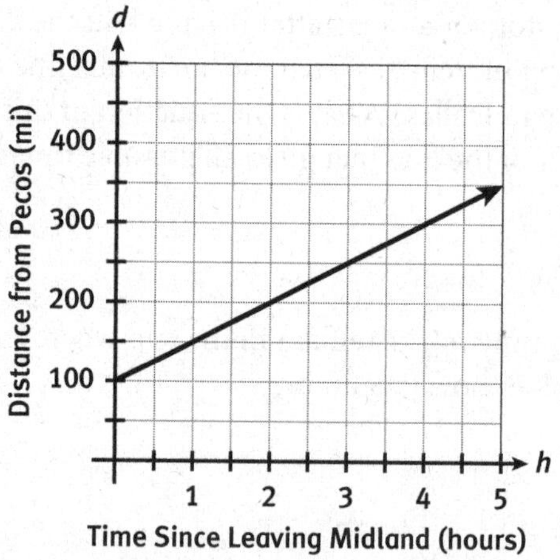

13. At what average speed does Travis expect to travel?

14. Travis hopes he will be able to maintain an average speed of 60 mi/h once he leaves his sister's house. Copy the graph above and draw the graph of Travis's distance from Pecos, based on an average speed of 60 mi/h.

15. Make sense of problems. Write an equation for the line you graphed in Item 14. Interpret the meanings of the constant and the coefficient of h in terms of the context.

16. How far might Travis be from Pecos after 3 hours of travel? Write your answer as an inequality.

My Notes

Learning Targets:

- Write a linear equation given a verbal description.
- Graph and analyze functions on the same coordinate plane.

SUGGESTED LEARNING STRATEGIES: Look for a Pattern, Create Representations, Identify a Subtask

1. Suppose that the speed limit on all parts of Interstate 20 has been changed to 70 mi/h. Travis finds that he can now average between 50 mi/h and 70 mi/h on the trip between Pecos and Dallas.

a. Write an equation that expresses Travis's distance d from Dallas as a function of the hours h since he left Pecos if his average speed is 50 mi/h.

b. Write an equation that expresses Travis's distance d from Dallas as a function of the hours h since he left Pecos if his average speed is 70 mi/h.

c. Reason abstractly. On the grid below, graph the two equations that you found in parts a and b. Describe how changing the coefficient of h in the equation affects the graph. Explain why this makes sense.

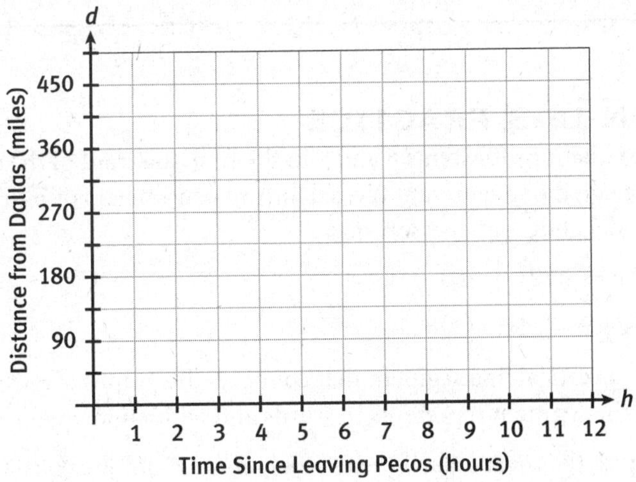

d. Shade the region of the graph above for the ordered pairs (h, d) such that h represents all the possible times and d represents all the possible distances from Dallas after h hours of travel.

My Notes

e. Describe Travis's possible distances from Dallas after 4 hours of travel.

f. Write an inequality to represent Travis's possible distances from Dallas after 4 hours.

Check Your Understanding

Sumi and Jerome start riding their bikes at the same time, traveling in the same direction on the same bike path. Sumi rides at a constant speed of 10 ft/s. Jerome rides at a constant speed of 15 ft/s.

2. Write an equation that expresses the distance d that Sumi has traveled as a function of the time s in seconds since she started riding her bike.

3. Write an equation that expresses the distance d that Jerome has traveled as a function of the time s in seconds since he started riding his bike.

4. Graph the equations from Items 2 and 3 on the same coordinate plane.

5. Explain how to use your graph to find the distance between Sumi and Jerome after they have ridden their bikes for 5 seconds.

LESSON 15-3 PRACTICE

Graph the equations for Items 6 and 7 in the first quadrant of the same graph. The equations represent the height y in centimeters of Ellie's and James's model gliders after x seconds.

6. Ellie: $y = -20x + 220$

7. James: $y = -15x + 220$

8. Draw a vertical line segment that connects the graphs at $x = 5$. Describe what the segment represents in words and with an inequality.

9. Interpret the meanings of the coefficients of x and the constants in terms of the context.

10. **Construct viable arguments.** Whose glider was in the air longer? Justify your response.

11. Julian's model glider begins at an initial height of 300 centimeters and descends at a rate of 18 cm per second. Write an equation that expresses the height y of Julian's glider after x seconds.

ACTIVITY 15 PRACTICE

Write your answers on notebook paper.
Show your work.

Lesson 15-1

1. The graph shows the number of gallons of water y remaining in a tub x minutes after the tub began draining.

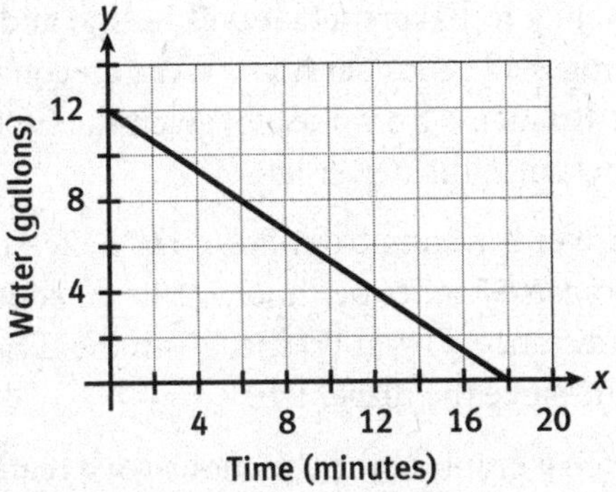

Which describes something that is being drained at nine times the rate of the tub?

A. A pool is drained at a rate of nine gallons per minute.
B. A rain barrel is drained at a rate of six gallons per minute.
C. A cooler is drained at a rate of 12 gallons per minute.
D. A pond is drained at a rate of 18 gallons per minute.

2. The equation $d = 30 - 60h$ gives the distance d in miles that a bus is from the station h hours after leaving an arena. How far is the station from the arena?

A. 0.5 miles **B.** 30 miles
C. 60 miles **D.** 120 miles

3. Claire walks home from a friend's house two miles away. She sketches a graph of her walk, showing her distance from home as a function of time in minutes. Is the slope of the graph positive or negative? Explain.

Use the table for Items 4–9.

Minutes Since Leaving Store, *m*	Distance in Miles from Home, *d*
10	1.5
15	1.25
20	1

4. The table represents Holly's walk home from the store. What is Holly's rate in miles per minute?

5. How far is the store from Holly's home? Explain how you know.

6. Write an equation that relates distance d to the number of minutes m.

Lesson 15-2

7. Tim left school on his bike at the same time Holly left the store. The equation $d = 4 - \frac{m}{5}$ gives Tim's distance from Holly's house after m minutes. Sketch a graph of Holly's and Tim's trips on the same coordinate plane.

8. Compare the total time for Tim's trip to the total time for Holly's trip.

9. Part of Tim's trip includes the way Holly will walk. Use your graph to estimate when Tim will run into Holly.

10. Kane researched the cost of a taxi ride in a nearby city. He found conflicting information about the per-mile cost of a ride. The graph below shows his findings.

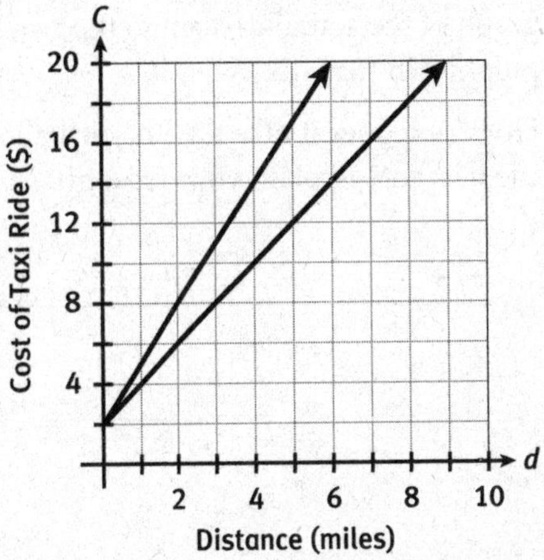

a. What can you conclude about the cost per mile of a taxi ride?

b. How much should Kane expect to pay for a five-mile taxi ride? Explain.

The graph shows the planned descent of a hot-air balloon. Use the graph for Items 11–16.

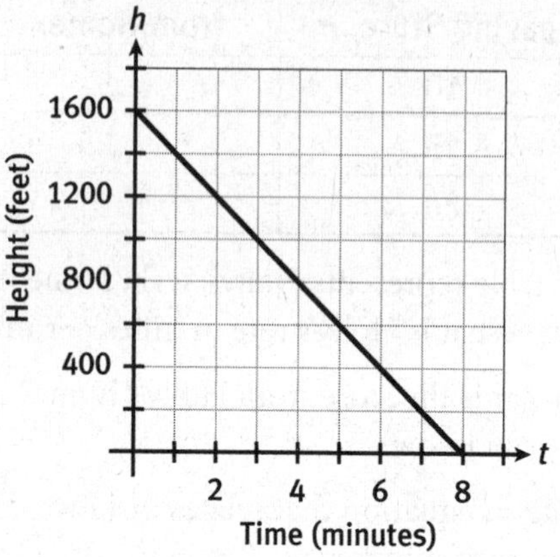

11. Write an equation that expresses the height of the balloon h as a function of the time t since the beginning of the descent.

12. State a reasonable domain and range for the function in Item 11 in this context.

13. The pilot thinks it may be possible to increase the rate of descent to a maximum of 400 ft/min. Copy the above graph and then draw a graph of the descent at 400 ft/min on the same coordinate plane.

14. On your graph from Item 13, draw the vertical line $t = 3$. Use the line to determine all the possible heights of the balloon after 3 minutes. Write your answer as an inequality.

15. The actual descent of the hot-air balloon is given by the equation $h = 1600 - 250t$. Compare the graph of the actual descent to the two lines graphed in Item 13.

16. How long does it take the hot-air balloon to make the descent? Explain your reasoning.

Lesson 15-3

17. Wendy leaves Sacramento and heads to San Jose. She averages 50 mi/h on the 120-mile trip. Which equation describes Wendy's distance d from San Jose h hours after she leaves Sacramento?

A. $d = 50h - 120$ **B.** $d = 120 - 50h$
C. $d = 120 + 50h$ **D.** $d = 50h$

18. Fresno is 150 miles from Bakersfield. A driver traveling to Bakersfield leaves Fresno and averages 62 miles per hour. Write an equation for the distance d from Fresno, given the time t in hours since the driver left.

19. A driver traveling from Fresno to Bakersfield averaged 57 miles per hour. Write an equation for the distance d from Fresno, given the time t in hours since the driver left.

20. Sketch a graph of the equations you created in Items 18 and 19 on the same coordinate axis.

21. If a driver traveling between Fresno and Bakersfield knew he could average between 57 and 62 miles per hour, sketch a graph of the region that represents all possible distances from Fresno that the driver could be at time t.

MATHEMATICAL PRACTICES
Make Sense of Problems and Persevere in Solving Them

22. The amount of water w in aquarium A, in gallons, is given by $w = 50 - 3.5m$, where m is the number of minutes since the aquarium started draining. The amount of water w in aquarium B, in gallons, m minutes after the aquarium started draining, is given by the table below. Which aquarium drains more quickly, and how long does it take until it is empty?

Minutes, m	Gallons of Water, w
0	50
5	40

My Notes

Learning Targets:

- Write linear inequalities in two variables.
- Read and interpret the graph of the solutions of a linear inequality in two variables.

SUGGESTED LEARNING STRATEGIES: Marking the Text, Look for a Pattern, Think-Pair-Share, Create Representations, Sharing and Responding

Axl and Aneeza pay for and share memory storage at a remote Internet site. Their plan allows them upload 250 terabytes (TB) or less of data each month. Let x represent the number of terabytes Axl uploads in a month, and let y represent the number of terabytes Aneeza uploads in a month.

1. Look at the data for four months below. Determine if Axl and Aneeza stayed within the limits of their monthly plan.

Axl's Uploads (x)	Aneeza's Uploads (y)	Within Plan? (Yes/No)
50 TB	120 TB	
135 TB	100 TB	
100 TB	150 TB	
162 TB	128 TB	

2. Write an inequality that models the plan's restriction on uploading.

3. **Reason quantitatively.** If Axl uploads 100 terabytes, write an inequality that shows how many terabytes Aneeza can upload.

4. If Aneeza uploads 75 terabytes, write an inequality that shows how many terabytes Axl can upload.

My Notes

The graph below represents the possible values for the number of terabytes that Axl and Aneeza can upload. An ordered pair (x, y) represents the number of terabytes that Axl and Aneeza upload. For example, the coordinates (100, 75) mean that Axl uploaded 100 TB and Aneeza uploaded 75 TB of data.

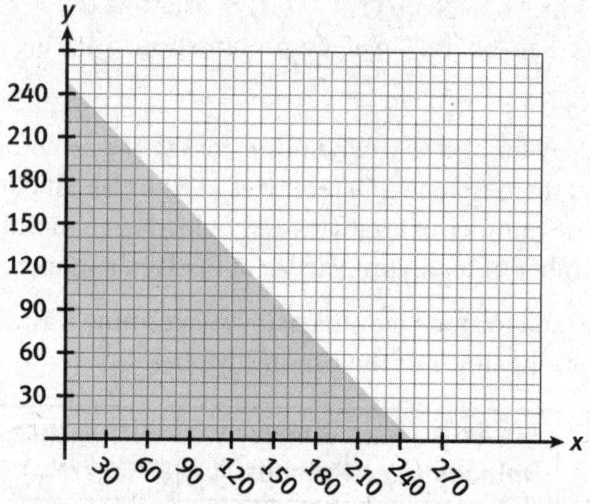

MATH TIP

Some equations and inequalities are graphed on the coordinate plane, but contain only one variable.

$x = 1$

$y < 3$

Other equations and inequalities that are graphed on the coordinate plane contain two variables.

$x + y = 10$

$y \geq 7 - x$

5. Graph each ordered pair on the graph above. Determine if it will allow Axl and Aneeza to remain within their plan, and explain your answers.

a. (200, 50)

b. (150, 150)

c. (225, 25)

d. (20, 120)

e. (120, 200)

MATH TERMS

A **linear inequality** is an inequality in the form $Ax + By > C$, $Ax + By \geq C$, $Ax + By < C$, or $Ax + By \leq C$, where A, B, and C are constants and A and B are not both zero.

A **solution of a linear inequality** is an ordered pair (x, y) that makes the inequality true.

The graph of Axl and Aneeza's storage plan above is an example of a graph of a ***linear inequality*** in two variables. All the points in the shaded region are ***solutions of the linear inequality***.

My Notes

Check Your Understanding

Axl and Aneeza decide to change to a 350-terabyte plan. Use this information for Items 6–8.

6. How does the graph in Item 5 change?

7. Graph the inequality that represents Axl and Aneeza's new plan. Let x represent the number of terabytes Axl can upload, and let y represent the number of terabytes Aneeza can upload.

8. Which ordered pairs will work with the new plan? Justify your response.

a. (200, 50) **b.** (150, 150)
c. (240, 180) **d.** (0, 300)
e. (210, 180) **f.** (360, −10)

9. Explain how to decide whether an ordered pair is a solution to a linear inequality.

LESSON 16-1 PRACTICE

You are on a treasure-diving ship that is hunting for gold and silver coins. You reel in a wire basket that contains gold and silver coins, among other things. The basket holds no more than 50 pounds of material. Each gold coin weighs about 0.5 ounce, and each silver coin weighs about 0.25 ounce. You want to know the different numbers of each type of coin that could be in the basket.

10. Write an inequality that models the weight in the basket.

11. Give three solutions of the inequality you wrote in Item 10.

12. Construct viable arguments. Explain why there cannot be 400 gold coins and 2800 silver coins in the basket.

13. Mitch and Mindy are in charge of purchasing art supplies for their local senior-citizen center. The monthly budget for supplies is $325, so the total they can spend in a month cannot exceed $325. This table displays how much each spent from January through April.

Month	Amount Mitch Spent (x)	Amount Mindy Spent (y)
January	150	155
February	215	110
March	60	252
April	175	150

Write an inequality in two variables that models how much Mitch and Mindy can spend together and stay within the budget.

My Notes

Learning Targets:

- Graph on a coordinate plane the solutions of a linear inequality in two variables.
- Interpret the graph of the solutions of a linear inequality in two variables.

SUGGESTED LEARNING STRATEGIES: Vocabulary Organizer, Interactive Word Wall, Create Representations, Identify a Subtask, Construct an Argument

The solutions of a linear inequality in two variables can be represented in the coordinate plane.

Example A

Graph the linear inequality $y \leq 2x + 3$.

Step 1: Graph the corresponding linear equation $y = 2x + 3$. The line you graphed is the ***boundary line***.

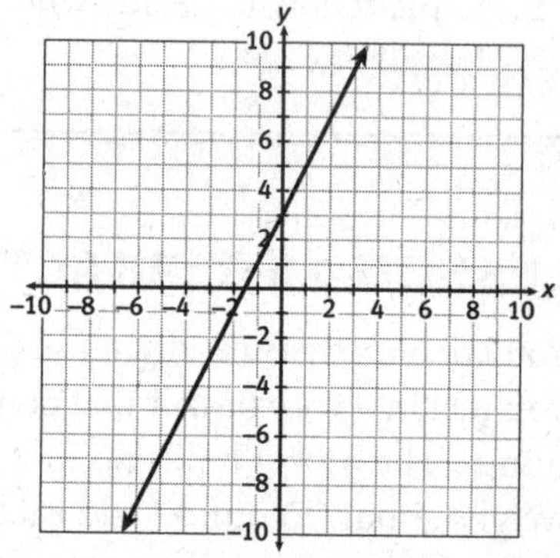

Step 2: Test a point in one of the ***half-planes*** to see if it is a solution of the inequality. Using (0, 0), $0 \leq 2(0) + 3$ is a true statement. So (0, 0) is a solution.

Step 3: If the point you tested is a solution, shade the half-plane in which it lies. If it is not, shade the other half-plane. (0, 0) is a solution, so shade the half-plane containing the solution point (0, 0).

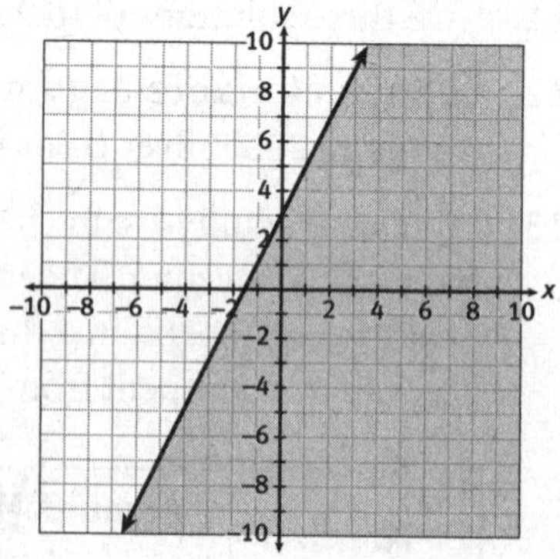

In Example A, the solution set includes the points on the boundary line. The solution set is a ***closed half-plane***. In an inequality containing $<$ or $>$ the solution set does not include the points on the boundary line, so the boundary line is dashed, and the solution set is an ***open half-plane***.

MATH TERMS

A line that separates the coordinate plane into two regions is a **boundary line**. The two regions are **half-planes**.

MATH TIP

The origin (0, 0) is usually an easy point to test if it is not on the boundary line.

MATH TERMS

A **closed half-plane** includes the points on the boundary line.

An **open half-plane** does not include the points on the boundary line.

My Notes

Example B

Graph $x + 2y > 8$.

Step 1: Solve the inequality for y.

$x - x + 2y > 8 - x$ Subtract x from each side.

$2y > -x + 8$

$\frac{2y}{2} > \frac{-x+8}{2}$ Divide each side by 2.

$y > -\frac{1}{2}x + 4$ Simplify.

Step 2: Graph the boundary line, $y = -\frac{1}{2}x + 4$. Since the inequality uses the < symbol, the solution set is an open half-plane. Draw a dashed line.

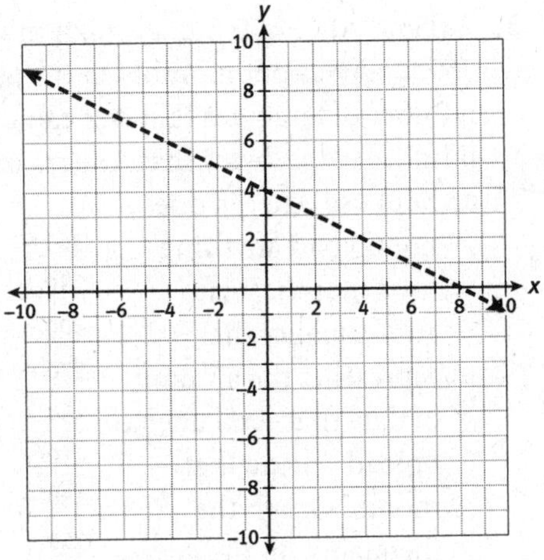

Step 3: Check the test point (0, 0):

$0 > -\frac{1}{2}(0) + 4$ is false.

Shade the half-plane that does **not** contain the point (0, 0).

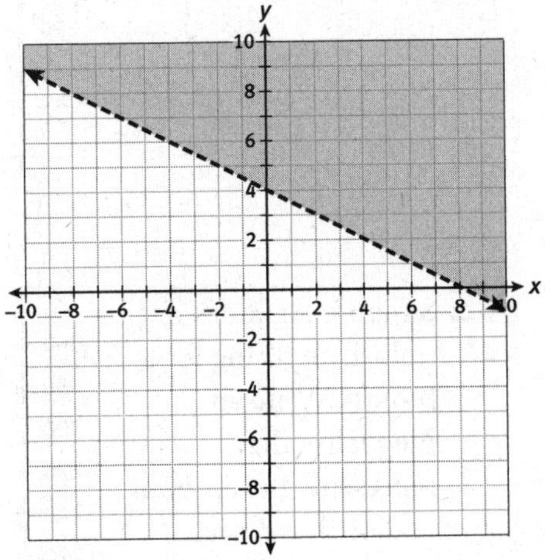

Try These A–B

Graph each inequality.

a. $3x - 4y \geq -12$

b. $y < 3x - 2$

c. $-2y > 6 + x$

d. $y \leq -4x + 5$

MATH TIP

When multiplying or dividing each side of an inequality by a negative number, you must reverse the inequality symbol.

1. When graphing a linear inequality, is it possible for a test point located on the boundary line to determine which half-plane should be shaded? Explain.

My Notes

2. Critique the reasoning of others. Axl tried to graph the inequality $2x + 4y < 8$. He first graphed the linear equation $2x + 4y = 8$. He then chose the test point (2, 1) and used his result to shade above the line. Explain Axl's mistake and how he should correct it.

3. Axl and Aneeza try a new plan in which they can upload no more than 350 terabytes per month. On the grids below, the x-axis represents the number of terabytes that Axl can upload and the y-axis represents the number of terabytes that Aneeza can upload.

a. Suppose Aneeza does not upload anything for a month. Write an inequality that represents the amount of data that Axl can upload during that month. Graph the inequality on the grid.

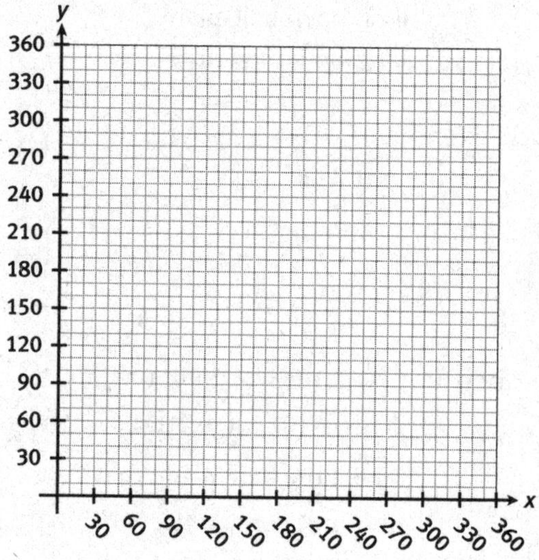

b. Suppose Axl does not upload anything for a month. Write an inequality that represents the amount of data that Aneeza can upload during that month. Graph the inequality on the grid.

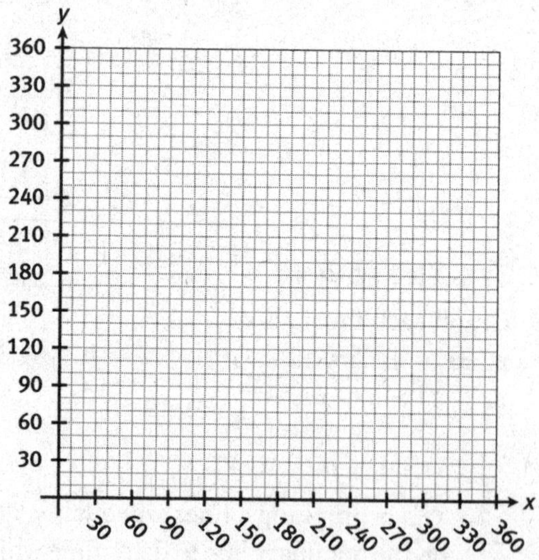

My Notes

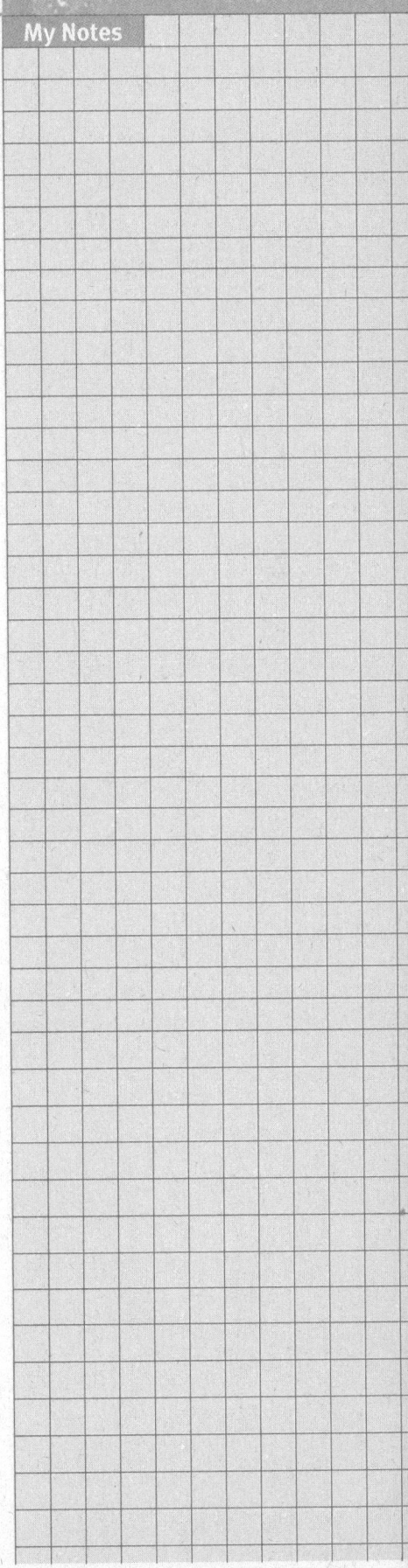

4. The graph below represents another plan that Axl and Aneeza considered. The x-axis represents the number of terabytes of data from photos that can be uploaded each month. The y-axis represents the number of terabytes of data from text files.

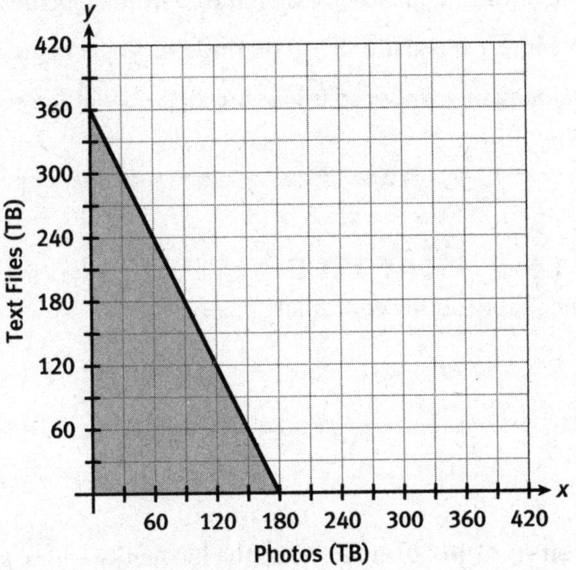

a. Identify the x-intercept and y-intercept. What do they represent in this context?

b. Determine the equation for the boundary line of the graph. Justify your response.

c. Model with mathematics. Write a linear inequality to represent this situation. Then describe the plan in your own words.

My Notes

Check Your Understanding

5. Graph the linear inequality $x < -3$ on the coordinate plane.
6. Graph the linear inequality $y > 2$ on the coordinate plane.
7. Graph the linear inequality $x \geq 0$ on the coordinate plane.
8. Write an inequality whose solutions are all points in the second and third quadrants.

LESSON 16-2 PRACTICE

Graph each inequality on the coordinate plane.

9. $x - y \leq 4$
10. $2x - y > 1$
11. $y \geq 3x + 7$
12. $-x + 6 > y$
13. **Make sense of problems.** Write the inequality whose solutions are shown in the graph.

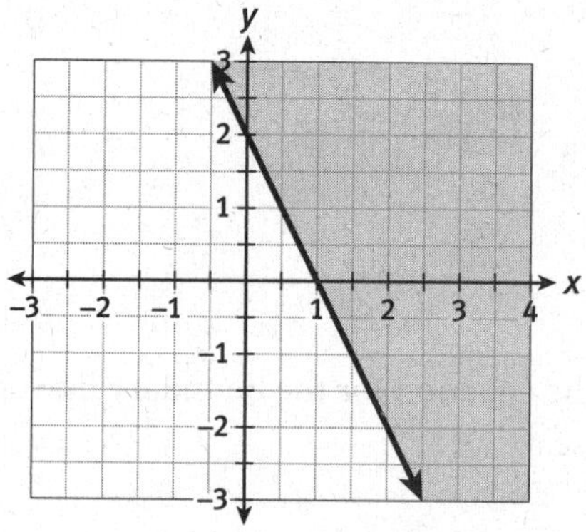

ACTIVITY 16 PRACTICE

Write your answers on notebook paper.
Show your work.

Lesson 16-1

1. Which ordered pairs are solutions of the inequality $5y - 3x \leq 7$?
   A. (0, 0) **B.** (3, 5)
   C. (−2, −5) **D.** (1, 2.5)
   E. (5, −3)

2. Apple juice costs $2 per bottle, and cranberry juice costs $3 per bottle. Tamiko has at most $18 with which to buy drinks for a club picnic. She lets x represent the number of bottles of apple juice and lets y represent the number of bottles of cranberry juice. Then she graphs the inequality $2x + 3y \leq 18$, as shown below.

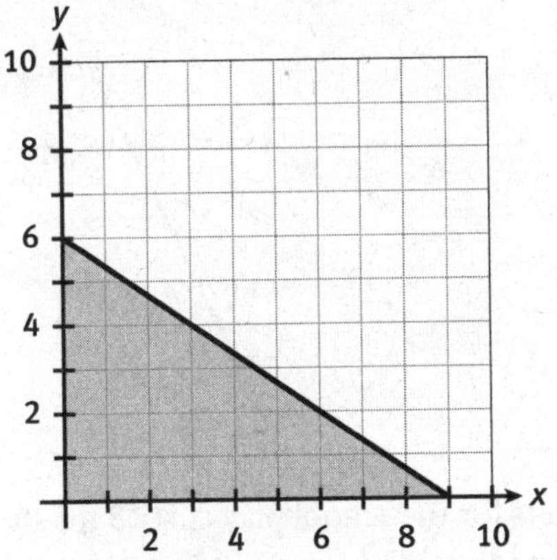

   a. Tamiko states that the graph does not help her decide how many bottles of each type of juice to buy, because there are infinitely many solutions. Do you agree or disagree? Why?
   b. Suppose Tamiko decides to buy two bottles of apple juice. Explain how she can use the graph to determine the possible numbers of bottles of cranberry juice she can buy.

3. Describe a real-world situation that can be represented by the inequality shown in the graph.

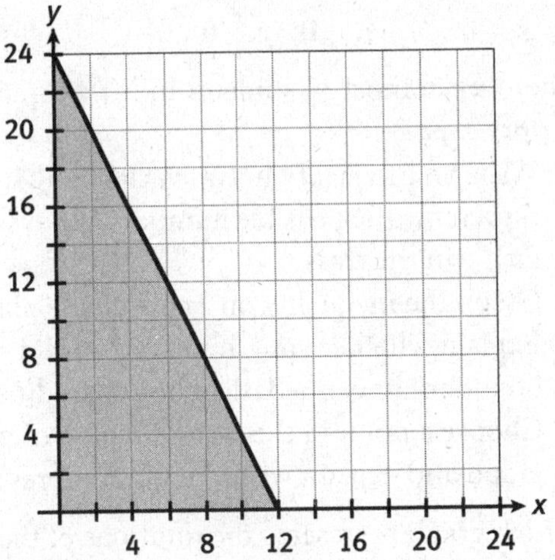

Lesson 16-2

4. Write an inequality for the half-plane. Is the half-plane open or closed?

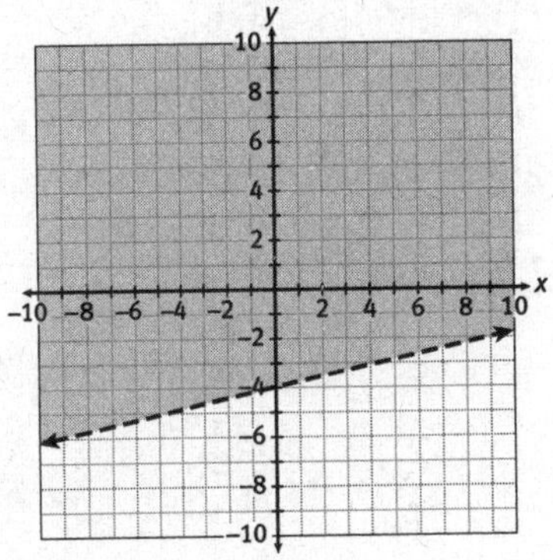

5. Write an inequality for the half-plane. Is the half-plane open or closed?

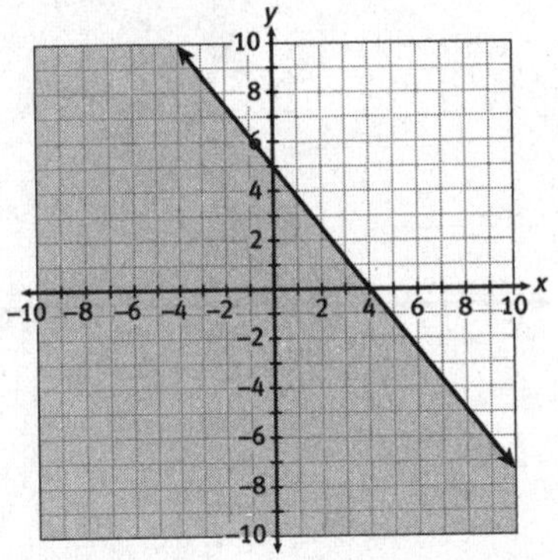

6. Sketch a graph of the inequality $y \geq -\frac{2}{5}x + 2$.

7. Sketch a graph of the inequality $3y > 7x - 15$.

8. Which inequality represents all of the points in the first and fourth quadrants?
   A. $x < 0$ **B.** $x > 0$
   C. $y < 0$ **D.** $y > 0$

9. There are at most 30 students in Mr. Moreno's history class.
   a. Write an inequality in two variables that represents the possible numbers of boys b and girls g in the class.
   b. Graph the inequality on a coordinate plane.
   c. Explain whether your graph has a solid boundary line or a dashed boundary line.
   d. Choose a point in the shaded region of your graph and explain what the point represents.

10. Which graph represents the solutions of the inequality $2x - y \geq 6$?

A.

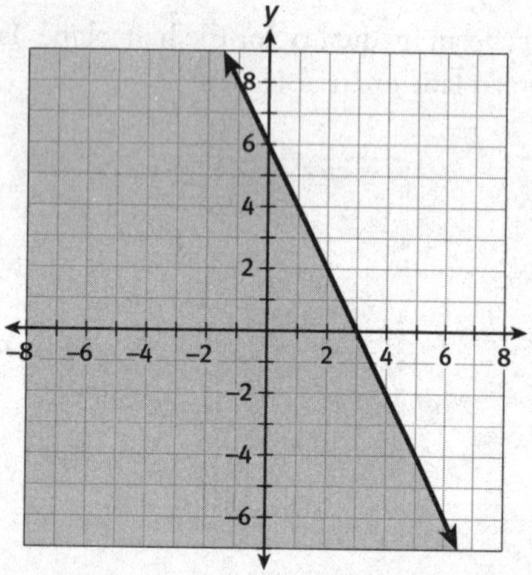

B.

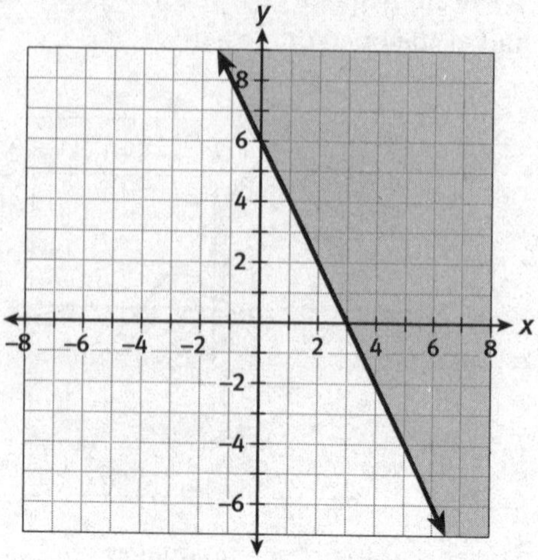

C.

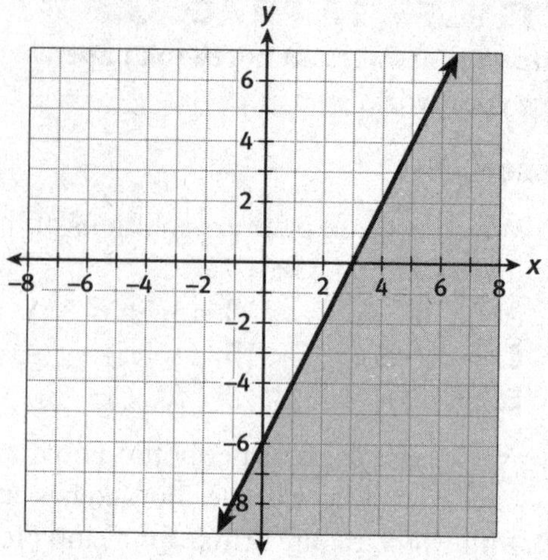

D.

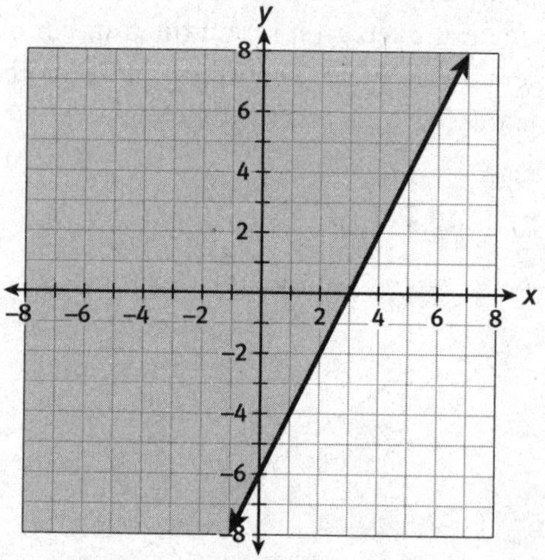

11. Tickets for the school play cost \$3 for students and \$6 for adults. The drama club hopes to bring in at least \$450 in sales. The auditorium has 120 seats. Let a represent the number of adult tickets and s represent the number of student tickets.
   a. Write an inequality in two variables that represents the desired ticket sales.
   b. Write an inequality in two variables that represents the possible numbers of tickets that can be sold.
   c. Sketch both inequalities on the same grid. What does the intersection of the two graphs represent?

12. When is the boundary line of the graph of an inequality in two variables part of the solution?

MATHEMATICAL PRACTICES
Look For and Make Use of Structure

13. Graph the inequality $x < 3$ on a number line and on the coordinate plane. Describe the differences in the graphs.

Graphing Inequalities and Piecewise-Defined Functions

EARNINGS ON A GRAPH

1. Steve works at a restaurant. He earns $8.50 per hour.

a. Write an equation that indicates the amount of money *m* in dollars that Steve can earn as a function of the hours *h* that he worked.

b. Steve works at least 10 hours and not more than 30 hours per week. Describe the reasonable domain and range for your function from part a.

c. The cost of any food that Steve buys while working is deducted from his earnings. Write an inequality that represents the possible amounts of money he can earn after buying food.

d. Copy the grid below. Graph the inequality you wrote in part c.

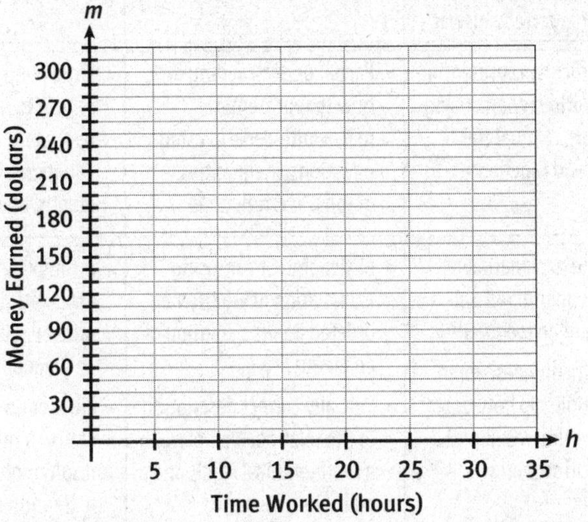

e. Determine whether the ordered pair (16, 126) is a solution of the inequality you wrote in part c. If so, interpret its meaning. If not, explain why not.

Bob has been working at the restaurant longer than Steve. He earns $9 per hour. During some weeks he works more than 40 hours. The hours he works beyond 40 are considered overtime. For overtime pay, Bob earns double time, or $18 per hour.

2. a. Write a function $B(h)$ that will give Bob's pay for working 40 hours or less.

b. Identify a reasonable domain and range for the function in this context.

3. a. Write a function $B(h)$ that will give Bob's pay for working more than 40 hours.

b. Identify a reasonable domain and range for the function in this context.

4. Write a piecewise-defined function $B(h)$ that gives Bob's pay for any number of hours *h*.

5. Graph your function from Item 4.

Scoring Guide	Exemplary	Proficient	Emerging	Incomplete
	The solution demonstrates the following characteristics:			
Mathematics Knowledge and Thinking (Items 1b, 2b, 3b)	• Clear understanding and accurate identification of reasonable domain and range	• Adequate understanding and accurate identification of reasonable domain and range	• Partial understanding and partially accurate identification of reasonable domain and range	• No understanding and inaccurate identification of reasonable domain and range
Problem Solving (Item 1e)	• Appropriate and efficient strategy that results in a correct answer	• Strategy that may include unnecessary steps but results in a correct answer	• Strategy that results in some incorrect answers	• No clear strategy when solving problems
Mathematical Modeling / Representations (Items 1a, 1c, 1d, 2a, 3a, 4, 5)	• Effective understanding of how to represent a real-world scenario using equations, inequalities, graphs, and functions	• Little difficulty representing a real-world scenario using equations, inequalities, graphs, and functions	• Partial understanding of how to represent a real-world scenario using equations, inequalities, graphs, and functions	• Little or no understanding of how to represent a real-world scenario using equations, inequalities, graphs, and functions
Reasoning and Communication (Item 1e)	• Precise use of appropriate math terms and language to explain whether an ordered pair is a solution of an inequality • Ease and accuracy describing the relationship between a mathematical result and a real-world scenario	• Adequate explanation of whether an ordered pair is a solution of an inequality • Little difficulty describing the relationship between a mathematical result and a real-world scenario	• Misleading or confusing explanation of whether an ordered pair is a solution of an inequality • Partially correct description of the relationship between a mathematical result and a real-world scenario	• Incomplete or inaccurate explanation of whether an ordered pair is a solution of an inequality • Little or no understanding of how a mathematical result might relate to a real-world scenario

Solving Systems of Linear Equations

A Tale of Two Truckers

Lesson 17-1 The Graphing Method

My Notes

Learning Targets:

- Solve a system of linear equations by graphing.
- Interpret the solution of a system of linear equations.

SUGGESTED LEARNING STRATEGIES: Summarizing, Paraphrasing, Marking the Text, Look for a Pattern, Create Representations

Travis Smith and his brother, Roy, are co-owners of a trucking company. The company needs to transport two truckloads of fruit grown in Pecos, Texas, to a distributing plant in Dallas, Texas. If the fruit does not get to Dallas quickly, it will spoil. The farmers offer Travis a bonus if he can get both truckloads to Dallas within 24 hours.

Due to road construction, Travis knows it will take 10 hours to drive from Pecos to Dallas. The return trip to Pecos will take only 7.5 hours. He estimates it will take 1.5 hours to load the fruit onto the truck and 1 hour to unload it.

1. Why is it impossible for Travis to earn the bonus by himself?

2. Travis wants to earn the bonus so he asks his brother, Roy, if he will help. With Roy's assistance, can the brothers meet the deadline and earn the bonus? Explain why or why not.

Roy is in Dallas ready to leave for Pecos. To meet the deadline and earn the bonus, Travis will leave Pecos first and meet Roy somewhere along the interstate to give him a key to the storage area in Pecos.

3. From Pecos to Dallas, Travis averages 45 mi/h. If Dallas is 450 mi from Pecos, write an equation that expresses Travis's distance d in miles from Dallas as a function of the hours h since he left Pecos.

My Notes

4. Graph the equation you wrote in Item 3.

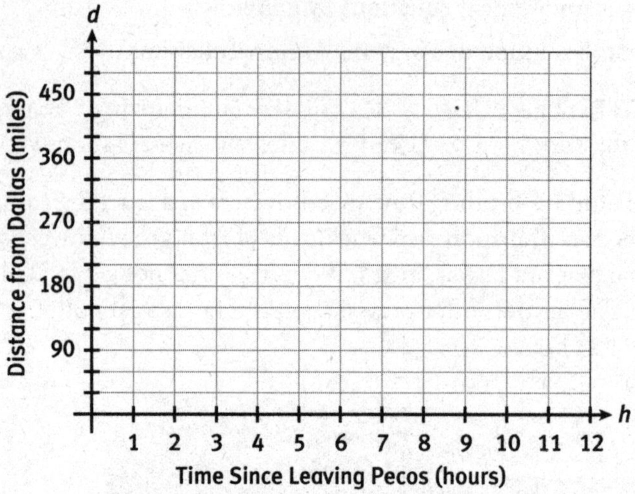

5. Roy leaves Dallas one-half hour before Travis leaves Pecos. In terms of the hours *h* since Travis left Pecos, write an expression that represents the time since Roy left Dallas.

6. Roy travels 60 mi/h from Dallas to Pecos. Write an equation that expresses Roy's distance *d* from Dallas as a function of the hours *h* since Travis left Pecos.

7. Graph the equation from Item 6 on the grid in Item 4.

8. Identify the intersection point of the two lines. Describe the information these coordinates provide.

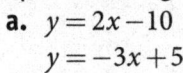

The two equations you wrote in Items 3 and 6 form a ***system of linear equations***.

To determine the solution of a system of linear equations, you must identify all the ordered pairs that make both equations true. One method is to graph each equation and determine the intersection point.

9. Graph each system of linear equations. Give each solution as an ordered pair. Check that the point of intersection is a solution of both equations by substituting the solution values into the equations.

a. $y = 2x - 10$
$y = -3x + 5$

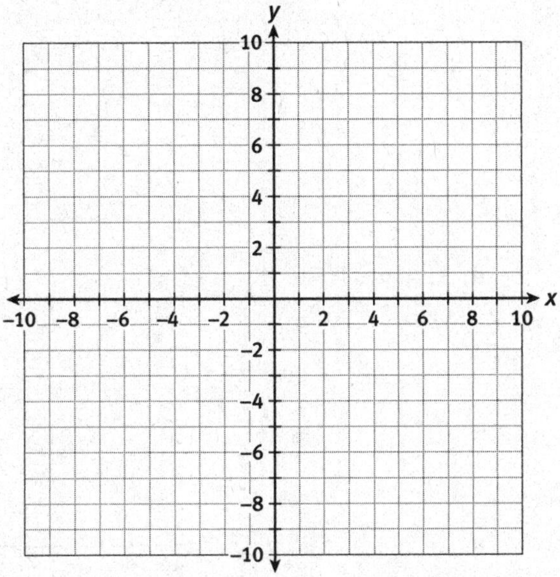

b. Reason abstractly. Edgar has nine coins in his pocket. All of the coins are nickels or dimes and are worth a total of $0.55. The system shown below represents this situation. How many of each type of coin does Edgar have in his pocket?

$n + d = 9$
$n + 2d = 11$

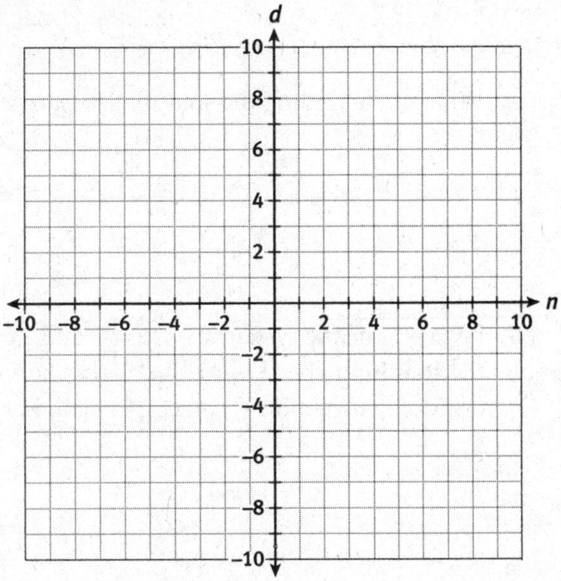

My Notes

MATH TERMS

Two or more linear equations with the same variables form a **system of linear equations.**

TECHNOLOGY TIP

You can graph each equation on a graphing calculator and use the built-in commands to determine the intersection point.

My Notes

c. $y = 3x - 5$
$y = -2x + 4$

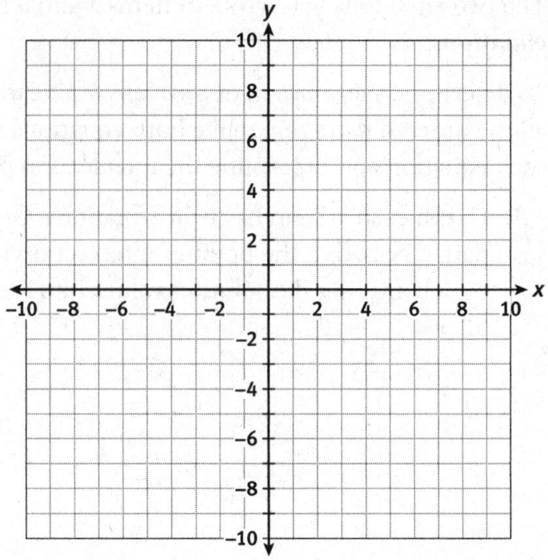

d. $3x + y = 1$
$6x + 2y = 10$

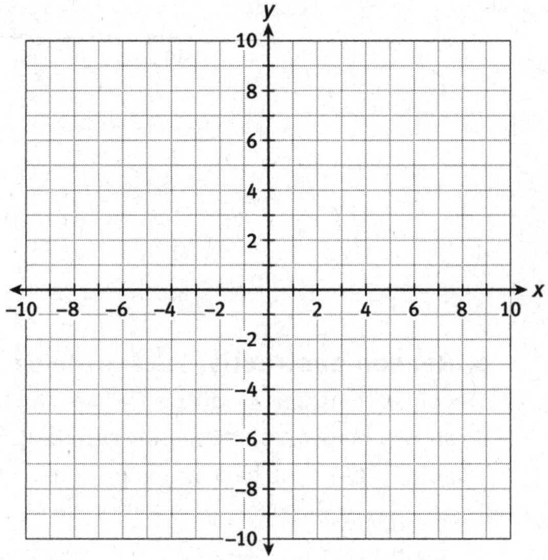

e. What made finding the solutions to parts c and d challenging?

10. Lena graphs the system $x - 2y = 3$ and $2x - y = -3$ and determines the solution to be $(1, -1)$. She checks her solution algebraically and decides the solution is correct. Explain her error.

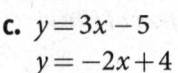

My Notes

Check Your Understanding

11. Solve each system.

a. $y = -2x + 5$
$y = \frac{1}{8}x - \frac{7}{2}$

b. $3x - y = 5$
$4x - 2y = 4$

c. $y = -2x + 3$
$y = x$

d. $2x + y = 5$
$4x - 1 = y$

12. Roberto has eight coins that are all dimes or nickels. They are worth \$0.50. The system $n + d = 8$ and $n + 2d = 10$ represents this situation. Graph the system to determine how many of each coin Roberto has.

LESSON 17-1 PRACTICE

13. Solve each system.

a. $y = 2x + 2$
$y = -2x - 6$

b. $y = \frac{1}{3}x - 2$
$y = -x + 2$

c. $3x + 2y = 6$
$x - y = -3$

d. $y = -2$
$2y = -x - 1$

14. Sandeep ordered peanuts and raisins for his bakery. He ordered a total of eight pounds of these ingredients. Peanuts cost \$1 per pound, and raisins cost \$2 per pound. He spent a total of \$10. The system $p + r = 8$ and $p + 2r = 10$ represents this situation. Graph the system to determine how many pounds of peanuts and raisins Sandeep ordered.

15. Critique the reasoning of others. Kyla was asked to solve the system of equations below. She made the graph shown and stated that the solution of the system is $(-4, -1)$. Is Kyla correct? Justify your response and identify Kyla's errors, if they exist.

$x + y = 3$
$x - 3y = -1$

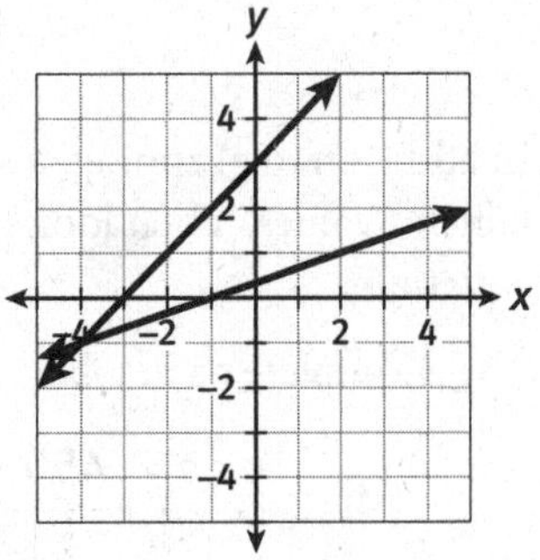

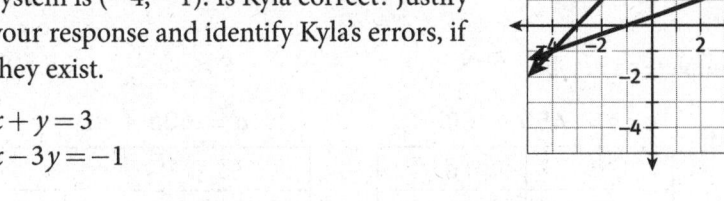

16. Write the system of equations represented by this graph.

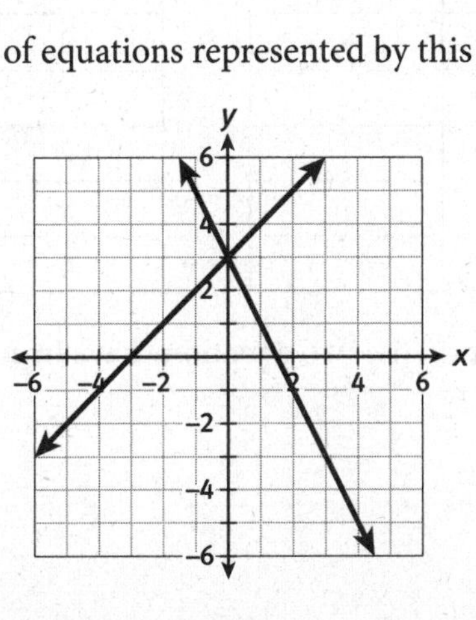

My Notes

Learning Targets:

- Solve a system of linear equations using a table or the substitution method.
- Interpret the solution of a system of linear equations.

SUGGESTED LEARNING STRATEGIES: Think-Pair-Share, Note Taking, Marking the Text, Guess and Check, Simplify the Problem

On another trip, Travis is traveling from Pecos to Dallas, and Roy is driving from Dallas to Pecos. They agree to meet for lunch along the way. Each driver averages 60 mi/h but Roy leaves 1.5 hours before his brother does. To determine when and where they will meet, you will solve this system of linear equations.

$$d = 450 - 60h$$
$$d = 60h + 90$$

1. What do the coefficient 60 and the constants 90 and 450 represent in the context of the problem?

2. Which equation represents Roy's distance from Dallas? How do you know?

In addition to graphing, a system of equations can be solved by first making a table of values. Then look for an ordered pair that is common to both equations.

3. Complete each table.

$d = 450 - 60h$

h	d
0	
1	
2	
3	
4	

$d = 60h + 90$

h	d
0	
1	
2	
3	
4	

4. What ordered pair do the two equations have in common?

My Notes

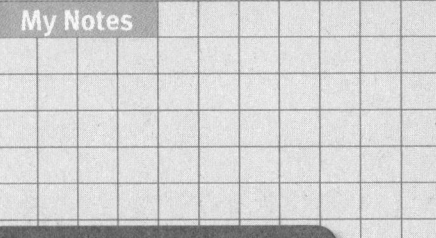

5. **Use appropriate tools strategically.** Graph the equations on a graphing calculator. Identify the solution of the linear system. Describe its meaning in terms of the situation.

6. Is it possible that the intersection point from a graph of a linear system does not show up on a table of values? How could the solution be determined if it is not shown in the table?

TECHNOLOGY TIP

You can use the table feature of a graphing calculator to quickly generate a table of values.

Check Your Understanding

7. Solve each system.

   a. $y = 2x + 6$
      $y = -3x + 16$

   b. $x + y = 8$
      $3x + 2y = 14$

   c. $y = 100 - 2x$
      $y = 20 + 6x$

   d. $2x + y = 6$
      $2x + 3y = 8$

8. What challenges did you encounter when solving these systems of linear equations?

Sometimes it is difficult to solve a system of equations by graphing or by using tables of values, and another solution method is necessary.

On another trip, Roy leaves from Pecos one hour before his brother and averages 55 mi/h. Travis leaves from Dallas during rush hour so he averages only 45 mi/h.

This system of linear equations represents each brother's distance d from Dallas, h hours after Roy leaves Pecos.

Roy: $d = 450 - 55h$
Travis: $d = 45(h - 1)$

Solve the system by finding when Travis's distance from Dallas is the same as Roy's distance.

Travis's distance = Roy's distance
$45(h - 1) = 450 - 55h$

9. Solve the equation for h and show your work.

My Notes

10. Reason quantitatively. What does the answer to Item 9 represent? Is your answer reasonable?

11. Use a graphing calculator to graph the system above.

a. What is the intersection point of the graphs? How would you use this point to answer Item 9?

b. The second coordinate of the intersection point represents the distance that the brothers are from Dallas. How could you confirm this using the equations?

Another method for solving systems of equations is the ***substitution method***, in which one equation is solved for one of the variables. Then the expression for that variable is substituted into the other equation.

Example A

For a Valentine's Day dance, tickets for couples cost $12 and tickets for individuals cost $8. Suppose 250 students attended the dance, and $1580 was collected from ticket sales. How many of each type of ticket was sold?

Step 1: Let $x =$ number of couples, and $y =$ number of individuals.

Step 2: Write an equation to represent the number of people attending.

$2x + y = 250$ — The number of attendees is 250.

Write another equation to represent the money collected.

$12x + 8y = 1580$ — The total ticket sales is $1580.

Step 3: Use substitution to solve this system.

$2x + y = 250$ — Solve the first equation for y.

$y = 250 - 2x$

$12x + 8(250 - 2x) = 1580$ — Substitute for y in the second equation.

$12x + 2000 - 16x = 1580$ — Solve for x.

$-4x = -420$

$x = 105$

My Notes

Step 4: Substitute the value of x into one of the original equations to find y.

$$2x + y = 250$$
$$2(105) + y = 250 \qquad \text{Substitute 105 for } x.$$
$$210 + y = 250$$
$$y = 40$$

Solution: For the dance, 105 couples' tickets and 40 individual tickets were sold.

Try These A

Solve each system using substitution.

a. $x + 2y = 8$ and $3x - 4y = 4$

b. $5x - 2y = 0$ and $3x + y = -1$

c. Patty and Toby live 345 miles apart. They decide to drive to meet one another. Patty leaves at noon, traveling at an average rate of 45 mi/h, and Toby leaves at 3:00 P.M., traveling at an average speed of 60 mi/h. At what time will they meet?

12. Write the system of equations represented by these tables of values. Then use any method to solve the system.

x	−2	0	3	4
y	−4	0	6	8

x	−1	1	3	5
y	7	5	3	1

Check Your Understanding

Solve the systems by any method. Explain why you chose the method you did.

13. $x - 5y = 2$ and $x + y = 8$

14. $2y = x$ and $x - 7y = 10$

15. Theo buys a slice of pizza and a bottle of water for \$3. Ralph buys 3 slices of pizza and 2 bottles of water for \$8. How much does a slice of pizza cost?

My Notes

LESSON 17-2 PRACTICE

16. Delano was asked to solve the system $y = 2x - 4.5$ and $y = -x + 6$. He made the tables shown below.

$y = 2x - 4.5$

x	y
0	−4.5
1	−2.5
2	−0.5
3	1.5
4	3.5
5	5.5

$y = -x + 6$

x	y
0	6
1	5
2	4
3	3
4	2
5	1

a. Are any of the ordered pairs in the tables solutions of the system? Why or why not?

b. What can you conclude about the solution of the system? Explain your reasoning.

c. Would you choose to solve the system graphically or algebraically? Justify your choice.

17. A rock-climbing gym called Rock-and-Roll charges $2.75 to rent shoes, and $3 per hour to climb. A competing gym, Climb the Walls, charges $4.25 to rent shoes, and $2.50 per hour to climb.

a. Write an equation that gives the cost y of renting shoes and climbing for x hours at Rock-and-Roll.

b. Write an equation that gives the cost y of renting shoes and climbing for x hours at Climb the Walls.

c. Solve the system of the two equations you wrote in parts a and b using any method. Why did you choose the solution method you did?

d. What does your solution to the system represent in the context of the problem?

e. **Construct viable arguments.** Suppose you plan to rent shoes and go rock climbing for 4 hours. Which gym offers a better deal in this case? Justify your response.

18. Write the system of equations represented by these tables of values. Then use any method to solve the system.

x	−2	−1	1	2
y	−2	0	4	6

x	−2	2	4	6
y	1	3	4	5

My Notes

Learning Targets:

- Use the elimination method to solve a system of linear equations.
- Write a system of linear equations to model a situation.

SUGGESTED LEARNING STRATEGIES: Note Taking, Discussion Groups, Critique Reasoning, Vocabulary Organizer, Marking the Text

Elimination is another algebraic method that may be used to solve a system of equations. Two equations can be combined to yield a third equation that is also true. The ***elimination method*** creates like terms that add to zero.

MATH TERMS

The **elimination method,** also called the linear combination method, for solving a system of two linear equations involves *eliminating* one variable. To eliminate one variable, multiply each equation in the system by an appropriate number so that the terms for one of the variables will combine to zero when the equations are added. Then substitute the value of the known variable to find the value of the unknown variable. The ordered pair is the *solution* of the system.

Example A

Solve the system using the elimination method: $4x - 5y = 30$
$3x + 4y = 7$

Step 1: To solve this system of equations by elimination, decide to eliminate the y variable.

Original system	Multiply the first equation by 4. Multiply the second equation by 5.	Add the two equations to eliminate y.
$4x - 5y = 30$ →	$4(4x - 5y) = 4(30)$ →	$16x - 20y = 120$
$3x + 4y = 7$ →	$5(3x + 4y) = 5(7)$ →	$\underline{15x + 20y = 35}$
		$31x = 155$
	Solve for x.	$x = 5$

Step 2: Find y by substituting the value of x into one of the original equations.

$$\begin{aligned} 4x - 5y &= 30 \\ 4(5) - 5y &= 30 \quad \text{Substitute 5 for } x. \\ 20 - 5y &= 30 \\ -5y &= 10 \\ y &= -2 \end{aligned}$$

Step 3: Check $(5, -2)$ in the second equation $3x + 4y = 7$.

$$\begin{aligned} 3x + 4y &= 7 \\ 3(5) + 4(-2) &\ ?\ 7 \\ 15 - 8 &\ ?\ 7 \\ 7 &= 7 \quad \text{check} \end{aligned}$$

Solution: The solution is $(5, -2)$.

Try These A

Solve each system using elimination.

a. $3x - 2y = -21$
$2x + 5y = 5$

b. $7x + 5y = 9$
$4x - 3y = 11$

My Notes

1. Phuong has \$2.00 in nickels and dimes in his bank. The number of dimes is five more than twice the number of nickels. How many of each type of coin are in his bank?

Example B

Noah is given two beakers of saline solution in chemistry class. One contains a 3% saline solution and the other an 8% saline solution. How much of each type of solution will Noah need to mix to create 150 mL of a 5% solution?

Step 1: To solve this problem, write and solve a system of linear equations.

Let x = number of mL of 3% solution.
Let y = number of mL of 8% solution.

Step 2: Write one equation based on the amounts of liquid being mixed. Write another equation on the amount of saline in the final solution.

$x + y = 150$ The amount of the mixture is 150 mL.
$0.03x + 0.08y = 0.05(150)$ The amount of saline in the mixture is 5%.

Step 3: To solve this system of equations by elimination, decide to eliminate the x variable.

$-3(x + y) = -3(150)$ Multiply the first equation by -3.
$100(0.03x + 0.08y) = 100(7.5)$ Multiply the second equation by 100 to remove decimals.

$-3x - 3y = -450$
$3x + 8y = 750$ Add the two equations to eliminate x.
$5y = 300$ Solve for y.
$y = 60$

Step 4: Find the value of the eliminated variable x by using one of the original equations.

$x + y = 150$
$x + 60 = 150$ Substitute 60 for y.
$x = 90$ Subtract 60 from both sides.

Step 5: Check your answers by substituting into the original second equation.

$0.03x + 0.08y = 0.05(150)$
$0.03(90) + 0.08(60) \ ? \ 0.05(150)$ Substitute 90 for x and 60 for y.
$2.7 + 4.8 \ ? \ 7.5$
$7.5 = 7.5$ check

Solution: Noah needs 90 mL of 3% solution mixed with 60 mL of 8% solution to make 150 mL of the 5% solution.

My Notes

Try These B

Solve each system using elimination.

a. $7x + 5y = -1$
$4x - y = -16$

b. Make sense of problems. Mary had \$25,000 to invest. She invested part of that amount at 3% annual interest and part at 5% annual interest for one year. The amount of interest she earned for both investments was \$1100. How much was invested at each rate?

MATH TIP

Recall the formula for simple interest,

$i = prt,$

where
i = interest earned
p = principal amount
r = annual interest rate
t = time in years

2. Sylvia wants to mix 100 pounds of Breakfast Blend coffee that will sell for \$25 per pound. She is using two types of coffee to create the mixture. Kona coffee sells for \$51 per pound and Columbian coffee sells for \$11 per pound. How many pounds of each type of coffee should Sylvia use?

Check Your Understanding

3. Fay wants to solve the system $2x - 3y = 5$ and $3x + 2y = -5$ using elimination. She multiplies the second equation by 1.5 to get $4.5x + 3y = -7.5$.

a. Do you think Fay's approach is correct? Explain why or why not.

b. Describe how Fay could have multiplied to avoid decimal coefficients.

LESSON 17-3 PRACTICE

For each situation, write and solve a system of equations.

4. Attend to precision. A pharmacist has a 10% alcohol solution and a 25% alcohol solution. How many milliliters of each solution will she need to mix together in order to have 200 mL of a 20% alcohol solution?

5. Alyssa invested a total of \$1500 in two accounts. One account paid 2% annual interest, and the other account paid 4% annual interest. After one year, Alyssa earned a total of \$44 interest. How much did she invest in each account?

6. Kendall is Jamal's older brother. The sum of their ages is 39. The difference of their ages is 9. How old are Kendall and Jamal?

7. Yolanda wants to solve the system shown below.

$$3x - 4y = 5$$
$$2x + 3y = -2$$

She decides to use the elimination method to eliminate the x variable. Describe how she can do this.

My Notes

Learning Targets:

- Explain when a system of linear equations has no solution.
- Explain when a system of linear equations has infinitely many solutions.

SUGGESTED LEARNING STRATEGIES: Create Representations, Marking the Text, Close Reading, Think-Pair-Share

MATH TERMS

Objects that are coincident lie in the same place. **Coincident** lines occupy the same location in the plane and pass through the same set of ordered pairs.

The graph of a system of linear inequalities does not always result in a unique intersection point. ***Parallel*** lines have graphs that do not intersect. ***Coincident*** lines have graphs that intersect infinitely many times.

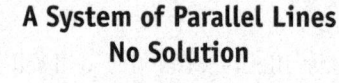

A System of Parallel Lines
No Solution

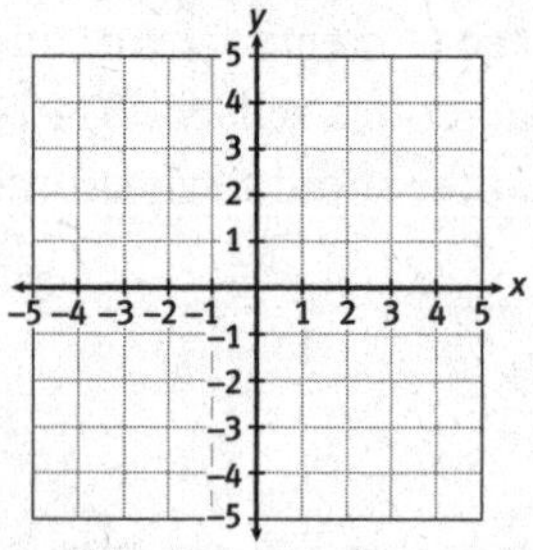

$y = -2x + 2$
$y = -2x - 2$

A System of Coincident Lines
Infinitely Many Solutions

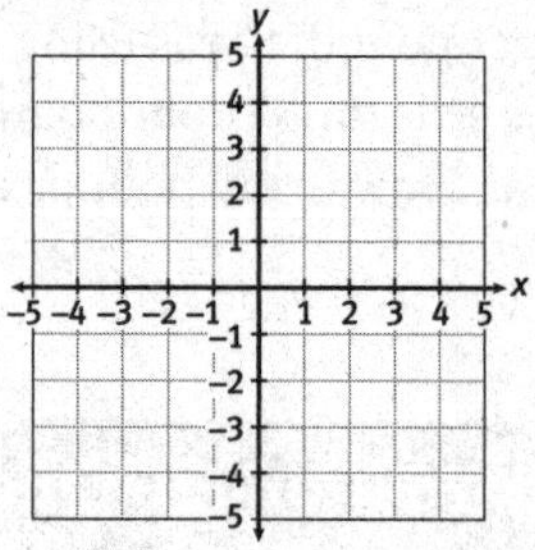

$y = -2x + 2$
$6x = 6 - 3y$

On a particular trip, Roy leaves from Pecos driving 55 mi/h. After 20 minutes, Travis realizes Roy forgot two cases of peaches. In 4 minutes, Travis loads the cases into his truck and heads out after Roy, traveling 55 mi/h.

This system of equations represents each brother's distance d from Dallas h hours after Roy leaves Pecos.

Roy: $d = 450 - 55h$ Travis: $d = 450 - 55(h - 0.4)$

1. **Model with mathematics.** Graph the system of equations. Describe the graph.

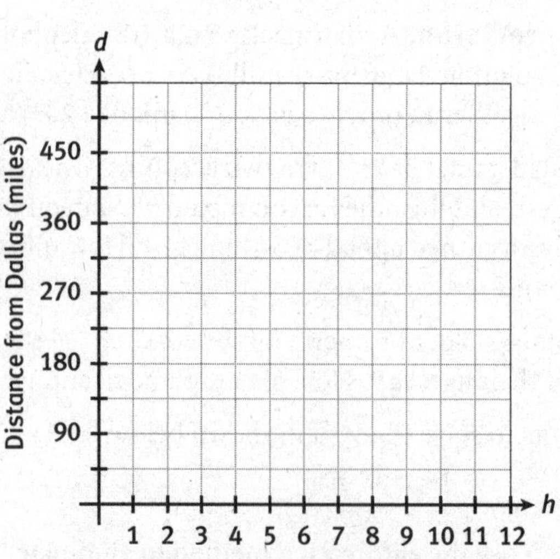

My Notes

2. What information about Travis and Roy does the graph provide?

3. How many solutions exist to the system of equations? How is this shown in the graph?

The system below represents a return trip from Dallas to Pecos, where d represents the distance from Pecos h hours after Roy leaves Dallas.

Travis: $d = 450 - 60h$
Roy: $d = -10(6h - 45)$

4. Model with mathematics. Graph the system of equations. Describe the graph.

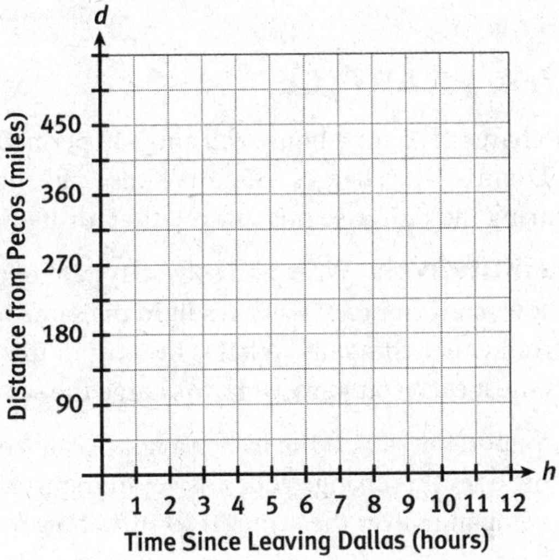

5. What information about Travis and Roy does the graph provide?

6. How many solutions exist to the system of equations? How is this shown in the graph?

My Notes

Check Your Understanding

7. Solve each system.

a. $y = 50 + 3x$
$y = 100 - 2x$

b. $x + 3y = 9$
$-3x + 2y = 8$

8. Tom leaves for Los Angeles averaging 65 mi/h. Michelle leaves for Los Angeles one hour later than Tom from the same location. She travels the same route averaging 70 mi/h. When will she pass Tom?

9. Juan bought a house for $200,000 and each year its value increases by $10,000. Tia bought a house for $350,000 and its value is decreasing annually by $5000. When will the two homes be worth the same amount of money?

10. Solve each system by graphing. If the lines are parallel, write *no solution.* If the lines are coincident, write *infinitely many solutions.*

a. $4x + 2y = 10$
$y = -2x + 5$

b. $y - 2x = 4$
$y = 5 + 2x$

LESSON 17-4 PRACTICE

Speedy Plumber charges $75 for a house call and $40 per hour for work done during the visit. Drains-R-Us charges $35 for a house call and $60 per hour for work done during the visit. Use this information for Items 11–13.

11. **Reason quantitatively.** Write and solve a system of equations to determine how many hours of work result in the same total cost for a house call from either company. What is the cost in this case? How many hours must each company work to charge the same amount?

12. Eliza has a coupon for $40 off the fee for a house call from Speedy Plumber. How does this change your answer to Item 11? Is the total cost for the two companies ever the same? If so, after how many hours?

13. Tyrone has a coupon that lowers the hourly rate for Drains-R-Us to $40 per hour. How does this change your answer to Item 11? Is the total cost for the two companies ever the same? If so, after how many hours?

My Notes

Learning Targets:

- Determine the number of solutions of a system of equations.
- Classify a system of linear equations as independent or dependent and as consistent or inconsistent.

SUGGESTED LEARNING STRATEGIES: Summarizing, Paraphrasing, Sharing and Responding, Interactive Word Wall, Predict and Confirm

When a system of two linear equations in two variables is solved, three possible relationships can occur.

- Two distinct lines that intersect produce one ordered pair as the solution.
- Two distinct parallel lines that do not intersect produce no solutions.
- Two lines that are coincident produce the same solution set—an infinite set of ordered pairs that satisfy both equations.

The three systems in the chart represent each of the possible relationships described above.

Relationship of Lines	Sketch	Number of Solutions
two intersecting lines		one solution
two parallel lines	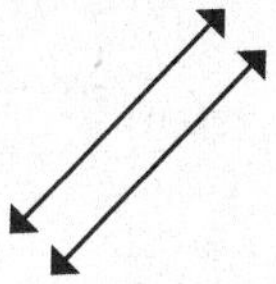	no solution
two coincident lines	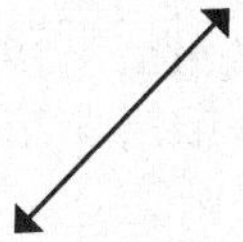	infinitely many solutions

1. Use the system $\begin{array}{l} 4x-2y=21 \\ y-2x=10 \end{array}$ to answer parts a–d below.

a. **Make use of structure.** Write each equation in the system in *slope-intercept form*. Compare the slopes and y-intercepts.

b. Make a conjecture about the graph and the solution of this system.

MATH TERMS

The **slope-intercept form** of a linear equation is $y = mx + b$; m is the slope of the line and $(0, b)$ is the y-intercept.

My Notes

c. Check the conjecture from part b by graphing the system on a graphing calculator. Revise the conjecture if necessary.

d. Solve the system using either the substitution method or the elimination method. Describe the result.

2. Use the system $\begin{array}{l} 2x = -y + 1 \\ 6x + 3y = 3 \end{array}$ to answer parts a–d below.

a. Write each equation in the system in slope-intercept form. Compare the slopes and y-intercepts.

b. Make a conjecture about the graph and the solution of this system.

c. Check the conjecture from part b by graphing the system on a graphing calculator. Revise the conjecture if necessary.

d. Solve the system using either the substitution method or the elimination method. Describe the result.

3. Write each equation in the system $\begin{array}{l} 2x - y = 6 \\ x = 6 - y \end{array}$ in slope-intercept form.

4. Construct viable arguments. Without graphing or solving, describe the solution of the system in Item 3. Justify your response.

5. Verify your answer to Item 4 by solving the system using any method. Revise your answer to Item 4 if necessary.

Systems of linear equations are classified by the relationships of their lines.

- Systems that produce two distinct lines when graphed are ***independent***. Systems that produce coincident lines are ***dependent***.
- Systems that have no solution are ***inconsistent***. Systems that have at least one solution are ***consistent***.

6. Classify the systems in Items 1, 2, and 3.

My Notes

7. For each system below, complete the table with the information requested.

The Nature of Solutions to a System of Two Linear Equations		
Equations in Standard Form:		
$2x+y=2$ $6x+3y=6$	$2x+y=2$ $x+y=3$	$2x+y=2$ $4x+2y=-4$
Graph Each System:		
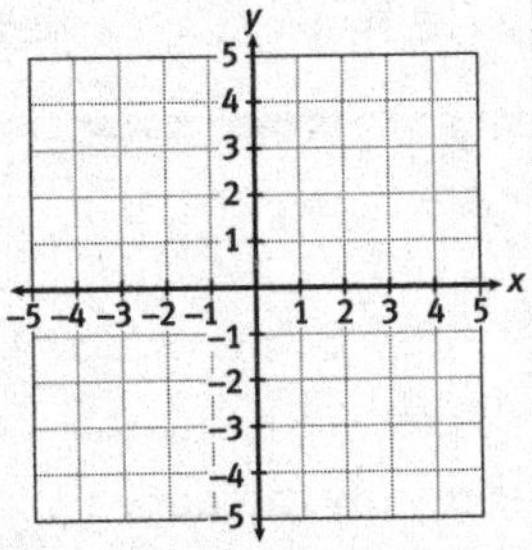	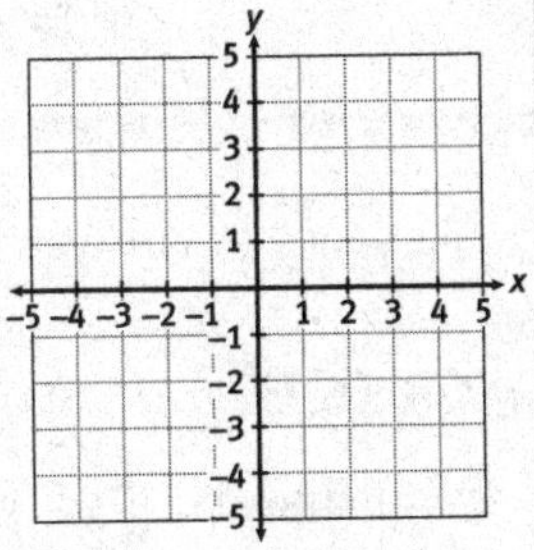	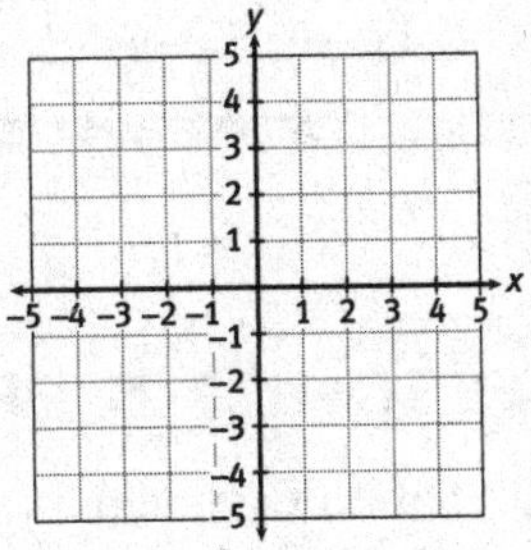
Write the Number of Solutions:		
Write the Relationship of the Lines:		
Solve Algebraically:		
Write the Equations in Slope-Intercept Form:		
Compare the Slopes and *y*-intercepts:		
Classify the System:		

My Notes

Check Your Understanding

For each system below:

a. Tell how many solutions the system has.
b. Describe the graph.
c. Classify the system.

8. $2x - 2y = 6$
$y - x = -3$

9. $y = 1.5x + 5$
$3x - 2y = 10$

10. $y = \frac{2}{3}x + 1$
$4x - 6y = -6$

11. $3x + 4y = 1$
$2x - 5y = 16$

LESSON 17-5 PRACTICE

12. Solve the system $3x + 4y = 8$ and $y = \frac{3}{4}x - 2$ using any method. Classify the system.

13. Approximate the point of intersection for the system of linear equations graphed below. Verify algebraically using substitution or elimination that the selected point is a solution for the system.

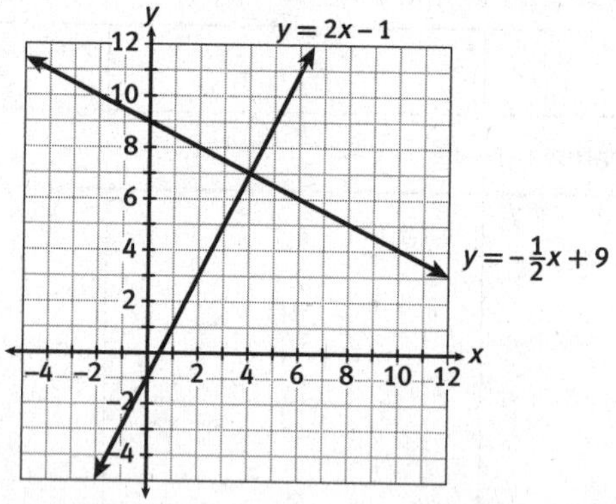

14. Monica claims that the system $3x + y = 16$ and $2x + 2y = 12$ has the same solutions as the system $6x + 2y = 32$ and $2x + 2y = 12$. Explain how you can tell whether Monica is correct without solving the systems.

15. Find the solution of the system $y = -\frac{2}{5}x + 1$ and $2x + 5y = 3$ by any method. Classify the system.

16. **Critique the reasoning of others.** Kristen graphed the system $y = 3x + 5$ and $10y = 30x + 51$ on her graphing calculator. She saw a single line on the screen and concluded that the system was dependent and consistent. Do you agree or disagree? Explain.

ACTIVITY 17 PRACTICE

Write your answers on notebook paper. Show your work.

Lesson 17-1

1. Solve each system of linear equations.

a. $y = 3x - 4$
$y = \frac{2}{5}x + 9$

b. $x + y = 7$
$x - 3y = -1$

2. Which ordered pair is a solution to the system shown at right? $y = \frac{2}{3}x + 3$
$y = -3x + 14$

A. $(-3, 1)$ **B.** $(5, -1)$
C. $(3, 5)$ **D.** $(5, 3)$

3. Which system's solution is represented by the graphs shown below?

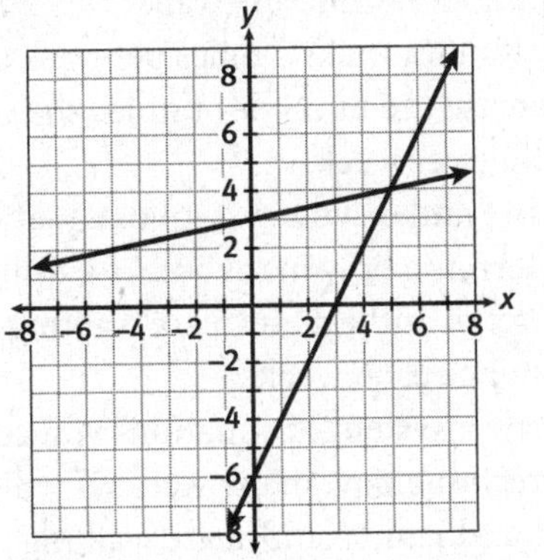

A. $y = 2x + 3$
$y = \frac{1}{5}x - 6$

B. $y = 2x - 6$
$y = \frac{1}{5}x + 3$

C. $y = 5x - 6$
$y = \frac{1}{2}x + 3$

D. $y = 5x + 3$
$y = \frac{1}{2}x - 6$

Lesson 17-2

4. A bushel of apples currently costs \$10 and th price is increasing by \$0.50 per week. A bushel o pears currently costs \$15 and the price is decreasing by \$0.25 per week. Which system of linear equations could be used to determine when the two fruits will cost the same amount per bushel?

A. $y = 0.5x + 10$
$y = -0.25x + 15$

B. $y = 0.25x + 15$
$y = -0.5x + 10$

C. $y = 10x + 0.5$
$y = 15x - 0.25$

D. $y = 5x + 3$
$y = -\frac{1}{2}x - 6$

5. Ray starts walking to school at a rate of 2 mi/h. Ten minutes later, his sister runs after him with his lunch, averaging 6 mi/h.

a. Write a system of linear equations to represent this situation.

b. Solve the system to determine how much time it took Ray's sister to catch up to him.

6. Colleen is in charge of ordering office supplies for her company. Last month she ordered ink cartridges and toner cartridges for the office printers. The cost of a toner cartridge is \$19.50 more than the cost of an ink cartridge. Colleen ordered 11 ink cartridges and 4 toner cartridges, and the total cost was \$460.50. Write and solve a system of equations to find the cost of each cartridge.

Rocking Horse Ranch, admission is $45 and rail rides are $22.50 per hour. At Saddlecreek Ranch, admission is $30 and trail rides are $35 per hour.

a. Write a system of equations that shows the total cost c for a trail ride that lasts h hours at each ranch.

b. Solve your system from part a. Interpret the meaning of the solution in the context of the problem.

c. Janelle has $125, and she wants to go on a three-hour trail ride. Which ranch should Janelle choose? Justify your answer.

Lesson 17-3

8. Lawrence has 10 coins in his pocket. One coin is a quarter, and the others are all nickels or dimes. The coins are worth 90 cents.

a. Write a system that represents this situation.

b. Solve the system to determine the number of dimes and nickels in Lawrence's pocket.

9. Pedro placed an order with an online nursery for 6 apple trees and 5 azaleas and the order came to $147. The next order for 3 apple trees and 4 azaleas came to $96. What was the unit cost for each apple tree and for each azalea?

10. Jeremiah scored 28 points in yesterday's basketball game. He made a total of 17 baskets. Some of the baskets were field goals (worth two points) and the rest were free throws (worth one point). Write and solve a system of equations to find the number of field goals and the number of free throws that Jeremiah made.

Lesson 17-4

11. Graph each system of linear equations and describe the solutions.

a. $x + y = 3$ and $2x + 2y = 6$

b. $2x + 3y = 6$ and $-x + y = -3$

c. $x - 4y = 1$ and $2x - 9 = 8y$

12. Ming graphs a system of linear equations on his calculator. When he looks at the result, he sees only one line. Assuming he graphed the equations correctly, what could this mean?

A. The system has infinitely many solutions.

B. The system has no solution.

C. The system has exactly one solution.

D. Every possible ordered pair (x, y) is a solution.

Lesson 17-5

13. Which is the best way to classify the system $y = 1.5x - 2$ and $3y = 4.5x - 7$?

A. Independent and inconsistent

B. Independent and consistent

C. Dependent and inconsistent

D. Dependent and consistent

14. Solve the system $3x + 4y = 8$ and $y = -\frac{3}{4}x + 2$ and classify the system.

15. The equation $3x - y = -1$ is part of a system of linear equations that is dependent and consistent. Write an equation that could be the other equation in the system.

16. Three friends decide to increase their exercise programs at the same time. Carolyn walks 7 miles per week and decides to increase the number of miles she walks by 1.5 miles per week. Eduardo walks 3 miles per week and decides to increase the number of miles he walks by 3.5 miles per week. Kendra walks 5 miles per week and decides to increase the number of miles she walks by 3.5 miles per week.

a. Write and solve a system of equations to determine how many weeks it will be until Carolyn and Eduardo are walking the same distance each week.

b. Write a system of equations you could use to determine how many weeks it will be until Eduardo and Kendra are walking the same distance each week.

c. Classify the systems you wrote in parts a and b.

d. Describe what would happen if you tried to solve the system you wrote in part b using the graphing method

MATHEMATICAL PRACTICES
Look for and Make Use of Structure

17. Explain how you can determine whether a system of two linear equations has a unique solution by examining the equations.

Which Region Is It?

Lesson 18-1 Representing the Solution of a System of Inequalities

My Notes

Learning Targets:

- Determine whether an ordered pair is a solution of a system of linear inequalities.
- Graph the solutions of a system of linear inequalities.

SUGGESTED LEARNING STRATEGIES: Create Representations, Look for a Pattern, Discussion Groups, Quickwrite, Graphic Organizer

1. Graph each inequality on the number lines and grids.

Inequality	Graph all x	Graph all (x, y)
$x < 2$	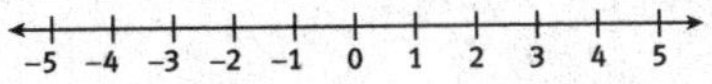	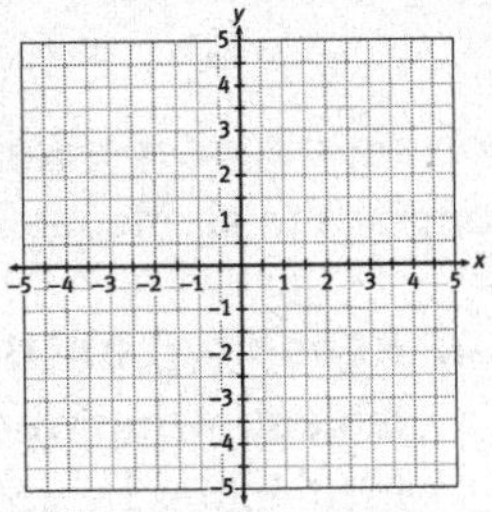
$x \geq -3$	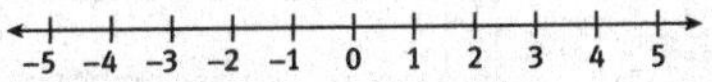	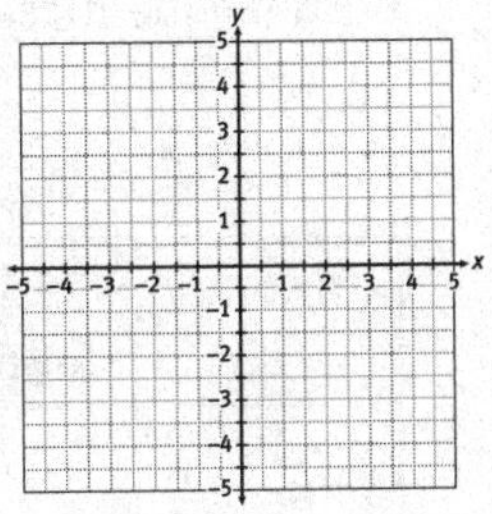
$x < 2$ and $x \geq -3$	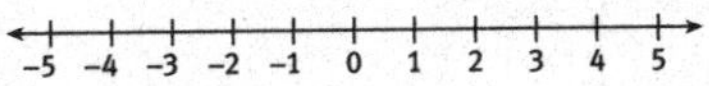	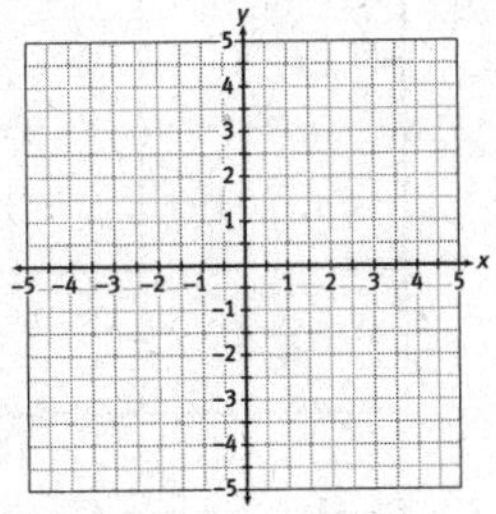

2. Compare and contrast the graphs you made in the third row of the table. In your explanation, compare the graphs to those in the first two rows and use the following words: dimension, half-line, half-plane, open, closed, and intersection.

My Notes

3. On the coordinate grid, graph the solutions common to the inequalities $y \leq 4$ and $y > 1$.

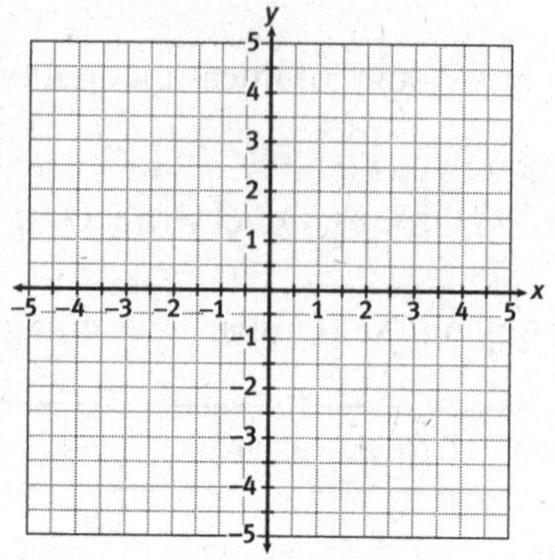

Solving a ***system of linear inequalities*** means finding all solutions that are common to all inequalities in the system.

MATH TERMS

A **system of linear inequalities** consists of two or more linear inequalities with the same variables.

4. Reason quantitatively. For the inequalities below, complete the table showing whether each ordered pair is a solution of the system of inequalities by deciding whether the ordered pair is a solution of **both** inequalities. Justify your responses.

$$x + y > 2$$
$$2x - y \geq -5$$

Ordered Pair	Is it a Solution?	Why or Why Not?
(−2, −3)		
(3, 2)		
(3, −1)		
(0, 5)		

5. Make use of structure. Find two more solutions of the system in Item 4.

My Notes

Since a system of inequalities has infinitely many solutions, you can represent all solutions using a graph. To solve a system of inequalities, graph each inequality on the same coordinate grid by shading a half-plane. The region that is the intersection of the two shaded half-planes, called the ***solution region***, represents all solutions of the system.

MATH TERMS

A **solution region** is the part of the coordinate plane in which the ordered pairs are solutions to all inequalities in a system.

Example A

Solve the system of inequalities.

$x + y \geq 1$
$x - 3y > 3$

Step 1: First, graph $x + y \geq 1$.

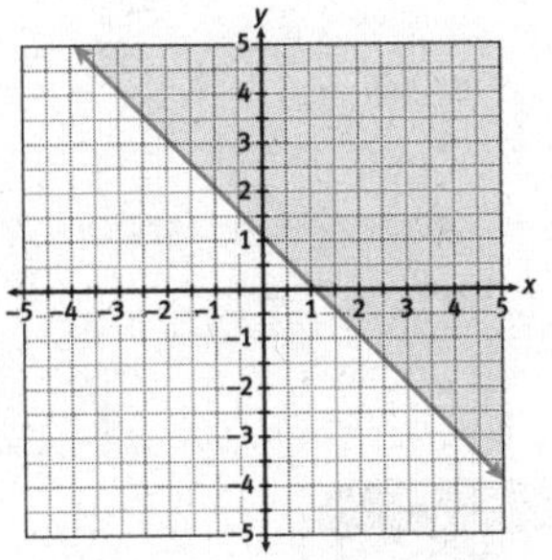

MATH TIP

To graph an inequality, it may be helpful to solve the inequality for y first.

Step 2: Next, graph $x - 3y > 3$.

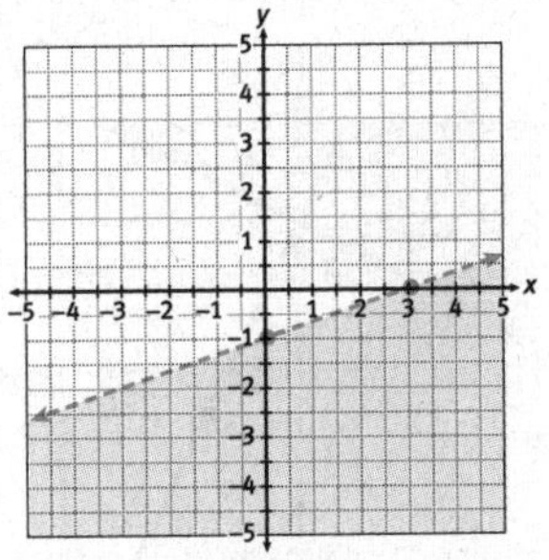

Step 3: Identify the solution region.

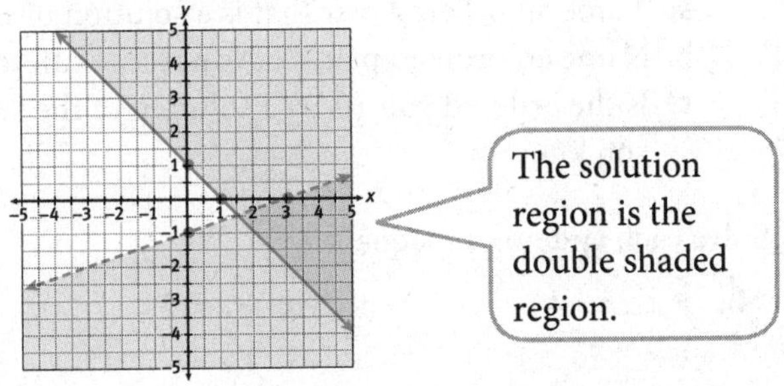
The solution region is the double shaded region.

MATH TIP

When finding the solution region for a system of inequalities, it may be helpful to shade each half-plane with a different pattern or color.

Solution: The solution region is the double shaded region shown in Step 3.

My Notes

a.

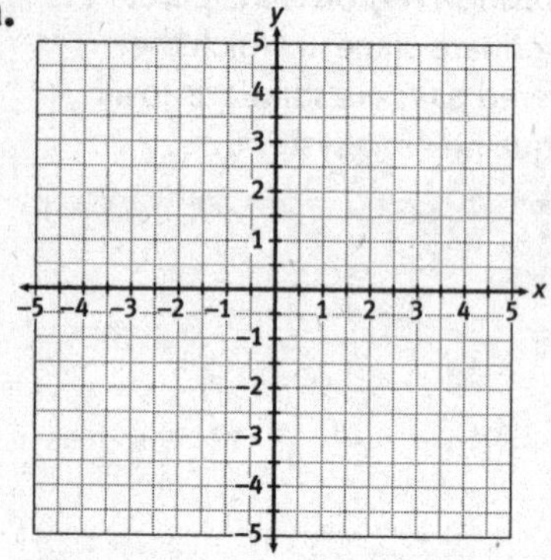

b.

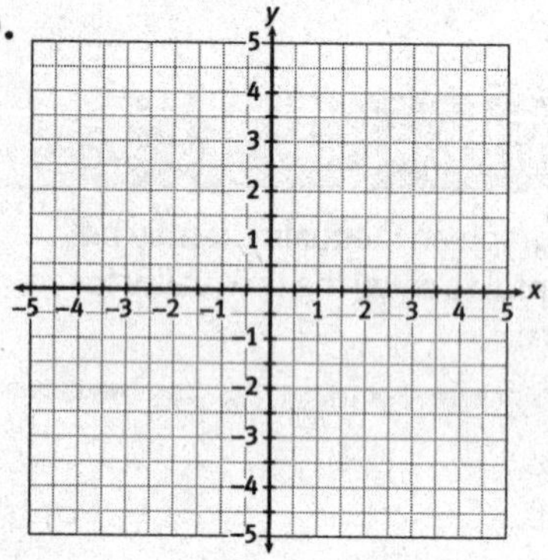

c.

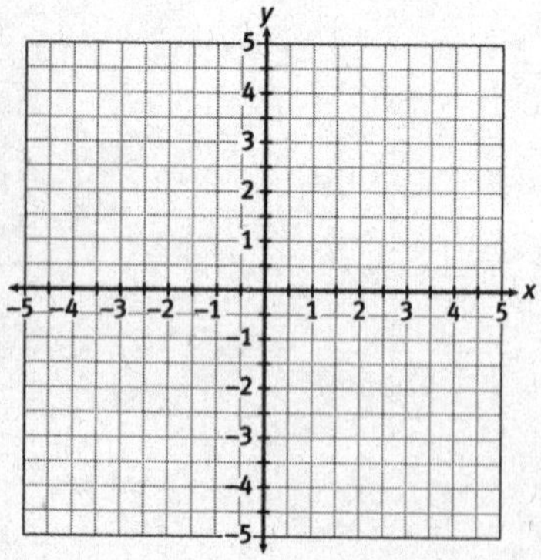

Try These A

Solve each system of inequalities using the coordinate grids in the *My Notes* section.

a. $x + y > 2$
$2x - y \geq -5$

b. $y \geq x - 1$
$y \leq -\frac{1}{2}x + 2$

c. $2x + 3y > 6$
$x - 2y < 4$

6. The system in Try These Part (a) is the same as the system in Item 4. Plot the points listed in Items 4 and 5 on the graph you made for Try These A part a. Where do the points that are solutions lie? Where do points that are not solutions lie? Give three additional points that are solutions.

Check Your Understanding

7. The graph below shows the solution of a system of inequalities.

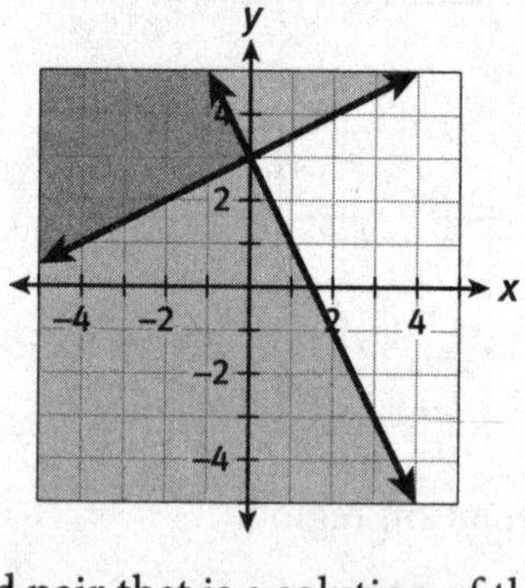

a. Name an ordered pair that is a solution of the system.
b. Name an ordered pair that is **not** a solution of the system.
c. Is the ordered pair (0, 3) a solution of the system? Explain how you know.

Solve each system of inequalities.

8. $y < 2x - 1$
$y \geq -x$

9. $3x + y < 3$
$x - y > 1$

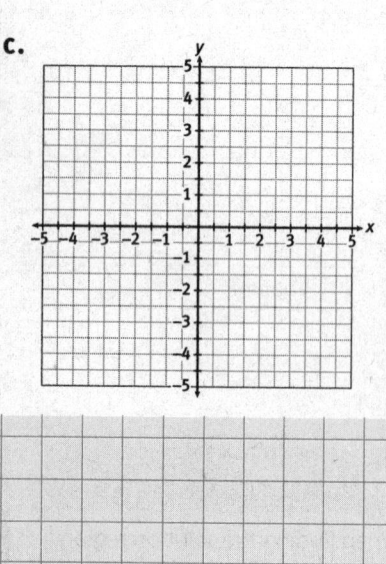

10. $y \geq \frac{2}{3}x + 4$
$x + 2y < 6$

My Notes

LESSON 18-1 PRACTICE

Solve each system of inequalities.

11. $y \geq x - 3$
$y \leq \frac{1}{2}x + 1$

12. $3x + 3y > 1$
$2y < 11$

13. $y < \frac{1}{3}x$
$x + y \geq -2$

14. Name three ordered pairs that are solutions of the system of inequalities shown below.

$$y < 4x + 4$$
$$x - y > 3$$

15. Write a system of inequalities whose solution is shown by the overlapping regions in the graph below.

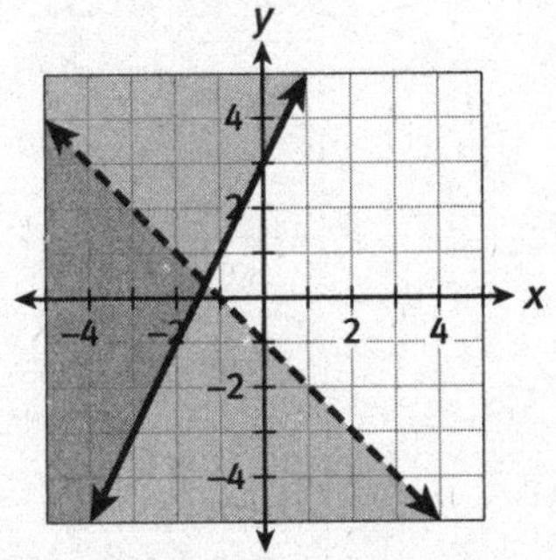

16. **Critique the reasoning of others.** A student was asked to solve the system $y \geq \frac{1}{2}x + 4$ and $x + 4y \leq -20$. She made the graph shown below. She noticed that the two shaded regions do not overlap, so she concluded that the system of inequalities has no solution. Do you agree or disagree? Justify your reasoning.

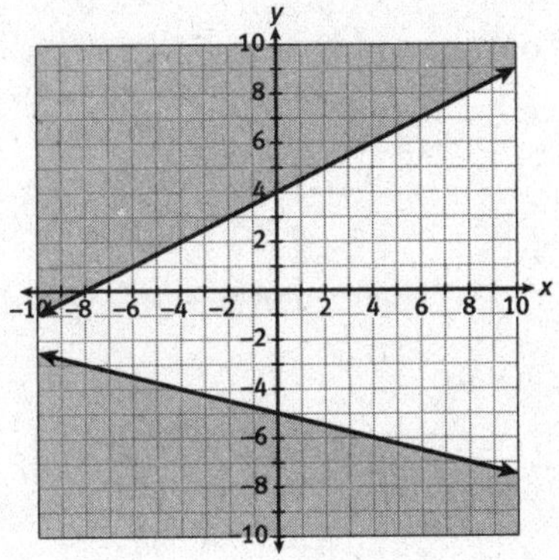

My Notes

Learning Targets:

- Identify solutions to systems of linear inequalities when the solution region is determined by parallel lines.
- Interpret solutions of systems of linear inequalities.

SUGGESTED LEARNING STRATEGIES: Create Representations, Look for a Pattern, Discussion Groups, Close Reading, Marking the Text

As you share your ideas, be sure to use mathematical terms and academic vocabulary precisely. Make notes to help you remember the meaning of new words and how they are used to describe mathematical concepts.

1. Determine the solutions to the systems of inequalities by graphing.

a. $y > \frac{1}{2}x + 2$

$x - 2y > 8$

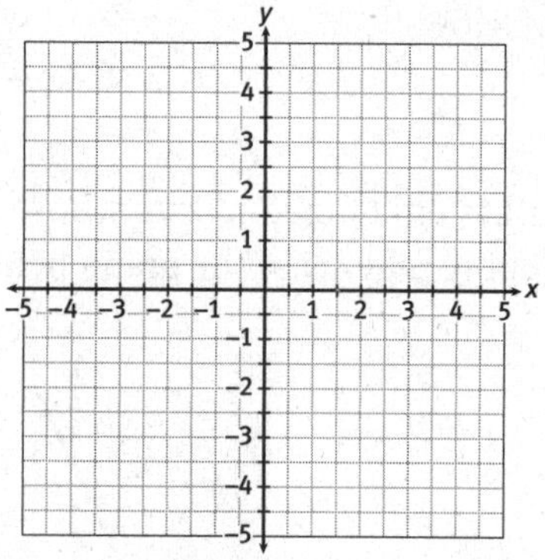

b. $3x + y < 3$

$y + 2 \geq -3x$

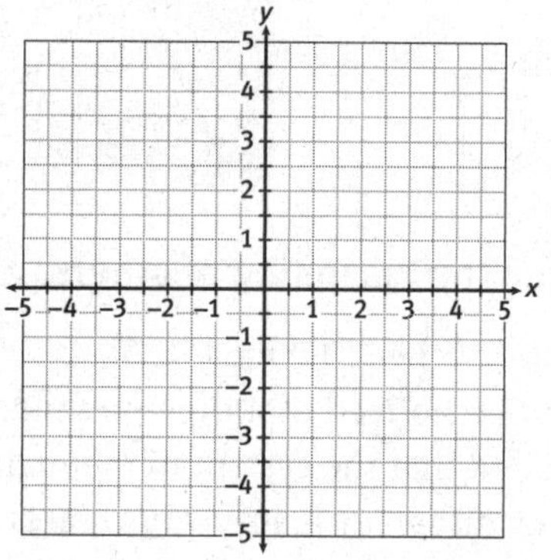

2. Compare and contrast the systems in Items 1a and 1b.

My Notes

3. Ray plays on the basketball team. Last week, he scored all of his points from free throws (worth one point each) and field goals (worth two points each). He has forgotten how many points he scored, but he remembers some facts from the game. Review with your group the background information that is given as you solve the items below.

a. Reason abstractly. Let f represent the number of free throws and g represent the number of field goals. Write an inequality for each of the facts below.

Ray scored fewer than 20 points.

Ray made fewer than six free throws.

At most, Ray made twice as many free throws as he made field goals.

b. Graph the solutions to the system represented by your inequalities from part a.

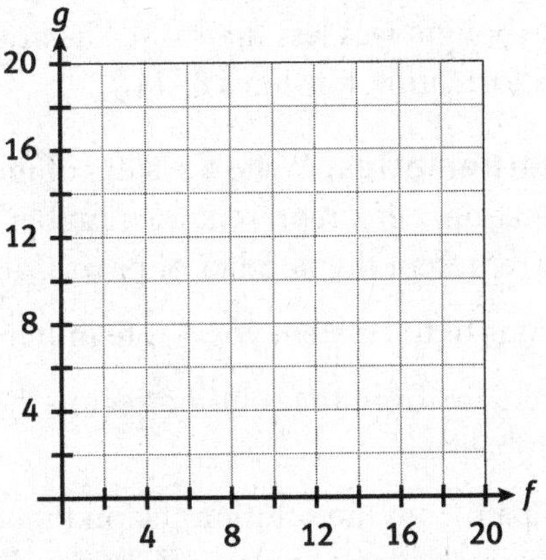

c. Reason quantitatively. In the solution region you graphed in part b not every point makes sense in this context. Give two solutions that make sense in the context of the problem and one that does not. Explain your reasoning.

My Notes

Check Your Understanding

4. Use guess and check to identify an ordered pair that is a solution to the system of inequalities.
$$3x + y \geq 6$$
$$x + 3y < 3$$
5. Solve the system in Item 4 by graphing. Confirm that the ordered pair you wrote in Item 4 is a solution. Explain.
6. Solve the system of linear inequalities.
$$y > 2x - 3$$
$$y < 4x - 3$$
7. Write a system of inequalities whose solution region is all the points in the third quadrant.

LESSON 18-2 PRACTICE

Catherine bought apples and pears for a school picnic. The total number of pieces of fruit that she bought was less than 100. She bought more apples than pears. Use this information for Items 8–11.

8. **Model with mathematics.** Write a system of inequalities that describes the situation. Let x represent the number of apples Catherine bought and let y represent the number of pears Catherine bought.
9. Graph the solutions of the system you wrote in Item 8.
10. Give two ordered pairs from the solution region that make sense in the context of the problem.
11. Give an ordered pair from the solution region that does not make sense in the context of the problem. Explain why the ordered pair does not make sense.
12. Describe the solution region of the system of the inequalities $-x + 2y \leq 6$ and $2y \geq x + 6$.

ACTIVITY 18 PRACTICE

Write your answers on notebook paper. Show your work.

Lesson 18-1

1. Determine which of the following ordered pairs are solutions to each of the given systems of inequalities.

   $\{(5, 3), (-2, 1), (1, 2), (2, -3), (3, 5), (-2, 3), (2, 0)\}$

   a. $y < -x + 3$
   $y \geq x - 2$

   b. $2x - y \leq 0$
   $y > -\frac{1}{2}x + 1$

   c. $3x - y > 6$
   $y \leq 3$

2. Graph each of the systems of inequalities in Item 1. Graph the points that you chose as solutions from Item 1 on the same coordinate grid to verify that they are solutions.

3. Identify four ordered pairs that are solutions to the following system of equations. Explain how you chose your points.

   $4x + 3y \geq -12$
   $2x - y < 4$

4. Solve the following systems of inequalities.

   a. $2x + 3y \geq 15$
   $5x - y \geq 3$

   b. $x - 4y \geq 4$
   $4y - x > 8$

5. Tickets for the school play cost $3 for students and $6 for adults. The drama club hopes to bring in at least $450 in sales, and the auditorium has 120 seats. Let a represent the number of adult tickets, and let s represent the number of student tickets.

   a. Write a system of inequalities representing this situation.

   b. Show the solutions to the system of inequalities by graphing.

   c. If the show sells out, what is the greatest number of student tickets that could be sold to get the desired amount of sales?

6. Which system of inequalities represents all of the points in the second quadrant?

   A. $x > 0, y < 0$ **B.** $x > 0, y > 0$
   C. $x < 0, y < 0$ **D.** $x < 0, y > 0$

7. Which graph represents the system of inequalities shown?

   $2x - y > 3$
   $x + y > 4$

A.

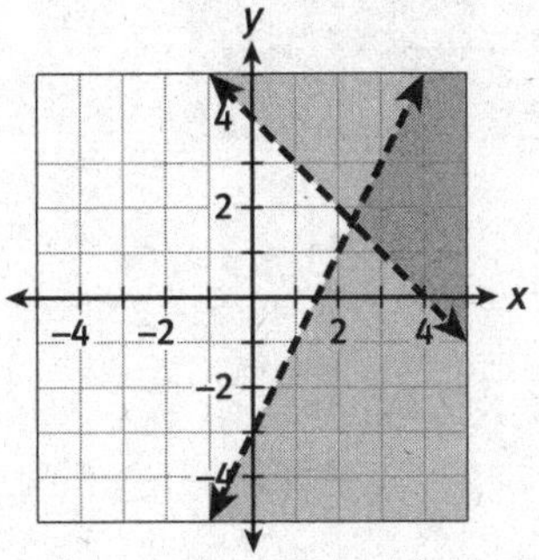

B.

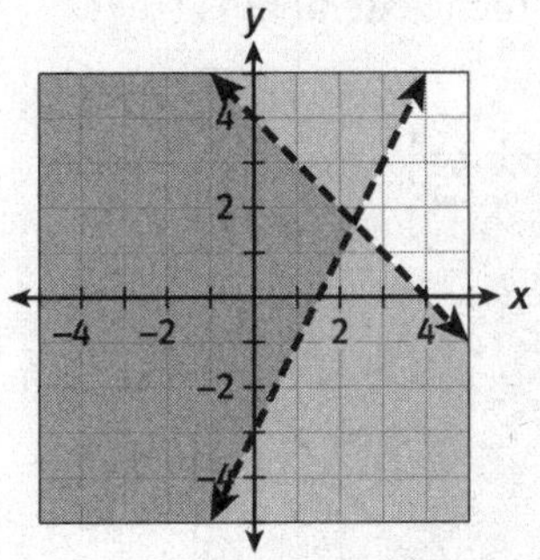

C.

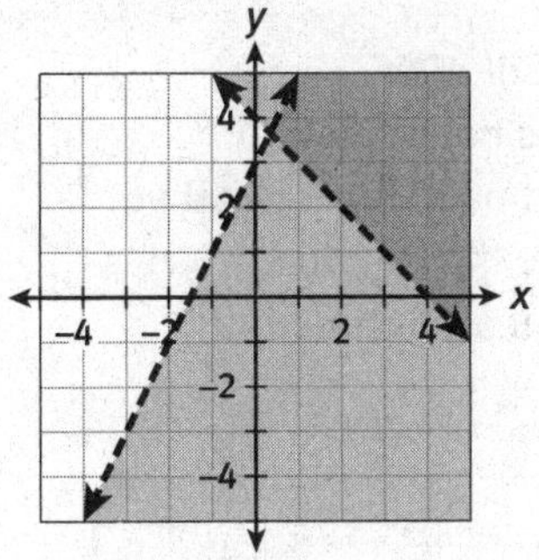

D.

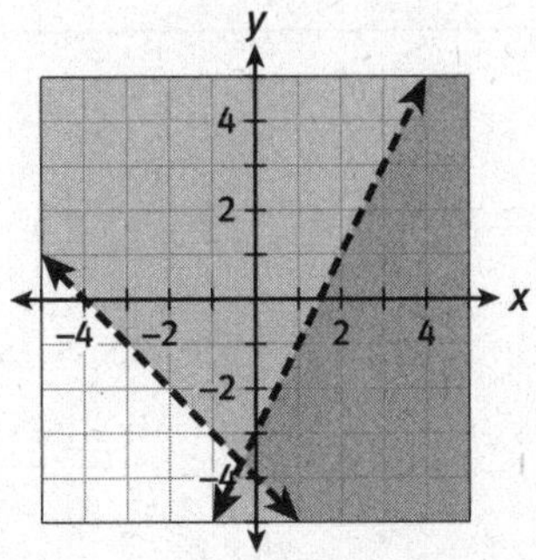

8. Which system of inequalities has the solution region shown in the graph below?

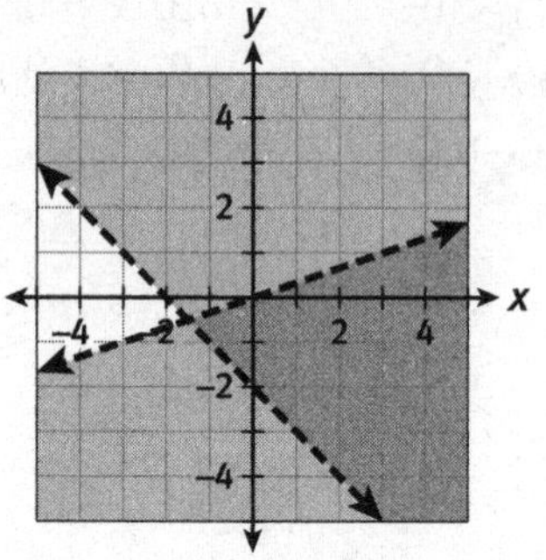

A. $3y < x$
$x + y > 2$

B. $3y < x$
$x + y > -2$

C. $3y > x$
$x + y > 2$

D. $3y > x$
$x + y > -2$

9. The graph shows the solution of a system of inequalities. Which statement is true?

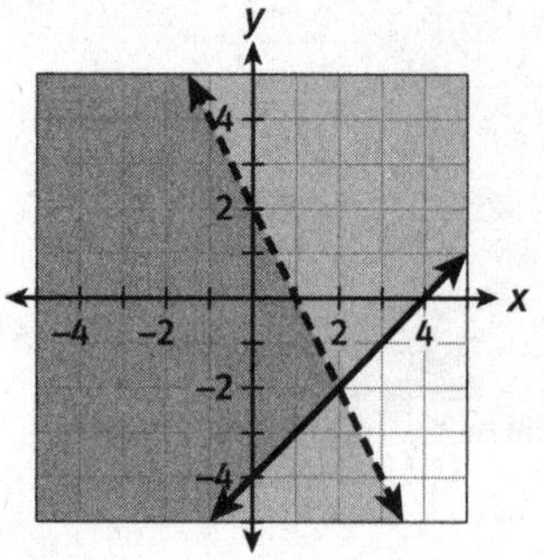

A. (2, −2) is a solution.
B. The origin is not a solution.
C. Any ordered pair with a negative x-coordinate is a solution.
D. (2, −4) is not a solution.

Lesson 18-2

10. Consider the system of inequalities shown below.

$$x + 2y \leq 6$$
$$x + 2y \geq -2$$

a. Graph the solution of the system of inequalities.
b. Describe the solution region.
c. Name three ordered pairs that are solutions of the system.
d. How would the solution be different if the inequality signs were reversed? That is, what is the solution of the system $x + 2y \geq 6$ and $x + 2y \leq -2$? Explain why your answer makes sense.

11. Connor bought used books from a Web site. The paperbacks cost \$2 each, and the hardcovers cost \$3 each. He spent no more than \$90 on the books, and he bought no more than 35 books.
a. Write a system of inequalities to represent the situation. State what the variables represent.
b. Graph the solution of the system of inequalities you wrote in part a.
c. Name two ordered pairs that are solutions. Interpret the meaning of each ordered pair in the context of the problem.
d. Suppose you know that Connor spent exactly \$90 and that he bought exactly 35 books. What can you conclude in this case? Why?

12. a. Write a system of linear inequalities that has no solutions.
b. Is it possible for a system of linear inequalities with no solutions to have nonparallel boundary lines? If so, explain why and give an example. If not, explain why not. (Hint: Consider coincident boundary lines.)

MATHEMATICAL PRACTICES
Attend to Precision

13. Explain why graphing is the preferred method of representing the solutions of a system of linear inequalities.

1. Rajesh and his brother Mohib are each mailing a birthday gift to a friend. Rajesh's package weighs three more pounds than twice the weight of Mohib's package. The combined weight of both packages is 15 pounds.
   a. Write a system to represent this situation. Define each variable that you use.
   b. Solve the system using substitution or elimination to determine the weight of each package. Justify the reasonableness of your solution.
   c. Rajesh and Mohib each graph the system that represents this situation. Who is correct? Explain why.

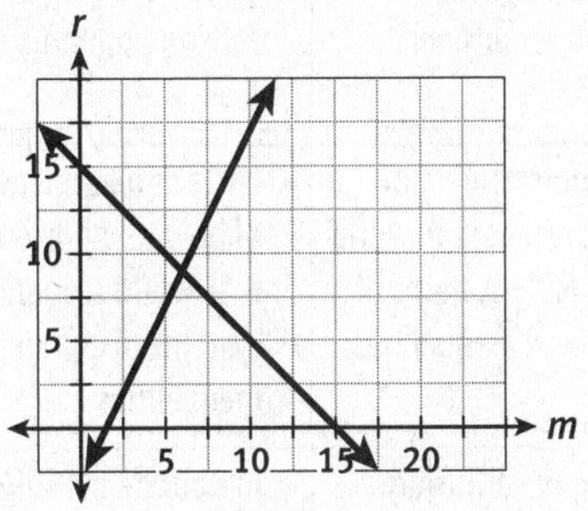

Rajesh's Graph

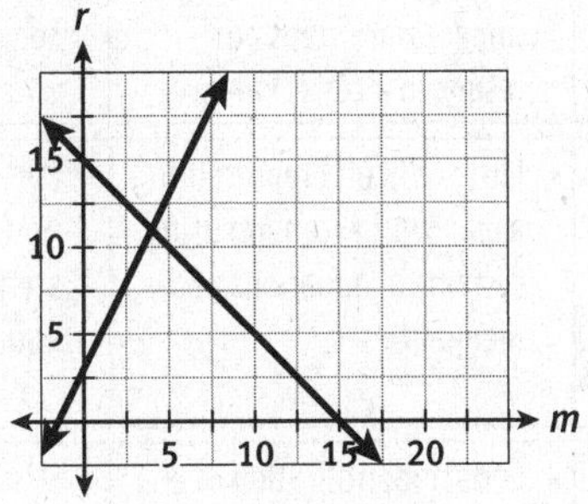

Mohib's Graph

2. Rajesh and Mohib will mail rectangular packages that meet the weight and height requirements of their delivery service. Their packages are described in the table.

	Height	Length	Width
Rajesh's package	7 in.	18 in.	15 in.
Mohib's package	8 in.	19 in.	13 in.

   a. The delivery service also requires that both the length and width of the packages be 20 inches or less and that the length plus the width be no more than 32 inches. Write a system of inequalities to represent this situation.
   b. Graph the system from part a to show all of the possible dimensions of length and width that the packages could have.
   c. Is either package unacceptable? Justify your answer using the table or the graph.

Systems of Equations and Inequalities

TILT THE SCALES

Scoring Guide	Exemplary	Proficient	Emerging	Incomplete
	The solution demonstrates the following characteristics:			
Mathematics Knowledge and Thinking (Items 1b, 2b)	• Effective understanding of and accuracy in solving systems of equations and inequalities	• Adequate understanding of and accuracy in solving systems of equations and inequalities	• Partial understanding of and some difficulty solving systems of equations and inequalities	• Incomplete understanding of and significant difficulty solving systems of equations and inequalities
Problem Solving (Items 1b, 2c)	• Appropriate and efficient strategy that results in a correct answer	• Strategy that may include unnecessary steps but results in a correct answer	• Strategy that results in some incorrect answers	• No clear strategy when solving problems
Mathematical Modeling / Representations (Items 1a, 2a)	• Fluency in representing real-world scenarios using systems of equations and inequalities	• Little difficulty representing a real-world scenario using systems of equations and inequalities	• Partial understanding of how to represent real-world scenarios using systems of equations and inequalities	• Little or no understanding of how to represent real-world scenarios using systems of equations and inequalities
Reasoning and Communication (Items 1c, 2c)	• Precise use of appropriate math terms and language to identify and explain an error • Ease and accuracy describing the relationship between a table or a graph and a real-world scenario	• Correct identification of an error with an adequate explanation • Little difficulty describing the relationship between a table or a graph and a real-world scenario	• Misleading or confusing explanation of an error • Partially correct description of the relationship between a table or a graph and a real-world scenario	• Inaccurate identification of an error with an incomplete or inaccurate explanation • Little or no understanding of how a table or a graph might relate to a real-world scenario

Exponents, Radicals, and Polynomials

4

Unit Overview

In this unit you will explore multiplicative patterns and representations of nonlinear data. Exponential growth and decay will be the basis for studying exponential functions. You will investigate the properties of powers and radical expressions. You will also perform operations with radical and rational expressions.

Key Terms

As you study this unit, add these and other terms to your math notebook. Include in your notes your prior knowledge of each word, as well as your experiences in using the word in different mathematical examples. If needed, ask for help in pronouncing new words and add information on pronunciation to your math notebook. It is important that you learn new terms and use them correctly in your class discussions and in your problem solutions.

Math Terms

- radical expression
- principal square root
- negative square root
- cube root
- rationalize
- tree diagram
- geometric sequence
- common ratio
- arithmetic sequence
- recursive formula
- exponential growth
- exponential function
- exponential decay
- compound interest
- exponential regression
- term
- polynomial
- coefficient
- constant term
- degree of a term
- degree of a polynomial
- standard form of a polynomial
- descending order
- leading coefficient
- monomial
- binomial
- trinomial
- like terms
- difference of two squares
- square of a binomial
- greatest common factor of a polynomial
- perfect square trinomial
- rational expression

ESSENTIAL QUESTIONS

How do multiplicative and exponential patterns model the physical world?

How are adding and multiplying polynomial expressions different from each other?

EMBEDDED ASSESSMENTS

This unit has four embedded assessments, following Activities 21, 23, 25, and 28. They will give you an opportunity to demonstrate what you have learned.

Getting Ready

Write your answers on notebook paper.
Show your work.

1. Find the greatest common factor of 36 and 54.
2. List all the factors of 90.
3. Which of the following is equivalent to $39 \cdot 26 + 39 \cdot 13$?

   A. 13^9 **B.** $13^4 \cdot 14$

   C. $13^2 \cdot 3^2 \cdot 2$ **D.** $13^2 \cdot 3^2$
4. Identify the coefficient, base, and exponent of $4x^5$.
5. Explain two ways to evaluate $15(90 - 3)$.
6. Complete the following table to create a linear relationship.

x	2	4	6	8	10
y	3	5			

7. Graph the function described in the table in Item 6.
8. Use ratios to model the following:

   a. 7.5

   b. Caleb receives 341 of the 436 votes cast for class president.

   Students in Mr. Bulluck's Class

Girls	Boys
12	19

   c. girls to boys

   d. boys to total class members
9. Tell whether each number is rational or irrational.

   a. $\sqrt{25}$ **b.** $\frac{4}{3}$

   c. 2.16 **d.** π
10. Calculate.

   a. $\frac{1}{2} + \frac{3}{8}$ **b.** $\frac{5}{12} - \frac{1}{3}$

   c. $\frac{3}{2} \cdot \frac{2}{5}$ **d.** $\frac{5}{8} \div \frac{3}{4}$

Exponent Rules

ACTIVITY 19

Icebergs and Exponents

Lesson 19-1 Basic Exponent Properties

My Notes

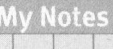

Learning Targets:

- Develop basic exponent properties.
- Simplify expressions involving exponents.

SUGGESTED LEARNING STRATEGIES: Create Representations, Predict and Confirm, Look for a Pattern, Think-Pair-Share, Discussion Groups, Sharing and Responding

An *iceberg* is a large piece of freshwater ice that has broken off from a glacier or ice shelf and is floating in open seawater. Icebergs are classified by size. The smallest sized iceberg is called a "growler."

A growler was found floating in the ocean just off the shore of Greenland. Its volume above water was approximately 27 cubic meters.

CONNECT TO GEOLOGY

Because ice is not as dense as seawater, about one-tenth of the volume of an iceberg is visible above water. It is difficult to tell what an iceberg looks like underwater simply by looking at the visible part. Growlers got their name because the sound they make when they are melting sounds like a growling animal.

1. **Reason quantitatively.** Two icebergs float near this growler. One iceberg's volume is 3^4 times greater than the growler. The second iceberg's volume is 2^8 times greater than the growler. Which iceberg has the larger volume? Explain.

GROUP DISCUSSION TIPS

Work with your peers to set rules for:
- discussions and decision-making
- clear goals and deadlines
- individual roles as needed

2. What is the meaning of 3^4 and 2^8? Why do you think ***exponents*** are used when writing numbers?

MATH TERMS

The expression 3^4 is a **power**. The **base** is 3 and the **exponent** is 4. The term **power** may also refer to the **exponent**.

3. Suppose the original growler's volume under the water is 9 times the volume above. How much of its ice is below the surface?

4. Write your solution to Item 3 using powers. Complete the equation below. Write the missing terms as a ***power*** of 3.

$$\text{volume above water} \cdot 3^2 = \text{volume below the surface}$$

$$\square \cdot 3^2 = \square$$

5. Look at the equation you completed for Item 4. What relationship do you notice between the exponents on the left side of the equation and the exponent on the right?

My Notes

6. Use the table below to help verify the pattern you noticed in Item 5. First write each product in the table in expanded form. Then express the product as a single power of the given base. The first one has been done for you.

Original Product	Expanded Form	Single Power
$2^2 \cdot 2^4$	$2 \cdot 2 \cdot 2 \cdot 2 \cdot 2 \cdot 2$	2^6
$5^3 \cdot 5^2$		
$x^4 \cdot x^7$		
$a^6 \cdot a^2$		

7. Express regularity in repeated reasoning. Based on the pattern you observed in the table in Item 6, write the missing exponent in the box below to complete the **Product of Powers Property** for exponents.

$$a^m \cdot a^n = a^{\square}$$

8. Use the Product of Powers Property to write $x^{\frac{3}{4}} \cdot x^{\frac{5}{4}}$ as a single power.

9. The density of an iceberg is determined by dividing its mass by its volume. Suppose a growler had a mass of 59,049 kg and a volume of 81 cubic meters. Compute the density of the iceberg.

CONNECT TO SCIENCE

The formula for density is

$$D = \frac{M}{V}$$

where *D* is density, *M* is mass, and *V* is volume.

10. Write your solution to Item 9 using powers of 9.

$$\frac{\boxed{\text{Mass}}}{\boxed{\text{Volume}}} = \boxed{\text{Density}}$$

11. What pattern do you notice in the equation you completed for Item 10?

My Notes

12. Use the table to help verify the patterns you noticed in Item 11. First write each quotient in the table below in expanded form. Then express the quotient as a single power of the given base. The first one has been done for you.

Original Quotient	Expanded Form	Single Power
$\frac{2^5}{2^2}$	$\frac{2\cdot2\cdot2\cdot2\cdot2}{2\cdot2}=\frac{\not{2}\cdot\not{2}\cdot2\cdot2\cdot2}{\not{2}\cdot\not{2}}$	2^3
$\frac{5^8}{5^6}$		
$\frac{a^3}{a^1}$		
$\frac{x^7}{x^3}$		

13. Based on the pattern you observed in Item 12, write the missing exponent in the box below to complete the **Quotient of Powers Property** for exponents.

$$\frac{a^m}{a^n}=a^{\square}, \text{ where } a \neq 0$$

14. Use the Quotient of Powers Property to write $\frac{a^{\frac{11}{3}}}{a^{\frac{2}{3}}}$ as a single power.

The product and quotient properties of exponents can be used to simplify expressions.

Example A

Simplify: $2x^5 \cdot 5x^4$

Step 1: Group powers with the same base.
$2x^5 \cdot 5x^4 = 2 \cdot 5 \cdot x^5 \cdot x^4$

Step 2: Product of Powers Property $= 10x^{5+4}$

Step 3: Simplify the exponent. $= 10x^9$

Solution: $2x^5 \cdot 5x^4 = 10x^9$

My Notes

Example B

Simplify: $\frac{2x^5y^4}{xy^2}$

Step 1: Group powers with the same base. $\frac{2x^5y^4}{xy^2} = 2 \cdot \frac{x^5}{x} \cdot \frac{y^4}{y^2}$

Step 2: Quotient of Powers Property $= 2x^{5-1} \cdot y^{4-2}$

Step 3: Simplify the exponents. $= 2x^4y^2$

Solution: $\frac{2x^5y^4}{xy^2} = 2x^4y^2$

Try These A–B

Simplify each expression.

a. $(4xy^4)(-2x^2y^5)$

b. $\frac{2a^2b^5c}{4ab^2c}$

c. $\frac{6y^3}{18x} \cdot 2xy$

Check Your Understanding

15. Simplify $3yz^2 \cdot 5y^2z$.

16. Simplify $\frac{21f^2g^{\frac{7}{4}}}{7fg^{\frac{3}{4}}}$.

17. A growler has a mass of 243 kg and a volume of 27 cubic meters. Compute the density of the iceberg by completing the following. Write your answer using powers of 3. $\frac{3^5}{3^3} =$

LESSON 19-1 PRACTICE

18. Which expression has the greater value? Explain your reasoning.

a. $2^3 \cdot 2^5$ **b.** $\frac{4^7}{4^3}$

19. The mass of an object is x^{15} grams. Its volume is x^9 cm^3. What is the object's density?

20. The density of an object is y^{10} grams/cm^3. Its volume is y^4 cm^3. What is the object's mass?

21. Simplify the expression $\frac{(3x)^{\frac{1}{3}} \cdot (3x)^{\frac{7}{3}}}{(3x)^{\frac{2}{3}}}$.

22. Make sense of problems. Tanika asks Toby to multiply the expression $8^7 \cdot 8^3 \cdot 8^2$. Toby says he doesn't know how to do it, because he believes the Product of Powers Property works with only two exponential terms, and this problem has three terms. Explain how Toby could use the Product of Powers Property with three exponential terms.

MATH TIP

Use a graphic organizer to record the properties of exponents you learn in this activity.

My Notes

Learning Targets:

- Understand what is meant by negative and zero powers.
- Simplify expressions involving exponents.

SUGGESTED LEARNING STRATEGIES: Look for a Pattern, Discussion Groups, Sharing and Responding, Think-Pair-Share, Close Reading, Note Taking

1. **Attend to precision.** Write each quotient in expanded form and simplify it. Then apply the Quotient of Powers Property. The first one has been done for you.

Original Quotient	Expanded Form	Single Power
$\frac{2^5}{2^8}$	$\frac{2\cdot2\cdot2\cdot2\cdot2}{2\cdot2\cdot2\cdot2\cdot2\cdot2\cdot2\cdot2}=\frac{\not2\cdot\not2\cdot\not2\cdot\not2\cdot\not2}{\not2\cdot\not2\cdot\not2\cdot\not2\cdot\not2\cdot2\cdot2\cdot2}=\frac{1}{2^3}$	$2^{5-8}=2^{-3}$
$\frac{5^3}{5^6}$		
$\frac{a^3}{a^8}$		
$\frac{x^4}{x^{10}}$		

2. Based on the pattern you observed in Item 1, write the missing exponent in the box below to complete the **Negative Power Property** for exponents.

$$\frac{1}{a^n}=a^{\square},\text{ where } a\neq 0$$

3. Write each quotient in expanded form and simplify it. Then apply the Quotient of Powers Property. The first one has been done for you.

Original Quotient	Expanded Form	Single Power
$\frac{2^4}{2^4}$	$\frac{2\cdot2\cdot2\cdot2}{2\cdot2\cdot2\cdot2}=\frac{\not2\cdot\not2\cdot\not2\cdot\not2}{\not2\cdot\not2\cdot\not2\cdot\not2}=1$	$2^{4-4}=2^0$
$\frac{5^6}{5^6}$		
$\frac{a^3}{a^3}$		

CONNECT TO AP

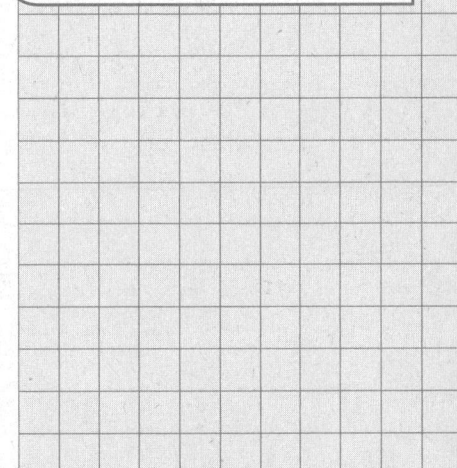

In calculus, an expression containing a negative exponent is often preferable to one written as a quotient. For example, $\frac{1}{x^3}$ is written x^{-3}.

My Notes

4. Based on the pattern you observed in Item 3, fill in the box below to complete the **Zero Power Property** of exponents.

$$a^0 = \square\text{, where } a \neq 0$$

5. Use the properties of exponents to evaluate the following expressions.

a. 2^{-3} **b.** $\dfrac{10^2}{10^{-2}}$ **c.** $3^{-2} \cdot 5^0$ **d.** $(-3.75)^0$

When evaluating and simplifying expressions, you can apply the properties of exponents and then write the answer without negative or zero powers.

Example A

Simplify $5x^{-2}yz^0 \cdot \dfrac{3x^4}{y^4}$ and write without negative powers.

Step 1: Commutative Property $5x^{-2}yz^0 \cdot \dfrac{3x^4}{y^4}$

$$= 5 \cdot 3 \cdot x^{-2} \cdot x^4 \cdot y^1 \cdot y^{-4} \cdot z^0$$

Step 2: Apply the exponent rules.

$$= 5 \cdot 3 \cdot x^{-2+4} \cdot y^{1-4} \cdot z^0$$

Step 3: Simplify the exponents.

$$= 15 \cdot x^2 \cdot y^{-3} \cdot 1$$

Step 4: Write without negative exponents.

$$= \frac{15x^2}{y^3}$$

Solution: $5x^{-2}yz^0 \cdot \dfrac{3x^4}{y^4} = \dfrac{15x^2}{y^3}$

Try These A

Simplify and write without negative powers.

a. $2a^2b^{-3} \cdot 5ab$ **b.** $\dfrac{10x^2y^{-4}}{5x^{-3}y^{-1}}$ **c.** $(-3xy^{-5})^0$

My Notes

Check Your Understanding

Simplify each expression. Write your answer without negative exponents.

6. $(z)^{-3}$ **7.** $12(xyz)^0$ **8.** $\frac{6^{-4}}{6^{-2}}$

9. $2^3 \cdot 2^{-6}$ **10.** $\frac{4x^{-2}}{x^3}$ **11.** $\frac{-5}{(ab)^0}$

LESSON 19-2 PRACTICE

12. For what value of v is $a^v = 1$, if $a \neq 0$?

13. For what value of w is $b^{-w} = \frac{1}{b^9}$, if $b \neq 0$?

14. For what value of y is $\frac{3^3}{3^y} = \frac{1}{9}$?

15. For what value of z is $5^8 \cdot 5^z = 1$?

16. Determine the values of n and m that would make the equation $7^n \cdot 7^m = 1$ a true statement. Assume that $n \neq m$.

17. For what value of x is $\frac{3^x \cdot 2^2}{3^4} = \frac{4}{3}$?

18. **Reason abstractly.** What is the value of $2^0 \cdot 3^0 \cdot 4^0 \cdot 5^0$? What is the value of any multiplication problem in which all of the factors are raised to a power of 0? Explain.

My Notes

Learning Targets:

- Develop the Power of a Power, Power of a Product, and the Power of a Quotient Properties.
- Simplify expressions involving exponents.

SUGGESTED LEARNING STRATEGIES: Note Taking, Look for a Pattern, Create Representations, Think-Pair-Share, Sharing and Responding, Close Reading

1. Write each expression in expanded form. Then write the expression using a single exponent with the given base. The first one has been done for you.

Original Expression	Expanded Form	Single Power
$(2^2)^4$	$2^2 \cdot 2^2 \cdot 2^2 \cdot 2^2 = 2 \cdot 2 \cdot 2 \cdot 2 \cdot 2 \cdot 2 \cdot 2 \cdot 2$	2^8
$(5^5)^3$		
$(x^3)^4$		

2. Based on the pattern you observed in Item 1, write the missing exponent in the box below to complete the **Power of a Power Property** for exponents.

$$(a^m)^n = a^{\square}$$

3. Use the Power of a Power Property to write $\left(x^{\frac{6}{5}}\right)^{25}$ as a single power.

4. Write each expression in expanded form and group like terms. Then write the expression as a product of powers. The first one has been done for you.

Original Expression	Expanded Form	Product of Powers
$(2x)^4$	$2x \cdot 2x \cdot 2x \cdot 2x =$ $2 \cdot 2 \cdot 2 \cdot 2 \cdot x \cdot x \cdot x \cdot x$	2^4x^4
$(-4a)^3$		
$(x^3y^2)^4$		

My Notes

5. Based on the pattern you observed in Item 4, write the missing exponents in the boxes below to complete the **Power of a Product Property** for exponents.

$$(ab)^m = a^{\square} \cdot b^{\square}$$

6. Use the Power of a Product Property to write $\left(c^{\frac{1}{2}}d^{\frac{1}{4}}\right)^8$ as a product of powers.

7. **Make use of structure.** Use the patterns you have seen. Predict and write the missing exponents in the boxes below to complete the **Power of a Quotient Property** for exponents.

$$\left(\frac{a}{b}\right)^m = \frac{a^{\square}}{b^{\square}}, \text{ where } b \neq 0$$

8. Use the Power of a Quotient Property to write $\left(\frac{x^3}{y^6}\right)^{\frac{1}{3}}$ as a quotient of powers.

You can apply these power properties and the exponent rules you have already learned to simplify expressions.

MATH TIP

Create an organized summary of the properties used to simplify and evaluate expressions with exponents.

Example A

Simplify $(2x^2y^5)^3\,(3x^2)^{-2}$ and write without negative powers.

Step 1: Power of a Power Property

$$(2x^2y^5)^3\,(3x^2)^{-2} = 2^3x^{2\cdot 3}\,y^{5\cdot 3} \cdot 3^{-2} \cdot x^{2\cdot -2}$$

Step 2: Simplify the exponents and the numerical terms.

$$= 8 \cdot x^6y^{15} \cdot \frac{1}{3^2} \cdot x^{-4}$$

Step 3: Commutative Property

$$= 8 \cdot \frac{1}{9}x^6 \cdot x^{-4}y^{15}$$

Step 4: Product of Powers Property

$$= \frac{8}{9}x^{6-4}y^{15}$$

Step 5: Simplify the exponents.

$$= \frac{8}{9}x^2y^{15}$$

Solution: $(2x^2y^5)^3(3x^2)^{-2} = \frac{8}{9}x^2y^{15}$

My Notes

Example B

Simplify $\left(\frac{x^2y^{-3}}{z}\right)^2$.

Step 1: Power of a Quotient Property $\left(\frac{x^2y^{-3}}{z}\right)^2 = \frac{x^{2\cdot 2}y^{-3\cdot 2}}{z^2}$

Step 2: Simplify the exponents. $= \frac{x^4y^{-6}}{z^2}$

Step 3: Negative Power Property $= \frac{x^4}{y^6z^2}$

Solution: $\left(\frac{x^2y^{-3}}{z}\right)^2 = \frac{x^4}{y^6z^2}$

Try These A–B

Simplify and write without negative powers.

a. $(2x^2y)^3(-3xy^3)^2$ **b.** $-2ab(5b^2c)^3$ **c.** $\left(\frac{4x}{y^3}\right)^{-2}$

d. $\left(\frac{5x}{y}\right)^2\left(\frac{y^3}{10x^2}\right)$ **e.** $(3xy^{-2})^2(2x^3yz)(6yz^2)^{-1}$

Check Your Understanding

Simplify each expression. Write your answer without negative exponents.

9. $(4x^3y^{-1})^2$

10. $\left(\frac{5x}{y^2}\right)^3$

11. $(-2a^2b^{-2}c)^3(3ab^4c^5)(xyz)^0$

12. $(4fg^3)^{-2}(-4fg^3h)^2(3gh^4)^{-1}$

13. $\left(\frac{2ab}{a^2b^{-2}}\right)^{-3}$

14. $\left[\left(-7nm^2\right)^{-3}\right]^0$

LESSON 19-3 PRACTICE

Simplify.

15. a. $\left(\frac{2}{3}\right)^2$ **b.** $\left(\frac{2}{3}\right)^{-2}$

16. a. $(3x)^3$ **b.** $(3x)^{-3}$

17. a. $(2^5)^4$ **b.** $(2^5)^{-4}$

18. Model with mathematics. The formula for the area of a square is $A = s^2$, where s is the side length. A square garden has a side length of x^4y. What is the area of the garden?

ACTIVITY 19 PRACTICE

Write your answers on notebook paper.
Show your work.

Lesson 19-1

For Items 1–5, evaluate the expression. Write your answer without negative powers.

1. $x^8 \cdot x^7$

2. $\dfrac{6a^{10}b^9}{3ab^3}$

3. $(6a^2b)(-3ab^3)$

4. $\dfrac{7x^2y^5}{14xy^4}$

5. $\dfrac{2xy^2}{x^5y^3} \cdot \dfrac{5xy^3}{-30y^{-2}}$

6. The volume of an iceberg that is below the water line is 2^5 cubic meters. The volume that is above the water line is 2^2 cubic meters. How many times greater is the volume below the water line than above it?
   A. $2^{2.5}$
   B. 2^3
   C. 2^7
   D. 2^{10}

7. A megabyte is equal to 2^{20} bytes, and a gigabyte is equal to 2^{30} bytes. How many times larger is a gigabyte than a megabyte?

8. A jackpot is worth 10^5 dollars. The contestant who wins the jackpot has the opportunity to put it all on the line with the single spin of a prize wheel. If the contestant spins the number 7 on the wheel, she will win 10^2 times more money. How many dollars will the contestant win if she risks her prize money and spins a 7?

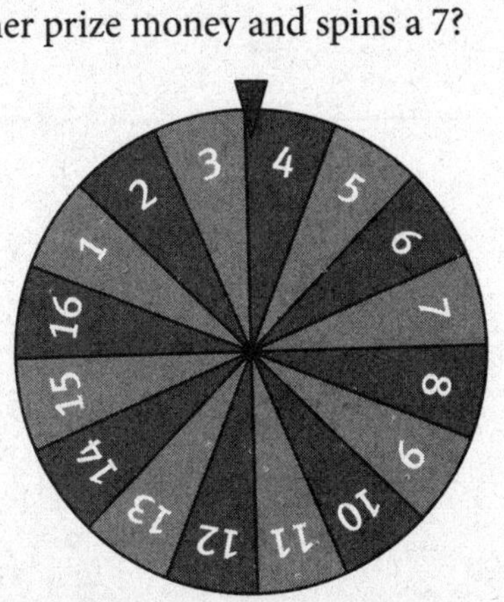

The number of earthquakes of a given magnitude that are likely to occur in any given year is represented by the formula $10^{(8-M)}$, where M is the magnitude. Use this formula for Items 9 and 10.

9. How many earthquakes of magnitude 8 are likely to occur next year?

10. If an earthquake of magnitude 10 occurred last year, how many years will it be before another one of that magnitude is likely to occur?

Lesson 19-2

11. Which of the following expressions is **not** equal to 1?
    A. $x^3 \cdot x^{-3}$ **B.** 1001^0
    C. $\dfrac{a^2b}{ba^2}$ **D.** $\dfrac{y^2}{y^{-2}}$

12. Which of the following expressions is equal to $\dfrac{y}{x^2}$?
    A. $x^{-2}y^3 \cdot y^{-2}$ **B.** $xy^2 \cdot x^{-3}y^{-2}$
    C. $\dfrac{y^2x}{yx^{-3}}$ **D.** $\dfrac{x^2y}{y^{-2}}$

Determine whether each statement is always, sometimes, or never true.

13. For $a \neq 0$, the value of a^{-1} is positive.

14. If n is an integer, then $3^n \cdot 3^{-n}$ equals 1.

15. If $6^p > 0$, then $p > 0$.

16. 4^{-x} equals $\dfrac{1}{4^x}$.

17. If m is an integer, then the value of 2^m is negative.

18. For what value of a is $w^{a-2} = 1$, if $w \neq 0$?

19. For what value of b is $p^{b-1} = \frac{1}{p^5}$, if $p \neq 0$?

For each of the following, give the value of the expression or state that the expression is undefined.

20. x^0 when $x = 0$

21. 2^{-a} when $a = 0$

22. $\frac{1}{x^p}$ when $x = 0$ and $p > 0$

23. $0^n \cdot 0^{-n}$ when n is an integer

Lesson 19-3

24. The area of a square is given by the formula $A = s^2$, where s is the length of the side. What is the area of the square shown?

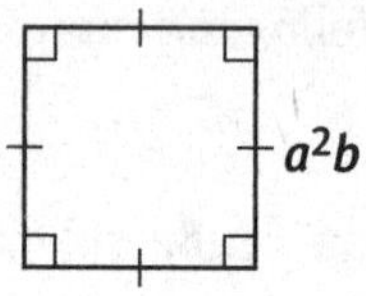

The volume of a cube is given by the formula $V = s^3$, where s is the length of the side. Use this formula for Items 25–27.

25. What is the volume of the cube shown?

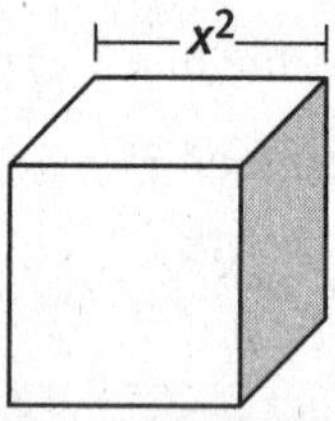

26. What is the volume of the cube shown?

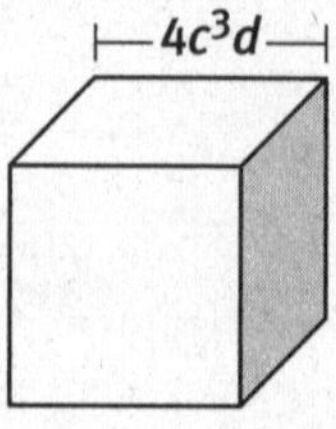

27. The volume of a cube is x^{27} cubic inches. What expression represents the length of one side of the cube? Justify your reasoning.

Simplify each expression. Write your answer without negative exponents.

28. $(-5x^2y^{-1})^4$

29. $\left(\frac{c^2d^{-2}}{c}\right)^5$

30. $(x^2y^2z^{-1})^3(xyz^4)(x^3y)$

31. $(m^2n^{-5})^0m^{-7}$

32. $\left(\frac{2x^{-2}}{3}\right)\left(\frac{3x}{4}\right)^2$

33. Which of the following is a true statement about the expression $a^4\left(\frac{1}{a}\right)^2$, given that $a \neq 0$?

A. The expression is always equal to 1.
B. The value of the expression is positive.
C. If a is negative, then the value of the expression is also negative.
D. The expression cannot be simplified any further.

MATHEMATICAL PRACTICES
Construct Viable Arguments and Critique the Reasoning of Others

34. Alana says that $(ab)^3 \cdot (ab)^4$ is the same as $[(ab)^3]^4$. Is Alana correct? Justify your response.

Operations with Radicals

Go Fly a Kite

Lesson 20-1 Radical Expressions

Learning Targets:

- Write and simplify radical expressions.
- Understand what is meant by a rational exponent.

SUGGESTED LEARNING STRATEGIES: Create Representations, Close Reading, Discussion Groups, Sharing and Responding, Note Taking, Think-Pair-Share

The frame of a box kite has four "legs" of equal length and four pairs of crossbars, all of equal length. The legs of the kite form a square base. The crossbars are attached to the legs so that each crossbar is positioned as a diagonal of the square base.

1. a. Label the legs of the kite pictured to the right. How many legs are in a kite? How many crossbars?

b. Label the points on the top view where the ends of the crossbars are attached to the legs *A*, *B*, *C*, and *D*. Begin at the bottom left and go clockwise.

c. Use one color to show the sides of the square and another color to show crossbar *AC*. What two figures are formed by two sides of the square and one diagonal?

My Notes

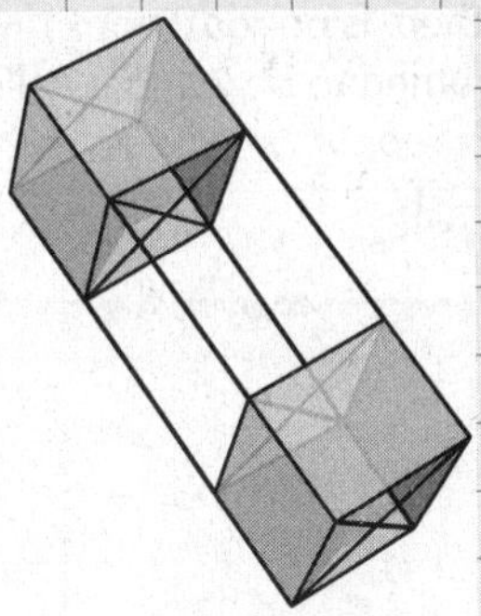

Top View

Members of the Windy Hill Science Club are building kites to explore aerodynamic forces. Club members will provide paper, plastic, or lightweight cloth for the covering of their kite. The club will provide the balsa wood for the frames.

2. Model with mathematics. The science club advisor has created the chart below to help determine how much balsa wood he needs to buy.

a. For each kite, calculate the exact length of one crossbar that will be needed to stabilize the kite. Use your drawing from Item 1c as a guide for the rectangular base of these box kites.

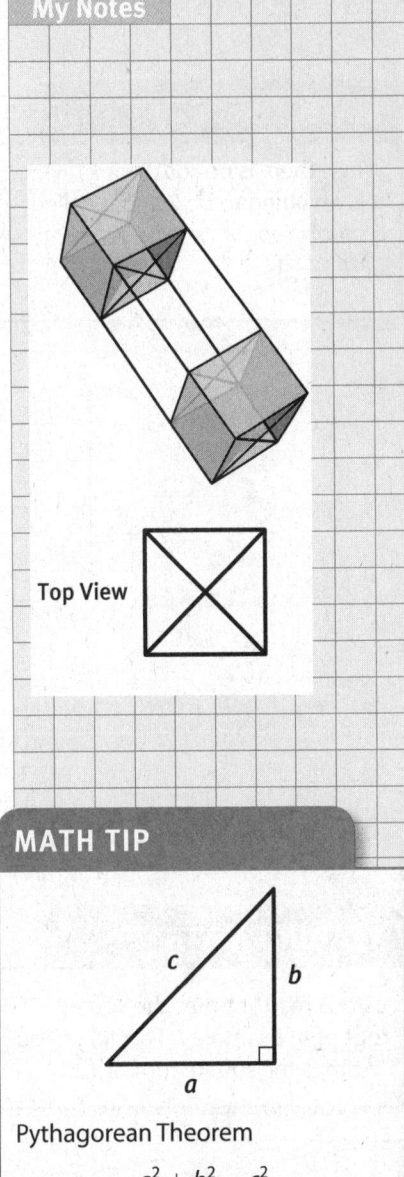

Kite	Dimensions of Base (in feet)	Exact Length of One Crossbar (in feet)	Kite	Dimensions of Base (in feet)	Exact Length of One Crossbar (in feet)
A	1 by 1		D	1 by 2	
B	2 by 2		E	2 by 4	
C	3 by 3		F	3 by 6	

b. How much wood would you recommend buying for the crossbars of Kite A? Explain your reasoning.

MATH TIP

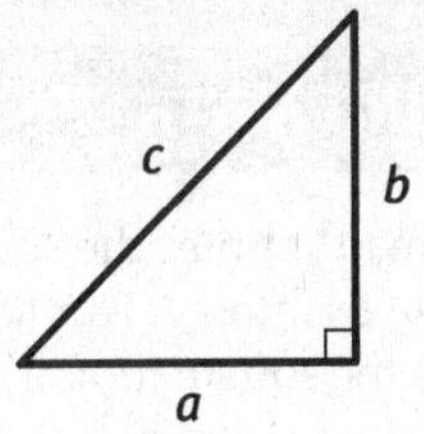

Pythagorean Theorem

$$a^2 + b^2 = c^2$$

MATH TIP

If you take the square root of a number that is not a perfect square, the result is a decimal number that does not terminate or repeat and is called an **irrational number**. The exact value of an irrational number must be written using a radical sign.

My Notes

MATH TIP

When there is no root index given, it is assumed to be 2 and is called a *square root*.
$\sqrt{36} = \sqrt[2]{36}$

READING MATH

$a\sqrt{b}$ is read "a times the square root of b." Example A Part (c) is read "7 times the square root of 12."

Each amount of wood in the table in Item 2 is a ***radical expression***.

Radical Expression	
An expression of the form $\sqrt[n]{a}$, where a is the radicand, $\sqrt{}$ is the radical symbol, and n is the root index.	
$\sqrt[n]{a} = b$, if $b^n = a$.	b is the nth root of a.

Finding the square root of a number or expression is the inverse operation of squaring a number or expression.

$$\sqrt{25} = 5, \text{ because } (5)(5) = 25$$

$$\sqrt{81} = 9, \text{ because } (9)(9) = 81$$

$$\sqrt{x^2} = x, \text{ because } (x)(x) = x^2, x \geq 0$$

Notice also that $(-5)(-5) = (-5)^2 = 25$. The ***principal square root*** of a number is the positive square root value. The expression $\sqrt{25}$ simplifies to 5, the principal square root. The ***negative square root*** is the negative root value, so $-\sqrt{25}$ simplifies to -5.

To simplify square roots in which the radicand is not a perfect square:

Step 1: Write the radicand as a product of numbers, one of which is a perfect square.

Step 2: Find the square root of the perfect square.

Example A

Simplify each expression.

a. $\sqrt{75} = \sqrt{25 \cdot 3} = 5\sqrt{3}$

b. $\sqrt{72} = \sqrt{36 \cdot 2} = 6\sqrt{2}$

$\sqrt{72} = \sqrt{9 \cdot 4 \cdot 2} = (3 \cdot 2)\sqrt{2} = 6\sqrt{2}$

c. $7\sqrt{12} = 7\sqrt{4 \cdot 3} = 7(2\sqrt{3}) = 14\sqrt{3}$

d. $\sqrt{c^3} = \sqrt{c^2 \cdot c} = c\sqrt{c}, c \geq 0$

Try These A

Simplify each expression.

a. $\sqrt{18}$ **b.** $5\sqrt{48}$ **c.** $\sqrt{126}$

d. $\sqrt{24y^2}$ **e.** $\sqrt{45b^3}$

My Notes

3. Copy the lengths of the crossbars from the chart in Item 1. Then express the lengths of the crossbars in simplified form.

Kite	Dimensions of Base (feet)	Exact Length of One Crossbar (feet)	Simplified Form of Length of Crossbar
A	1 by 1		
B	2 by 2		
C	3 by 3		
D	1 by 2		
E	2 by 4		
F	3 by 6		

The process of finding roots can be expanded to ***cube roots***. Finding the cube root of a number or an expression is the inverse operation of cubing that number or expression.

$$\sqrt[3]{125} = 5, \text{ because } (5)(5)(5) = 125$$
$$\sqrt[3]{y^3} = y, \text{ because } (y)(y)(y) = y^3$$

To simplify cube roots in which the radicand is not a perfect cube, follow the same two-step process that you used for square roots.

Step 1: Write the radicand as a product of numbers, one of which is a perfect cube.

Step 2: Find the cube root of the perfect cube.

MATH TERMS

The root index n can be any integer greater than or equal to 2. A **cube root** has $n = 3$. The cube root of 8 is $\sqrt[3]{8} = 2$ because $2 \cdot 2 \cdot 2 = 8$.

Example B

Simplify each expression.

a. $\sqrt[3]{16} = \sqrt[3]{8 \cdot 2} = \sqrt[3]{8} \cdot \sqrt[3]{2} = 2\sqrt[3]{2}$

b. $3\sqrt[3]{128} = 3\sqrt[3]{64 \cdot 2} = 3\sqrt[3]{64} \cdot \sqrt[3]{2} = 3(4\sqrt[3]{2}) = 12\sqrt[3]{2}$

c. $\sqrt[3]{x^5} = \sqrt[3]{x^3 \cdot x^2} = x\sqrt[3]{x^2}$

Try These B

Simplify each expression.

a. $\sqrt[3]{24}$ **b.** $\sqrt[3]{54z^3}$ **c.** $\sqrt[3]{40b^4}$

My Notes

Check Your Understanding

4. A kite has a base with dimensions of 2 feet by 3 feet. What is the length of one crossbar that will be needed to stabilize the kite?
5. Simplify.
   a. $\sqrt{124}$ **b.** $\sqrt{125d^4}$ **c.** $\sqrt[3]{250}$ **d.** $\sqrt[3]{81m^7}$

Another way to write radical expressions is with fractional exponents.

6. **Make use of structure.** Use the definition of a radical and the properties of exponents to simplify the expressions of each row of the table. The first row has been done for you.

Radical Form	Simplified Form	Fractional Exponent Form	Simplified Form
$\sqrt{16}\cdot\sqrt{16}$	$4\cdot 4=16$	$16^{\frac{1}{2}}\cdot 16^{\frac{1}{2}}$	$16^{\frac{1}{2}+\frac{1}{2}}=16^1=16$
$\sqrt[3]{8}\cdot\sqrt[3]{8}\cdot\sqrt[3]{8}$		$8^{\frac{1}{3}}\cdot 8^{\frac{1}{3}}\cdot 8^{\frac{1}{3}}$	
$\sqrt[4]{81}\cdot\sqrt[4]{81}\cdot\sqrt[4]{81}\cdot\sqrt[4]{81}$		$81^{\frac{1}{4}}\cdot 81^{\frac{1}{4}}\cdot 81^{\frac{1}{4}}\cdot 81^{\frac{1}{4}}$	
$\sqrt{a}\cdot\sqrt{a}$		$a^{\frac{1}{2}}\cdot a^{\frac{1}{2}}$	

7. Identify and describe any patterns in the table. Write $a^{\frac{1}{n}}$ as a radical expression.

The general rule for fractional exponents when the numerator is not 1 is $a^{\frac{m}{n}}=\sqrt[n]{a^m}=\left(\sqrt[n]{a}\right)^m$.

Example C

Write $6^{\frac{2}{3}}$ as a radical expression.

Method 1: $6^{\frac{2}{3}}=6^{2\cdot\frac{1}{3}}=(6^2)^{\frac{1}{3}}=\sqrt[3]{6^2}$

Method 2: $6^{\frac{2}{3}}=6^{\frac{1}{3}\cdot 2}=\left(6^{\frac{1}{3}}\right)^2=\left(\sqrt[3]{6}\right)^2$

Try These C

Write each of the following as a radical expression.

a. $13^{\frac{1}{4}}$ **b.** $7^{\frac{3}{5}}$ **c.** $x^{\frac{3}{2}}$

My Notes

Check Your Understanding

8. a. What is the value of $16^{\frac{1}{4}}$?

b. What is the value of $16^{\frac{3}{4}}$?

9. For each radical expression, write an equivalent expression with a fractional exponent.

a. $\sqrt{15}$ **b.** $\sqrt[3]{21}$

LESSON 20-1 PRACTICE

10. A square has an area of 72 square inches. What is its side length *s*? Give the exact answer using simplified radicals.

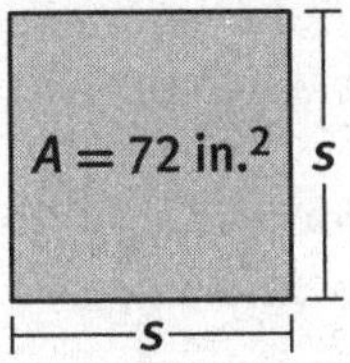

11. A cube has a volume of 216 cubic centimeters. What is its edge length *s*? Give the exact answer using simplified radicals.

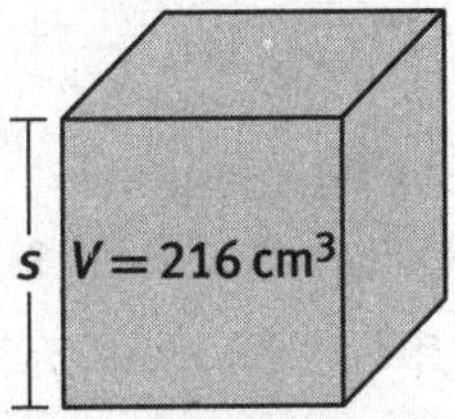

12. A square has an area of $12x^2$ square feet. What is the length of its sides?

13. A cube has a volume of $128y^3$ cubic millimeters. What is its edge length?

14. A kite has a square base with dimensions of 4 feet by 4 feet. What is the length of one of the diagonal crossbars that will be needed to stabilize the kite?

15. For each radical expression, write an equivalent expression with a fractional exponent.

a. $\sqrt{6}$ **b.** $\sqrt{10}$ **c.** $\sqrt[3]{5}$ **d.** $\sqrt[3]{18}$

16. Reason abstractly. Devise a plan for simplifying the fourth root of a number that is not a perfect fourth power. Explain to a friend how to use your plan to simplify the fourth root. Be sure to include examples.

My Notes

Learning Targets:

- Add radical expressions.
- Subtract radical expressions.

SUGGESTED LEARNING STRATEGIES: Discussion Groups, Close Reading, Note Taking, Think-Pair-Share, Identify a Subtask

The Windy Hill Science Club advisor wants to find the total length of the balsa wood needed to make the frames for the kites. To do so, he will need to add radicals.

Addition Property of Radicals
$a\sqrt{b} \pm c\sqrt{b} = (a \pm c)\sqrt{b}$, where $b \geq 0$.

To add or subtract radicals, the index and radicand must be the same.

MATH TIP

The rational numbers are **closed** under addition and subtraction. This means that the sum or difference of two rational numbers is rational.

Example A

Add or subtract each expression and simplify. State whether the sum or difference is rational or irrational.

a. $3\sqrt{5} + 7\sqrt{5}$
$= (3+7)\sqrt{5}$
$= 10\sqrt{5}$
irrational

←Add or subtract→ the coefficients.

b. $10\sqrt[3]{3} - 4\sqrt[3]{3}$
$= (10-4)\sqrt[3]{3}$
$= 6\sqrt[3]{3}$
irrational

c. $2\sqrt{5} + 8\sqrt{3} + 6\sqrt{5} - 3\sqrt{3}$

Step 1: Group terms with like radicands. $2\sqrt{5} + 6\sqrt{5} + 8\sqrt{3} - 3\sqrt{3}$

Step 2: Add or subtract the coefficients. $= (2+6)\sqrt{5} + (8-3)\sqrt{3}$
$= 8\sqrt{5} + 5\sqrt{3}$

Solution: $2\sqrt{5} + 8\sqrt{3} + 6\sqrt{5} - 3\sqrt{3} = 8\sqrt{5} + 5\sqrt{3}$; irrational

Try These A

Add or subtract each expression and simplify. State whether the sum or difference is rational or irrational.

a. $2\sqrt{7} + 3\sqrt{7} + \frac{2}{3}$

b. $5\sqrt{6} + 2\sqrt{5} - \sqrt{6} + 7\sqrt{5}$

c. $2 + 2\sqrt{2} + \sqrt{8} + 3\sqrt{2}$

My Notes

1. The club advisor also needs to know how much wood to buy for the legs of the kites. Each kite will be 3 feet tall.

a. Complete the table below.

Kite	Dimensions of Base (feet)	Length of One Crossbar (feet)	Length of One Leg (feet)	Wood Needed for Legs (feet)	Wood Needed for Crossbars (feet)
A	1 by 1				
B	2 by 2				
C	3 by 3				
D	1 by 2				
E	2 by 4				
F	3 by 6				

b. Reason quantitatively. How much balsa wood should the club advisor buy if the club is going to build the six kites described above? Is the result rational or irrational?

c. Explain how you reached your conclusion.

2. Use appropriate tools strategically. Approximately how much balsa wood, in decimal notation, will the club advisor need to buy?

a. Use your calculator to approximate the amount of balsa wood, and then decide on a reasonable way to round.

b. Explain why the club advisor would need this approximation rather than the exact answer expressed as a radical.

My Notes

Check Your Understanding

Perform the indicated operations. Be sure to completely simplify your answer. State whether each sum or difference is rational or irrational.

3. a. $7\sqrt{6}+9\sqrt{6}$ **b.** $12\sqrt{4}-5\sqrt{4}+2\sqrt{4}$

4. a. $8\sqrt{3}-\sqrt{12}+3\sqrt{3}$ **b.** $\sqrt{18}+\sqrt{8}+\sqrt{32}$

5. a. $\sqrt[3]{16}-\sqrt[3]{2}$ **b.** $\sqrt[3]{3}+\sqrt[3]{24}+16$

6. a. $3\sqrt{9}+5\sqrt{9}$ **b.** $3\sqrt{10}+5\sqrt{10}$

7. Is the sum of a rational number and an irrational number rational or irrational? Support your response with an example.

LESSON 20-2 PRACTICE

Use the figures of a rectangle and kite for Items 8–12.

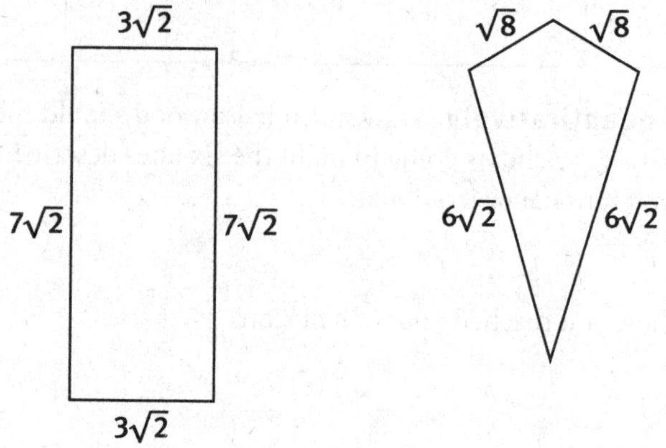

8. Determine the perimeter of the rectangle.

9. Determine the perimeter of the kite.

10. How much longer is the long side of the rectangle than the longer side of the kite?

11. How much greater is the perimeter of the rectangle than the perimeter of the kite?

12. **Make sense of problems.** How much wood would be required to insert diagonal crossbars in the rectangle?

My Notes

Learning Targets:

- Multiply and divide radical expressions.
- Rationalize the denominator of a radical expression.

SUGGESTED LEARNING STRATEGIES: Think-Pair-Share, Predict and Confirm, Discussion Groups, Close Reading, Marking the Text, Note Taking

1. a. Complete the table below and simplify the radical expressions in the third and fifth columns.

a	b	$\sqrt{a} \cdot \sqrt{b}$	ab	$\sqrt{ab}$
4	9			
100	25			
9	16			

b. Express regularity in repeated reasoning. Use the patterns you observe in the table above to write an equation that relates $\sqrt{a}$, $\sqrt{b}$, and $\sqrt{ab}$.

c. All the values of *a* and *b* in part a are perfect squares. In the table below, choose some values for *a* and *b* that are *not* perfect squares and use a calculator to show that the equation you wrote in Part (b) is true for those numbers as well.

TECHNOLOGY TIP

Approximate values of square roots that are not perfect squares can be found using a calculator.

a	b	$\sqrt{a} \cdot \sqrt{b}$	ab	$\sqrt{ab}$

d. Simplify the products in Columns A and B below.

A	Simplified Form	B	Simplified Form
$(2\sqrt{4})(\sqrt{9})$		$2\sqrt{4 \cdot 9}$	
$(3\sqrt{4})(5\sqrt{16})$		$(3 \cdot 5)\sqrt{4 \cdot 16}$	
$(2\sqrt{7})(3\sqrt{14})$		$(2 \cdot 3)\sqrt{7 \cdot 14}$	

e. Which products in the table in Part (d) are rational and which are irrational?

My Notes

MATH TIP

The rational numbers are **closed** under multiplication. This means that the product of two rational numbers is rational. Since the coefficients *a* and *c* are rational, their product will also be rational.

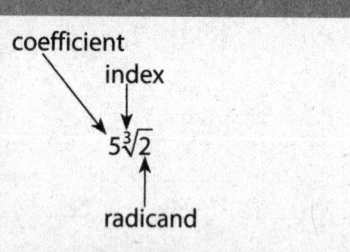

MATH TIP

coefficient → $5\sqrt[3]{2}$ ← index

radicand

CONNECT TO AP

Later in this course, you will study another system of numbers, called the complex numbers. In the complex number system, $\sqrt{-1}$ is defined as the imaginary number *i*.

f. Write a verbal rule that explains how to multiply radical expressions.

Multiplication Property of Radicals
$\left(a\sqrt{b}\right)\left(c\sqrt{d}\right)=ac\sqrt{bd}$,   where $b \geq 0$, $d \geq 0$.

To multiply radical expressions, the index must be the same. Find the product of the coefficients and the product of the radicands. Simplify the radical expression.

Example A

Multiply each expression and simplify.

a. $\left(3\sqrt{6}\right)\left(4\sqrt{5}\right)=(3\cdot 4)\left(\sqrt{6\cdot 5}\right)=12\sqrt{30}$

b. $\left(2\sqrt{10}\right)\left(3\sqrt{6}\right)$

c. $\left(2x\sqrt{6x}\right)\left(5\sqrt{3x^2}\right)$

Step 1: Multiply.

b. $=(2\cdot 3)\sqrt{10\cdot 6}$

$=6\sqrt{60}$

c. $=10x\sqrt{6x\cdot 3x^2}$

$=10x\left(\sqrt{18x^3}\right)$

Step 2: Simplify.

b. $=6\left(\sqrt{4\cdot 15}\right)$

$=(6\cdot 2)\sqrt{15}$

$=12\sqrt{15}$

c. $=10x\left(\sqrt{9x^2\cdot 2x}\right)$

$=(10x)(3x)\left(\sqrt{2x}\right)$

$=30x^2\sqrt{2x}$

Try These A

Multiply each expression and simplify.

a. $\left(2\sqrt{10}\right)\left(5\sqrt{3}\right)$

b. $\left(3\sqrt{8}\right)\left(2\sqrt{6}\right)$

c. $\left(4\sqrt{12}\right)\left(5\sqrt{18}\right)$

d. $\left(3\sqrt{5a}\right)\left(2a\sqrt{15a^2}\right)$

Division Property of Radicals
$\dfrac{a\sqrt{b}}{c\sqrt{d}}=\dfrac{a}{c}\sqrt{\dfrac{b}{d}}$ where $b \geq 0$, $d \geq 0$.

To divide radical expressions, the index must be the same. Find the quotient of the coefficients and the quotient of the radicands. Simplify the expression.

My Notes

Example B

Divide each expression and simplify.

a. $\frac{\sqrt{6}}{\sqrt{2}} = \sqrt{\frac{6}{2}} = \sqrt{3}$

b. $\frac{2\sqrt{10}}{3\sqrt{2}} = \frac{2}{3}\sqrt{\frac{10}{2}} = \frac{2}{3}\sqrt{5}$

c. $\frac{8\sqrt{24x^2}}{2\sqrt{3}} = \frac{8}{2}\sqrt{\frac{24x^2}{3}} = 4\sqrt{8x^2} = 4\sqrt{4 \cdot 2 \cdot x^2}$

$= 4\left(2x\sqrt{2}\right) = 8x\sqrt{2}$

Try These B

Divide each expression and simplify.

a. $\frac{4\sqrt{42}}{5\sqrt{6}}$

b. $\frac{10\sqrt{54}}{2\sqrt{2}}$

c. $\frac{12\sqrt{75}}{3\sqrt{3}}$

d. $\frac{16\sqrt[3]{8x^{11}}}{8\sqrt[3]{x^2}}$

A radical expression in simplified form does not have a radical in the denominator. Most frequently, the denominator is ***rationalized***. You ***rationalize the denominator*** by simplifying the expression to get a perfect square under the radicand in the denominator.

$$\frac{\sqrt{a}}{\sqrt{b}} \cdot 1 = \left(\frac{\sqrt{a}}{\sqrt{b}}\right)\left(\frac{\sqrt{b}}{\sqrt{b}}\right) = \frac{\sqrt{ab}}{\sqrt{b^2}} = \frac{\sqrt{ab}}{b}$$

MATH TERMS

Rationalize means to make rational. You can **rationalize the denominator** without changing the value of the expression by multiplying the fraction by an appropriate form of 1.

Example C

Rationalize the denominator of $\frac{\sqrt{5}}{\sqrt{3}}$.

Step 1: Multiply the numerator and denominator by $\sqrt{3}$.

$\frac{\sqrt{5}}{\sqrt{3}} = \frac{\sqrt{5}}{\sqrt{3}} \cdot \frac{\sqrt{3}}{\sqrt{3}} = \frac{\sqrt{15}}{\sqrt{9}}$

Step 2: Simplify.

$= \frac{\sqrt{15}}{3}$

Solution: $\frac{\sqrt{5}}{\sqrt{3}} = \frac{\sqrt{15}}{3}$

CONNECT TO AP

In calculus, both numerators and denominators are rationalized. The procedure for rationalizing a numerator is similar to that for rationalizing a denominator.

Try These C

Rationalize the denominator in each expression.

a. $\frac{\sqrt{11}}{\sqrt{6}}$

b. $\frac{2\sqrt{7}}{\sqrt{5}}$

c. $\frac{3\sqrt{5}}{\sqrt{8}}$

My Notes

Check Your Understanding

Express each expression in simplest radical form. State whether each result in Items 2–5 is rational or irrational.

2. $(4\sqrt{7})(2\sqrt{3})$
3. $\sqrt{2}(\sqrt{2}+3\sqrt{6})$
4. $\frac{\sqrt{75}}{\sqrt{5}}$
5. $\sqrt{\frac{5}{8}}$
6. $(3\sqrt{32y})(4\sqrt{25y})$
7. $\frac{6\sqrt{98x^4}}{\sqrt{2x}}$
8. Is the product of a nonzero rational number and an irrational number rational or irrational? Support your response with an example.

LESSON 20-3 PRACTICE

Express each expression in simplest radical form. State whether each result in Items 9–12 is rational or irrational.

9. $\left(\sqrt{\frac{1}{2}}\right)\left(\sqrt{\frac{3}{5}}\right)$
10. $\sqrt{27} \bullet \sqrt{\frac{1}{27}}$
11. $\frac{2\sqrt{7}}{\sqrt{3}}$
12. $\frac{4\sqrt{5}}{3\sqrt{2}}$
13. $(4\sqrt[3]{4m^2})(5m\sqrt[3]{2m^2})$
14. $\frac{2\sqrt{52x^9}}{\sqrt{13x}}$
15. **Attend to precision.** What conditions must be satisfied for a radical expression to be in simplified form?

ACTIVITY 20 PRACTICE

Write your answers on notebook paper.
Show your work.

Lesson 20-1

Write each expression in simplest radical form.

1. $\sqrt{40}$
2. $\sqrt{128}$
3. $\sqrt{162}$

Use the Pythagorean Theorem and the triangle below for Items 4 and 5. Recall that the Pythagorean Theorem states that for all right triangles, $a^2 + b^2 = c^2$.

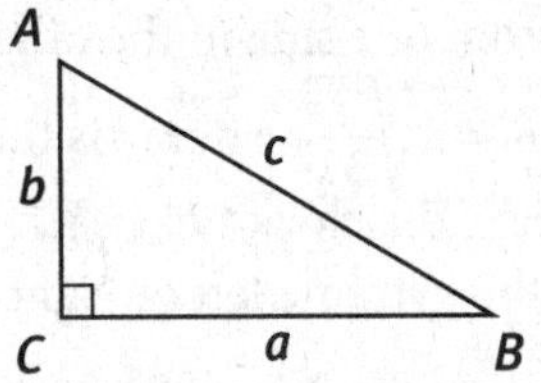

4. In the right triangle, if $a = 3$ and $b = 6$, what is the value of c?
   A. $3\sqrt{5}$ **B.** 9
   C. $9\sqrt{5}$ **D.** 45
5. In the right triangle, if $a = 12$ and $b = 15$, what is the value of c?

Simplify each expression.

6. $\sqrt{4m^7}$
7. $3\sqrt[4]{16n^8}$
8. $\sqrt[3]{16x^4}$

Write each of the following as a radical expression.

9. $15^{\frac{2}{5}}$
10. $(2p)^{\frac{1}{3}}$
11. $16x^{\frac{3}{4}}$
12. Which of the following expressions is **not** equivalent to $(8x)^{\frac{2}{3}}$?
    A. $4\sqrt[3]{x^2}$ **B.** $\sqrt[3]{8x^2}$
    C. $4x^{\frac{2}{3}}$ **D.** $8^{\frac{2}{3}}x^{\frac{2}{3}}$

Lesson 20-2

Write each expression in simplest radical form. State whether each result is rational or irrational.

13. $4\sqrt{27} + 6\sqrt{12}$
14. $8\sqrt{6} + 2\sqrt{12} + 5\sqrt{3} - \sqrt{54}$
15. $3\sqrt{36} - 5\sqrt{16} + 4$
16. Which of the following is the difference of $9\sqrt{20}$ and $2\sqrt{5}$?
    A. $7\sqrt{15}$ **B.** $16\sqrt{5}$
    C. $9\sqrt{5}$ **D.** $7\sqrt{5}$

The figure below is composed of a rectangle and a right triangle. Use the figure for Items 17–19.

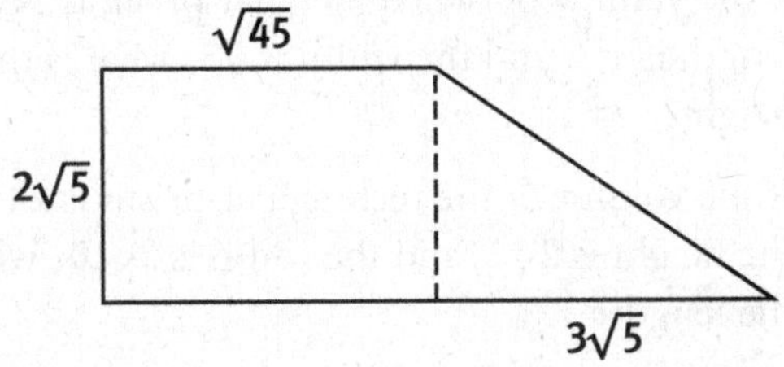

17. Determine the perimeter of the rectangle.
18. Determine the perimeter of the triangle.
19. Determine the perimeter of the composite figure.
20. A student was asked to completely simplify the expression $3\sqrt{3} + \sqrt{12} + 2\sqrt{3}$. The student wrote $5\sqrt{3} + \sqrt{12}$. Do you agree with the student's answer? Explain.

Lesson 20-3

The figure shows a rectangular prism. The volume of the rectangular prism is the product of the length, width, and height. Use the figure for Items 21–24.

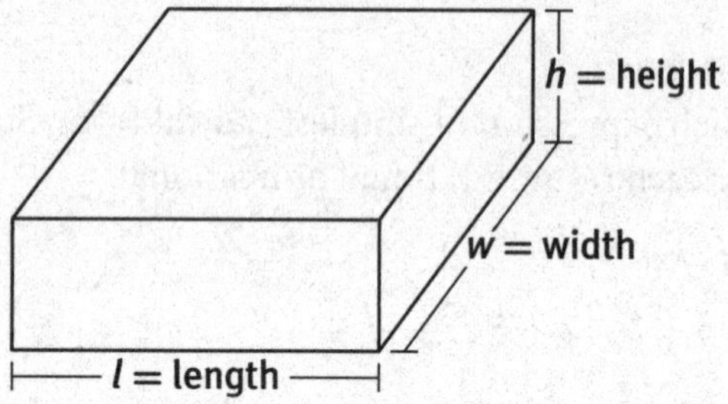

21. If $l = \sqrt{3}$, $w = \sqrt{2}$, and $h = \sqrt{6}$, what is the volume of the rectangular prism? Is the volume rational or irrational?

22. If $l = 3\sqrt{3}$, $w = 2\sqrt{2}$, and $h = 5\sqrt{10}$, what is the volume of the rectangular prism? Is the volume rational or irrational?

23. If the volume of the rectangular prism is 20, the length is $\sqrt{3}$, and the width is $\sqrt{5}$, what is the height?

24. If the volume of the rectangular prism is $24\sqrt{3}$, the height is $2\sqrt{2}$, and the width is $3\sqrt{10}$, what is the length?

Write each expression in simplest form.

25. $\left(2\sqrt{2x^2}\right)\left(3x\sqrt{x^2}\right)$

26. $\left(6p\sqrt{p^3}\right)\left(0.2\sqrt{16p}\right)$

27. $\left(4\sqrt[3]{8m}\right)\left(7m\sqrt[3]{m^5}\right)$

28. $\dfrac{3\sqrt{32}}{\sqrt{2}}$

29. $\dfrac{\sqrt[3]{81x}}{\sqrt[3]{3}}$

30. $\dfrac{\sqrt{x^2y^5}}{\sqrt{y}}$

31. Which of the following expressions cannot be simplified any further?

A. $\sqrt{\dfrac{5}{2}}$ **B.** $\dfrac{\sqrt{5}}{\sqrt{2}}$

C. $\sqrt{52}$ **D.** $5\sqrt{2}$

32. Elena was asked to simplify the expression $\left(2\sqrt{12x}\right)\left(4x\sqrt{3}\right)$. Her answer was $48x\sqrt{x}$.

a. Explain how Elena can use her calculator to check whether her answer is reasonable.

b. Is Elena's answer correct? If not, explain Elena's mistake and give the correct answer.

The time, T, in seconds, it takes the pendulum of a clock to swing from one side to the other side is given by the formula $T = \pi\sqrt{\dfrac{l}{32}}$, where l is the length of the pendulum, in feet. The clock ticks each time the pendulum is at the extreme left or right point.

Use this information for Items 33–36.

33. If the pendulum is 4 feet long, how long does it take the pendulum to swing from left to right? Give an exact value in terms of π.

34. If the pendulum is 8 feet long, how long does it take the pendulum to swing from left to right? Give an exact value in terms of π.

35. If the pendulum is shortened, will the clock tick more or less often? Explain how you arrived at your conclusion.

36. Approximately what length of the pendulum will result in its swinging from one side to the other every second?

MATHEMATICAL PRACTICES
Construct Viable Arguments and Critique the Reasoning of Others

37. Amil knows that the formula for the area of a circle is $A = \pi r^2$. He says that the area of a circle with a radius of $2\sqrt{5}$ feet is $4\sqrt{5}\pi$ square feet. Is he correct? If not, describe his error.

Geometric Sequences

Go Viral!

Lesson 21-1 Identifying Geometric Sequences

My Notes

Learning Targets:

- Identify geometric sequences and the common ratio in a geometric sequence.
- Distinguish between arithmetic and geometric sequences.

SUGGESTED LEARNING STRATEGIES: Visualization, Look for a Pattern, Create Representations, Think-Pair-Share, Sharing and Responding

For her Electronic Communications class, Keisha has been tasked with investigating the effects of social media. She decides to post a video in cyberspace to see if she can make it go viral.

To get things started, Keisha e-mails the video link to three of her friends. In the message, she asks each of the recipients to forward the link to three of his or her friends. Whenever a recipient forwards the link, Keisha asks him or her to attach the following message: *After watching, please forward this video link to three of your friends who have not yet received it.*

One way to visually represent this situation is with a tree diagram. A ***tree diagram*** shows all the possible outcomes of an event.

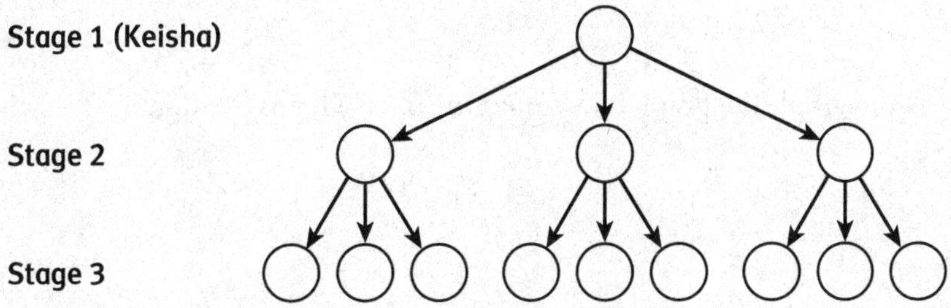

1. Use the tree diagram to help you complete the table. (Assume that everyone who receives the video link watches the video.)

Stage	Number of People Who Watch the Video
1	1
2	
3	
4	

2. Express regularity in repeated reasoning. Describe any patterns you notice in the table.

My Notes

3. Use the table of values to graph the viral video situation.

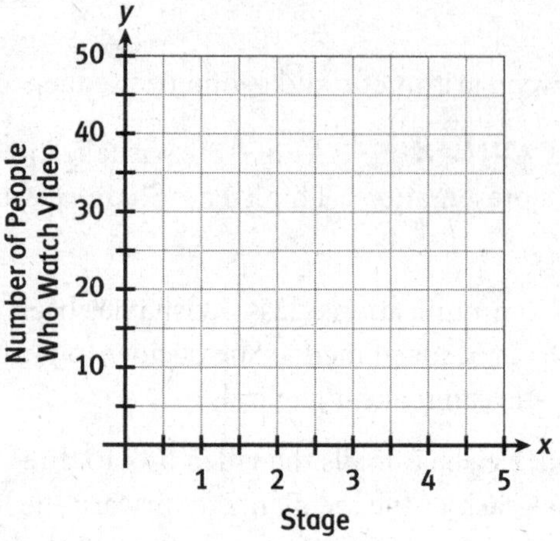

4. Is the relationship a linear relationship? Justify your response.

5. Is the graph the graph of a ***function***? If so, what is the domain?

MATH TERMS

A **function** is a relation in which every input is paired with exactly one output.

The number of people who have received the video link at each stage form a *geometric sequence.* A ***geometric sequence*** is a sequence of values in which a nonzero constant ratio exists between consecutive terms. The constant ratio is called the common ratio and is typically denoted by the letter *r*. The common ratio is the value that each term is multiplied by to get the next term.

6. a. Write the numbers of people who have received the video as a sequence.

b. Reason quantitatively. Identify the common ratio. Justify your response.

My Notes

7. Identify each sequence as arithmetic, geometric, or neither. If it is arithmetic, state the common difference. If it is geometric, state the common ratio.

a. 5, 8, 11, 14, …

b. $18, 6, 2, \frac{2}{3}, \ldots$

c. 1, 4, 9, 16, …

d. $-1, 4, -16, 64, \ldots$

e. $16, -8, 4, -2, \ldots$

MATH TERMS

An **arithmetic sequence** is a sequence in which the difference between consecutive terms is constant. This difference is called the **common difference.**

Check Your Understanding

Identify each sequence as arithmetic, geometric, or neither. If it is arithmetic, state the common difference. If it is geometric, state the common ratio.

8. 10, 8, 6, 4, . . . **9.** $2, \frac{1}{2}, \frac{1}{8}, \frac{1}{32}, \ldots$ **10.** $9, -3, 1 -\frac{1}{3}, \ldots$

LESSON 21-1 PRACTICE

A cell divides in two every day. The tree diagram shows the first few stages of this process. Use the tree diagram for Items 11–13.

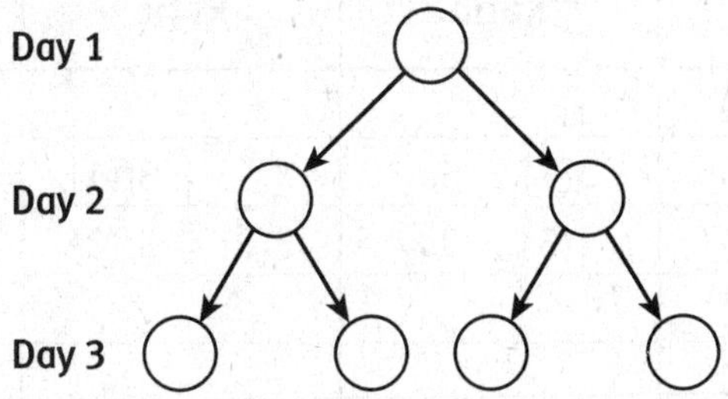

11. Make a table of values to represent the scenario shown in the tree diagram.

12. Does the tree diagram represent a geometric sequence? If so, what is the common ratio?

13. If the diagram were extended to a sixth day, how many circles would there be on Day 6?

14. Reason abstractly. Can a geometric sequence ever have a term equal to 0? Explain.

My Notes

Learning Targets:

- Write a recursive formula for a geometric sequence.
- Write an explicit formula for a geometric sequence.
- Use a formula to find a given term of a geometric sequence.

SUGGESTED LEARNING STRATEGIES: Create Representations, Look for a Pattern, Discussion Groups, Think-Pair-Share, Construct an Argument, Sharing and Responding

Remember that the numbers in a sequence are called *terms*, and you can use sequence notation a_n or function notation $f(n)$ to refer to the *n*th term.

1. a. Use the above notation to rewrite the first four terms of the viral video sequence. Also write the common ratio.

$a_1 = f(1) =$

$a_2 = f(2) =$

$a_3 = f(3) =$

$a_4 = f(4) =$

$r =$

b. What is the value of the term following $a_4 = f(4)$? Write an expression to represent this term using a_4 and the common ratio.

MATH TERMS

A **recursive formula** is a formula that gives any term as a function of preceding terms.

Just as with arithmetic sequences, you can use a ***recursive formula*** to represent a geometric sequence.

2. Complete the table below for the viral video sequence.

Term	Sequence Representation Using Common Ratio	Function Representation Using Common Ratio	Numerical Value (number of people who have seen the video)
$a_1 = f(1)$	1	1	1
$a_2 = f(2)$	$3(1) = 3a_1$	$3(1) = 3f(1)$	3
$a_3 = f(3)$	$3(3) = 3a_2$	$3(3) = 3f(2)$	
$a_4 = f(4)$			
$a_5 = f(5)$			
...	...	...	...
$a_n = f(n)$			—

3. The recursive formulas for the viral video sequence are partially given below. Complete the formulas by writing the expressions for a_n and $f(n)$.

$\begin{cases} a_1 = 1 \\ a_n = \end{cases}$ $\begin{cases} f(1) = 1 \\ f(n) = \end{cases}$

My Notes

Check Your Understanding

Write a recursive formula for each geometric sequence. Include the recursive formula in function notation.

4. 64, 16, 4, 1, . . .

5. 64, −16, 4, −1, . . .

6. −1, 1, −1, 1, . . .

7. Use the recursive formula to find a_6, a_7, and a_8 for the viral video sequence. Explain your results.

8. Why might it be difficult to find the 100th term of the viral video sequence using the recursive formula?

As with arithmetic sequences, geometric sequences can be represented with explicit formulas. The terms in a geometric sequence can be written as the product of the first term and a power of the common ratio.

9. For the viral video sequence, identify a_1 and r. Then fill in the missing exponents and blanks.

$a_1 =$ ________ $r =$ ________

$a_2 = 1 \cdot 3^{\square} = 3$

$a_3 = 1 \cdot 3^{\square} = 9$

$a_4 = 1 \cdot 3^{\square} =$ ________

$a_5 = 1 \cdot 3^{\square} =$ ________

$a_6 = 1 \cdot 3^{\square} =$ ________

$a_{10} = 1 \cdot 3^{\square} =$ ________

10. Express regularity in repeated reasoning. Describe any patterns you observe in your responses to Item 9. Then use a_1, r, and n to write a formula for the nth term of any geometric sequence.

11. Write the explicit formula for the viral video sequence. Use the formula to determine the 12th term of the sequence. What does the 12th term represent?

My Notes

12. The explicit formula for a geometric sequence can be thought of as a function.

a. What is the input? What is the output?

b. State the domain of the function.

c. Rewrite the explicit formula for the viral video sequence using function notation.

d. Use appropriate tools strategically. Use a graphing calculator to graph your function from Part (c). Is the function linear or nonlinear? Justify your response.

13. Consider the geometric sequence 5, 10, 20, 40, . . .

a. Write the explicit formula for the sequence.

b. How can you check that your formula is correct?

c. Determine the 16th term in the sequence.

d. Use function notation to write the explicit formula for the sequence.

e. What is the value of $f(10)$? What does it represent?

14. a. Write the explicit formula for the geometric sequence 32, 16, 8, 4, . . .

b. Determine the 9th term in the sequence.

My Notes

15. The explicit formula for a geometric sequence is $a_n = 6 \cdot 3^{n-1}$. State the recursive formula for the sequence. Include the recursive formula in function notation.

Check Your Understanding

16. How can you use the recursive formula for a geometric sequence to write the explicit formula?

Write the explicit formula for each geometric sequence. Then determine the 6th term of each sequence.

17. 1, 5, 25, . . .

18. 48, −24, 12, . . .

19. $\begin{cases} a_1 = -81 \\ a_n = -\frac{1}{3}a_{n-1} \end{cases}$

20. Make sense of problems. Revisit the viral video scenario at the beginning of the activity. How many stages will it take until 1 million new people receive the link to the viral video? Explain how you found your answer.

Check Your Understanding

21. Write a recursive formula for the geometric sequence whose explicit formula is $a_n = 1 \cdot (-2)^{n-1}$. Include the recursive formula in function notation.

22. Write an explicit formula for the sequence $\begin{cases} a_1 = 3 \\ a_n = 2a_{n-1} \end{cases}$.

My Notes

LESSON 21-2 PRACTICE

The diagram below shows a square repeatedly divided in half. The entire square has an area of 1 square unit. The number in each region is the area of the region. Use the diagram for Items 23–25.

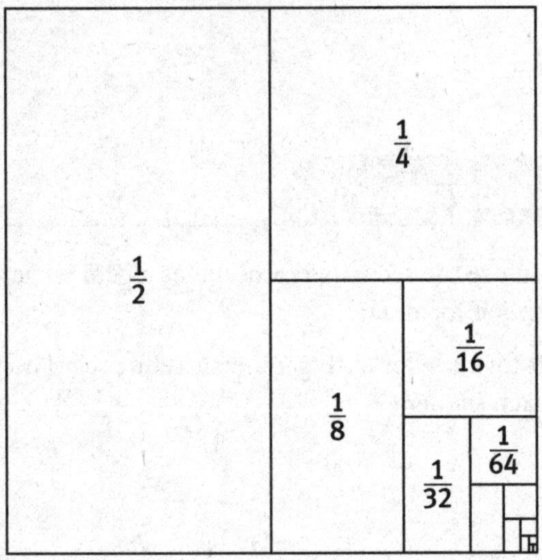

23. Write a geometric sequence to describe the areas of successive regions.

24. Write an explicit formula for the geometric sequence that you wrote in Item 23.

25. **Model with mathematics.** What is the 10th term of the sequence? What does it represent?

26. The explicit formula for a geometric sequence is $f(n) = 5(-2)^{n-1}$. Give the recursive formula for the sequence.

ACTIVITY 21 PRACTICE

Write your answers on notebook paper.
Show your work.

Lesson 21-1

In Items 1 and 2, assume that the first term of a sequence is −3.

1. Write the first four terms of the sequence if it is an arithmetic sequence with common difference $-\frac{1}{3}$.
2. Write the first four terms of the sequence if it is a geometric sequence with common ratio $-\frac{1}{3}$.

The tree diagram below shows the number of possible outcomes when tossing a coin a number of times. For example, if you toss a coin once (Stage 1), there are two possible outcomes: heads (H) and tails (T). If you toss a coin twice (Stage 2), there are four possible outcomes for the two tosses: HH, HT, TH, and TT.

Use the tree diagram for Items 3–5.

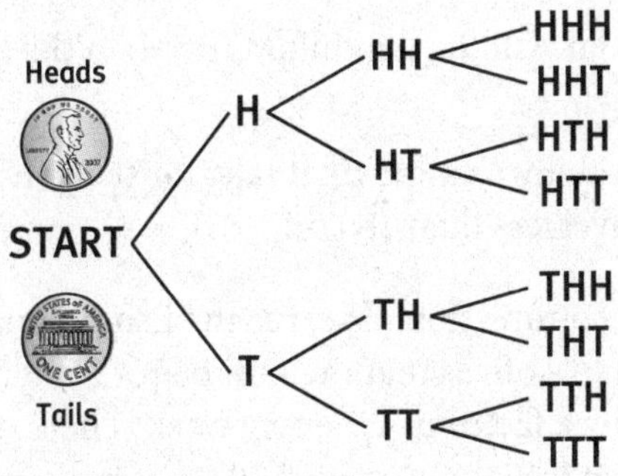

3. How many possible outcomes are there when you toss a coin 4 times?
4. Identify the common ratio of the sequence represented by the tree diagram.
5. How many possible outcomes are there when you toss a coin 23 times? Express your answer using exponents.

For Items 6–10, identify each sequence as arithmetic, geometric, or neither. If it is arithmetic, state the common difference. If it is geometric, state the common ratio.

6. 17, 25, 33, 41, . . .
7. 1, 3, 6, 10, 15, . . .
8. −27, −9, −3, −1, . . .
9. 0.1, 0.5, 0.9, 1.3, . . .
10. $\frac{1}{2}, \frac{1}{3}, \frac{1}{4}, \frac{1}{5}, \ldots$

11. A geometric sequence begins with the value 1 and has a common ratio of −2. Identify the eighth term in the sequence.
    A. −128 B. 128
    C. −256 D. 256
12. A geometric sequence begins with the value 2 and has a common ratio of $\frac{1}{2}$. Identify the fifth term in the sequence.
    A. 4 B. $\frac{1}{4}$
    C. $\frac{1}{8}$ D. $\frac{1}{16}$
13. Which of the following is a *false* statement about the sequence 2, 4, 8, 16, 32, . . . ?
    A. The common ratio of the sequence is 2.
    B. The tenth term of the sequence is 2^{10}.
    C. Every term of the sequence is even.
    D. The number 216 appears in the sequence.
14. Give an example of a geometric sequence with a common ratio of 0.2. Write at least the first four terms of the sequence.

Lesson 21-2

Write a recursive formula for each geometric sequence. Include the recursive formula in sequence notation.

15. 7, 21, 63, 189, ...

16. 100, 10, 1, 0.1, ...

17. −10, 20, −40, 80, ...

Write the explicit formula for each geometric sequence. Then determine the 8th term of each sequence.

18. 4, 16, 64, 256, ...

19. $\frac{1}{2}, \frac{1}{6}, \frac{1}{18}, \ldots$

20. $\begin{cases} a_1 = 5 \\ a_n = -3a_{n-1} \end{cases}$

A contestant on a game show wins $100 for answering a question correctly in Round 1. In each subsequent round, the contestant's winnings are doubled if she gives a correct answer. If the contestant gives an incorrect answer, she loses everything. Use this information for Items 21–23.

21. Write an explicit formula that gives the contestant's winnings in round *n*, assuming she answers all questions correctly.

22. How much does the contestant win in Round 10, assuming she answers all questions correctly?

23. How many rounds does a contestant need to play in order to answer a question worth at least $1,000,000?

24. A geometric sequence is given by the recursive formula $\begin{cases} f(1) = -6 \\ f(n) = -\frac{1}{2} f(n-1) \end{cases}$. Which of the following is a term in the sequence?

A. $\frac{3}{4}$ **B.** $-\frac{3}{4}$

C. $\frac{3}{2}$ **D.** -3

Each time a pendulum swings, the distance it travels decreases, as shown in the figure.

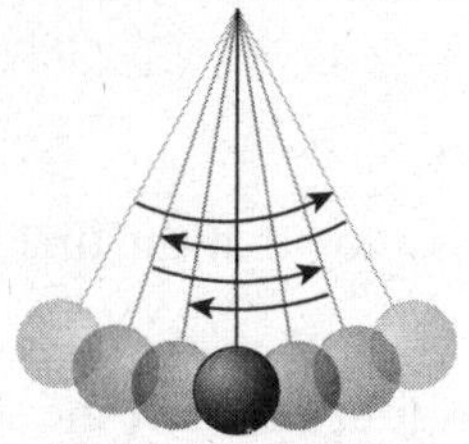

Pendulum Swing

The table shows how far the pendulum travels with each swing. Use this table for Items 25–27.

Swing Number	Distance Traveled (cm)
1	80
2	60
3	45
4	33.75

25. Write the explicit formula for the pendulum situation.

26. How far will the pendulum travel on the seventh swing?

27. How many swings will it take for the pendulum to travel less than 10 cm?

The game commission observes the fish population in a stream and notices that the number of trout increases by a factor of 1.5 every week. The commission initially observed 80 trout in the stream. Use this information for Items 28–31.

28. Write the explicit formula for the trout situation.

29. Make a graph of the population growth.

30. If this pattern continues, how many trout will be in the stream on the fifth week?

31. If this pattern continues, on what week will the trout population exceed 500?

MATHEMATICAL PRACTICES
Construct Viable Arguments and Critique the Reasoning of Others

32. Samir says that it is possible for a sequence to be both an arithmetic sequence and a geometric sequence. Do you agree or disagree? Explain.

Exponents, Radicals, and Geometric Sequences

TAKING STOCK

Stocking a lake is the process of adding fish to the lake. Once the fish have been added to the lake, their population growth depends upon many factors, such as the species of the fish, the number of predators in the lake, the quality of the water, and the lake's food supply.

Sapphire Lake was stocked with five species of fish several years ago. Michelle, a new employee at the parks and recreation commission, wants to analyze the population growth of the fish. She is able to find only the information shown below.

Species	Population Model	Meaning of Variables
A	$(3x^2y)^2$	x = average length of species in centimeters y = year of project (Year 1 is the year in which fish are first added to the lake.)
B	$6(xy)^3$	
C	$20\sqrt{xy}$	
D	$4\sqrt{x^2y}$	

Species E	
Year	**Population (hundreds)**
1	4
2	6
3	9
4	13.5

1. Michelle wants to know the ratio of the population of species A to species B.

a. Write the ratio as a fraction in simplified form without negative exponents.

b. Write the ratio using negative exponents.

2. Next, Michelle analyzes the population of species C and D.

a. Write the ratio of the population of species C to species D as a fraction in simplified form.

b. Michelle needs to know the total population of species C and D in Year 1. She learns that the average length of both species is 8 centimeters. Write a simplified expression for the approximate total population of species C and D in Year 1.

c. Is the expression you wrote in Part (b) rational or irrational? Explain your reasoning.

3. Michelle assumes that the population of species E continues to grow as shown in the table.

a. Write an explicit formula for the sequence.

b. Write a recursive formula for the sequence. Include the recursive formula in function notation.

c. According to the model, what was the approximate population of species E in year 7?

Exponents, Radicals, and Geometric Sequences

TAKING STOCK

Scoring Guide	Exemplary	Proficient	Emerging	Incomplete
	The solution demonstrates the following characteristics:			
Mathematics Knowledge and Thinking (Items 1a, 1b, 2a, 2b, 3c)	• Clear and accurate understanding of how to write and simplify exponential and radical expressions, including expressions with negative exponents • Fluency in determining a specified term of a geometric sequence	• Adequate understanding of how to write and simplify exponential and radical expressions, including expressions with negative exponents • Correct identification of a specified term of a geometric sequence	• Partial understanding of how to write and simplify exponential and radical expressions, including expressions with negative exponents • Partial understanding of and some difficulty determining a specified term of a geometric sequence	• Little or no understanding of how to write and simplify exponential and radical expressions, including expressions with negative exponents • Incomplete understanding of and significant difficulty determining a specified term of a geometric sequence
Problem Solving (Item 3)	• Appropriate and efficient strategy that results in a correct answer	• Strategy that may include unnecessary steps but results in a correct answer	• Strategy that results in some incorrect answers	• No clear strategy when solving problems
Mathematical Modeling/ Representations (Items 3a, 3b)	• Effective understanding of how to describe a real-world data set using explicit and recursive formulas	• Little difficulty describing a real-world data set using explicit and recursive formulas	• Partial understanding of how to describe a real-world data set using explicit and recursive formulas	• Inaccurate or incomplete understanding of how to describe a real-world data set using explicit and recursive formulas
Reasoning and Communication (Items 2c, 3c)	• Precise use of appropriate math terms and language to explain whether an expression is rational or irrational • Ease and accuracy describing the relationship between a sequence and a real-world scenario	• Correct identification of an expression as rational or irrational with an adequate explanation • Little difficulty describing the relationship between a sequence and a real-world scenario	• Misleading or confusing explanation of whether an expression is rational or irrational • Partially correct description of the relationship between a sequence and a real-world scenario	• Incomplete or inaccurate explanation of whether an expression is rational or irrational • Little or no understanding of how a sequence might relate to a real-world scenario

Exponential Functions

Protecting Your Investment

Lesson 22-1 Exponential Functions and Exponential Growth

My Notes

Learning Targets:

- Understand the definition of an exponential function.
- Graph and analyze exponential growth functions.

SUGGESTED LEARNING STRATEGIES: Marking the Text, Create Representations, Look for a Pattern, Interactive Word Wall, Predict and Confirm, Think-Pair-Share

The National Association of Realtors estimates that, on average, the price of a house doubles every ten years. Tony's grandparents bought a house in 1960 for $10,000. Assume that the trend identified by the National Association of Realtors applies to Tony's grandparents' house.

1. What was the value of Tony's grandparents' house in 1970 and in 1980?

2. Compute the difference in value from 1960 to 1970.

3. Compute the ratio of the 1970 value to the 1960 value.

MATH TIP

The ratio of the quantity a to the quantity b is evaluated by dividing a by b (ratio of a to $b = \frac{a}{b}$).

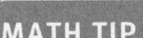

4. Complete the table of values for the years 1960 to 2010.

House Value				
Year	Decades Since 1960	Value of House	Difference Between Values of Consecutive Decades	Ratio of Values of Consecutive Decades
1960	0	$10,000	—	—
1970				
1980				
1990				
2000				
2010				

5. What patterns do you recognize in the table?

My Notes

6. Write the house values as a sequence. Identify the sequence as arithmetic or geometric and justify your answer.

7. Using the data from the table, graph the ordered pairs (decades since 1960, house value) on the coordinate grid below.

House Value

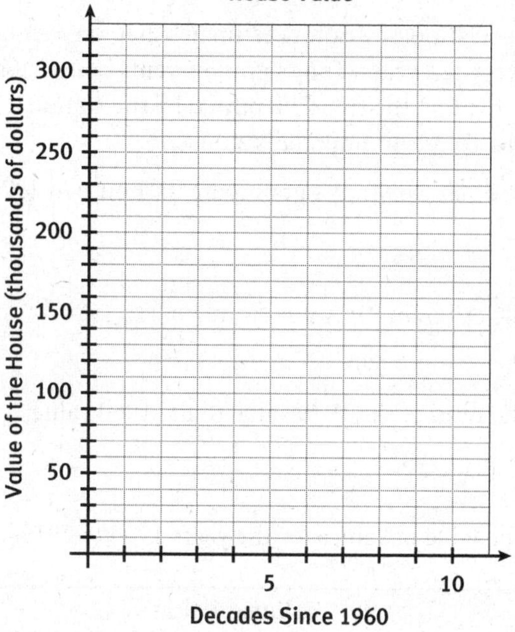

8. The data comparing the number of decades since 1960 and value of the house are not linear. Explain why using the table and the graph.

9. **Make use of structure.** Using the information that you have regarding the house value, predict the value of the house in the year 2020. Explain how you made your prediction.

10. Tony would like to know what the value of the house was in 2005. Using the same data, predict the house value in 2005. Explain how you made your prediction.

DISCUSSION GROUP TIPS

As you discuss ideas about the housing scenario, make notes and listen to what your group members have to contribute. Ask and answer questions to clearly aid comprehension and to ensure understanding of all group members' ideas.

The increase in house value for Tony's grandparents' house is an example of ***exponential growth***. Exponential growth can be modeled using an ***exponential function***.

Exponential Function
A function of the form $f(x) = a \cdot b^x$, where x is the domain value, $f(x)$ is the range value, $a \neq 0$, $b > 0$, and $b \neq 1$.

My Notes

In exponential growth, a quantity is multiplied by a constant factor greater than 1 during each time period.

11. The value of Tony's grandparents' house is growing exponentially because it is multiplied by a constant factor for each decade. What is this constant factor?

A function that can be used to model the house value is $h(t) = 10{,}000 \cdot (2)^t$. Use this function for Items 12–17.

12. Identify the meaning of $h(t)$ and t. What are the reasonable domain and range?

13. Describe how your answer to Item 11 is related to the function $h(t) = 10{,}000 \cdot (2)^t$.

14. Complete the table of values for t and $h(t)$. Then graph the function $h(t)$ on the grid below.

t	$h(t)$

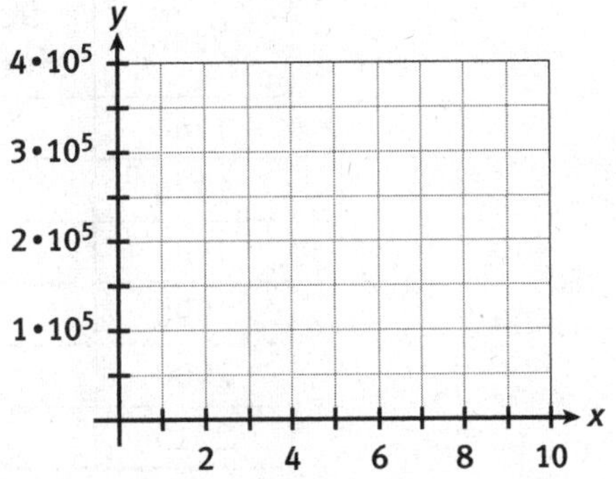

15. What was the value of the house in 1960? Describe how this value is related to the function $h(t) = 10{,}000 \cdot (2)^t$ and to the graph.

16. Calculate the value of the house in the year 2020. How does the value compare with your prediction in Item 9?

17. Calculate the value of the house in the year 2005. How does the value compare with your prediction in Item 10?

TECHNOLOGY TIP

Graph $h(t)$ on a graphing calculator. Find the y-coordinate when x is about 4.5. The value should be close to your calculated value in Item 17.

My Notes

Check Your Understanding

18. Copy and complete the table for the exponential function $g(x) = 3^x$.

x	$g(x)$
0	
1	
2	
3	
4	

19. Use your table to make a graph of $g(x)$.

20. Identify the constant factor for this exponential function.

Isaac evaluates the function modeling Tony's grandparents' house value, $h(t) = 10{,}000 \cdot (2)^t$, at $t = 2.5$. The variable t represents the number of decades since 1960.

21. What is $h(2.5)$?

22. For which year is Isaac estimating the house's value?

LESSON 22-1 PRACTICE

The value of houses in different locations can grow at different rates. The table below shows the value of Maddie's house from 1960 until 2010. Use the table for Items 23–25.

Year	Decades Since 1960	Value of House
1960	0	$10,000
1970	1	$15,000
1980	2	$22,500
1990	3	$33,750
2000	4	$50,625
2010	5	$75,938

23. Create a graph showing the value of Maddie's house from 1960 until 2010.

24. Explain how you know that the value of Maddie's house is growing exponentially.

25. What was the approximate value of Maddie's house in 1995?

The function $f(t) = 20{,}000 \cdot (1.2)^t$ can be used to find the value of Eduardo's house between 1970 and 2010, where the initial value of the function is the value of Eduardo's house in 1970.

26. Model with mathematics. Describe what the domain and range of the function mean in the context of Eduardo's house value.

27. What was the value of Eduardo's house in 1970?

28. Approximately how much was the house worth in 2000?

My Notes

Learning Targets:

- Describe characteristics of exponential decay functions.
- Graph and analyze exponential decay functions.

SUGGESTED LEARNING STRATEGIES: Look for a Pattern, Create Representations, Predict and Confirm, Discussion Groups, Visualization

Radon, a naturally occurring radioactive gas, was identified as a health hazard in some homes in the mid 1980s. Since radon is colorless and odorless, it is important to be aware of the concentration of the gas. Radon has a *half-life* of approximately four days.

Tony's grandparents' house was discovered to have a radon concentration of 400 pCi/L. Renee, a chemist, isolated and eliminated the source of the gas. She then wanted to know the quantity of radon in the house in the days following so that she could determine when the house would be safe.

CONNECT TO SCIENCE

All radioactive elements have a *half-life*. A half-life is the amount of time in which a radioactive element decays to half of its original quantity.

CONNECT TO SCIENCE

The US Environmental Protection Agency (EPA) recommends that the level of radon be below 4 pCi/L (picoCuries per liter) in any home. The EPA recommends that all homes be tested for radon.

1. **Make sense of problems.** What is the amount of the radon in the house four days after the source was eliminated? Explain your reasoning.

2. Compute the difference in the amount of radon from Day 0 to Day 4.

3. Determine the ratio of the amount of radon on Day 4 to the amount of radon on Day 0.

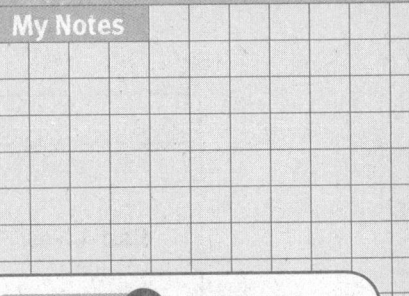

My Notes

4. Complete the table for the radon concentration.

Radon Concentration				
Half-Lives	Days After Radon Source Was Eliminated	Concentration of Radon in pCi/L	Difference Between Concentration of Consecutive Half-Lives	Ratio of Concentrations of Consecutive Half-Lives
0	0	400	—	—
1				
2				
3				

5. **Express regularity in repeated reasoning.** What patterns do you recognize in the table?

6. Graph the data in the table as ordered pairs in the form (half-lives, concentration).

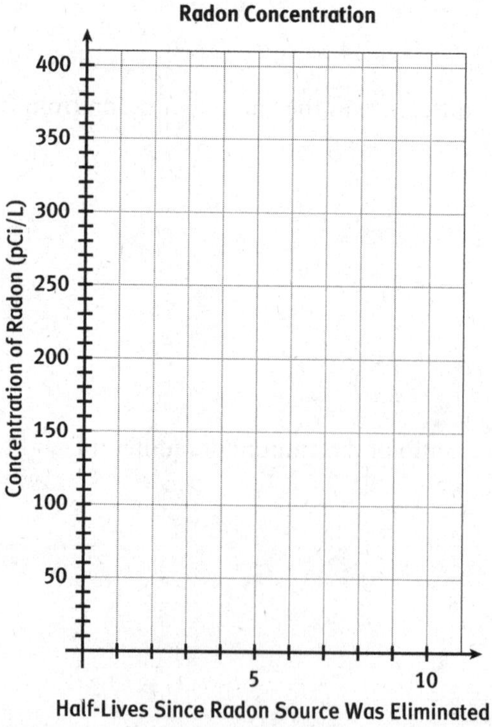

7. The data that compares the number of half-lives and the concentration of radon are not linear. Explain why using the table of values and the graph.

My Notes

8. Renee needs to know the concentration of radon in the house after 20 days. How many radon half-lives are in 20 days? What is the concentration after 20 days?

9. How many radon half-lives are in 22 days? Predict the concentration after 22 days.

The decrease in radon concentration in Tony's grandparents' house is an example of ***exponential decay***. Exponential decay can be modeled using an exponential function.

In exponential decay, a quantity is multiplied by a constant factor that is greater than 0 but less than 1 during each time period.

10. a. The concentration of radon is multiplied by a constant factor for each half-life. What is this constant factor?

b. Write an exponential function $r(t)$ for the table in Item 4, using 400 as the initial concentration of radon and the constant factor from part a.

Use the function from Item 10b for Items 11–17.

11. Identify the meaning of $r(t)$ and t. What are the reasonable domain and range?

12. Describe how your answer to Item 10a is related to the function $r(t)$.

13. Graph the function $r(t)$.

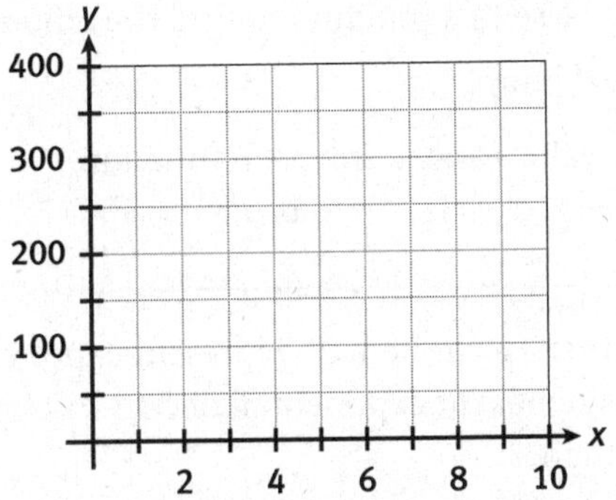

14. Describe how the original concentration of radon is related to the function and to the graph.

My Notes

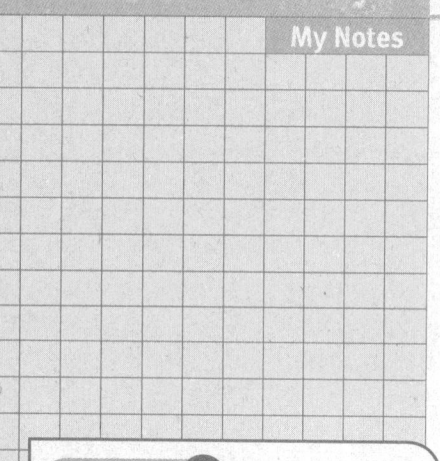

15. Use the function to identify the concentration of radon after 20 days. How does the concentration compare with your prediction in Item 8?

16. Use the function to calculate the concentration of radon after 22 days. How does the concentration compare with your prediction in Item 9?

CONNECT TO AP

In calculus, you will discover what happens as functions approach 0.

17. Construct viable arguments. Will the concentration of radon ever be 0? Explain your reasoning.

Check Your Understanding

18. Copy and complete the table for the exponential function $g(x) = \left(\frac{1}{4}\right)^x$.

19. Identify the constant factor for this exponential function.

x	$g(x)$
0	
1	
2	
3	

LESSON 22-2 PRACTICE

20. The amount of medication in a patient's bloodstream decreases exponentially from the time the medication is administered. For a particular medication, a function that gives the amount of medication in a patient's bloodstream t hours after taking a 100 mg dose is $A(t) = 100\left(\frac{7}{10}\right)^t$. Use this function to find the amount of medication remaining after 2 hours.

21. Make a table of values and graph each function.

a. $h(x) = 2^x$

b. $l(x) = 3^x$

c. $m(x) = \left(\frac{1}{2}\right)^x$

d. $p(x) = \left(\frac{1}{3}\right)^x$

22. Which of the functions in Item 21 represent exponential growth? Which of the functions represent exponential decay? Explain using your table of values and graph.

23. Reason abstractly. How can you identify which of the functions represent growth or decay by looking at the function?

24. Express regularity in repeated reasoning. Write an exponential function and identify its constant factor.

My Notes

Learning Targets:

- Describe key features of graphs of exponential functions.
- Compare graphs of exponential and linear functions.

SUGGESTED LEARNING STRATEGIES: Predict and Confirm, Discussion Groups, Create Representations, Look for a Pattern, Sharing and Responding, Summarizing

Recall that an exponential function is a function of the form $f(x) = ab^x$, where $a \neq 0$, $b > 0$, and $b \neq 1$.

1. Use a graphing calculator to graph each function. Sketch each graph on the coordinate grid provided.

	$a > 0$	$a < 0$
$b > 1$	**a.** $y = 3 \cdot 2^x$ 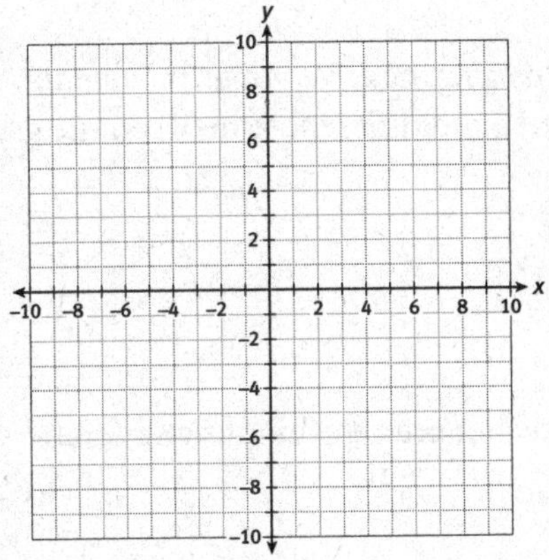	**b.** $y = (-3) \cdot 2^x$ 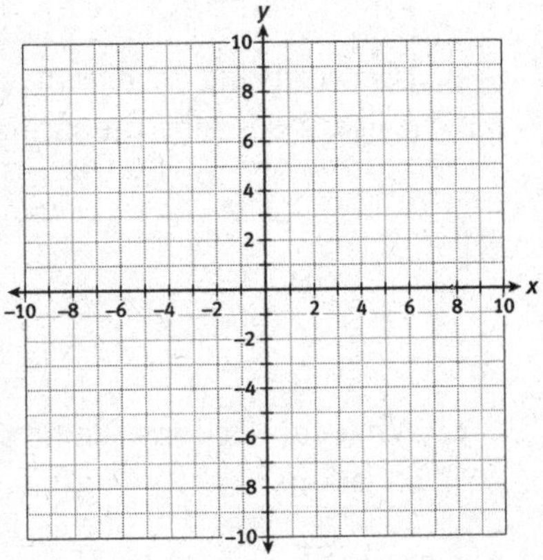
$0 < b < 1$	**c.** $y = 3 \cdot (0.5^x)$ 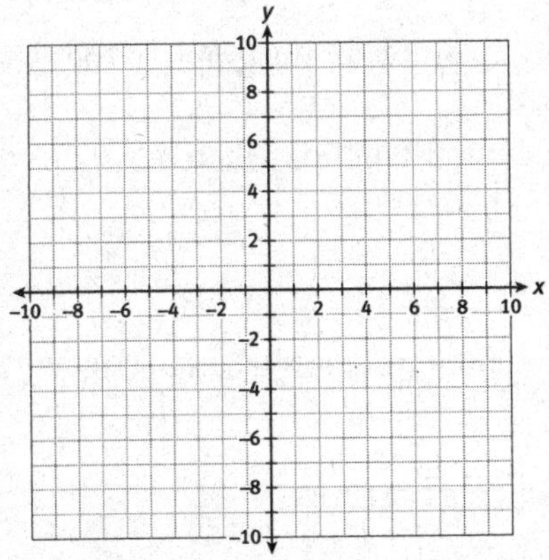	**d.** $y = -2 \cdot (0.5^x)$ 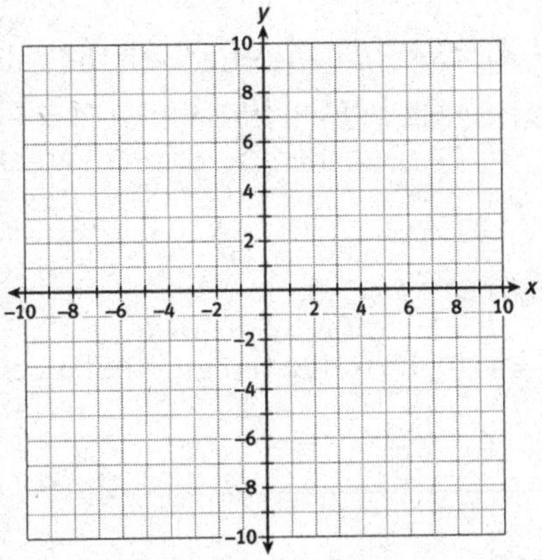

2. Compare and contrast the graphs and equations in Items 1a and 1b above.

a. How are the equations similar and different?

My Notes

b. Use words like *increasing, decreasing, positive, negative, domain,* and *range* to describe the similarities and differences in the graphs.

c. What connections can be made between the graphs and their equations?

3. Compare and contrast the graphs and equations in Items 1c and 1d.

a. How are the equations similar and different?

b. Use words like *increasing, decreasing, positive, negative, domain,* and *range* to describe the similarities and differences between the graphs.

c. What connections can be made between the graphs and their equations?

4. Describe the effects of the values of a and b on the graph of the exponential function $f(x) = ab^x$.

a. Describe the graph of an exponential function when $a > 0$.

b. Describe the graph of an exponential function when $a < 0$.

c. Describe the graph of an exponential function when $b > 1$.

d. Describe the graph of an exponential function when $0 < b < 1$.

My Notes

Check Your Understanding

5. Describe the values of a and b for which the exponential function $f(x) = ab^x$ is always positive.

6. Describe the values of a and b for which the exponential function $f(x) = ab^x$ is increasing.

7. Let $f(x) = 2x$ and $g(x) = 2^x$. Complete the tables below for each function.

x	0	1	2	3	4	5
$f(x)$						

x	0	1	2	3	4	5
$g(x)$						

8. Graph $f(x)$ and $g(x)$ below.

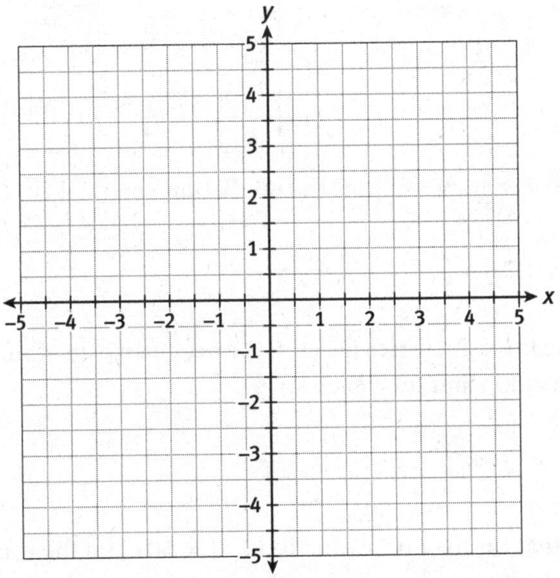

9. Examine the graphs of $f(x)$ and $g(x)$. Compare the values of each function from $x = -2$ through $x = 2$. Which function is greater on this interval?

10. Examine the values of $f(x)$ and $g(x)$ for $x > 2$.

a. Which function is greater on this interval?

My Notes

b. Do you think this will continue to be true as x continues to increase? Explain your reasoning.

11. To take a closer look at the graphs of $f(x)$ and $g(x)$ for larger values of x, regraph the two functions below. Note the new scale.

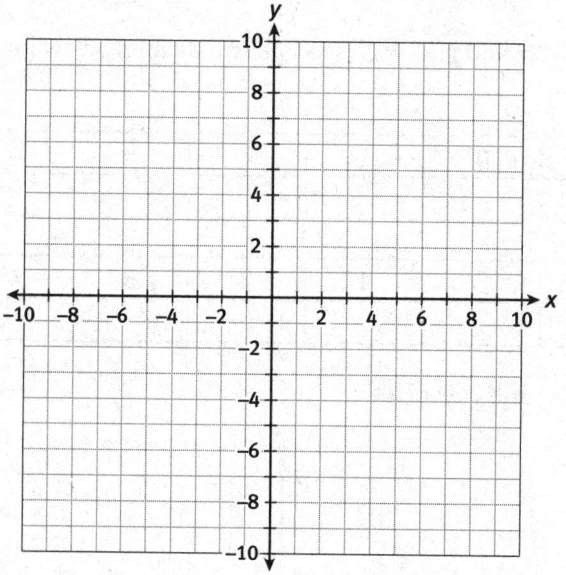

12. Does the new graph support the prediction you made in Item 10b?

13. Which function increases faster, $f(x)$ or $g(x)$? Explain your reasoning using the graph and the tables.

Alex believes that for the linear function $f(x) = 50x$ and the exponential function $g(x) = 2^x$, the value of $f(x)$ is always greater than the value of $g(x)$.

Glenda believes that for a linear function $f(x)$ to always be greater than an exponential function $g(x)$, the graph of $f(x)$ must be very steep while the graph of $g(x)$ must be very flat. She proposes graphing $f(x) = 50x$ and $g(x) = 1.1^x$ to test her conjecture.

14. a. Test Alex's conjecture by graphing $f(x) = 50x$ and $g(x) = 2^x$ on your graphing calculator. Do you agree or disagree with Alex's conjecture? Explain your reasoning.

b. Use appropriate tools strategically. Now adjust your viewing window to match the coordinate plane below. Sketch the graphs of $f(x)$ and $g(x)$.

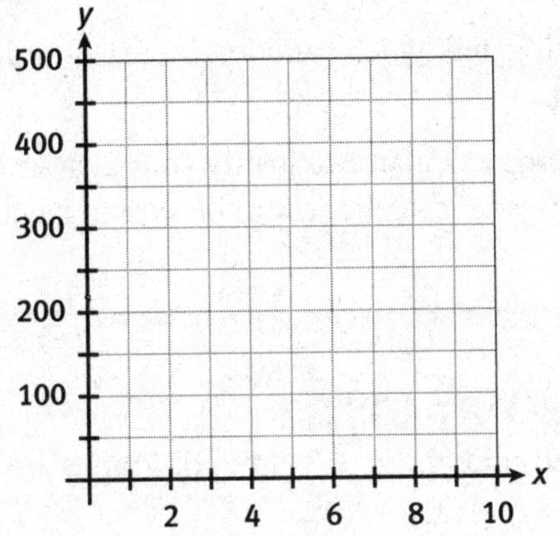

15. Should Alex revise his conjecture? Use the graph in Item 14b to explain.

16. a. Test Glenda's conjecture by graphing $f(x) = 50x$ and $g(x) = 1.1^x$ on your graphing calculator. Do you agree with Glenda's conjecture?

b. Now adjust your viewing window to match the coordinate plane below. Sketch the graphs of $f(x)$ and $g(x)$.

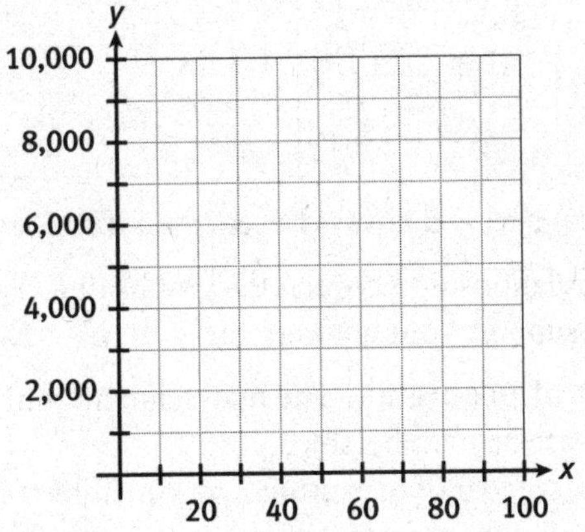

17. Should Glenda revise her conjecture? Use the graph in Item 16b to support your response.

18. Attend to precision. Is an exponential function always greater than a linear function? Explain your reasoning.

My Notes

TECHNOLOGY TIP

Use the axis labels on the blank grid to determine how to set the viewing window on your calculator.

My Notes

Check Your Understanding

Use the functions $a(x) = 25x$ and $b(x) = 5 \cdot 3^x$ for Items 19 and 20.

19. Without graphing, tell which function increases more quickly. Explain your reasoning.

20. Use your graphing calculator to justify your answer to Item 19. Sketch the graphs from your calculator, and be sure to label your viewing window.

LESSON 22-3 PRACTICE

Isaac graphs $f(x) = 9x$ and $g(x) = 0.5 \cdot 4^x$. His graphs are shown below.

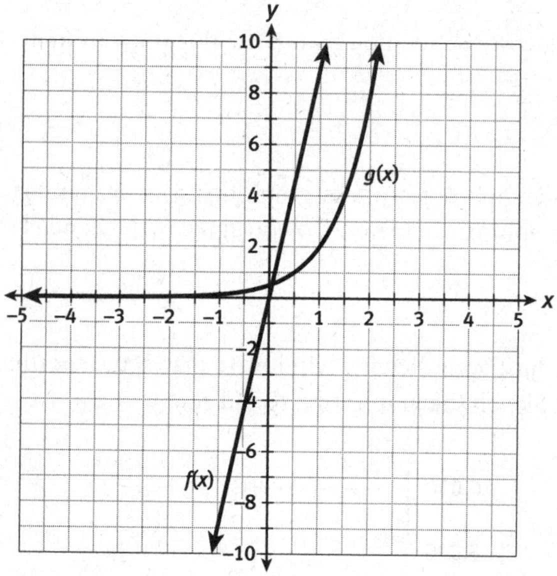

21. Isaac states that $g(x)$ will always be less than $f(x)$. Explain Isaac's error.

22. Describe the relationship between the graphs of $f(x)$ and $g(x)$. Make a new graph to support your answer.

23. Make sense of problems. The math club has only 10 members and wants to increase its membership.

- Julia proposes a goal of recruiting 2 new members each month. If the club meets this goal, the function $y = 2x + 10$ will give the total number of members y after x months.
- Jorge proposes a goal to increase membership by 10% each month. If the club meets this goal, the function $y = 10 \cdot 1.1^x$ will give the total number of members y after x months.

Club members want to choose the goal that will cause the membership to grow more quickly. Assume that the club will meet the recruitment goal that they choose. Which proposal should they choose? Use a graph to support your answer.

ACTIVITY 22 PRACTICE

Write your answers on notebook paper. Show your work.

Lesson 22-1

In January of this year, a clothing store earned \$175,000. Since then, earnings have increased by 10% each month. A function that models the store's earnings after m months is $e(m) = 175{,}000 \cdot (1.1)^m$. Use this information for Items 1–3.

1. Copy and complete the table.

Months After January (m)	Earnings $e(m)$
0	\$175,000
1	
2	
3	
4	

2. Make a graph of the function.

3. Predict the store's earnings after 9 months.

4. A scientist studying a bacteria population recorded the data in the table below.

Time (min)	0	1	2	3
Number of Bacteria	8	20	50	125

Is the number of bacteria growing exponentially? Justify your response.

5. The function $f(x) = 3 \cdot b^x$ is an exponential growth function. Which statement about the value of b is true?

A. Because $f(x)$ is an exponential growth function, b must be positive.

B. Because $f(x)$ is an exponential growth function, b must be greater than 1.

C. Because $f(x)$ is an exponential growth function, b must be between 0 and 1.

D. The function represents exponential growth because $3 > 1$, so b can have any value.

Lesson 22-2

A new car *depreciates*, or loses value, each year after it is purchased. A general rule is that a car loses 15% of its value each year.

Christopher bought a new car for \$25,000. A function that models the value of Christopher's car after t years is $v(t) = 25{,}000 \cdot (0.85)^t$. Use this information for Items 6–8.

6. Copy and complete the table.

Years After Purchase (t)	Value of Car $v(t)$
0	\$25,000
1	
2	
3	
4	

7. Make a graph of the function.

8. Predict the value of Christopher's car after 10 years.

For Items 9–12, graph each function and tell whether it represents exponential growth, exponential decay, or neither.

9. $y = (2.5)^x$

10. $y = -0.75x$

11. $y = 3(1.5)^x$

12. $y = 80(0.25)^x$

For Items 13–15, tell whether each function represents exponential growth, exponential decay, or neither. Justify your responses.

13.

x	0	1	2	3
y	2	60	118	176

14.

x	0	1	2	3
y	25	5	1	0.2

15.

x	0	2	4	6
y	1	9	81	729

16. A wildlife biologist is studying an endangered species of salamander in a particular region. She finds the following data.

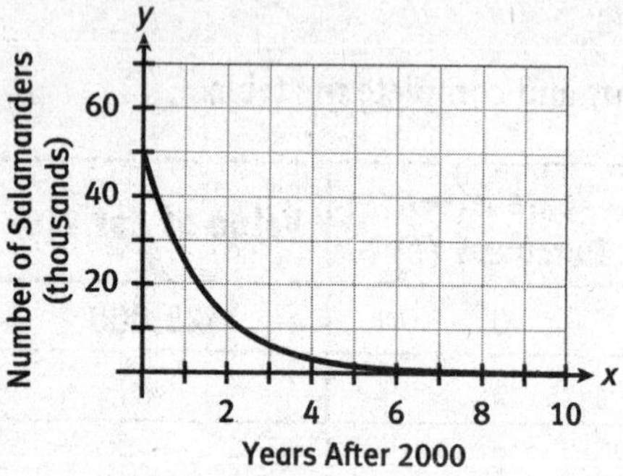

What was the initial number of salamanders in 2000?

17. Write a function that represents exponential decay. Explain how you know that your function represents exponential decay.

Lesson 22-3

Graph each of the following functions. Identify the values of a and b, and describe how these values affect the graphs.

18. $y = -4(2)^x$

19. $y = -1(2.5)^x$

20. $y = 1.5(2)^x$

21. $y = 0.5(0.2)^x$

For Items 22–29, use a graphing calculator to graph each function. Compare each function to $a(x) = 2^x$, graphed below. Describe the similarities and differences between the graphs.

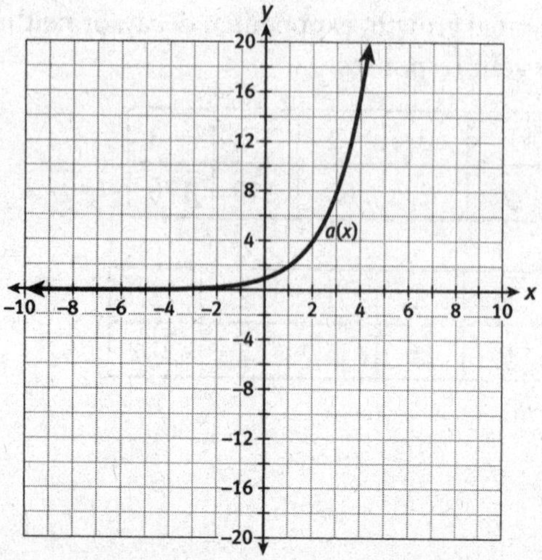

22. $f(x) = 0.5 \cdot 5^x$

23. $f(x) = 2 \cdot (1.1)^x$

24. $f(x) = 12^x$

25. $f(x) = 0.25 \cdot 4^x$

26. $f(x) = -3 \cdot 6^x$

27. $f(x) = -1 \cdot (0.3)^x$

28. $f(x) = 0.1 \cdot 2^x$

29. $f(x) = (0.5)^x$

30. Which function increases the fastest?

A. $y = 104x$
B. $y = -2 \cdot 15^x$
C. $y = 12^x$
D. $y = -220x$

31. Examine the graphs of $f(x) = 3^x$ and $g(x) = 5x$, shown below.

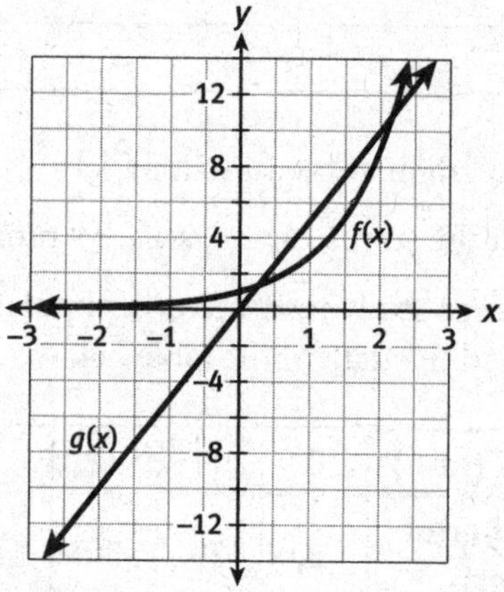

a. Estimate the values of x for which $f(x)$ is greater than $g(x)$.
b. Estimate the values of x for which $g(x)$ is greater than $f(x)$.
c. As the values of x decrease, the graph of $f(x)$ gets closer and closer to 0, or the x-axis. Will the graph ever intersect the x-axis? Explain.

MATHEMATICAL PRACTICES
Reason Abstractly and Quantitatively

32. Why can't the value of a in an exponential function be 0? Why can't the value of b be equal to 1?

Modeling with Exponential Functions

Growing, Growing, Gone

Lesson 23-1 Compound Interest

My Notes

Learning Target:

- Create an exponential function to model compound interest.

SUGGESTED LEARNING STRATEGIES: Create Representations, Look for a Pattern, Predict and Confirm, Discussion Groups, Think-Pair-Share, Critique Reasoning

Madison received $10,000 in gift money when she graduated from college. She deposits the money into an account that pays 5% ***compound interest*** annually.

MATH TERMS

Compound interest is interest, or money paid by a bank to an account holder, that is earned on both initial account funds, or *principal*, and previously earned interest.

1. To find the total amount of money in her account after the first year, Madison must add the interest earned in the first year to the initial amount deposited.

a. Calculate the earned interest for the first year by multiplying the amount of Madison's deposit by the interest rate of 5%.

b. Including interest, how much money did Madison have in her account at the end of the first year?

2. Madison wants to record the amount of money she will have in her account at the end of each year. Complete the table. Round amounts to the nearest cent.

Year	Account Balance
0	$10,000.00
1	$10,500.00
2	$11,025.00
3	
4	
5	
6	
7	
8	
9	
10	

My Notes

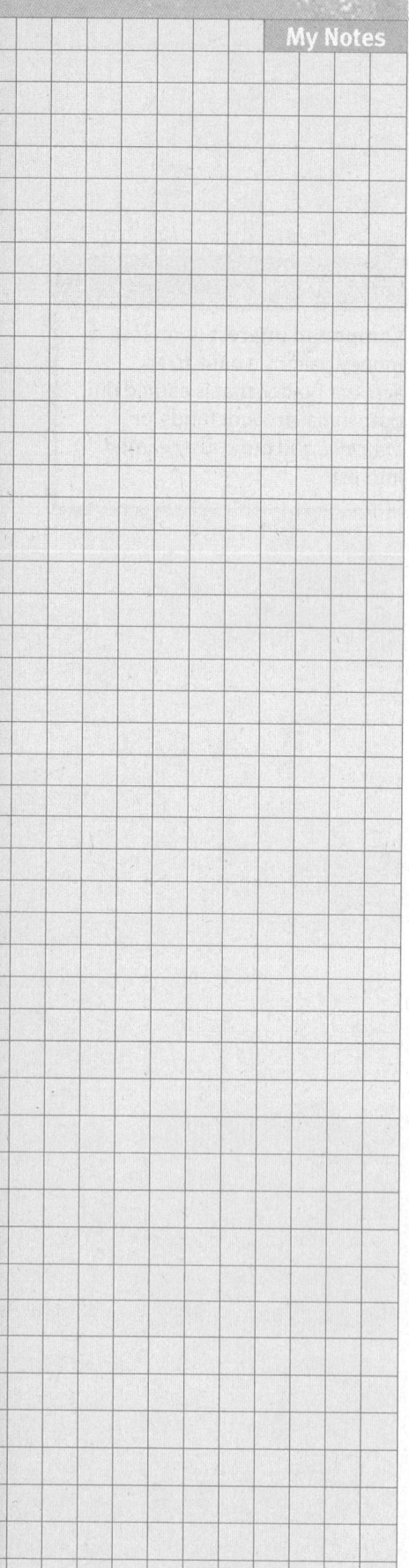

The amount of money in the account increases by a constant growth factor each year.

3. Identify the constant growth factor to the nearest hundredth.

4. How is the interest rate on Madison's account related to the constant growth factor in Item 3?

5. Instead of calculating the amount of money in the account after each year, write an expression for each amount of money using $10,000 and repeated multiplication of the constant factor. Then rewrite each expression using exponents.

Year	Account Balance
0	$10,000.00
1	$\$10{,}000 \cdot 1.05 = \$10{,}000 \cdot 1.05^1$
2	$(\$10{,}000 \cdot 1.05) \cdot 1.05 = \$10{,}000 \cdot 1.05^2$
3	
4	
5	
6	

6. Describe the relationship between the year number and the exponential expression.

7. Write an expression to represent the amount of money in the account at the end of Year 8.

8. Let t equal the number of years.

a. Express regularity in repeated reasoning. Write an expression to represent the amount of money in the account after t years.

b. Evaluate the expression for $t = 6$ to confirm that the expression is correct.

c. Evaluate the expression for $t = 10$.

9. Write your expression as a function $m(t)$, where $m(t)$ is the total amount of money in Madison's account after t years.

My Notes

en graph
Item 15

table in Item 2 to graph the function.

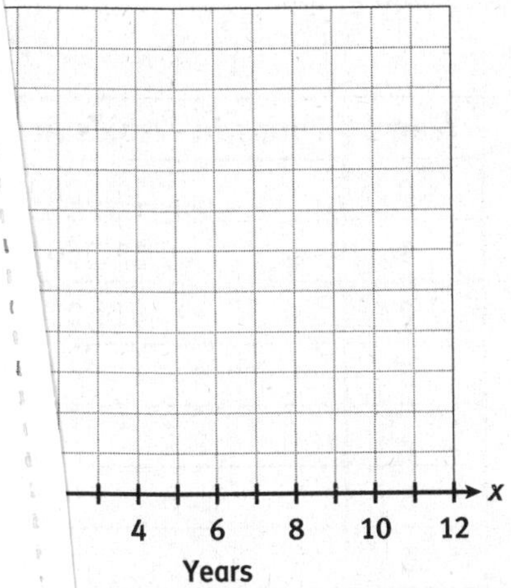

ear or non-linear. Justify your response.

in and range. Explain your reasoning.

other
deposits

w function

ɔurchasing a home. She estimates that
a down payment. Determine the year
ıgh funds in her account for the down

since she
nk, she will
n, even
our response

er account, her friend Frank deposits
ɔmpound interest rate of 6%.

he total funds in Frank's account,

account balance will differ from
e.

My Notes

16. Create a table of values for $f(t)$, rounding to the nearest dollar. Th
$f(t)$ on the grid in Item 10. Confirm or revise your prediction in
using the table and graph.

Year, t	Funds in Frank's Account, $f(t)$
0	\$10,000
1	\$10,600
2	\$11,236
3	
4	
5	
6	
7	
8	
9	
10	

At the same time that Madison and Frank open their accounts, ar
friend, Kasey, opens a savings account in a different bank. Kasey
\$12,000 at an annual compound interest rate of 4%.

17. How does Kasey's situation change the function? Write a ne
$k(t)$ to represent Kasey's account balance at any year t.

18. Critique the reasoning of others. Kasey believes that
started her account with more money than Madison or Fra
always have more money in her account than either of the
though her interest rate is lower. Is Kasey correct? Justify y
using a table, graph, or both.

My Notes

10. Use the data from the table in Item 2 to graph the function.

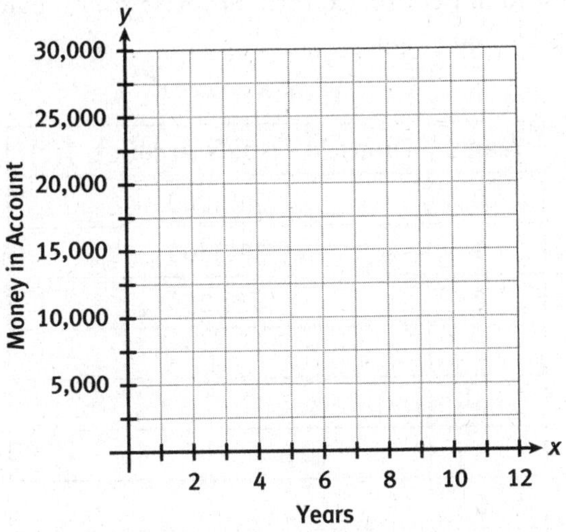

11. Describe the function as linear or non-linear. Justify your response.

12. Identify the reasonable domain and range. Explain your reasoning.

13. Madison's future plans include purchasing a home. She estimates that she will need at least $20,000 for a down payment. Determine the year in which Madison will have enough funds in her account for the down payment.

At the same time that Madison opens her account, her friend Frank deposits $10,000 in an account with an annual compound interest rate of 6%.

14. Write a new function to represent the total funds in Frank's account, $f(t)$, after t years.

15. Predict how the graph of Frank's bank account balance will differ from the graph of Madison's account balance.

My Notes

16. Create a table of values for $f(t)$, rounding to the nearest dollar. Then graph $f(t)$ on the grid in Item 10. Confirm or revise your prediction in Item 15 using the table and graph.

Year, t	Funds in Frank's Account, $f(t)$
0	\$10,000
1	\$10,600
2	\$11,236
3	
4	
5	
6	
7	
8	
9	
10	

At the same time that Madison and Frank open their accounts, another friend, Kasey, opens a savings account in a different bank. Kasey deposits \$12,000 at an annual compound interest rate of 4%.

17. How does Kasey's situation change the function? Write a new function $k(t)$ to represent Kasey's account balance at any year t.

18. **Critique the reasoning of others.** Kasey believes that since she started her account with more money than Madison or Frank, she will always have more money in her account than either of them, even though her interest rate is lower. Is Kasey correct? Justify your response using a table, graph, or both.

My Notes

19. Over a long period of time, does the initial deposit or the interest rate have a greater effect on the amount of money in an account that has interest compounded yearly? Explain your reasoning.

Most savings institutions offer compounding intervals other than annual compounding. For example, a bank that offers *quarterly compounding* computes interest on an account every quarter; that is, at the end of every 3 months. Instead of computing the interest once each year, interest is computed four times each year. If a bank advertises that it is offering 8% interest compounded quarterly, 8% is not the actual growth factor. Instead, the bank will use $\frac{8\%}{4} = 2\%$ to determine the quarterly growth factor.

CONNECT TO FINANCE

Interest can be compounded semiannually (every 6 months), quarterly (every 3 months), monthly, and daily.

20. What is the quarterly interest rate for an account with an annual interest rate of 5%, compounded quarterly?

21. Suppose that Madison invested her $10,000 in the account described in Item 20.

a. In the table below, determine Madison's account balance after the specified times since her initial deposit.

Time Since Initial Deposit	Number of Times Interest Has Been Compounded	Account Balance
3 months		
6 months		
9 months		
1 year		
4 years		
t years		

b. Write a function $A(t)$ to represent the balance in Madison's account after t years.

c. Calculate the balance in Madison's account after 20 years.

My Notes

MATH TIP

For a given annual interest rate, properties of exponents can be used to approximate equivalent semi-annual, quarterly, monthly, and daily interest rates. For example, the function $f(t) = 5000(1.03)^t$ is used to approximate the balance in an account with an initial deposit of \$5000 and an annual interest rate of 3%. $f(t)$ can be rewritten as

$$f(t) = 5000\left(1.03^{\frac{1}{12}}\right)^{12t}$$ and is equivalent to the function $g(t) = 5000(1.0025)^{12t}$, which reveals that the approximate equivalent monthly interest rate is 0.25%.

$$f(t) = 5000\left(1.03^{\frac{1}{4}}\right)^{4t}$$ is equivalent to $h(t) = 5000(1.0074)^{4t}$, which reveals that the approximate equivalent quarterly interest rate is 0.74%.

22. For the compounding periods given below, write a function to represent the balance in Madison's account after t years. Then calculate the balance in the account after 20 years. She is investing \$10,000 at a rate of 5% annual compound interest.

a. Yearly:

b. Quarterly:

c. Monthly:

d. Daily (assume there are 365 days in a year):

23. What is the effect of the compounding period on the amount of money in the account after 20 years as the number of times the interest is compounded each year increases?

Check Your Understanding

24. Write a function that gives the amount of money in Frank's account after t years when 6% annual interest is compounded monthly.

25. Create a table and a graph for the function in Item 24. Be sure to label the units on the x-axis correctly.

LESSON 23-1 PRACTICE

Model with mathematics. Nick deposits \$5000 into an account with a 4% annual interest rate, compounded annually.

26. Create a table showing the amount of money in Nick's account after 0–8 years.

27. Write a function that gives the amount of money in Nick's account after t years. Identify the reasonable domain and range.

28. Create a graph of your function.

29. Explain how Nick's account balance would be different if he deposited his money into an account that pays 2% annual interest, compounded annually. Graph this situation on the same coordinate plane that you used in Item 28. Describe the similarities and differences between the graphs.

Learning Targets:

- Create an exponential function to fit population data.
- Interpret values in an exponential function.

SUGGESTED LEARNING STRATEGIES: Create Representations, Look for a Pattern, Predict and Confirm, Think Aloud, Sharing and Responding, Construct Arguments

The population of Nevada since 1950 is shown in the table in the *My Notes* section.

1. Graph the data from the table.

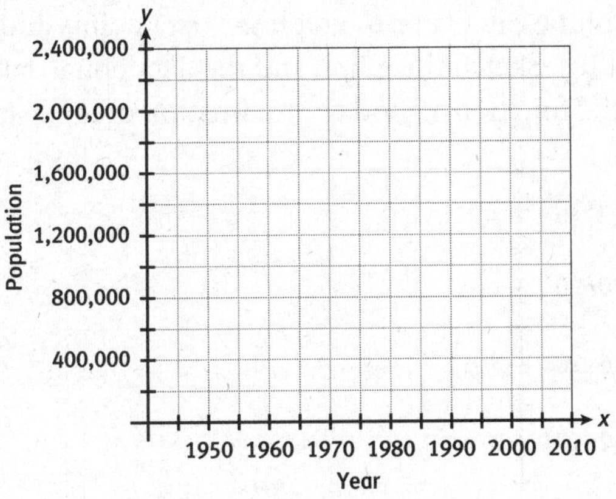

2. Use the table and the graph to explain why the data are not linear.

3. a. Complete the table by finding the approximate ratio between the populations in each decade.

Decades Since 1950	Resident Population	Ratio
0	160,083	--
1	285,278	$\frac{285,278}{160,083} \approx 1.782$
2	488,738	$\frac{488,738}{285,278} \approx$
3	800,508	
4	1,201,833	
5	1,998,257	

b. Explain how the table shows that the data are not exponential.

The data are not exactly exponential, but the shape of the graph resembles an exponential curve. Also, the table in Item 3a shows a near-constant factor. These suggest that the data are approximately exponential. Use ***exponential regression*** to find an exponential function that models the data.

My Notes

Year	Resident Population
1950	160,083
1960	285,278
1970	488,738
1980	800,508
1990	1,201,833
2000	1,998,257

MATH TERMS

Exponential regression is a method used to find an exponential function that models a set of data.

My Notes

TECHNOLOGY TIP

On a Texas Instruments (TI) calculator, perform exponential regression using the ExpReg function.

4. Use a graphing calculator to determine the exponential regression equation to model the relationship between the decades since 1950 and the population.

a. The calculator returns two values, a and b. Write these values below. Round a to the nearest whole number and b to the nearest thousandth, if necessary.

$a =$ $b =$

b. The general form of an exponential function is $y = ab^x$. Use this general form and the values of a and b from Part (a) to write an exponential function that models Nevada's population growth.

c. Use a graphing calculator to graph the data points and the function from Part (b). Sketch the graph and the data points below. Is the exponential function a good approximation of the data? Explain.

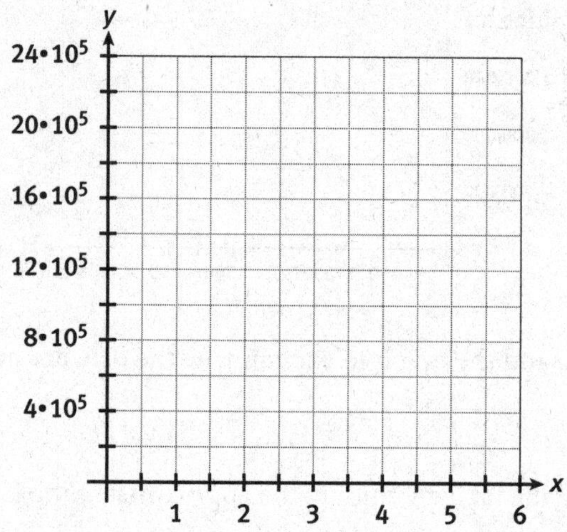

5. Reason abstractly. What does the value of b tell you about Nevada's population growth?

6. Interpret the value of a in terms of Nevada's population. How is this value related to the graph?

7. What do the domain values represent?

8. What would the x-intercept represent in terms of Nevada's population? Does the graph have an x-intercept? Explain.

My Notes

TECHNOLOGY TIP

Use the table function on a graphing calculator to determine the value of the function when $x = 4.5$.

9. **Make sense of problems.** Describe how to estimate the population of Nevada in 1995 using each of the following:
   a. the function identified in Item 4b

   b. the graph of the function

   c. a table

10. Estimate the population in 1995. Which method did you use, and why?

11. **a.** Estimate Nevada's population in 2010.

    b. Construct viable arguments. Which estimate do you think is likely to be more accurate, your estimate of the population in 1995 or in 2010? Explain.

The function for the growth rate of Nevada's population estimates the growth per decade. You can use this rate to estimate the growth per year, or the annual growth rate.

12. Let n be the number of years since 1950. Write an equation that gives the number of years n in x decades. Solve your equation for x.

13. Rewrite the function that models Nevada's population from Item 4b. Then write the function again, but replace x with the equivalent expression for x from Item 12.

14. Simplify to write the function in the form $y = ab^n$.

$$y = 170{,}377 \cdot (1.645)^{\frac{n}{10}} = 170{,}377 \cdot \left(1.645^{\frac{1}{10}}\right)^{\square}$$

$$\approx 170{,}377 \cdot (\quad)^n$$

My Notes

15. What is the approximate annual growth rate of Nevada's population? How do you know?

16. To find the approximate population of Nevada in 2013, what value should you use for n? Explain.

17. Use the function from Item 14 to find the approximate population of Nevada for the year 2013.

18. Compare the approximate population for 2013 that you found in Item 17 to the approximate population you found for 2010 in Item 11. Does your estimate for 2013 seem reasonable? Why or why not?

Check Your Understanding

19. Create a graph showing the annual growth of Nevada's population.

20. Describe the similarities and differences between the graph in Item 19 and the previous graph of Nevada's population from Item 4c.

LESSON 23-2 PRACTICE

The population of Texas from 1950 to 2000 is shown in the table below.

Year	Resident Population
1950	7,711,194
1960	9,579,677
1970	11,198,655
1980	14,225,513
1990	16,986,510
2000	20,851,820

21. **Use appropriate tools strategically.** Use a graphing calculator to find a function that models Texas's population growth.

22. Create a graph showing the actual population from the table and the approximate population from the function in Item 21.

23. Is the function a good fit for the data? Why or why not?

24. Describe the meanings of the domain, range, y-intercept, and x-intercept in the context of Texas's population growth.

ACTIVITY 23 PRACTICE

Write your answers on notebook paper.
Show your work.

Lesson 23-1

1. Four friends deposited money into savings accounts. The amount of money in each account is given by the functions below.

Marisol: $m(t) = 100 \cdot (1.01)^t$
Iris: $i(t) = 200 \cdot (1.04)^t$
Brenda: $b(t) = 300 \cdot (1.05)^t$
José: $j(t) = 400 \cdot (1.03)^t$

Which statement is correct?

A. José has the greatest interest rate.
B. Brenda has the greatest initial deposit.
C. The person with the least initial deposit also has the least interest rate.
D. The person with the greatest initial deposit also has the greatest interest rate.

Darius makes an initial deposit into a bank account, and then earns interest on his account. He records the amount of money in his account each year in the table below. Use this table for Items 2–5.

Year	Amount
0	$4000.00
1	$4120.00
2	$4243.60
3	$4370.91
4	$4502.04

2. Make a graph showing the amount of money in Darius's account each year.

3. Identify the constant factor. Round to the nearest hundredth.

4. Identify the reasonable domain and range. Explain your answers.

5. What is the annual interest rate? How do you know?

The amount of money y in Jesse's checking account t years after the account was opened is given by the function $j(t) = 15{,}000 \cdot (1.02)^t$. Use this information for Items 6–10.

6. What was the initial amount of money deposited in Jesse's account?

7. What is the annual interest rate?

8. Create a graph of the amount of money in Jesse's checking account.

9. Interpret the meaning of the y-intercept in the context of Jesse's account.

10. Find the amount of money in the account after 4 years.

The two graphs on the coordinate grid below represent the amounts of money in two different savings accounts. Graph a represents the amount of money in Allison's account, and graph b represents the amount of money in Boris's account. Use the graph for Items 11–13.

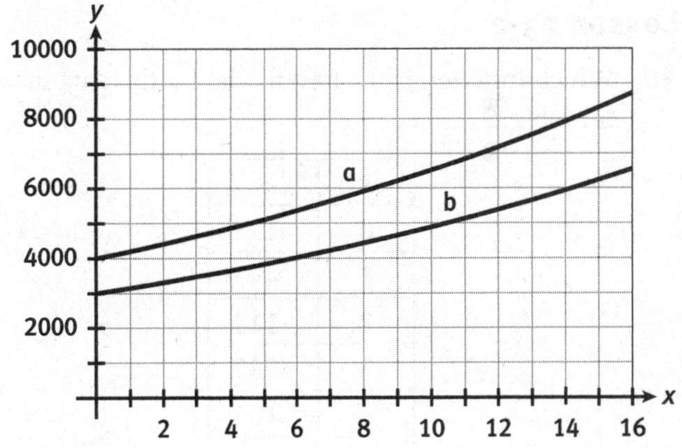

11. Whose account had a higher initial deposit? Use the graph to justify your answer.

12. What was the amount of Allison's initial deposit?

13. Identify the reasonable domain and range for each function, and explain your answers.

Maria's bank offers two types of savings accounts. The first has an annual interest rate of 8% compounded annually. The second also has an annual interest rate of 8%, but it is compounded monthly. She is going to open an account by depositing $1000. Use this information for Items 14–19.

14. If Maria chooses the first account, determine the amount of money she will have in the account after 3 years.

15. Write a function that gives the amount of money in the first account after *t* years.

16. Write a function that gives the amount of money in the second account after *t* years.

17. What is the *monthly* interest rate for the second account?

18. If Maria chooses the second account, determine the amount of money she will have in the account after 1 year.

19. After 10 years, which account will have the higher balance?

Lesson 23-2

20. Which function is the best model for the data in the table?

x	y
0	19
1	44.5
2	112
3	282
4	704

A. $y = 172x + 18$ **B.** $y = 44x^2$
C. $y = 44x$ **D.** $y = 172 \cdot 18^x$

For Items 21 and 22, tell whether an exponential function would be a good model for each data set. Explain your answers.

21.

x	0	1	2	3	4
y	33	58	120	247	506

22.

x	0	3	6	9	4
y	16.5	12	8.5	4.6	0

The head circumference of an infant is measured and recorded to track the infant's growth and development. Nathan's head circumferences from age 3 months through 12 months are recorded in the table below. Use the table for Items 23–27.

Age (months)	Head Circumference (cm)
3	43
4	44
5	44.7
6	45.2
7	45.8
8	46.3
9	47.1
10	47.6
11	48.0
12	48.3

23. Use a graphing calculator to find an exponential function to approximate Nathan's head circumference.

24. Identify the reasonable domain and range for the function in Item 23. Explain your answers.

25. Create a graph showing Nathan's head circumference.

26. Determine the growth rate, and explain how you found your answer.

27. Interpret the meaning of the *y*-intercept in the context of Nathan's head circumference.

MATHEMATICAL PRACTICES
Attend to Precision

28. Explain why the *x*-intercept can have a meaning in the context of a situation, such as population growth, but cannot be shown on the graph.

Mr. Davis has just become a grandfather! He wants to invest money for his new granddaughter's college education.

Mr. Davis has done some research on savings bonds. He has learned that you buy a savings bond from the government or from a bank. After one year, you can cash in your bond and get back the money you paid for it. However, if you wait at least five years, you will get back your money plus interest.

Mr. Davis has also learned that he can buy paper bonds or electronic bonds. While there are many similarities and differences between the bonds, Mr. Davis has summarized the most important information below.

Paper Bond	Electronic Bond
Current rate of interest: 1.8% annual, but the interest rate may change over the life of the bond.	Current rate of interest: 1.4% annual; this rate will not change.
For both bonds, the interest is compounded semiannually (every 6 months, or twice per year) for 30 years or until the bond is cashed in, whichever comes first.	

Mr. Davis decides to buy a $5000 bond that he will give to his granddaughter on her 18th birthday. He will use the current interest rates to decide which bond he will purchase.

1. Using the current interest rate, a function that gives the value of a $5000 paper bond after t years is $p(t) = 5000 \cdot (1.009)^{2t}$.
   a. How is the interest rate of 1.8% related to the function?
   b. Why is the exponent $2t$ instead of t?
   c. Use the function to determine the value of the bond in 18 years. Round your answer to the nearest cent.

2. **a.** Write a function $e(t)$ that gives the value of a $5000 electronic bond after t years.
   b. Use your function to determine the value of the bond in 18 years. Round your answer to the nearest cent.

3. Identify the reasonable domain and range for each function. Explain your answers.

4. Use a graphing calculator to graph both functions on the same coordinate plane. Sketch the graphs and label each function.

5. Explain to Mr. Davis which bond you think he should purchase and why.

6. Mr. Davis's accountant has more information about electronic bonds. She tells Mr. Davis that if you keep an electronic bond for 20 years, the value becomes double what you paid for it. Would this change your advice to Mr. Davis? Explain.

Scoring Guide	Exemplary	Proficient	Emerging	Incomplete
	The solution demonstrates the following characteristics:			
Mathematics Knowledge and Thinking (Items 1c, 2b)	• Effective understanding of and accuracy in evaluating an exponential function	• Largely correct understanding of and accuracy in evaluating an exponential function	• Partial understanding of and some difficulty in evaluating an exponential function	• Incomplete understanding of and significant difficulty in evaluating an exponential function
Problem Solving (Items 5, 6)	• Appropriate and efficient strategy that results in a correct answer	• Strategy that may include unnecessary steps but results in a correct answer	• Strategy that results in some incorrect answers	• No clear strategy when solving problems
Mathematical Modeling/ Representations (Item 1a, 1b, 2a, 3, 4)	• Fluency in representing a real-world scenario using an exponential function, including identification of a reasonable domain and range • Clear and accurate graphs of exponential functions	• Adequate understanding of how to represent a real-world scenario using an exponential function, including identification of a reasonable domain and range • Little difficulty graphing exponential functions	• Partial understanding of how to represent a real-world scenario using an exponential function, including identification of a reasonable domain and range • Partially accurate graphs of exponential functions	• Inaccurate or incomplete understanding of how to represent a real-world scenario using an exponential function, including identification of a reasonable domain and range • Inaccurate or incomplete graphs of exponential functions
Reasoning and Communication (Items 5, 6)	• Precise use of appropriate math terms and language to make and justify a recommendation	• Recommendation with an adequate justification	• Recommendation with a misleading or confusing justification	• No recommendation or a recommendation with an inaccurate or incomplete justification

Adding and Subtracting Polynomials

ACTIVITY 24

Polynomials in the Sun

Lesson 24-1 Polynomial Terminology

My Notes

Learning Targets:

- Identify parts of a polynomial.
- Identify the degree of a polynomial.

SUGGESTED LEARNING STRATEGIES: Create Representations, Vocabulary Organizer, Interactive Word Wall, Think-Pair-Share, Close Reading

A solar panel is a device that collects and converts solar energy into electricity or heat. The solar panel consists of interconnected solar cells. The panels can have differing numbers of solar cells and can come in square or rectangular shapes.

CONNECT TO SCIENCE

Solar panels, also known as photovoltaic panels, are made of semiconductor materials. A panel has both positive and negative layers of semiconductor material. When sunlight hits the semiconductor, electrons travel across the intersection of the two different layers of materials, creating an electric current.

1. How many solar cells are in the panel below?

2. **Reason abstractly.** If a solar panel has four rows as the picture does, but can be extended to have an unknown number of columns, x, write an expression to give the number of solar cells that could be in the panel.

3. Write an expression that would give the total number of cells in the panel for a solar panel having x rows and x columns.

4. If there were 5 panels like those found in Item 3, write an expression to represent the total number of solar cells.

All the answers in Items 1–4 are called *terms*. A ***term*** is a number, variable, or the product of a number and/or variable(s).

5. Write an expression to represent the sum of your answers from Items 1, 2, and 4.

My Notes

Expressions like the answer to Item 5 are called polynomials. A ***polynomial*** is a single term or the sum of two or more terms with *whole-number powers*.

6. List the terms of the polynomial you wrote in Item 5.

7. What are the ***coefficients*** of the polynomial in Item 5? What is the ***constant term***?

MATH TERMS

A **coefficient** is the numeric factor of a term.

A **constant term** is a term that contains only a number, such as the answer to Item 1. The constant term of a polynomial is a term of degree zero.

Check Your Understanding

8. Tell whether each expression is a polynomial. Explain your reasoning.

a. $3x^{-2} - 5$ **b.** $6x + 4x^2$ **c.** 15 **d.** $2 + x^{\frac{1}{2}}$

9. For the expressions in Item 8 that are polynomials, identify the terms, coefficients, and constant terms.

DISCUSSION GROUP TIPS

As needed, refer to the Glossary to review definitions and pronunciations of key terms. Incorporate your understanding into group discussion to confirm your knowledge and use of key mathematical language.

The ***degree of a term*** is the sum of the exponents of the variables contained in the term.

10. Identify the degree and coefficient of each term in the polynomial $4x^5 + 12x^3 + x^2 - x + 5$.

Term	Degree	Coefficient
$4x^5$	5	
$12x^3$		12
x^2		
$-x$		

11. Make use of structure. For the polynomial $2x^3y - 6x^2y^2 + 9xy - 13y^5 + 5x + 15$, list each term and identify its degree and coefficient. Identify the constant term.

My Notes

The ***degree of a polynomial*** is the greatest degree of any term in the polynomial.

12. Identify the degree and constant term of each polynomial.

Polynomial	Degree of Polynomial	Constant Term
$2x^2 + 3x + 7$	2	
$-5y^3 + 4y^2 - 8y - 3$		
$36 + 12x + x^2$		36

The ***standard form of a polynomial*** is a polynomial whose terms are written in ***descending order*** of degree. The ***leading coefficient*** is the coefficient of a polynomial's leading term when the polynomial is written in standard form.

MATH TERMS

Descending order of degree means that the term that has the highest degree is written first, the term with the next highest degree is written next, and so on.

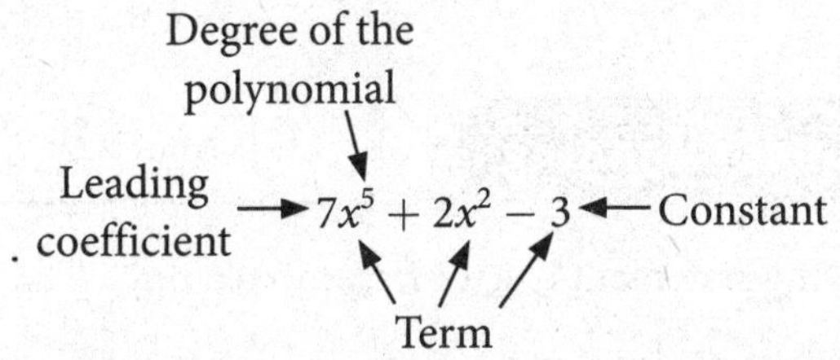

A polynomial can be classified by the number of terms it has when it is in simplest form.

Name	Number of Terms n	Examples
monomial	1	8 or $-2x$ or $3x^2$
binomial	2	$3x + 2$ or $4x^2 - 7x$
trinomial	3	$-x^2 - 3x + 9$
polynomial	$n > 3$	$9x^4 - x^3 - 3x^2 + 7x - 2$

READING MATH

The prefixes mono (one), bi (two), tri (three), and poly (many) appear in many math terms such as bisect (cut in half), triangle (three-sided figure), and polygon (many-sided figure).

My Notes

13. Fill in the missing information in the table below.

Polynomial	Number of Terms	Name	Leading Coefficient	Constant Term	Degree
$3x^2 - 5x$					
$-2x^2 + 13x + 6$					
$15x^2$					
$5p^3 + 2p^2 - p - 7$					
$a^2 - 25$					
$0.23x^3 + 0.54x^2 - 0.58x + 0.0218$					
$-9.8t^2 - 20t + 150$					

Check Your Understanding

14. Is the following statement true or false? Explain.

"All polynomials are binomials."

15. Describe your first step for writing $3x - 5x^2 + 7$ in standard form.

LESSON 24-1 PRACTICE

For Items 16–20, use the polynomial $4x^3 + 3x^2 - 9x + 7$.

16. Name the coefficients of the terms in the polynomial that have variables.

17. List the terms, and give the degree of each term.

18. What is the degree of the polynomial?

19. Identify the leading coefficient of the polynomial.

20. Identify the constant term of the polynomial.

Write each polynomial in standard form.

21. $9 + 8x^2 + 2x^3$

22. $y^2 + 1 + 4y^3 - 2x$

23. **Construct viable arguments.** Is the expression $5x^2 + \sqrt{2}x$ a polynomial? Justify your response.

My Notes

Learning Targets:

- Use algebra tiles to add polynomials.
- Add polynomials algebraically.

SUGGESTED LEARNING STRATEGIES: Discussion Groups, Use Manipulatives, Create Representations, Close Reading, Note Taking

Notice that in the solar panels at the right, there are 4^2 or 16 cells. Each column has 4 cells.

1. If a square solar panel with an unknown number of cells along the edge can be represented by x^2, how many cells would be in one column of the panel?

A square solar panel with x rows and x columns can be represented by the algebra tile:

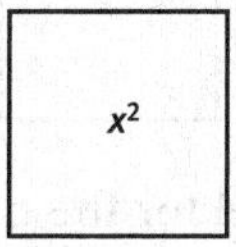

A column of x cells can be represented by using the tile 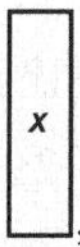

, and a single solar cell can be represented by [+1].

Suppose there were 3 square solar panels that each had x columns and x rows, 2 columns with x cells, and 3 single solar cells. You can represent $3x^2 + 2x + 3$ using algebra tiles.

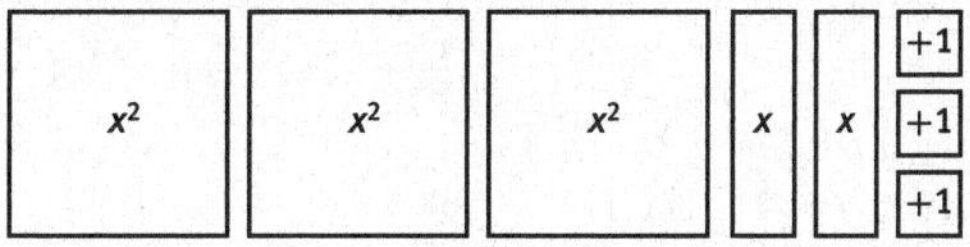

2. Represent $2x^2 - 3x + 2$ using algebra tiles. Draw a picture of the representation below.

MATH TIP

The additive inverse of the x^2, x, and 1 algebra tiles can be represented with another color, or the flip side of the tile.

My Notes

Adding polynomials using algebra tiles can be done by:

- modeling each polynomial
- identifying and removing zero pairs
- writing the new polynomial

Example A

Add $(3x^2 - 3x - 5) + (2x^2 + 5x + 3)$ using algebra tiles.

Step 1: Model the polynomials.

$3x^2 - 3x - 5$ $\qquad$ $2x^2 + 5x + 3$

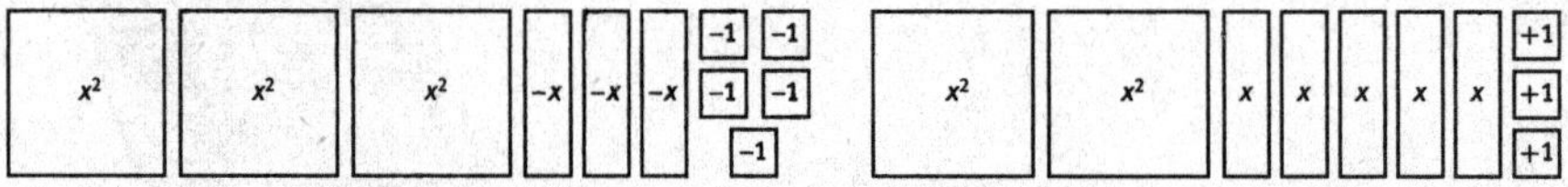

Step 2: Identify and remove zero pairs.

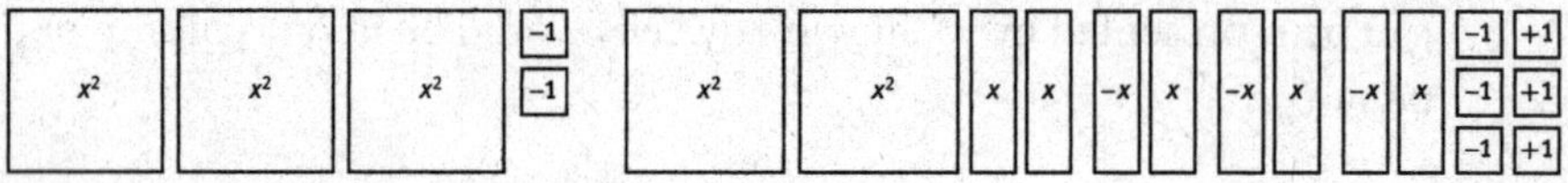

Step 3: Combine like tiles.

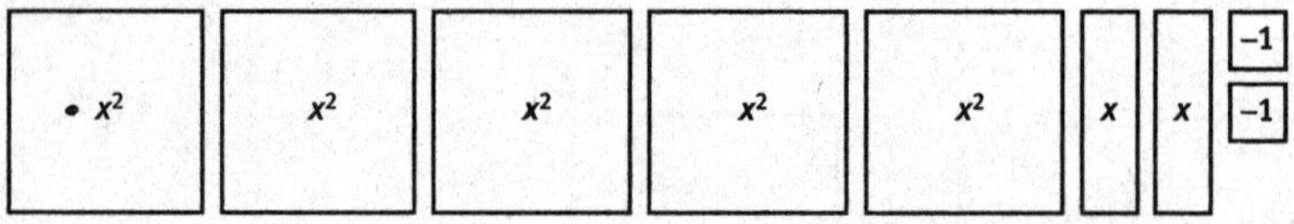

Step 4: Write the polynomial for the model in Step 3.

$5x^2 + 2x - 2$

Solution: $(3x^2 - 3x - 5) + (2x^2 + 5x + 3) = 5x^2 + 2x - 2$

Try These A

Add using algebra tiles.

a. $(x - 2) + (2x + 5)$

b. $(2y^2 + 3y + 6) + (3y^2 - 4)$

c. $(2x^2 + 3x + 9) + (-x^2 - 4x - 6)$

d. $(5 - 3x + x^2) + (2x + 4 - 3x^2)$

My Notes

3. **Use appropriate tools strategically.** Can you use algebra tiles to add $(4x^4 + 3x^2 + 15) + (x^4 + 10x^3 - 4x^2 + 22x - 23)$? If so, model the polynomials and add. If not, explain why.

Like terms in an expression are terms that have the same variable and exponent for that variable. All constants are like terms.

4. State whether the terms are like or unlike terms. Explain.

   a. $2x$; $2x^3$

   b. 5; $5x$

   c. $-3y$; $3y$

   d. x^2y; xy^2

   e. 14; -0.6

5. **Attend to precision.** Using vocabulary from this activity, describe a method that could be used to add polynomials without using algebra tiles.

My Notes

Use the properties of real numbers to add polynomials algebraically.

Example B

Add $(3x^3 + 2x^2 - 5x + 7) + (4x^3 + 2x - 3)$ horizontally and vertically. Write your answer in standard form.

Horizontally

Step 1: Identify like terms. $(3x^3 + 2x^2 - 5x + 7) + (4x^3 + 2x - 3)$

Step 2: Group like terms. $(3x^3 + 4x^3) + (2x^2) + (-5x + 2x) + (7 - 3)$

Step 3: Add the coefficients of like terms. $7x^3 + 2x^2 - 3x + 4$

Solution: $(3x^3 + 2x^2 - 5x + 7) + (4x^3 + 2x - 3) = 7x^3 + 2x^2 - 3x + 4$

Vertically

Step 1: Vertically align like terms.

Step 2: Add the coefficients of like terms.

$$\begin{array}{r} 3x^3 + 2x^2 - 5x + 7 \\ +4x^3 \qquad\quad + 2x - 3 \\ \hline 7x^3 + 2x^2 - 3x + 4 \end{array}$$

Solution: $(3x^3 + 2x^2 - 5x + 7) + (4x^2 + 2x - 3) = 7x^3 + 2x^2 - 3x + 4$

MATH TIP

The Commutative and Associative Properties of Addition allow you to re-order and group like terms.

Try These B

Add. Write your answers in standard form.

a. $(4x^2 + 3) + (x^2 - 3x + 5)$

b. $(10y^2 + 8y + 6) + (17y^2 - 11)$

c. $(9x^2 + 15x + 21) + (-13x^2 - 11x - 26)$

6. Are the answers to Try These B polynomials? Justify your response.

7. Explain why the sum of two polynomials will always be a polynomial.

MATH TIP

Polynomials are *closed* under addition. A set is closed under addition if the sum of any two elements in the set is also an element of the set.

My Notes

Check Your Understanding

Write a polynomial for each expression represented by algebra tiles.

8. 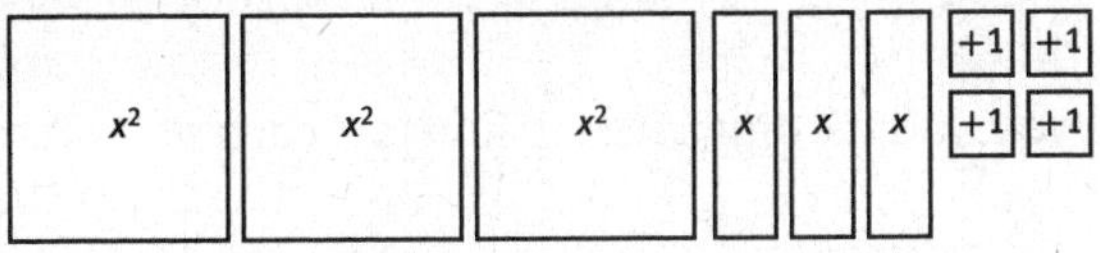

9. 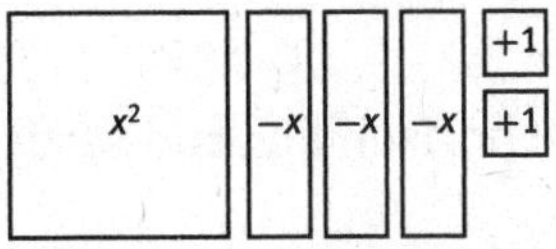

10. Add the expressions you wrote for Items 8 and 9.

11. What property or properties justify Steps 1 and 2 below?

$(2x^2 + x + 1) + (x^2 + 2x - 1)$

Step 1: $(2x^2 + x^2) + (x + 2x) + (1 - 1)$

Step 2: $(2 + 1)x^2 + (1 + 2)x + (1 - 1)$

Step 3: $3x^2 + 3x$

LESSON 24-2 PRACTICE

Add. Write your answers in standard form.

12. $(3x^2 + x + 5) + (2x^2 + x - 5)$

13. $(-4x^2 + 2x -1) + (x^2 - x + 9)$

14. $(7x^2 - 2x + 3) + (3x^2 + 2x + 7)$

15. $(-x^2 + 5x + 2) + (-3x^2 + x - 9)$

Write the perimeter of each figure as a polynomial in standard form.

16. 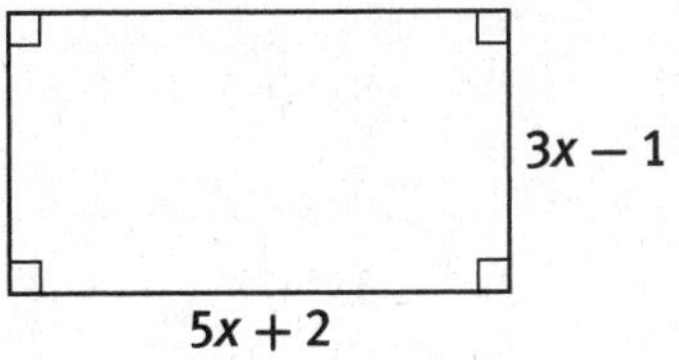

17. 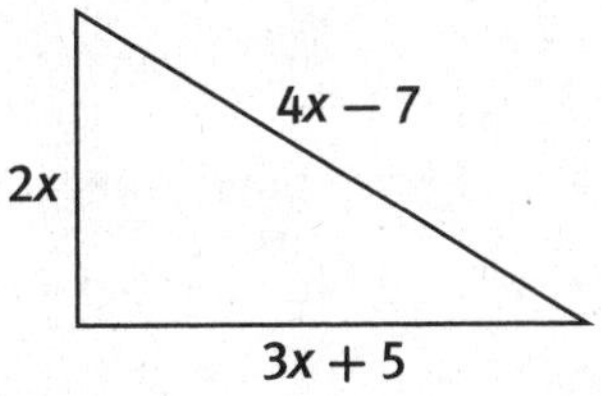

18. **Critique the reasoning of others.** A student added the expressions $x^4 + 5x^2 - 2x + 1$ and $2x^4 + x^3 + 2x - 7$. Identify and correct the student's error.

$$\begin{array}{r} x^4 + 5x^2 - 2x + 1 \\ \underline{2x^4 + \ x^3 + 2x - 7} \\ 3x^4 + 6x^2 \qquad - 6 \end{array}$$

My Notes

MATH TERMS

The **opposite** of a number or a polynomial is its additive inverse.

Learning Target:

- Subtract polynomials algebraically.

SUGGESTED LEARNING STRATEGIES: Note Taking, Close Reading, Think-Pair-Share

To subtract a polynomial you add its ***opposite***, or subtract each of its terms.

Example A

Subtract $(2x^3 + 8x^2 + x + 10) - (5x^2 - 4x + 6)$ horizontally and vertically. Write the answer in standard form.

Horizontally

Step 1: Distribute the negative. $(2x^3 + 8x^2 + x + 10) - (5x^2 - 4x + 6)$

Step 2: Identify like terms. $2x^3 + 8x^2 + x + 10 - 5x^2 + 4x - 6$

Step 3: Group like terms. $2x^3 + (8x^2 - 5x^2) + (x + 4x) + (10 - 6)$

Step 4: Combine coefficients of like terms. $2x^3 + 3x^2 + 5x + 4$

Solution: $(2x^3 + 8x^2 + x + 10) - (5x^2 - 4x + 6) = 2x^3 + 3x^2 + 5x + 4$

Vertically

Step 1: Vertically align like terms.

$$\begin{array}{r} 2x^3 + 8x^2 + x + 10 \\ \underline{-(5x^2 - 4x + 6)} \end{array}$$

Step 2: Distribute the negative.

$$\begin{array}{r} 2x^3 + 8x^2 + x + 10 \\ \underline{-5x^2 + 4x - 6} \end{array}$$

Step 3: Combine coefficients of like terms. $2x^3 + 3x^2 + 5x + 4$

Solution: $(2x^3 + 8x^2 + x + 10) - (5x^2 - 4x + 6) = 2x^3 + 3x^2 + 5x + 4$

My Notes

Try These A

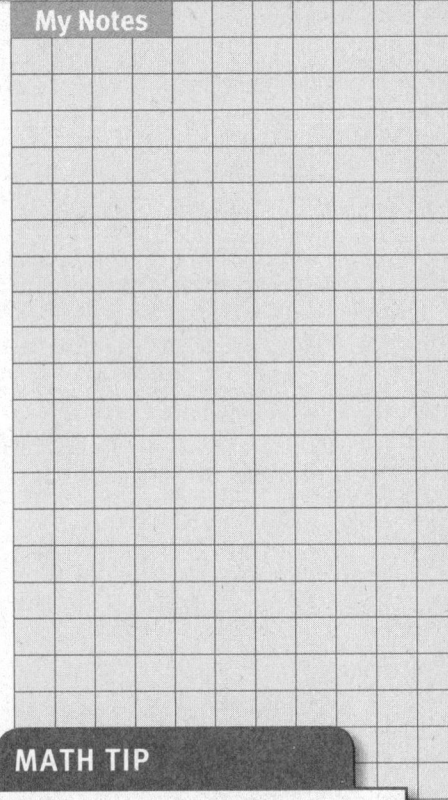

Subtract. Write your answers in standard form.

a. $(5x - 5) - (x + 7)$

b. $(2x^2 + 3x + 2) - (-5x^2 - 2x - 9)$

c. $(y^2 + 3y + 8) - (4y^2 - 9)$

d. $(12 + 5x + 14x^2) - (8x + 15 - 7x^2)$

1. Are the answers to Try These A polynomials? Justify your response.

MATH TIP

Polynomials are *closed* under subtraction. A set is closed under subtraction if the difference of any two elements in the set is also an element of the set.

2. Explain why the difference of two polynomials will always be a polynomial.

My Notes

Check Your Understanding

Rewrite each difference as addition of the opposite, or additive inverse, of the second polynomial.

3. $(x^2 + 2x + 3) - (4x^2 - x + 5)$

4. $(5y^2 + y - 2) - (-y^2 - 3y + 4)$

5. **Critique the reasoning of others.** Gil used the vertical method to subtract $(3x^2 - 5x + 2) - (x^2 + 2x + 4)$ as shown below. Identify Gil's error.

$$\begin{array}{r} 3x^2 - 5x + 2 \\ \underline{-\, x^2 + 2x + 4} \\ 2x^2 - 3x + 6 \end{array}$$

LESSON 24-3 PRACTICE

Subtract. Write your answers in standard form.

6. $(2x^2 + 4x + 1) - (7x^2 - 3x - 4)$

7. $(x^2 + 3x - 9) - (x^2 + 2x - 8)$

8. $(9x^2 + x - 12) - (14x^2 - 7x - 2)$

9. $(x^2 + 3x - 6) - (5x - 6)$

10. $(y^4 + y^2 + 2y) - (-y^4 + 3)$

11. Write two polynomials whose difference is $6x + 3$.

12. **Model with mathematics.** A rectangular piece of paper has area $4x^2 + 3x + 2$. A square is cut from the rectangle and the remainder of the rectangle is discarded. The area of the discarded paper is $3x^2 + x + 1$. What is the area of the square?

ACTIVITY 24 PRACTICE

Write your answers on notebook paper.
Show your work.

Lesson 24-1

For Items 1–5, use the polynomial $5x^4 - 2x^2 + 8x - 3$.

1. Identify the coefficients of the variable terms of the polynomial.
2. List the terms, and give the degree of each term.
3. State the degree of the polynomial.
4. Identify the leading coefficient of the polynomial.
5. Identify the constant term of the polynomial.
6. Consider the expressions $3x^2 + 2x - 7$ and $3x^2 + 2x - 7x^0$. Are the expressions equivalent? Explain.

Write each polynomial in standard form.

7. $5x^2 - 2x^3 - 10$
8. $11 + y^2 - 8y$
9. $y^2 - 12 - y^3 + y^4$
10. $9x + 7 - 5x^3$

Lesson 24-2

Add. Write your answers in standard form.

11. $(4x + 9) + (3x - 5)$
12. $(2x^2 + 3x - 1) + (x^2 - 5x + 2)$
13. $(x^3 + 5x^2 + 3) + (2x^3 - 5x^2)$
14. $(7y^3 - 2y^2 + 5) + (4y^3 - 3y)$
15. Which expression represents the perimeter of the trapezoid?

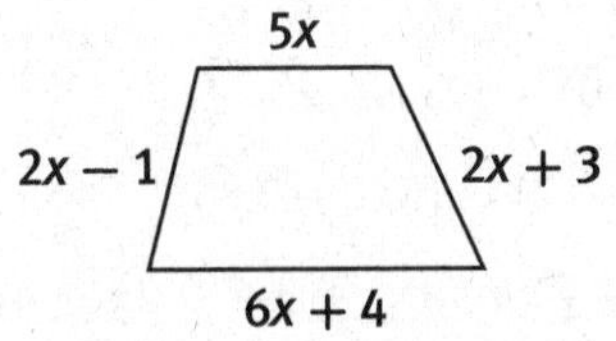

A. $11x + 4$
B. $4x + 2$
C. $15x + 6$
D. $13x + 8$

16. The length of each side of a square is $4y + 5$. Draw and label the square, and write an expression to represent its perimeter.

Lesson 24-3

17. Which expression is equivalent to $10x - (7x - 1)$?
A. $3x - 1$
B. $3x + 1$
C. $17x - 1$
D. $17x + 1$

Subtract. Write your answers in standard form.

18. $(5x - 4) - (3x + 2)$
19. $(3x^2 - 2x + 7) - (2x^2 + 2x - 7)$
20. $(8y^2 - 3y + 6) - (-2y^2 - 3)$
21. $(x^2 - 5x) - (4x - 6)$

Determine the sum or difference. Write your answers in standard form.

22. $(5y^2 + 3y + 7) + (7y - 2)$

23. $(3x^2 + x + 9) - (2x^2 + x + 2)$

24. $(x^3 + 3x^2 + 12) - (5x^3 + 7x^2)$

25. $(8x^3 - 5x + 7) + (4x^3 + 3x^2 - 3x - 4)$

26. $(-4y^2 - 2y + 1) - (7y^3 + y^2 - y + 5)$

27. $(3x + 7y) + (-4x + 3y)$

28. $(5x^2 + 8xy + y^2) - (-x^2 + 4xy - 5y^2)$

29. A playground has a sidewalk border around a play area.

The total area of the playground, the larger rectangle, is $16x^2 - 5x + 2$. The area of the play area, the smaller rectangle, is $10x^2 + 3x - 1$. Write an expression to represent the area of the sidewalk.

30. To make a box, four corners of a rectangular piece of cardboard are cut out and the box is folded and taped.

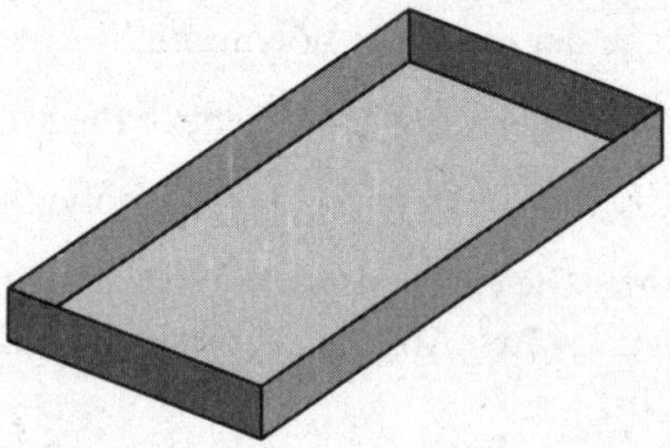

The area of the cardboard, after the corners are cut out, is $28x^2 + 12x + 32$. The area of each cut-out corner is $2x^2 + 3$. Write an expression to represent the area of the original piece of cardboard.

MATHEMATICAL PRACTICES
Reason Abstractly and Quantitatively

31. The set of polynomials is closed under the operations of addition and subtraction. This means that when you add or subtract two polynomials, the result is also a polynomial.

a. Are the integers closed under addition and subtraction? In other words, when you add or subtract two integers, is the result always an integer? Justify your response.

b. Give a counterexample to show that the whole numbers are **not** closed under subtraction.

Multiplying Polynomials

Tri-Com Computers

Lesson 25-1 Multiplying Binomials

My Notes

Learning Targets:

- Use a graphic organizer to multiply expressions.
- Use the Distributive Property to multiply expressions.

SUGGESTED LEARNING STRATEGIES: Think-Pair-Share, Look for a Pattern, Discussion Groups, Create Representations, Graphic Organizer

Tri-Com Computers is a company that sets up local area networks in offices. In setting up a network, consultants need to consider not only where to place computers, but also where to place peripheral equipment, such as printers.

Tri-Com typically sets up local area networks of computers and printers in square or rectangular offices. Printers are placed in each corner of the room. The primary printer A serves the first 25 computers and the other three printers, B, C, and D, are assigned to other regions in the room. Below is an example.

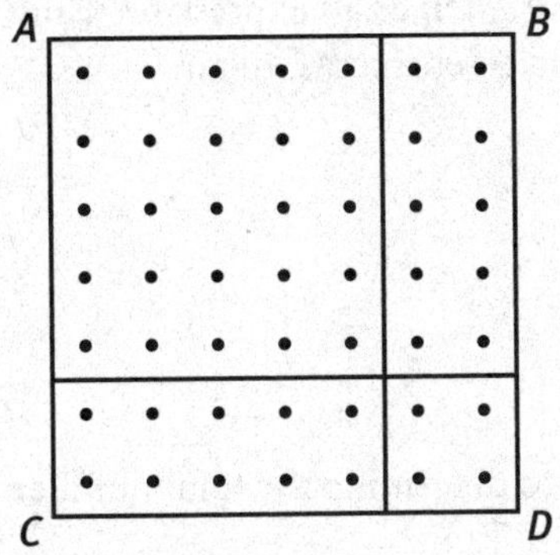

1. If each dot represents a computer, how many computers in this room will be assigned to each of the printers?

2. What is the total number of computers in the room? Describe two ways to find the total.

My Notes

Another example of an office in which Tri-Com installed a network had 9 computers along each wall. The computers are aligned in an array with the number of computers in each region determined by the number of computers along the wall.

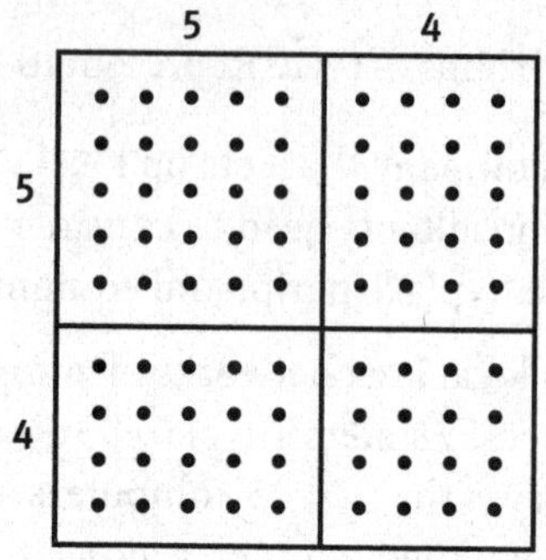

3. A technician claimed that since $9 = 5 + 4$, the number of computers in the office could be written as an expression using only the numbers 5 and 4. Is the technician correct? Explain.

4. Show another way to determine the total number of computers in the office.

5. Rewrite the expression $(5 + 4)(5 + 4)$ using the Distributive Property.

6. Make sense of problems. Explain why $(5 + 4)(5 + 4)$ could be used to determine the total number of computers.

My Notes

7. The office to the right has 8^2 computers. Fill in the number of computers in each section if it is split into a $(5 + 3)^2$ configuration.

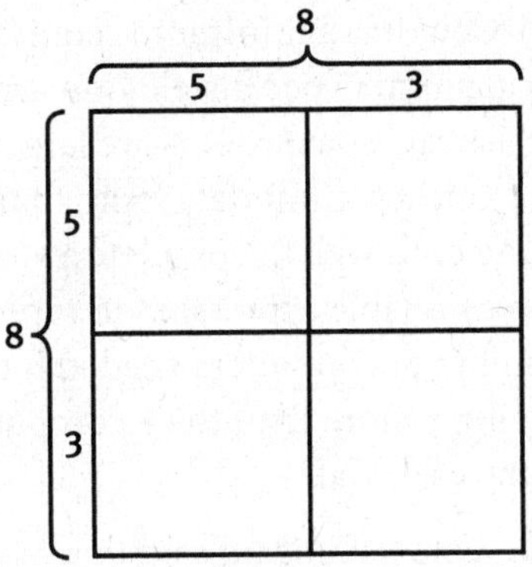

8. What is the total number of computers? Describe two ways to find the total.

9. For each possible office configuration below, draw a diagram like the one next to Item 7. Label the number of computers on the edge of each section and determine the total number of computers in the room by adding the number of computers in each section.

a. $(2 + 3)^2$

b. $(4 + 1)^2$

c. $(3 + 7)^2$

My Notes

Tri-Com has a minimum requirement of 25 computers per installation arranged in a 5 by 5 array. Some rooms are larger than others and can accommodate more than 5 computers along each wall to complete a square array. Use a variable expression to represent the total number of computers needed for any office having x more than the 5 computer minimum along each wall.

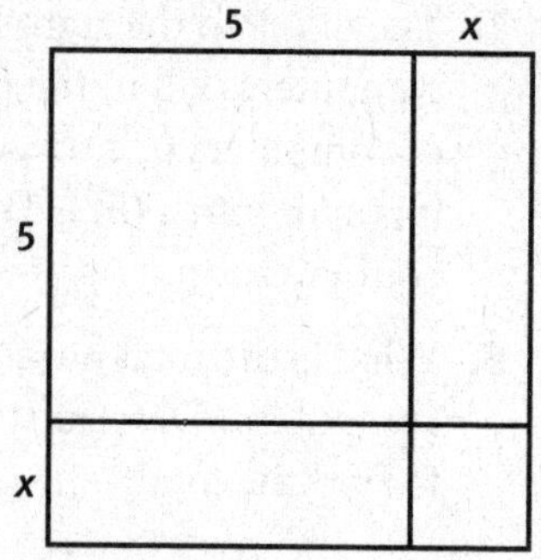

10. One technician said that $5^2 + x^2$ would be the correct way to represent the total number of computers in the office space. Use the diagram to explain how the statement is incorrect.

11. Model with mathematics. Write an expression for the sum of the number of computers in each region in Item 10.

12. For each of the possible room configurations, determine the total number of computers in the room.

a. $(2 + x)^2$

b. $(x + 3)^2$

c. $(x + 6)^2$

My Notes

The graphic organizer below can be used to help arrange the multiplications of the Distributive Property. It does not need to be related to the number of computers in an office. For example, this graphic organizer shows $5 \cdot 7 = (3 + 2)(4 + 3)$.

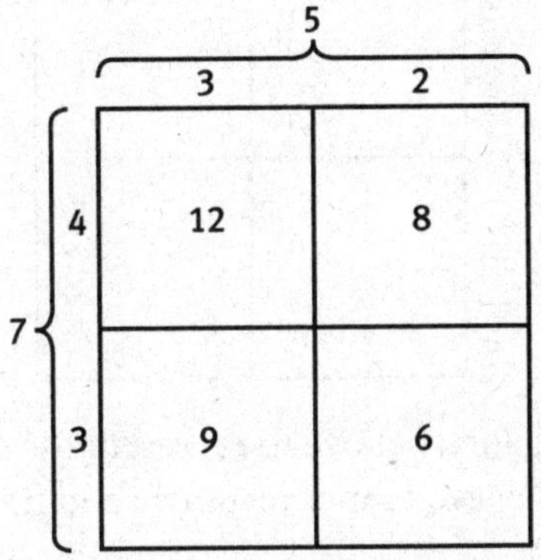

13. Draw a graphic organizer to represent the expression $(5 + 2)(2 + 3)$. Label each inner rectangle and find the sum.

14. Draw a graphic organizer to represent the expression $(6 - 3)(4 - 2)$. Label each inner rectangle and find the sum.

15. Multiply the binomials in Item 14 using the Distributive Property. What do you notice?

My Notes

You can use the same graphic organizer to multiply binomials that contain variables. The following diagram represents $(x-2)(x-3)$.

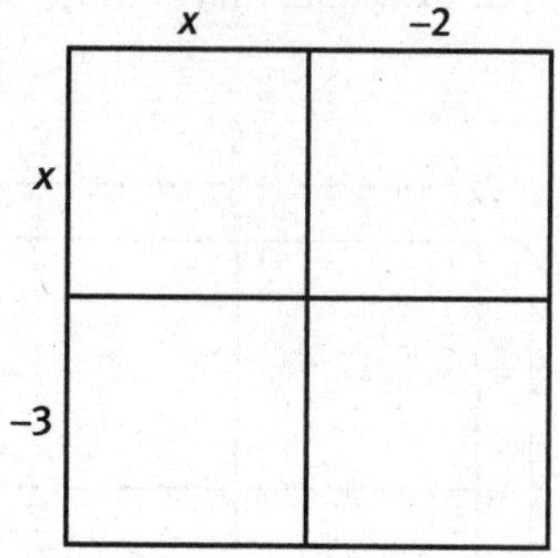

16. Use the graphic organizer above to represent the expression $(x-2)(x-3)$. Label each inner rectangle and find the sum.

17. Multiply the binomials in Item 16 using the Distributive Property. What do you notice?

18. Determine the product of the binomials.

a. $(x-7)(x-5)$

b. $(x-7)(x+5)$

c. $(x+7)(x+5)$

d. $(x+7)(x-5)$

e. $(4x+1)(x+3)$

f. $(2x-1)(3x+2)$

19. Reason abstractly. Examine the products in Item 18. How can you predict the sign of the last term?

My Notes

Check Your Understanding

20. Use a graphic organizer to calculate $(6 + 2)^2$. Explain why the product is not $6^2 + 2^2$.

Determine the product of the binomials using a graphic organizer or by using the Distributive Property.

21. $(x + 7)(x + 2)$

22. $(x + 7)(3x - 2)$

23. Compare the use of the graphic organizer and the use of the Distributive Property to find the product of two binomials.

LESSON 25-1 PRACTICE

Determine each product.

24. $(2 + 1)(3 + 5)$

25. $(2 + 3)(2 + 7)$

26. $(x + 9)(x + 3)$

27. $(x + 5)(x + 1)$

28. $(x - 3)(x + 4)$

29. $(x + 1)(x - 5)$

30. $(x + 3)(x - 3)$

31. $(x + 3)(x + 3)$

32. $(2x - 3)(x - 1)$

33. $(x + 7)(3x - 5)$

34. $(4x + 3)(2x + 1)$

35. $(6x - 2)(5x + 1)$

36. **Critique the reasoning of others.** A student determined the product $(x - 2)(x - 4)$. Identify and correct the student's error.

$$(x - 2)(x - 4)$$
$$x(x - 4) - 2(x - 4)$$
$$x^2 - 4x - 2x - 8$$
$$x^2 - 6x - 8$$

My Notes

Learning Targets:

- Multiply binomials.
- Find special products of binomials.

SUGGESTED LEARNING STRATEGIES: Think-Pair-Share, Look for a Pattern

1. Determine each product.

a. $(x + 1)(x - 1)$

b. $(x + 4)(x - 4)$

c. $(x - 7)(x + 7)$

d. $(2x - 3)(2x + 3)$

2. Describe any patterns in the binomials and products in Item 1.

3. Express regularity in repeated reasoning. The product of binomials of the form $(a + b)(a - b)$, has a special pattern called a ***difference of two squares***. Use the patterns you found in Items 1 and 2 to explain how to find the product $(a + b)(a - b)$.

MATH TERMS

A binomial of the form $a^2 - b^2$ is known as the **difference of two squares**.

DISCUSSION GROUP TIP

As you read and define new terms, discuss their meanings with other group members and make connections to prior learning.

My Notes

4. Determine each product.

a. $(x + 1)^2$

b. $(4 + y)^2$

c. $(x + 7)^2$

d. $(2y + 3)^2$

e. $(x - 5)^2$

f. $(4 - x)^2$

g. $(y - 7)^2$

h. $(2x - 3)^2$

5. Describe any patterns in the binomials and products in Item 4.

6. Reason abstractly. The ***square of a binomial***, $(a + b)^2$ or $(a - b)^2$, also has a special pattern. Use the pattern you found in Items 4 and 5 to explain how to determine the square of any binomial.

MATH TERMS

A binomial of the form $(a + b)^2$ or $(a - b)^2$ is known as the **square of a binomial.**

My Notes

Check Your Understanding

7. Use the difference of two squares pattern to find the product $(p + k)(p - k)$.
8. Use the square of a binomial pattern to determine $(p + k)^2$.
9. Can you use a special products pattern to determine $(x + 1)(x - 2)$? Explain your reasoning.

LESSON 25-2 PRACTICE

Determine each product.

10. $(x - 4)(x + 4)$
11. $(x + 4)^2$
12. $(y + 10)(y - 10)$
13. $(y - 10)^2$
14. $(2x - 3)^2$
15. $(2x - 3)(2x + 3)$
16. $(5x + 1)^2$
17. $(2y - 1)(2y - 1)$
18. **Construct viable arguments.** Explain why the products $(x - 3)^2$ and $(x + 3)(x - 3)$ have a different number of terms.

My Notes

Learning Targets:

- Use a graphic organizer to multiply polynomials.
- Use the Distributive Property to multiply polynomials.

SUGGESTED LEARNING STRATEGIES: Graphic Organizer, Create Representations, Think-Pair-Share, Look for a Pattern

A graphic organizer can be used to multiply polynomials that have more than two terms, such as a binomial times a trinomial. The graphic organizer at right can be used to multiply $(x + 2)(x^2 + 2x + 3)$.

	x	2
x^2	x^3	$2x^2$
$2x$	$2x^2$	$4x$
3	$3x$	6

$= x^3 + 4x^2 + 7x + 6$

1. Draw a graphic organizer in the space provided in the *My Notes* section to represent $(x - 3)(x^2 + 5x + 6)$. Label each inner rectangle and find the sum.

2. How many boxes would you need to represent the multiplication of $(x^3 + 5x^2 + 3x - 3)(x^4 - 6x^3 - 7x^2 + 5x + 6)$ using the graphic organizer?

   a. Explain how you determined your answer.

   b. **Use appropriate tools strategically.** Would you use the graphic organizer for other multiplications with this many terms? Explain your reasoning.

The Distributive Property can be used to multiply any polynomial by another. Multiply each term in the first polynomial by each term in the second polynomial.

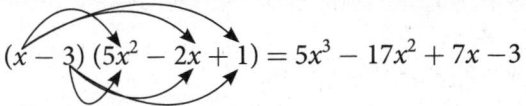

$$(x - 3)(5x^2 - 2x + 1) = 5x^3 - 17x^2 + 7x - 3$$

3. Determine each product.

   a. $x(x + 5)$

   b. $(x - 3)(x + 6)$

   c. $(x + 7)(3x^2 - x - 1)$

   d. $(3x - 7)(4x^2 + 4x - 3)$

4. How can you predict the number of terms the product will have before you combine like terms?

My Notes

MATH TIP

Polynomials are *closed* under multiplication. A set is closed under multiplication if the product of any two elements in the set is also an element of the set.

5. Are all of the answers to Item 3 polynomials? Justify your response.

6. Explain why the product of two polynomials will always be a polynomial.

7. You can find the product of more than two polynomials, such as $(x + 3)(2x + 1)(3x - 2)$.

a. To multiply $(x + 3)(2x + 1)(3x - 2)$, first determine the product of the first two polynomials, $(x + 3)(2x + 1)$.

$(x + 3)(2x + 1) =$

b. Multiply your answer to Part (a) by the third polynomial, $(3x - 2)$.

8. Determine each product.

a. $(x - 2)(x + 1)(2x + 2)$

b. $(x + 3)(3x + 1)(2x - 1)$

c. $(x - 1)(3x - 2)(x + 4)$

d. $(2x - 4)(4x + 1)(3x + 3)$

Check Your Understanding

Determine each product.

9. $a(b + c)$

10. $(a + b)(a + c)$

11. $(a + b)(a^2 + b + c)$

12. $(a + b)(a + c)(b + c)$

LESSON 25-3 PRACTICE

Determine each product.

13. $x(x + 7)$

14. $x(2x - 5)$

15. $(y + 3)(y + 6)$

16. $(y + 3)(y - 6)$

17. $x(2x^2 - 5x + 1)$

18. $(x - 1)(2x^2 - 5x + 1)$

19. $(2x - 7)(5x^2 - 1)$

20. $(2x - 7)(5x^2 - 3x - 1)$

21. $(x + 2)(x - 3)(x + 1)$

22. $(x + 2)(2x - 3)(2x + 1)$

23. **Attend to precision.** A binomial of degree 2 and variable x and a trinomial of degree 4 and variable x are multiplied. What will be the degree of the product? Explain your reasoning.

ACTIVITY 25 PRACTICE

Write your answers on notebook paper.
Show your work.

Lesson 25-1

Determine each product.

1. $(10 - 3)(10 - 8)$
2. $(x - 3)(x - 8)$
3. $(y - 7)(y + 2)$
4. $(x + 5)(x - 9)$
5. $(2y - 6)(3y - 8)$
6. $(4x + 3)(x - 11)$
7. Which expression represents the area of the rectangle?

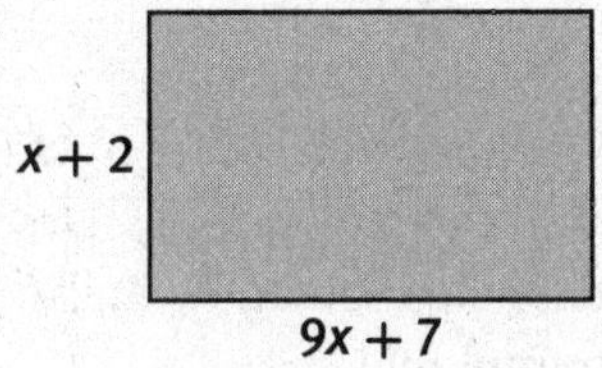

A. $10x + 9$
B. $9x^2 + 14$
C. $20x + 18$
D. $9x^2 + 25x + 14$

Lesson 25-2

Determine each product.

8. $(x - 7)(x + 7)$
9. $(y + 6)(y - 6)$
10. $(2x - 5)(2x + 5)$
11. $(3y + 1)(3y - 1)$
12. $(x - 11)^2$
13. $(x + 8)^2$
14. $(2y - 3)^2$
15. $(3y - 2)^2$
16. Which expression represents the area of the square?

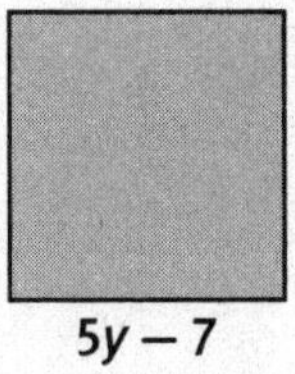

A. $20y^2 + 28$
B. $20y - 28$
C. $25y^2 - 70y + 49$
D. $25y^2 - 49$

Lesson 25-3

Determine each product.

17. $x(x^2 - 7)$

18. $2x(x^2 - 3x + 2)$

19. $(x + 2)(4x^2 - 7x + 5)$

20. $(y - 5)(4y^2 + 5y + 2)$

21. $(5x - 9)^2$

22. $(3x - 4)(3x + 4)$

23. $(2y + 1)(y^2 + 3y - 5)$

24. Which expression represents the area of the triangle? Use the formula $A = \frac{1}{2}bh$.

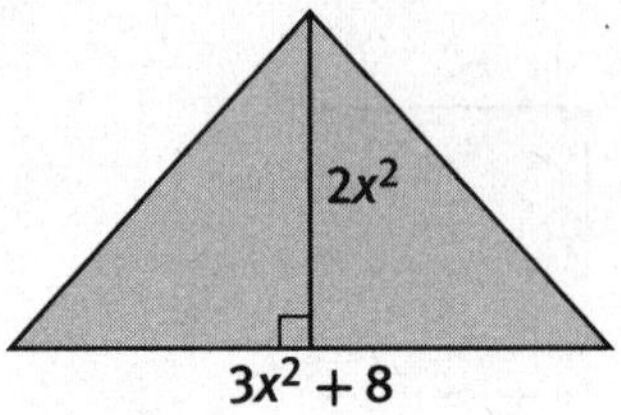

A. $3x^4 + x^2$
B. $3x^4 + 8x^2$
C. $6x^2 + 8$
D. $6x^4 + 16x$

Determine each product.

25. $(x - 1)(7x - 1)(x + 2)$

26. $(x + 5)(4x - 1)(2x + 3)$

27. $(y + 1)^3$

28. Devise a plan for finding the product of four polynomials.

MATHEMATICAL PRACTICES
Look for and Make Use of Structure

29. Determine each product and describe any patterns you observe.

$(x - 1)(x + 1)$

$(x - 1)(x^2 + x + 1)$

$(x - 1)(x^3 + x^2 + x + 1)$

From the patterns you see, predict the product of $(x - 1)(x^4 + x^3 + x^2 + x + 1)$. Describe the pattern that helps you know the answer without needing to multiply.

Employees at Ship-It-Quik must perform computations involving volume and surface area. As part of the job application, potential employees must take a test that involves surface area, volume, and algebraic skills.

1. The surface area of a figure is the total area of all faces. The areas of the faces of a rectangular prism are shown. The surface area of this prism is $18x^2 + 12x + 22$. Complete the first part of the job application by finding the area of the missing face.

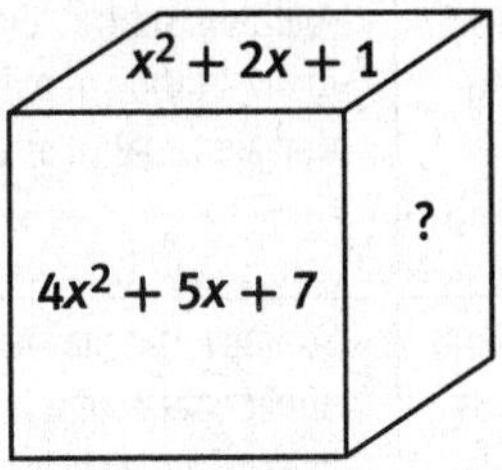

The formula for the volume of a rectangular prism is $V = lwh$, where l, w, and h are length, width, and height, respectively. The formula for the surface area is $SA = 2lw + 2wh + 2lh$.

2. Complete the second part of the job application by verifying whether or not the following computations are correct. Explain your reasoning by showing your work.

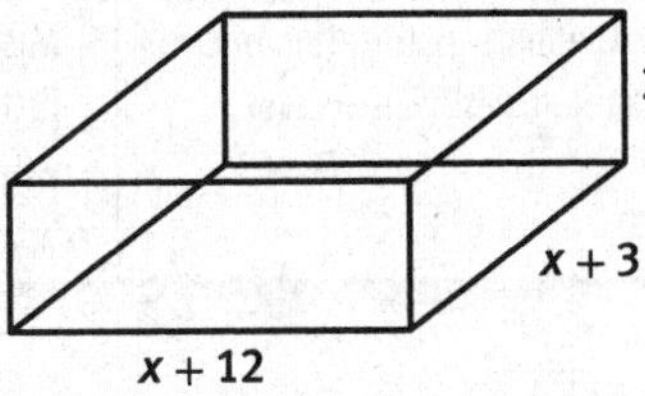

Volume:
$2x^2 + 15x + 36$

Surface Area:
$10x^2 + 90x + 72$

3. Complete the final part of the job application by writing an expression for the volume of a cylinder with radius $3x^2y$ and height $2xy$. Use the formula $V = \pi r^2 h$ where r is the radius and h is the height. Simplify your answer as much as possible.

Scoring Guide	Exemplary	Proficient	Emerging	Incomplete
	The solution demonstrates the following characteristics:			
Mathematics Knowledge and Thinking (Items 1–3)	• Effective understanding of and accuracy in adding, subtracting, and multiplying polynomials	• Addition, subtraction, and multiplication of polynomials that are usually correct	• Difficulty adding, subtracting, and multiplying polynomials	• Inaccurate addition, subtraction, and multiplication of polynomials
Problem Solving (Item 1)	• Appropriate and efficient strategy that results in a correct answer	• Strategy that may include unnecessary steps but results in a correct answer	• Strategy that results in some incorrect answers	• No clear strategy when solving problems
Mathematical Modeling/ Representations (Items 1–3)	• Clear and accurate understanding of geometric formulas	• Functional understanding of geometric formulas that results in correct answers	• Partial understanding of geometric formulas that results in some incorrect answers	• Little or no understanding of geometric formulas
Reasoning and Communication (Item 2)	• Precise use of appropriate math terms and language to justify each step in verifying an answer	• Adequate justification of each step to verify an answer	• Misleading or confusing justification of the steps to verify an answer	• Incomplete or inaccurate justification of the steps to verify an answer

Factoring

Factors of Construction

Lesson 26-1 Factoring by Greatest Common Factor (GCF)

Learning Targets:

- Identify the GCF of the terms in a polynomial.
- Factor the GCF from a polynomial.

SUGGESTED LEARNING STRATEGIES: Look for a Pattern, Think-Pair-Share, Discussion Groups, Note Taking

Factor Steele Buildings is a company that manufactures prefabricated metal buildings that are customizable. All the buildings come in square or rectangular designs. Most office buildings have an entrance area or great room, large offices, and cubicles. The diagram below shows the front face of one of their designs. The distance c represents space available for large offices, and p represents the space available for the great room.

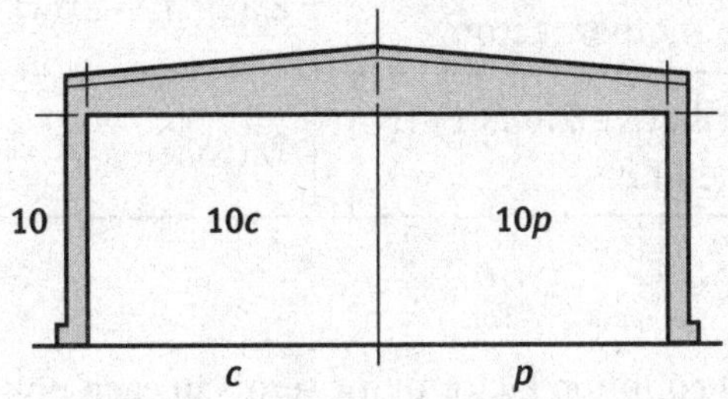

1. To determine how much material is needed to cover the front wall of the building, represent the total area as a product of a ***monomial*** and a ***binomial***.

2. Represent the same area from Item 1 as a sum of two monomials.

3. **Make use of structure.** What property can be used to show that the two quantities in Items 1 and 2 are equal?

4. Factor Steele Buildings inputs the length of the large office space c into an expression that gives the area of an entire space: $6c^2 + 12c - 9$. Determine the ***greatest common factor (GCF)*** of the terms in this polynomial. Explain your choice.

My Notes

MATH TERMS

A **monomial** is a number, a variable, or a product of numbers and variables with whole-number exponents. For example, 4, $-9x$, and $5xy^2$ are all monomials. A **binomial** is a sum or difference of two monomials.

MATH TERMS

The **greatest common factor (GCF)** of the terms in a polynomial is the greatest monomial that divides into each term without a remainder.

My Notes

MATH TERMS

A **factor** is any of the numbers or symbols that when multiplied together form a product. For example, 2 and x are factors of $2x$ because 2 and x are multiplied to get $2x$. *Factor* can be used as a noun or a verb.

To ***factor*** a number or expression means to write the number or expression as a product of its ***factors***.

Example A

To Factor a Monomial (the GCF) from a Polynomial	
Steps to Factoring	**Example**
• Determine the GCF of all terms in the polynomial.	$6x^3 + 2x^2 - 8x$ GCF $= 2x$
• Write each term as the product of the GCF and another factor.	$2x(3x^2) + 2x(x) + 2x(-4)$
• Use the Distributive Property to factor out the GCF.	$2x(3x^2 + x - 4)$

Try These A

Find the greatest common factor of the terms in each polynomial. Then write each polynomial with the GCF factored out.

a. $36y - 24$

b. $4x^5 - 6x^3 + 10x^2$

c. $15t^2 + 10t - 5$

Check Your Understanding

5. Identify the GCF of the terms in the polynomial $21x^3 + 14x^2 + 35x$.

Factor a monomial (the GCF) from each polynomial.

6. $36x + 9$

7. $6x^4 + 12x^2 - 18x$

8. $125n^6 + 250n^5 + 25n^3$

9. $3x^3 + 9x^2 + 6x$

10. $\frac{2}{3}y^4 + \frac{1}{3}y^3 - \frac{4}{3}y^2$

11. $4x^2y^2 + 12xy^2 - 8x^2y - 4xy$

My Notes

LESSON 26-1 PRACTICE

Use the cylinder for Items 12–15.

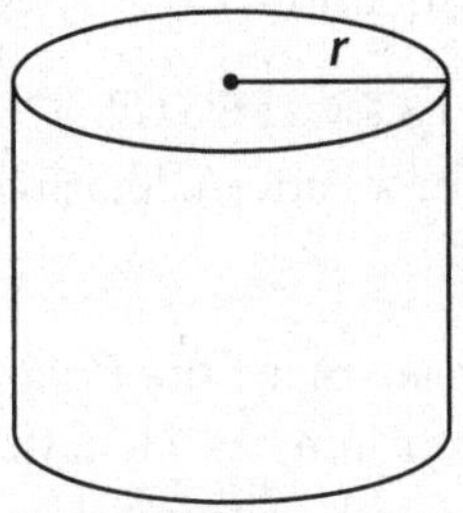

The surface area of a cylinder is given by the formula $SA = 2\pi r^2 + 2\pi rh$, where r is the radius and h is the height.

12. Factor a monomial (the GCF) from the formula.

13. Suppose the radius of the cylinder is y and the height is $y + 2$. Rewrite the formula in this case, using multiplication and exponent rules as needed to simplify the expression.

14. Factor the expression from Item 13 completely.

15. **Construct viable arguments.** Answer each of the following questions and justify your responses.

a. If the radius of a cylinder doubles, what happens to the GCF of its surface area?

b. What happens to the GCF of the cylinder's surface area if its radius is squared?

My Notes

Learning Targets:

- Factor a perfect square trinomial.
- Factor a difference of two squares.

SUGGESTED LEARNING STRATEGIES: Create Representations, Discussion Groups, Look for a Pattern, Sharing and Responding, Think-Pair-Share

Factor Steele Buildings can create many floor plans with different size spaces. In the diagram below the great room has a length and width of x units, and each cubicle has a length and width of 1 unit. Use the diagram below for Items 1–3.

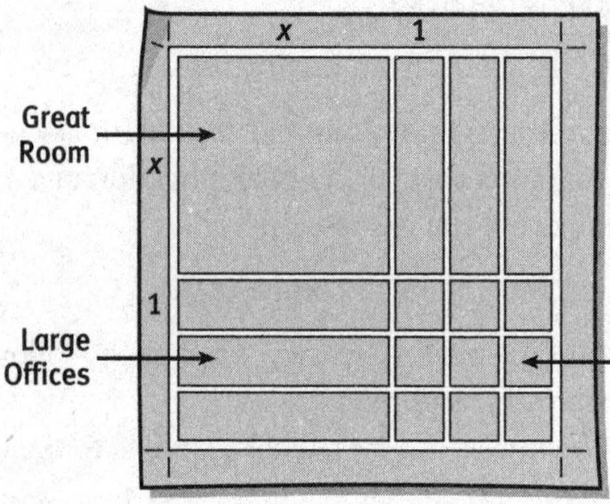

1. **Model with mathematics.** Represent the area of the entire office above as a sum of the areas of all the rooms.

2. Write the area of the entire office as a product of two binomials.

3. What property can you use to show how the answers to Items 1 and 2 are related? Show this relationship.

4. For each of the following floor plans, write the area of the office as a sum of the areas of all the rooms and as a product of binomials.

a.

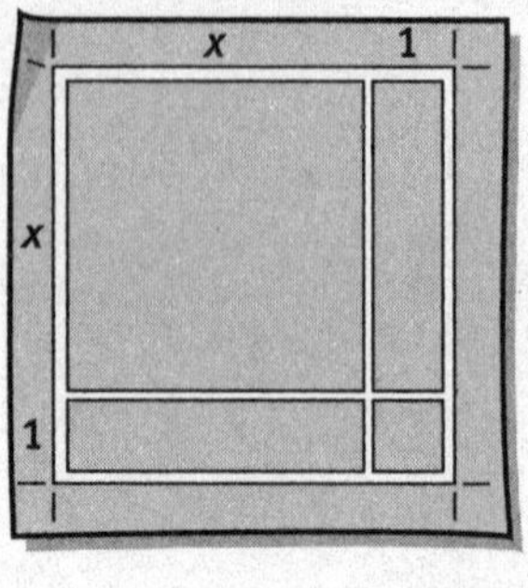

b.

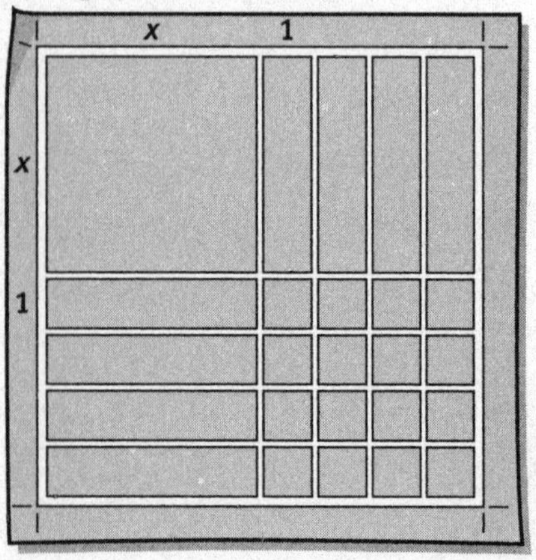

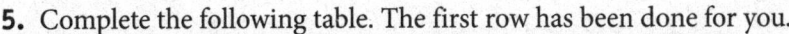

c. What patterns do you observe?

My Notes

5. Complete the following table. The first row has been done for you.

Polynomial	1st Factor	2nd Factor	First Term in Each Factor	Second Term in Each Factor
$x^2 + 6x + 9$	$(x + 3)$	$(x + 3)$	x	3
	$(x - 3)$	$(x - 3)$		
	$(x + 4)$	$(x + 4)$		
	$(x - 4)$	$(x - 4)$		
	$(x + 5)$	$(x + 5)$		
	$(x - 5)$	$(x - 5)$		

6. Express regularity in repeated reasoning. Describe any patterns that you observe in the table from Item 5.

7. Explain how to factor polynomials of the form $a^2 + 2ab + b^2$ and $a^2 - 2ab + b^2$.

Polynomials of the form $a^2 + 2ab + b^2$ and $a^2 - 2ab + b^2$ are called ***perfect square trinomials***.

Check Your Understanding

Factor each perfect square trinomial.

8. $x^2 - 14x + 49$ **9.** $m^2 + 20m + 100$ **10.** $y^2 - 16y + 64$

11. Complete the table by finding the polynomial product of each pair of binomial factors. The first row has been done for you.

1st Factor	2nd Factor	Polynomial
$(x + 3)$	$(x - 3)$	$x^2 - 9$
$(x + 4)$	$(x - 4)$	
$(x - 5)$	$(x + 5)$	
$(9 - x)$	$(9 + x)$	
$(2x - 7)$	$(2x + 7)$	
$(6x - 2y)$	$(6x + 2y)$	

My Notes

12. Describe any patterns you observe in the table from Item 11.

13. a. One factor of $36 - y^2$ is $6 + y$. What is the other factor?

b. One factor of $p^2 - 144$ is $p - 12$. What is the other factor?

c. Describe any patterns you observe.

14. Factor each of the following.

a. $49 - x^2$ **b.** $n^2 - 9$ **c.** $64w^2 - 25$

d. Describe any patterns you observe.

15. Explain how to factor a polynomial of the form $a^2 - b^2$.

A polynomial of the form $a^2 - b^2$ is referred to as the ***difference of two squares***.

Check Your Understanding

Factor each difference of two squares.

16. $x^2 - 121$ **17.** $16m^2 - 81$ **18.** $9 - 25p^2$

LESSON 26-2 PRACTICE

Identify each polynomial as a perfect square trinomial, a difference of two squares, or neither. Then factor the polynomial if it is a perfect square trinomial or a difference of two squares.

19. $z^2 + 6z + 12$ **20.** $4x^2 - 121$ **21.** $y^2 - 8y + 16$

22. $y^2 - 8y - 16$ **23.** $n^2 + 25$ **24.** $169 - 9x^2$

25. What factor would you need to multiply by $(4c + 7)$ to get $16c^2 - 49$?

26. What factor would you need to multiply by $(3d + 1)$ to get $9d^2 + 6d + 1$?

Factor completely. (*Hint*: First look for a GCF.)

27. $2x^2 + 8x + 8$ **28.** $3y^2 - 75$ **29.** $12x^2 - 12x + 3$

30. Use appropriate tools strategically. Explain how you can use your calculator to check that you have factored a polynomial correctly.

ACTIVITY 26 PRACTICE

Write your answers on notebook paper.
Show your work.

Lesson 26-1

1. What is the greatest common factor of the terms in the polynomial $24x^8 + 6x^5 + 9x^2$?
   A. 3 **B.** $3x^2$
   C. $6x$ **D.** $6x^2$

Factor a monomial (the GCF) from each polynomial.

2. $15x^4 + 20x^3 + 35x$
3. $12m^3 - 8m^2 + 16m + 8$
4. $32y^2 + 48y - 16$
5. $x^5 + x^4 + 3x^3 + 3x^2$
6. Which of these polynomials cannot be factored by factoring out the GCF?
   A. $7x^2 + 14x + 21$ **B.** $49x^3 + 21x^2 + x$
   C. $x^2 + 14x + 7$ **D.** $35x^3 + 28x^2 + 7x$
7. The figure shows the dimensions of a garden plot in the shape of a trapezoid. Write and simplify a polynomial for the perimeter of the plot. Then factor the polynomial completely.

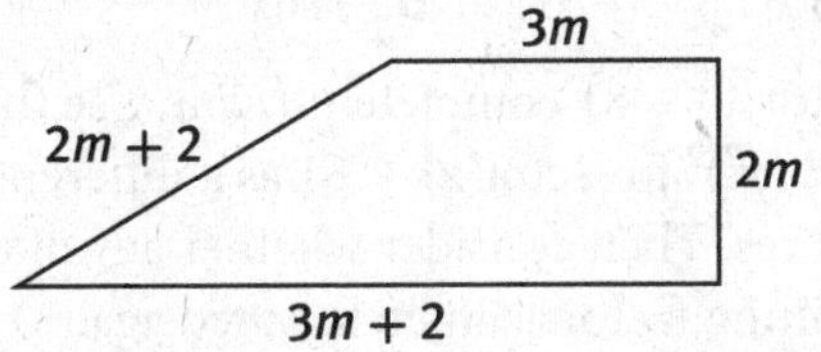

8. The area of the rectangle shown below is $6x^2 + 9x$ square feet. The width of the rectangle is given in the figure. What is the length of the rectangle? Justify your answer.

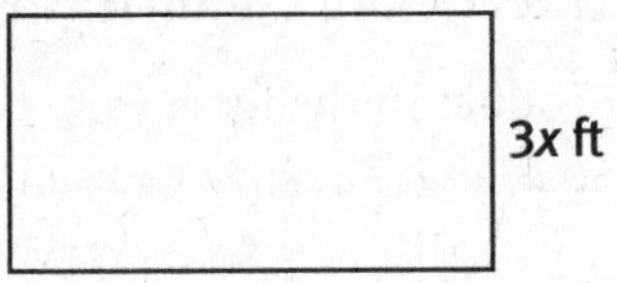

9. Marcus saw the factorization shown below in his textbook, but part of the factorization was covered by a drop of ink. What expression was covered by the drop of ink?

   $-24x^5 - 16x^3 = -8x^3($  $+ 2)$

10. Write a polynomial with four terms that has a GCF of $4x^2$.

Lesson 26-2

Identify each polynomial as a perfect square trinomial, a difference of two squares, or neither. Then factor the polynomial if it is a perfect square trinomial or a difference of two squares.

11. $9x^2 - 121$
12. $m^2 - 16m + 64$
13. $y^2 + 12y - 36$
14. $16z^2 + 25$
15. $25 - 144p^2$
16. $x^2 + 50x + 625$

Factor completely.

17. $2x^2 - 32$
18. $32 - 8p$
19. $3x^3 + 12x^2 + 12x$
20. $4y^3 - 32y^2 + 64y$
21. $5x^4 - 125x^2$
22. What factor would you need to multiply by $(4x - 1)$ to get $16x^2 - 8x + 1$?
   A. $4x - 1$ **B.** $4x + 1$
   C. $4x^2$ **D.** $4x$

Use the rectangle for Items 23–25.

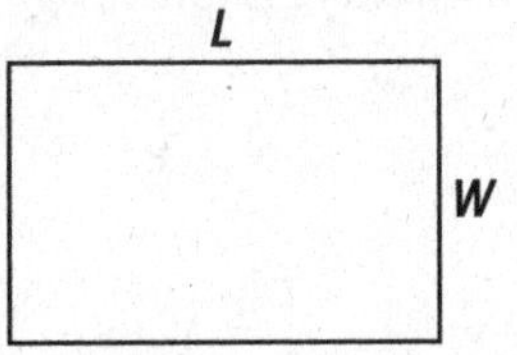

23. The area of a rectangle is $64b^2 - 4$ and $W = 8b - 2$. What is L?
24. The area of another rectangle is $144c^2 - 4$ and $L = 12c + 2$. What is W?

25. Suppose the area of a rectangle is $4x^2 - 4x + 1$ and $L = 2x - 1$.

a. What is W?

b. What must be true about the rectangle in this case? Explain.

26. The area of a square window is given by the expression $m^2 - 16m + 64$. Which expression represents the length of one side of the window?

A. $m - 4$ **B.** $m + 4$
C. $m - 8$ **D.** $m + 8$

27. What value of k makes the polynomial $x^2 + 6x + k$ a perfect square trinomial?

A. 3 **B.** 6
C. 9 **D.** 36

28. Consider the following values of c in the polynomial $36x^2 + c$.

I. $c = -25$
II. $c = 25$
III. $c = -36$

Which value or values of c make it possible to factor the polynomial?

A. I only
B. I and II only
C. I and III only
D. I, II, and III

29. Write a perfect square trinomial that includes the term $9x^2$.

30. The polynomial $x^2 + bx + 25$ is a perfect square trinomial. What is the value of b? Is there more than one possibility? Explain.

31. Sasha and Pedro were asked to factor the polynomial $9x^2 - 9$ completely and explain their process. Their work is shown below. Has either student factored the polynomial completely? Explain. If not, give the complete factorization.

Sasha's Work
$9x^2 - 9 = (3x + 3)(3x - 3)$
I used the fact that $9x^2 - 9$ is a difference of two squares.

Pedro's Work
$9x^2 - 9 = 9(x^2 - 1)$
I factored out the GCF.

32. Which of the following polynomials has $m - 4$ as a factor?

A. $m^2 - 4$ **B.** $m^2 + 16$
C. $m^2 - 8m + 16$ **D.** $m^2 - 8m - 16$

33. Given that $x^2 + \square + 100$ is a perfect square trinomial, which of these could be the missing term?

A. $10x$ **B.** $20x$
C. $50x$ **D.** $100x$

34. Factor $x^4 - 81$ completely. (*Hint*: Use the fact that $x^4 = (x^2)^2$ to factor $x^4 - 81$ as a difference of two squares. Then consider whether any of the resulting factors can be factored again.)

35. Use the method in Item 34 to factor $y^8 - 625$ completely.

MATHEMATICAL PRACTICES
Reason Abstractly and Quantitatively

36. Could a product in the form $(a + b)(a - b)$ ever be equal to $a^2 + b^2$? Justify your answer.

Factoring Trinomials

Deconstructing Floor Plans

Lesson 27-1 Factoring $x^2 + bx + c$

My Notes

Learning Targets:

- Use algebra tiles to factor trinomials of the form $x^2 + bx + c$.
- Factor trinomials of the form $x^2 + bx + c$.

SUGGESTED LEARNING STRATEGIES: Marking the Text, Create Representations, Think-Pair-Share, Look for a Pattern, Discussion Groups

Recall that Factor Steele Buildings can create many floor plans with different-size spaces. Custom Showrooms has asked Factor Steele Buildings for a floor plan with one great room, five large offices, and six cubicles. Each great room has a length and width equal to x units, each large office has a width of x units and a length of 1 unit, and each cubicle has a length and width of 1 unit.

Factor Steele Buildings proposes the rectangular floor plan shown below.

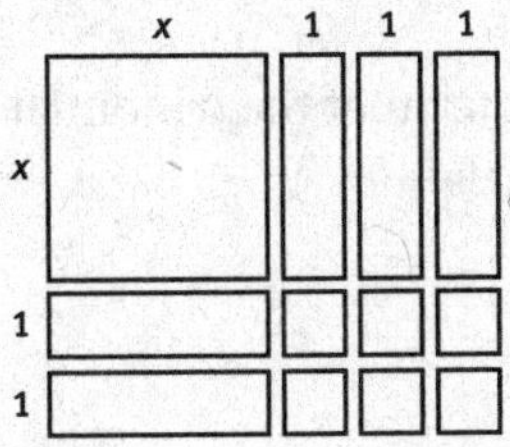

1. Represent the area of the entire office as a sum of the areas of all the rooms.

2. Write the area of the entire office as a product of two binomials by multiplying the length of the entire office by the width of the entire office.

3. **Make use of structure.** Multiply the binomials in Item 2 to check that their product is the expression you wrote in Item 1. Justify your steps and name any properties you use to multiply the binomials.

My Notes

Items 1 through 3 show how to use algebra tiles to factor a trinomial. However, drawing tiles to factor a trinomial can become time-consuming. Analyzing patterns and using graphic organizers can help factor a trinomial of the form $x^2 + bx + c$ without using tiles.

4. Consider the binomials $(x - 5)$ and $(x + 3)$.

a. Determine their product.

b. How is the coefficient of the trinomial's middle term related to the constant terms of the binomials?

c. How is the constant term of the trinomial related to the constant terms of the binomials?

5. Consider the binomials $(x + 6)$ and $(x + 1)$.

a. Determine their product.

b. How is the coefficient of the trinomial's middle term related to the constant terms of the binomials?

c. How is the constant term of the trinomial related to the constant terms of the binomials?

6. Express regularity in repeated reasoning. Use the patterns you observed in Items 4 and 5 to analyze a trinomial of the form $x^2 + bx + c$. Describe how the numbers in the binomial factors are related to the constant term c, and to b, the coefficient of x.

My Notes

Example A

Factor $x^2 + 12x + 32$.

Step 1: Create a graphic organizer as shown. Place the first term in the upper left region. Place the last term in the lower right region.

x^2	
	32

Step 2: Identify the factors of c that add to b. Use a table to help you test factors.

Factors of 32		Sum of the Factors		
32	1	32 + 1	=	33
16	2	16 + 2	=	18
8	4	8 + 4	=	12✓

Step 3: Fill in the missing factors and products in the graphic organizer.

	x	8
x	x^2	$8x$
4	$4x$	32

Step 4: Write the original trinomial as the product of two binomials.

$$x^2 + 12x + 32 = (x + 4)(x + 8)$$

My Notes

Try These A

a. Fill in the missing sections of the graphic organizer for the trinomial $x^2 - 6x + 8$. Express the trinomial as a product of two binomials.

	___	___
___	x^2	
___	$-4x$	8

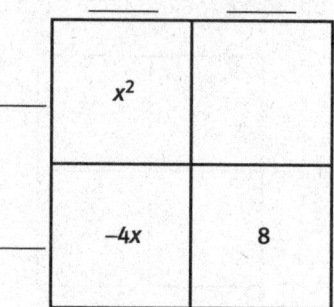

b. Make a graphic organizer like the one above for the trinomial $x^2 + 14x + 45$. Express the trinomial as a product of two binomials.

c. Factor $x^2 + 6x - 27$.

d. Factor $x^2 + 10x + 1$.

MATH TIP

If there are no factors of *c* that add to *b*, the trinomial cannot be factored. A polynomial that cannot be factored is called *unfactorable* or a *prime polynomial*.

My Notes

Check Your Understanding

Factor each trinomial. Then multiply your factors to check your work.

7. $x^2 + 15x + 56$

8. $x^2 + 22x + 120$

9. $x^2 + 6x - 27$

10. $x^2 - 14x + 48$

11. $x^2 - x + 1$

LESSON 27-1 PRACTICE

Factor each trinomial.

12. $x^2 + 8x + 15$

13. $x^2 - 5x - 14$

14. $x^2 - 5x + 3$

15. $x^2 - 16x + 48$

16. $24 + 10x + x^2$

17. Custom Showrooms has expanded and now wants Factor Steele Buildings to create a floor plan with one great room, 15 large offices, and 50 cubicles.

a. Write the area of the new floor plan as a trinomial.

b. Factor the trinomial.

c. Multiply the binomials in Part (b) to check your work.

18. **Reason abstractly.** Suppose $x^2 + bx + c$ is a factorable trinomial in which c is a positive prime number.

a. Write an expression to represent the value of b.

b. Write $x^2 + bx + c$ as the product of two factors using only c as an unknown constant.

MATH TIP

A prime number has only itself and 1 as factors. For example, the numbers 3 and 11 are prime numbers.

My Notes

Learning Targets:

- Factor trinomials of the form $ax^2 + bx + c$ when the GCF is 1.
- Factor trinomials of the form $ax^2 + bx + c$ when the GCF is not 1.

SUGGESTED LEARNING STRATEGIES: Think-Pair-Share, Note Taking, Guess and Check, Look for a Pattern, Work Backward

Custom Showrooms now wants Factor Steele Buildings to create a floor plan with more than one great room. Instead, Custom Showrooms wants two great rooms, seven large offices, and six cubicles.

The trinomial $2x^2 + 7x + 6$ can be factored to determine the length and width of the entire office space.

1. **Attend to precision.** How is the trinomial $2x^2 + 7x + 6$ different from the trinomials you factored in Lesson 27-1?

MATH TIP

The factors of c will both have the same sign if $c > 0$. If $b < 0$, both factors will be negative. If $b > 0$, both factors will be positive.

Example A

Factor $2x^2 + 7x + 6$ using a guess and check method.

Possible Binomial Factors	Reasoning
$(2x \quad)(x \quad)$	$a = 2$ can be factored as $2 \cdot 1$.
$(2x + \quad)(x + \quad)$	$c = 6$, so both factors have the same sign. $b = 7$, so both factors are positive. 6 can be factored as $1 \cdot 6$, $6 \cdot 1$, $2 \cdot 3$, or $3 \cdot 2$.
$(2x + 1)(x + 6)$	Product: $2x^2 + 13x + 6$, incorrect
$(2x + 6)(x + 1)$	Product: $2x^2 + 8x + 6$, incorrect
$(2x + 2)(x + 3)$	Product: $2x^2 + 8x + 6$, incorrect
$(2x + 3)(x + 2)$	Product: $2x^2 + 7x + 6$, correct factors

Example B

Factor $3x^2 + 8x - 11$ using a guess and check method.

Possible Binomial Factors	Reasoning
$(3x \quad)(x \quad)$	$a = 3$ can be factored as $3 \cdot 1$.
$(3x + \quad)(x - \quad)$ or $(3x - \quad)(x + \quad)$	$c = -11$, so the factors have different signs. 11 can be factored as $11 \cdot 1$ or $1 \cdot 11$.
$(3x + 11)(x - 1)$	Product: $3x^2 + 8x - 11$, correct factors
$(3x - 11)(x + 1)$	Product: $3x^2 - 8x - 11$, incorrect
$(3x + 1)(x - 11)$	Product: $3x^2 - 32x - 11$, incorrect
$(3x - 1)(x + 11)$	Product: $3x^2 + 32x - 11$, incorrect

My Notes

Try These A–B

Factor the trinomials.

a. $3x^2 + 5x + 2$

b. $2x^2 + 5x - 18$

c. $2x^2 + 6x - 7$

Example C

Factor $4x^2 - 4x - 15$ using a guess and check method.

Possible Binomial Factors	Reasoning
$(4x \quad)(x \quad)$ or $(2x \quad)(2x \quad)$	$a = 4$ can be factored as $4 \cdot 1$ or $2 \cdot 2$.
$(4x - \quad)(x + \quad)$ or $(4x + \quad)(x - \quad)$ or $(2x - \quad)(2x + \quad)$ or $(2x + \quad)(2x - \quad)$	$c = -15$, so the factors have different signs. 15 can be factored as $1 \cdot 15$, $15 \cdot 1$, $3 \cdot 5$, or $5 \cdot 3$.
$(4x - 1)(x + 15)$	Product: $4x^2 + 59x - 15$, incorrect
$(4x + 1)(x - 15)$	Product: $4x^2 - 59x - 15$, incorrect
$(4x - 15)(x + 1)$	Product: $4x^2 - 11x - 15$, incorrect
$(4x + 15)(x - 1)$	Product: $4x^2 + 11x - 15$, incorrect
$(4x - 3)(x + 5)$	Product: $4x^2 + 17x - 15$, incorrect
$(4x + 3)(x - 5)$	Product: $4x^2 - 17x - 15$, incorrect
$(4x - 5)(x + 3)$	Product: $4x^2 + 7x - 15$, incorrect
$(4x + 5)(x - 3)$	Product: $4x^2 - 7x - 15$, incorrect
$(2x - 1)(2x + 15)$	Product: $4x^2 + 28x - 15$, incorrect
$(2x + 1)(2x - 15)$	Product: $4x^2 - 28x - 15$, incorrect
$(2x - 3)(2x + 5)$	Product: $4x^2 + 4x - 15$, incorrect
$(2x + 3)(2x - 5)$	Product: $4x^2 - 4x - 15$, correct factors

Try These C

Factor the trinomials.

a. $6x^2 - 11x - 2$

b. $6x^2 - 13x - 4$

c. $4x^2 - 20x + 21$

My Notes

Example D

Factor $9x^2 - 24x + 12$.

Step 1: The coefficients 9, −24, and 12 are all divisible by 3. Factor out the GCF.

$$9x^2 - 24x + 12 = 3(3x^2 - 8x + 4)$$

Step 2: Factor $3x^2 - 8x + 4$ using a guess and check method.

Possible Binomial Factors	Reasoning
$(3x \quad)(x \quad)$	$a = 3$ can be factored as $3 \cdot 1$.
$(3x - \quad)(x - \quad)$	$c = 4$, so the factors have the same sign. $b = -8$, so both factors are negative. 4 can be factored as $4 \cdot 1$, $1 \cdot 4$, or $2 \cdot 2$.
$(3x - 4)(x - 1)$	Product: $3x^2 - 7x + 4$, incorrect
$(3x - 1)(x - 4)$	Product: $3x^2 - 13x + 4$, incorrect
$(3x - 2)(x - 2)$	Product: $3x^2 - 8x + 4$, correct factors

Solution: Write the complete factorization, including the GCF from Step 1: $3(3x - 2)(x - 2)$

Check: Multiply to check your answer.
$3(3x - 2)(x - 2)$
$= 3(3x^2 - 6x - 2x + 4) = 3(3x^2 - 8x + 4) = 9x^2 - 24x + 12$

Try These D

Factor the trinomials completely. Check your work by multiplying the factors.

a. $10x^2 + 19x + 6$ **b.** $8x^2 + 20x - 28$ **c.** $8x^3 - 14x^2 + 6x$

Check Your Understanding

Factor each trinomial completely. Check your work by multiplying the factors.

2. $5x^2 + x - 4$ **3.** $49x^2 - 126x + 56$ **4.** $9x^3 - 39x^2 - 30x$

LESSON 27-2 PRACTICE

Model with mathematics. Factor Steele Buildings has received several floor plan requests. For Items 5–8, factor each floor plan scenario completely to help Factor Steele Buildings determine the space's dimensions.

5. 3 great rooms, 23 large offices, 14 cubicles

6. 10 great rooms, 31 large offices, 15 cubicles

7. 8 great rooms, 92 large offices, 180 cubicles

8. 12 great rooms, 38 large offices, 20 cubicles

9. Suppose $ax^2 + bx + c$ is a factorable trinomial in which both a and c are positive prime numbers. Write an expression to represent the value of b.

ACTIVITY 27 PRACTICE

Write your answers on notebook paper. Show your work.

Lesson 27-1

Factor each trinomial.

1. $x^2 + 11x + 30$
2. $x^2 + 22x + 121$
3. $x^2 + x - 30$
4. $x^2 - 7x - 18$
5. $x^2 - 169$
6. $x^2 + 9x - 36$

Mrs. Harbrook can choose from two rectangular pool sizes. The pool manufacturer provides her with the area of the pool, but she needs to find the dimensions in order to determine if the pool will fit in her yard. Use the rectangle for Items 7 and 8.

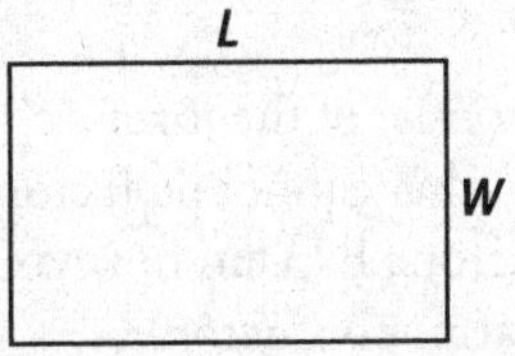

7. If the area of the pool is $x^2 - 17x + 72$, what are possible expressions to represent the length L and the width W?
8. **a.** If the area of the pool is $x^2 + 24x + 144$, what are possible expressions to represent the length L and the width W?
   b. What do these dimensions tell you about the shape of the pool?

The area of a parallelogram is given by the formula $A = bh$, where b is the base and h is the height. Use this information for Items 9 and 10.

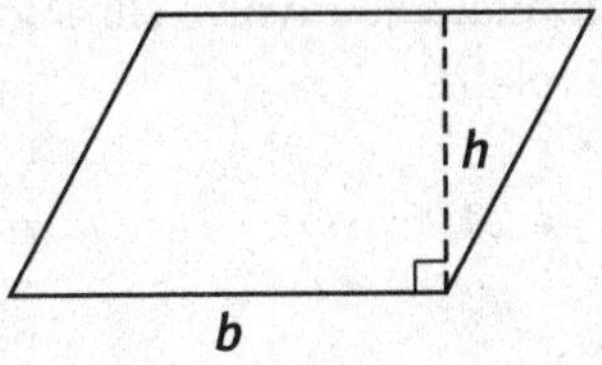

9. If the area of the parallelogram is $x^2 + x - 42$, what are possible expressions to represent the base b and the height h?
10. If the area of the parallelogram is $x^2 + 4x - 117$, what are possible expressions to represent the base b and the height h?
11. Which of the following trinomials cannot be factored?
    A. $x^2 + 3x + 2$ **B.** $x^2 + 3x - 2$
    C. $x^2 - 3x + 2$ **D.** $x^2 + 2x - 3$
12. Which of the following binomials is a factor of the trinomial $y^2 - y - 20$?
    A. $y - 4$ **B.** $y + 4$
    C. $y - 10$ **D.** $y + 10$

For Items 13–15, consider the trinomial $x^2 + 2x + c$. Determine whether each statement is always, sometimes, or never true.

13. If c is a prime number, then the trinomial cannot be factored.
14. If c is an even number, then the GCF of the terms in the trinomial is 2.
15. If $c < 0$, then the trinomial can be factored.
16. Write a trinomial that can be factored such that one of the binomial factors is $x - 5$. Explain how you found the trinomial.

Lesson 27-2

Factor each trinomial completely.

17. $3x^2 + 8x - 11$
18. $5x^2 - 7x + 2$
19. $2x^2 - 9x - 5$
20. $3x^2 + 17x - 28$
21. $7x^2 + 9x + 2$
22. $6x^2 - 11x - 7$
23. $12x^2 - 11x + 2$
24. $8x^2 + 16x + 6$

25. Which of the following is **not** a factor of the trinomial $24x^3 - 6x^2 - 9x$?
A. $3x$ **B.** $4x - 3$
C. $2x + 1$ **D.** $2x - 1$

26. Which binomial is a factor of $4x^2 + 12x + 5$?
A. $2x + 5$ **B.** $2x - 5$
C. $4x + 1$ **D.** $4x - 1$

The volume of a rectangular prism is found using the formula $V = lwh$, where l is the length, w is the width, and h is the height. Use the rectangular prism for Items 27–29.

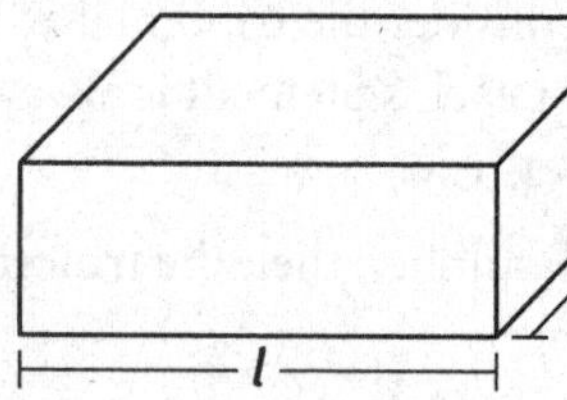

27. If the volume of a rectangular prism is $6x^3 + 3x^2 - 18x$, what are possible expressions to represent the length, width, and height?

28. If the volume of a rectangular prism is $10x^2 - 55x + 60$, what are possible expressions to represent the length, width, and height?

29. If the volume of a rectangular prism is $12x^2 + 22x + 6$, what are possible expressions to represent the length, width, and height?

30. For which value of k is it possible to factor the trinomial $2x^2 + 3x + k$?
A. -1 **B.** 1
C. 2 **D.** 3

31. Which of the following trinomials has the binomial $x + 1$ as a factor?
A. $2x^2 - x - 1$
B. $2x^2 - 3x + 1$
C. $3x^2 - 5x + 2$
D. $3x^2 + x - 2$

32. Mayumi was asked to completely factor the trinomial $4x^2 + 10x + 4$. Her work is shown below. Is her solution correct? Justify your response.

> 4 can be factored as $4 \cdot 1$ or $2 \cdot 2$.
> Try $(2x + \quad)(2x + \quad)$.
> $(2x + 2)(2x + 2) = 4x^2 + 8x + 4$; incorrect
> $(2x + 4)(2x + 1) = 4x^2 + 10x + 4$; correct!
> The factorization is $(2x + 4)(2x + 1)$.

33. Given that the trinomial $5x^2 + bx + 10$ can be factored, which of the following statements must be true?
A. The value of b must be positive.
B. The value of b must be negative.
C. The value of b cannot be 3.
D. The value of b cannot be -7.

34. What is the factorization of the trinomial $p^2x^2 - 2pqx + q^2$?

35. Write a trinomial of the form $ax^2 + bx + c$ (with $a \neq 1$) that cannot be factored into binomial factors. Explain how you know the trinomial cannot be factored.

36. The area of a rectangular carpet is $6x^2 - 11x + 4$ square yards. The length of the carpet is $3x - 4$ yards. Which of the following is the width?
A. $2x - 1$ yards **B.** $2x + 1$ yards
C. $3x - 1$ yards **D.** $3x + 1$ yards

MATHEMATICAL PRACTICES
Construct Viable Arguments and Critique the Reasoning of Others

37. Guillaume is asked to factor a trinomial of the form $x^2 + bx - 8$. He says that because the constant term is negative, both binomial factors of the trinomial will involve subtraction. Is he correct? Explain.

Simplifying Rational Expressions

Totally Rational

Lesson 28-1 Simplifying Rational Expressions

My Notes

Learning Targets:

- Simplify a rational expression by dividing a polynomial by a monomial.
- Simplify a rational expression by dividing out common factors.

SUGGESTED LEARNING STRATEGIES: Think-Pair-Share, Note Taking, Identify a Subtask

A field trips costs \$800 for the charter bus plus \$10 per student for x students. The cost per student is represented by the expression $\frac{10x + 800}{x}$.

The cost-per-student expression is a rational expression. A ***rational expression*** is a ratio of two polynomials.

Like fractions, rational expressions can be simplified and combined using the operations of addition, subtraction, multiplication, and division.

When a rational expression has a polynomial in the numerator and a monomial in the denominator, it may be possible to simplify the expression by dividing each term of the polynomial by the monomial.

Example A

Simplify by dividing: $\frac{12x^5 + 6x^4 - 9x^3}{3x^2}$

Step 1: Rewrite the rational expression to indicate each term of the numerator divided by the denominator.

$$\frac{12x^5}{3x^2} + \frac{6x^4}{3x^2} - \frac{9x^3}{3x^2}$$

Step 2: Divide. Use the Quotient of Powers Property.

$$\frac{12x^5}{3x^2} + \frac{6x^4}{3x^2} - \frac{9x^3}{3x^2}$$
$$4x^{5-2} + 2x^{4-2} - 3x^{3-2}$$
$$4x^3 + 2x^2 - 3x^1$$

Solution: $4x^3 + 2x^2 - 3x$

Try These A

Simplify by dividing.

a. $\frac{5y^4 - 10y^3 - 5y^2}{5y^2}$

b. $\frac{32n^6 - 24n^4 + 16n^2}{-8n^2}$

My Notes

To simplify a rational expression, first factor the numerator and denominator. Remember that factors can be monomials, binomials, or even polynomials. Then, divide out the common factors.

Example B

Simplify $\frac{12x^2}{6x^3}$.

Step 1: Factor the numerator and denominator.

$$\frac{2 \cdot 6 \cdot x \cdot x}{6 \cdot x \cdot x \cdot x}$$

Step 2: Divide out the common factors.

$$\frac{2 \cdot \cancel{6} \cdot \cancel{x} \cdot \cancel{x}}{\cancel{6} \cdot x \cdot \cancel{x} \cdot \cancel{x}}$$

Solution: $\frac{2}{x}$

Example C

Simplify $\frac{2x^2 - 8}{x^2 - 2x - 8}$.

Step 1: Factor the numerator and denominator.

$$\frac{2(x+2)(x-2)}{(x+2)(x-4)}$$

Step 2: Divide out the common factors.

$$\frac{2\cancel{(x+2)}(x-2)}{\cancel{(x+2)}(x-4)}$$

Solution: $\frac{2(x-2)}{x-4}$

MATH TIP

If a, b, and c are polynomials, and b and c do not equal 0, then $\frac{ac}{bc} = \frac{a}{b}$, because $\frac{c}{c} = 1$.

Try These B–C

Simplify each rational expression.

a. $\frac{6x^4y}{15xy^3}$ **b.** $\frac{x^2 + 3x - 4}{x^2 - 16}$ **c.** $\frac{15x^2 - 3x}{25x^2 - 1}$

MATH TIP

The graph of $y = \frac{2}{x}$ will never cross the x-axis since x cannot equal 0.

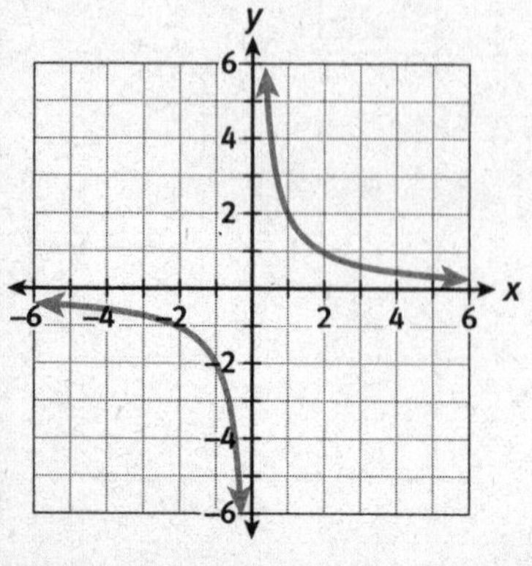

The value of the denominator in a rational expression cannot be zero because division by zero is undefined.

- In Example B, x cannot equal 0 because $6 \cdot (0)^3 = 0$.
- To find the excluded values of x in Example C, first factor the denominator. This shows that $x \neq -2$ because that would make the factor $x + 2 = 0$. Also, $x \neq 4$ because that would make the factor $x - 4 = 0$. Therefore, in Example C, x cannot equal -2 or 4.

My Notes

Example D

Divide $\frac{1-x}{x-1}$. Simplify your answer if possible.

Step 1: Factor the numerator. $-1(x-1)$

Step 2: Divide out the common factor. $\frac{-1\cancel{(x-1)}}{\cancel{x-1}}$

Solution: -1

Try These D

Divide. Simplify your answer if possible. Identify any excluded values of the variable.

a. $\frac{x-5}{5-x}$ **b.** $\frac{3x-3}{1-x}$

MATH TIP

The graph of the rational function $f(x) = \frac{1-x}{x-1}$ looks like:

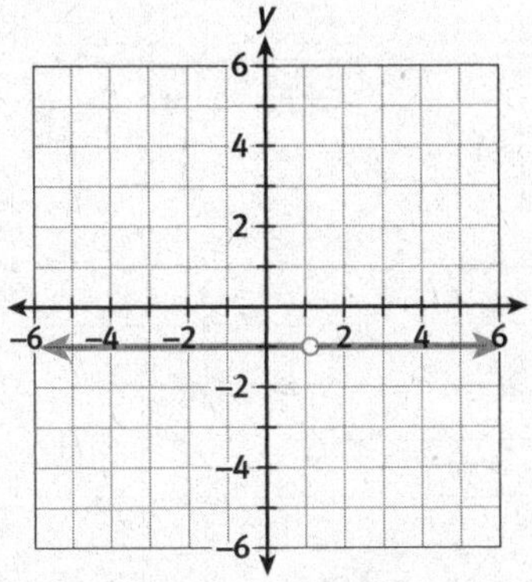

It looks like the graph of $y = -1$.

Check Your Understanding

Attend to precision. Describe the steps you would take to simplify each rational expression. Identify any excluded values of the variable.

1. $\frac{x^2-36}{6-x}$ **2.** $\frac{x^2-10x+24}{4x-16}$

LESSON 28-1 PRACTICE

Simplify by dividing.

3. $\frac{16x^5-8x^3+4x^2}{4x^2}$ **4.** $\frac{15x^6-20x^4}{-5x^3}$

5. $\frac{24x^6+18x^5-15x^3+12x^2}{3x^2}$

Simplify.

6. $\frac{3x^2yz}{12xyz^3}$ **7.** $\frac{25x^4y^3z^4}{-5x^5y^2z^3}$

8. $\frac{x^2-2x+1}{x^2+3x-4}$ **9.** $\frac{2x^2}{4x^3-16x}$

10. $\frac{x+1}{-4-4x}$ **11.** $\frac{x^2+6x+9}{x^2-9}$

12. Model with mathematics. The four algebra classes at Sanchez School are going on a field trip to a museum. Each class contains s students. The museum charges $8 per student for admission. There is also a flat fee of $200 for the buses.

a. Write an expression for the total cost of the buses and the museum admission fees for all four classes.

b. Write a rational expression for the cost per student. Simplify the expression as much as possible.

c. Use the expression you wrote in Part (b) to find the cost per student if each class has 20 students.

My Notes

Learning Targets:

- Divide a polynomial of degree one or two by a polynomial of degree one or two.
- Express the remainder of polynomial division as a rational expression.

SUGGESTED LEARNING STRATEGIES: Think-Pair-Share, Identify a Subtask, Close Reading, Note Taking, Discussion Groups

Division of polynomials is similar to long division of real numbers.

Example A

Divide $\frac{525}{25}$ using long division.

Step 1: Divide 52 by 25.

$$\begin{array}{r} 2 \\ 25\overline{)525} \\ -\underline{50} \\ 2 \end{array}$$

Step 2: Bring down 5.

$$\begin{array}{r} 2 \\ 25\overline{)525} \\ -\underline{50}\downarrow \\ 25 \end{array}$$

Step 3: Divide 25 by 25.

$$\begin{array}{r} 21 \\ 25\overline{)525} \\ -\underline{50}\downarrow \\ 25 \\ -\underline{25} \\ 0 \end{array}$$

Solution: The quotient is 21.

Division with polynomials can be done in the same way as long division with whole numbers.

My Notes

Example B

Simplify using long division: $\dfrac{12x^5 + 6x^4 - 9x^3}{3x^2}$

Step 1: Divide $12x^5$ by $3x^2$.

$$\begin{array}{r|l} & 4x^3 \\ 3x^2 & 12x^5 + 6x^4 - 9x^3 \\ & -\underline{12x^5} \\ & 0 \end{array}$$

Step 2: Bring down $6x^4$.

$$\begin{array}{r|l} & 4x^3 \\ 3x^2 & 12x^5 + 6x^4 - 9x^3 \\ & -\underline{12x^5} \quad \downarrow \\ & \quad\quad 6x^4 \end{array}$$

Step 3: Divide $6x^4$ by $3x^2$.

$$\begin{array}{r|l} & 4x^3 + 2x^2 \\ 3x^2 & 12x^5 + 6x^4 - 9x^3 \\ & -\underline{12x^5} \\ & \quad\quad 6x^4 \\ & \quad -\underline{6x^4} \\ & \quad\quad 0 \end{array}$$

Step 4: Bring down $-9x^3$.

$$\begin{array}{r|l} & 4x^3 + 2x^2 \\ 3x^2 & 12x^5 + 6x^4 - 9x^3 \\ & -\underline{12x^5} \\ & \quad\quad 6x^4 \\ & \quad -\underline{6x^4} \quad \downarrow \\ & \quad\quad\quad -9x^3 \end{array}$$

Step 5: Divide $-9x^3$ by $3x^2$.

$$\begin{array}{r|l} & 4x^3 + 2x^2 - 3x \\ 3x^2 & 12x^5 + 6x^4 - 9x^3 \\ & -\underline{12x^5} \\ & \quad\quad 6x^4 \\ & \quad -\underline{6x^4} \\ & \quad\quad\quad -9x^3 \\ & \quad\quad -\underline{(-9x^3)} \\ & \quad\quad\quad\quad 0 \end{array}$$

Solution: The quotient is $4x^3 + 2x^2 - 3x$.

My Notes

Example C

Simplify using long division: $\dfrac{x^2+9x+14}{x+7}$.

Step 1: Divide x^2 by x.

$$\begin{array}{r}
x \\
x+7\overline{)\,x^2+9x+14} \\
-(x^2+7x)
\end{array}$$

Step 2: Distribute the negative and subtract $x^2 + 7x$ from $x^2 + 9x$.

$$\begin{array}{r}
x \\
x+7\overline{)\,x^2+9x+14} \\
\underline{-x^2-7x} \\
2x
\end{array}$$

Step 3: Bring down the next term, 14.

$$\begin{array}{r}
x \\
x+7\overline{)\,x^2+9x+14} \\
\underline{-x^2-7x} \downarrow \\
2x+14
\end{array}$$

Step 4: Divide $2x$ by x.

$$\begin{array}{r}
x+2 \\
x+7\overline{)\,x^2+9x+14} \\
\underline{-x^2-7x} \\
2x+14 \\
\underline{-(2x+14)}
\end{array}$$

Step 5: Distribute the negative and subtract $2x + 14$ from $2x + 14$.

$$\begin{array}{r}
x+2 \\
x+7\overline{)\,x^2+9x+14} \\
\underline{-x^2-7x} \\
2x+14 \\
\underline{-2x-14} \\
0
\end{array}$$

Solution: The quotient is $x + 2$.

MATH TIP

You can check the quotient in a division problem by using multiplication. Multiply the quotient, $x + 2$, by the divisor, $x + 7$. If you have divided correctly, the product will be the dividend, $x^2 + 9x + 14$.

$(x+2)(x+7) =$
$x^2 + 7x + 2x + 14 =$
$x^2 + 9x + 14$

Try These A–B–C

Simplify using long division.

a. $\dfrac{24x^5 - 8x^4 + 12x^3 - 4x^2}{4x^2}$

b. $\dfrac{x^2 - x - 12}{x - 4}$

My Notes

Sometimes there are remainders when dividing integers. In a similar way, sometimes there are remainders when dividing polynomials.

MATH TIP

When dividing with integers, the remainder is often written as a fraction whose denominator is the divisor.

$$\begin{array}{r} 76\frac{1}{2} \\ 2\overline{)\,153} \\ \underline{-14} \\ 13 \\ \underline{-12} \\ 1 \end{array}$$

Example D

Simplify using long division: $\dfrac{2x^3 - 6x + 15}{x + 1}$.

Step 1: Divide. Add the term $0x^2$ to the dividend as a placeholder.

$$\begin{array}{r} 2x^2 - 2x - 4 \\ x + 1\overline{)\,2x^3 + 0x^2 - 6x + 15} \\ \underline{-(2x^3 + 2x^2)} \\ -2x^2 - 6x \\ \underline{-(-2x^2 - 2x)} \\ -4x + 15 \\ \underline{-(-4x - 4)} \\ 19 \end{array}$$

Step 2: Write the remainder as $\dfrac{19}{x+1}$.

$$\begin{array}{r} 2x^2 - 2x - 4 + \dfrac{19}{x+1} \\ x + 1\overline{)\,2x^3 + 0x^2 - 6x + 15} \\ \underline{-(2x^3 + 2x^2)} \\ -2x^2 - 6x \\ \underline{-(-2x^2 - 2x)} \\ -4x + 15 \\ \underline{-(-4x - 4)} \\ 19 \end{array}$$

Solution: The quotient is $2x^2 - 2x - 4 + \dfrac{19}{x+1}$.

Try These D

Simplify using long division.

a. $(3x^2 + 6x + 1) \div (3x)$

b. $(6x^2 + 5x - 20) \div (3x + 4)$

c. $(3x^3 + 5x - 10) \div (x - 4)$

My Notes

1. **Make sense of problems.** Consider the following polynomial division problem: $\frac{3x^2 - 8x + 15}{x^2 + 3x - 4}$.

a. How does this division problem differ from those in the examples?

b. Use long division to perform the division.

c. Write the remainder as a rational expression.

d. What is the quotient?

Check Your Understanding

2. **Make use of structure.** Describe how dividing polynomials using long division is similar to dividing whole numbers using long division.

3. Explain how to check a division problem involving whole numbers that has a remainder.

4. Explain how to check a division problem involving polynomials that has a remainder. To demonstrate, use $(3x^2 + x - 2) \div (x^2 + 2x + 3) = 3 + \frac{-5x - 11}{x^2 + 2x + 3}$.

LESSON 28-2 PRACTICE

Simplify using long division.

5. $\frac{4x^2 + 6x}{2x}$

6. $\frac{3x^4 - 9x^3 + 6x^2}{3x^2}$

7. $\frac{12x^5 + 24x^4 - 16x^3 - 12x^2}{-4x^2}$

8. $\frac{3x^2 - 6x - 24}{3x - 6}$

9. $\frac{5x^2 - 21x + 4}{5x - 1}$

10. $\frac{12x^2 - 15}{x + 5}$

11. $\frac{3x^2 + 6x - 9}{x + 1}$

12. $\frac{25x^2 + 20x - 15}{5x^2 + 5x + 5}$

13. **Reason abstractly.** The area of a rectangular swimming pool is $2x^2 + 11x + 4$ square feet. The width of the pool is $x - 2$ feet.

a. Write a rational expression that represents the length of the pool. Simplify the expression using long division.

b. What are the length, width, and area of the pool when $x = 19$?

My Notes

Learning Targets:

- Multiply rational expressions.
- Divide rational expressions.

SUGGESTED LEARNING STRATEGIES: Note Taking, Close Reading

To multiply rational expressions, first factor the numerator and denominator of each expression. Next, divide out any common factors. Then simplify, if possible.

Example A

Multiply $\frac{2x-4}{x^2-1} \cdot \frac{3x+3}{x^2-2x}$. Simplify your answer if possible.

Step 1: Factor the numerators and denominators.

$$\frac{2(x-2)}{(x+1)(x-1)} \cdot \frac{3(x+1)}{x(x-2)}$$

Step 2: Divide out common factors.

$$\frac{2(\cancel{x-2}) \cdot 3(\cancel{x+1})}{(\cancel{x+1})(x-1)(x)(\cancel{x-2})}$$

Solution: $\frac{6}{x(x-1)}$

Try These A

Multiply. Simplify your answer.

a. $\frac{y^2+5y+6}{y+2} \cdot \frac{y}{2y+6}$

b. $\frac{2x+2}{x^2-16} \cdot \frac{x^2-5x+4}{4x^2-4}$

To divide rational expressions, use the same process as dividing fractions. Write the division as multiplication of the reciprocal. Then simplify.

Example B

Divide: $\frac{x^2-5x+6}{x^2-9} \div \frac{2x-4}{x^2+2x-3}$. Simplify your answer.

Step 1: Rewrite the division as multiplication by the reciprocal.

$$\frac{x^2-5x+6}{x^2-9} \cdot \frac{x^2+2x-3}{2x-4}$$

Step 2: Factor the numerators and the denominators.

$$\frac{(x-2)(x-3)}{(x+3)(x-3)} \cdot \frac{(x+3)(x-1)}{2(x-2)}$$

Step 3: Divide out common factors.

$$\frac{(\cancel{x-2})(\cancel{x-3})(\cancel{x+3})(x-1)}{(\cancel{x+3})(\cancel{x-3})(2)(\cancel{x-2})}$$

Solution: $\frac{x-1}{2}$

MATH TIP

When dividing fractions, write the division as multiplication by the reciprocal.

$$\frac{a}{b} \div \frac{c}{d} = \frac{a}{b} \cdot \frac{d}{c} = \frac{ad}{bc}$$

If *a*, *b*, *c*, and *d* have any common factors, you can divide them out before you multiply.

$$\frac{4}{15} \div \frac{8}{3} = \frac{4}{15} \cdot \frac{3}{8}$$

$$= \frac{\cancel{4}}{\cancel{3} \cdot 5} \cdot \frac{\cancel{3}}{2 \cdot \cancel{4}} = \frac{1}{10}$$

My Notes

Try These B

Divide. Simplify your answer.

a. $\dfrac{w^2 - 2w - 3}{w^2 - 6w + 9} \div \dfrac{5}{w - 3}$

b. $\dfrac{3xy}{3x^2 - 12} \div \dfrac{xy + y}{x^2 + 3x + 2}$

Check Your Understanding

1. **Critique the reasoning of others.** A student was asked to divide the rational expressions shown below. Examine the student's solution, and then identify and correct the error.

$$\frac{a^2 - 9}{3a} \div \frac{a + 3}{a - 3} = \frac{3a}{a^2 - 9} \cdot \frac{a + 3}{a - 3}$$

$$= \frac{3a}{\cancel{(a + 3)}(a - 3)} \cdot \frac{\cancel{(a + 3)}}{a - 3} = \frac{3a}{(a - 3)^2}$$

2. What is the quotient when $\dfrac{2x + 6}{x + 5}$ is divided by $\dfrac{2}{x + 5}$?

LESSON 28-3 PRACTICE

Multiply or divide.

3. $\dfrac{x^2 - 5x - 6}{x^2 - 4} \cdot \dfrac{x + 2}{x^2 - 12x + 36}$

4. $\dfrac{x^3}{x^2 - 1} \cdot \dfrac{2x + 2}{4x}$

5. $\dfrac{x^2 - y^2}{12} \cdot \dfrac{36}{x + y}$

6. $(b^2 + 12b + 11) \cdot \dfrac{b + 9}{b^2 + 20b + 99}$

7. $\dfrac{x^2 + 4x + 4}{2x + 4} \div \dfrac{x^2 - 4}{x^2 - 6x + 5}$

8. $\dfrac{1}{x - 1} \div \dfrac{x}{x - 1}$

9. $\dfrac{2x + 4}{x^2 + 11x + 18} \div \dfrac{x + 1}{x^2 + 14x + 45}$

10. $\dfrac{m^2 + m - 6}{m^2 + 8m + 15} \div \dfrac{m^2 - m - 2}{m^2 + 9m + 20}$

11. **Make sense of problems.** The figure shows a rectangular prism. The area of the rectangular face $ABCD$ is $x^2 + 2x - 15$.

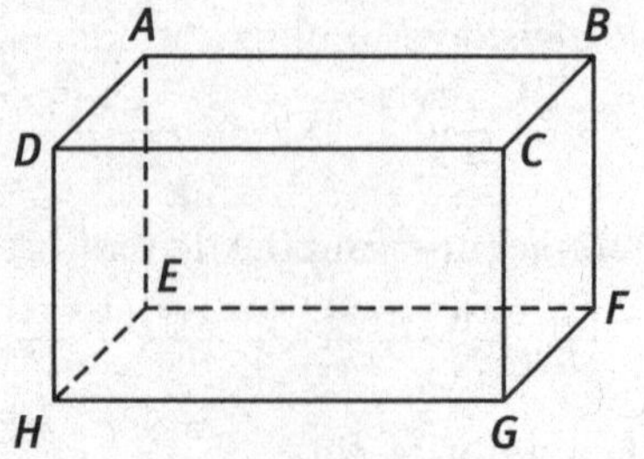

a. The length of edge $\overline{DC}$ is $x + 1$. Write a rational expression that represents the length of edge $\overline{BC}$.

b. The length of edge $\overline{BF}$ is $\dfrac{x^2 + 2x + 1}{x + 5}$. Write and simplify a product to find the area of face $BFGC$.

My Notes

Learning Targets:

- Identify the least common multiple (LCM) of algebraic expressions.
- Add and subtract rational expressions.

SUGGESTED LEARNING STRATEGIES: Note Taking, Close Reading, Sharing and Responding, Identify a Subtask

To add or subtract rational expressions with the same denominator, add or subtract the numerators and then simplify if possible.

Example A

Simplify $\frac{10}{x} - \frac{5}{x}$.

Step 1: Subtract the numerators. $\frac{10}{x} - \frac{5}{x} = \frac{10-5}{x}$

Solution: $\frac{5}{x}$

Example B

Simplify $\frac{2x}{x+1} + \frac{2}{x+1}$.

Step 1: Add the numerators. $\frac{2x}{x+1} + \frac{2}{x+1} = \frac{2x+2}{x+1}$

Step 2: Factor. $= \frac{2(x+1)}{x+1}$

Step 3: Divide out common factors. $= \frac{2(\cancel{x+1})}{\cancel{x+1}}$

Solution: 2

Try These A–B

Add or subtract. Simplify your answer.

a. $\frac{3}{x^2} - \frac{x}{x^2}$

b. $\frac{2}{x+3} - \frac{6}{x+3} + \frac{x}{x+3}$

c. $\frac{x}{x^2-x} + \frac{4x}{x^2-x}$

To add or subtract rational expressions with unlike denominators, first identify a common denominator. The ***least common multiple*** (LCM) of the denominators is used for the common denominator.

One way to determine the LCM is to factor each expression. The LCM is the product of each factor common to the expressions as well as any non-common factors.

MATH TERMS

The **least common multiple** is the smallest multiple that two or more numbers or expressions have in common.

The numbers 10 and 25 have many common multiples. The number 50 is the least common multiple.

My Notes

Example C

Determine the LCM of $x^2 - 4$ and $2x + 4$.

Step 1: Factor each expression.

$$x^2 - 4 = (x + 2)(x - 2)$$
$$2x + 4 = 2(x + 2)$$

Step 2: Identify the factors.
Common Factor: $(x + 2)$
Factors Not in Common: 2 and $(x - 2)$

Step 3: The LCM is the product of all of the factors in Step 2.

Solution: The LCM is $2(x + 2)(x - 2)$.

Try These C

a. Determine the LCM of $2x + 2$ and $x^2 + x$. Use the steps below.
Factor each expression:

Common Factor(s):

Factors Not in Common:

LCM:

b. Determine the LCM of $x^2 - 2x - 15$ and $3x + 9$.

Now add and subtract rational expressions with different denominators. First, determine the LCM of the denominators. Next, write each fraction with the LCM as the denominator. Then, add or subtract. Simplify if possible.

Example D

Subtract $\frac{2}{x} - \frac{3}{x^2 - 2x}$. Simplify your answer if possible.

Step 1: Determine the LCM.
Factor the denominators: x and $x(x - 2)$
The LCM is $x(x - 2)$.

Step 2: Multiply the numerator and denominator of the first term by $(x - 2)$. The denominator of the second term is the LCM.

$$\frac{2}{x} \cdot \frac{(x-2)}{(x-2)} - \frac{3}{x(x-2)}$$

Step 3: Use the Distributive Property in the numerator.

$$\frac{2x - 4}{x(x-2)} - \frac{3}{x(x-2)}$$

Step 4: Subtract the numerators. $\frac{2x - 7}{x(x-2)}$

Solution: $\frac{2x - 7}{x(x-2)}$

MATH TIP

Multiplying a fraction by a form of 1 gives an equivalent fraction.

$\frac{1}{3} = \frac{1 \cdot 2}{3 \cdot 2} = \frac{2}{6}$, because $\frac{2}{2} = 1$.

The same is true for rational expressions. Multiplying by $\frac{(x-2)}{(x-2)}$ gives an equivalent expression because $\frac{(x-2)}{(x-2)} = 1$ when $x \neq 2$.

My Notes

Example E

Add $\frac{-4}{5-p}+\frac{3}{p-5}$. Simplify your answer if possible.

Step 1: Determine a common denominator. $p-5$

Step 2: Multiply the numerator and denominator of the first term by −1.

$$\frac{-4}{5-p}\cdot\frac{-1}{-1}+\frac{3}{p-5}$$

Step 3: Multiply. $\frac{4}{p-5}+\frac{3}{p-5}$

Step 4: Add. $\frac{7}{p-5}$

Solution: $\frac{7}{p-5}$

MATH TIP

If you multiply $(5-p)$ by −1, the product is $p-5$.

Try These D–E

a. Add $\frac{1}{x^2-1}+\frac{2}{x+1}$. Use the steps below.

Factor each denominator:

Common Factors:

Factors Not in Common:

LCM:

Factor the denominator of the first term. Multiply the numerator and denominator of the second term by ____________________:

Add the numerators:

Use the Distributive Property:

Combine like terms:

Solution:

Add or subtract. Simplify your answer.

b. $\frac{3}{x+1}-\frac{x}{x-1}$ **c.** $\frac{2}{x}-\frac{3}{x^2-3x}$ **d.** $\frac{2}{x^2-4}+\frac{x}{x^2+4x+4}$

My Notes

Check Your Understanding

1. **Make use of structure.** Sometimes the denominator of one fraction or one rational expression works as a common denominator for all fractions or rational expressions in a set.
   a. Write two fractions (rational numbers) in which the denominator of one of the fractions is a common denominator.
   b. Write two rational expressions in which the denominator of one of the expressions is a common denominator.
   c. Show how to add the two rational expressions you wrote in Part (b).
2. List the steps you usually use to add or subtract rational expressions with unlike denominators.

LESSON 28-4 PRACTICE

Determine the least common multiple of each set of expressions.

3. $2x + 4$ and $x^2 - 4$
4. $2x - 8$ and $x - 4$
5. $x - 3$ and $x + 3$
6. $x + 6$, $x + 7$, and $x^2 + 7x + 6$
7. $x + 3$, $x^2 + 6x + 9$, and $x^2 - 7x - 30$

Perform the indicated operation.

8. $\frac{x}{x+1} - \frac{2}{x+3}$
9. $\frac{2}{3x-3} - \frac{x}{x^2-1}$
10. $\frac{x}{x+5} - \frac{2}{x+3}$
11. $\frac{3}{x-3} - \frac{x}{x+4}$
12. $\frac{x}{3x-2} + \frac{2}{x-5}$
13. $\frac{x-2}{x^2+4x+4} + \frac{x-2}{x+2}$
14. **Model with mathematics.** In the past week, Emilio jogged for a total of 7 miles and biked for a total of 7 miles. He biked at a rate that was twice as fast as his jogging rate.
    a. Suppose Emilio jogs at a rate of *r* miles per hour. Write an expression that represents the amount of time he jogged last week and an expression that represents the amount of time he biked last week. $\left(\textit{Hint:} \text{distance} = \text{rate} \times \text{time, so time} = \frac{\text{distance}}{\text{rate}}.\right)$
    b. Write and simplify an expression for the total amount of time Emilio jogged and biked last week.
    c. Emilio jogged at a rate of 5 miles per hour. What was the total amount of time Emilio jogged and biked last week?

ACTIVITY 28 PRACTICE

Write your answers on notebook paper. Show your work.

Lesson 28-1

1. Allison correctly simplified the rational expression shown below by dividing.

$$\frac{35x^7 + 15x^5 - 10x^3}{5x^3}$$

Which of these is a term in the resulting expression?

A. $3x^8$ **B.** $3x^4$
C. $-2x$ **D.** -2

For Items 2–5, simplify each expression.

2. $\frac{56x^2y}{70x^3y}$

3. $\frac{28x^2}{49xy}$

4. $\frac{x^2 - 25}{5x + 25}$

5. $\frac{x + 5}{x^2 + x - 20}$

6. Which of the following expressions is equivalent to a negative integer?

A. $\frac{5y + 5}{5y - 5}$ **B.** $\frac{6y - 6}{3 - 3y}$
C. $\frac{2 - 2y}{4y - 4}$ **D.** $\frac{8y - 8}{4y - 4}$

7. A rental car costs \$24 plus \$3 per mile.
   a. Write an expression that represents the total cost of the rental if you drive the car m miles.
   b. Write and simplify an expression that represents the cost per day if you keep the car for 3 days.
   c. What is the cost per day if you drive 50 miles?

8. The expression $\frac{x^2 + 8x + c}{x + 4}$ can be simplified to $x + 4$. What is the value of c?

A. -16 **B.** 0
C. 16 **D.** 64

Lesson 28-2

For Items 9–14, determine each quotient by using long division.

9. $(3x^2 + 6x + 2) \div 3x$

10. $(3x^2 - 7x - 6) \div (3x + 2)$

11. $\frac{2x^2 - 7x - 16}{2x + 3}$

12. $\frac{x^2 - 19x + 9}{x - 4}$

13. $\frac{4x^2 + 17x - 1}{4x + 1}$

14. $\frac{5x^3 + x - 2}{x - 1}$

15. The area A and length ℓ of a rectangle are shown below. Write a rational expression that represents the width w of the rectangle. Then simplify the expression using long division.

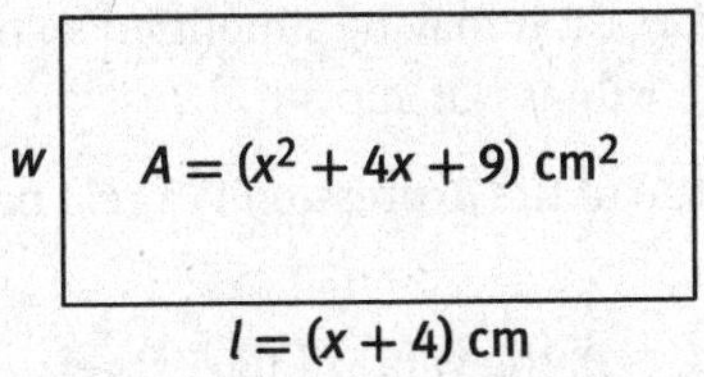

16. Greg was asked to simplify each expression below using long division. For which expression should he have a remainder?

A. $\frac{6x^2 + 9x + 3}{3x}$ **B.** $\frac{8x^4 + 12x^3 + 16x^2}{4x^2}$
C. $\frac{15x^2 + 5x + 25}{5}$ **D.** $\frac{6x^4 + 12x^3 + 6x^2}{6x}$

17. A student performed the long division shown below. Is the student's work correct? Justify your response.

$$\begin{array}{r} 4 \\ x^2 + 2x - 5 \overline{\big) 4x^2 - 6x + 11} \\ \underline{4x^2 + 8x - 20} \\ 2x \;\; - 9 \end{array}$$

The quotient is $4 + \frac{2x - 9}{x^2 + 2x - 5}$.

Lesson 28-3

Multiply or divide. Simplify your answer if possible.

18. $\dfrac{x+4}{3x} \bullet \dfrac{4x^2}{x^2+9x+20}$

19. $\dfrac{3x+9}{x} \bullet \dfrac{x^2}{x^2-9}$

20. $\dfrac{x^2-x-6}{x^2-9} \bullet \dfrac{x^2+7x+12}{x^2+4x+4}$

21. $\dfrac{x^2-25}{x^2-10x+25} \div \dfrac{x^2+10x+25}{2x-10}$

22. $\dfrac{n^2-4n-5}{n^2+2n+1} \div \dfrac{n^2-6n+5}{n^2-1}$

In the expression $\dfrac{1}{(x+5)^2} \div \dfrac{k}{(x+5)^2}$, k is a real number with $k \neq 0$. For Items 23–25, determine whether each statement is always, sometimes, or never true.

23. The expression may be simplified so that the variable x does not appear.

24. The value of the expression is a real number less than 1.

25. When $k > 0$, the value of the expression is also greater than 0.

26. A student was asked to divide the rational expressions shown below. Examine the student's solution, then identify and correct the error.

$$\frac{x^2-6x+9}{5x} \div \frac{x-3}{x+3} = \frac{(x+3)(x-3)}{5x} \bullet \frac{x+3}{x-3}$$

$$= \frac{(x+3)\cancel{(x-3)}}{5x} \bullet \frac{x+3}{\cancel{x-3}}$$

$$= \frac{(x+3)^2}{5x}$$

27. Which expression is equivalent to $\dfrac{3x+3}{x^2} \bullet \dfrac{x^2-x}{x^2-1}$?

A. 3 **B.** $\dfrac{3}{x}$

C. $3-x$ **D.** $\dfrac{3x+3}{x}$

Lesson 28-4

For Items 28–31, determine the least common multiple of each set of expressions.

28. $x^2 - 25$ and $x + 5$

29. $y + 3$, y, and y^2

30. $x^2 + 5x + 6$ and $x^2 + 7x + 12$

31. $x^2 - 4x + 4$, $x - 2$, and $(x - 2)^3$

32. Which pair of expressions has a least common multiple that is the product of the expressions?

A. $x + 7$ and $x^2 + 14x + 49$

B. $x + 7$ and $x - 7$

C. $x - 3$ and $x^2 - 9$

D. $x - 3$ and $(x - 3)^2$

Add or subtract. Express in simplest form.

33. $\dfrac{4}{x} + \dfrac{3}{x}$

34. $\dfrac{x}{2} + \dfrac{x}{2}$

35. $\dfrac{x}{x+1} + \dfrac{1}{x+1}$

36. $\dfrac{x}{x^2-4x} - \dfrac{5x}{x-4}$

37. $\dfrac{3x+2}{3x-6} + \dfrac{x+2}{x^2-4}$

38. $\dfrac{-18}{3-x} + \dfrac{7}{x-3}$

MATHEMATICAL PRACTICES
Reason Abstractly and Quantitatively

39. Justine lives one mile from the grocery store. While she was driving to the store, there was a lot of traffic. On her way home, there was no traffic at all, and her average rate (speed) was twice the average rate of her trip to the store.

a. Let r represent Justine's average rate on her way to the store. Write an expression for the time it took her to get to the store. $\left(\textit{Hint:}\ \text{distance} = \text{rate} \times \text{time, so time} = \dfrac{\text{distance}}{\text{rate}}.\right)$

b. Write an expression for the time it took Justine to drive home.

c. Write and simplify an expression for the total time of the round trip to and from the store.

d. If Justine drove at 30 miles per hour to the store, what was the total time for the round trip? Write your answer in minutes.

Factoring and Simplifying Rational Expressions
ROCK STAR DEMANDS

Rockstar Platforms sets up outdoor stages for rock concerts. The musician Fuchsia requires a square stage at all of her concerts. Rockstar Platforms lays down a stage with an area of $x^2 + 8x + 16$ square feet for her.

1. a. Draw a diagram to represent the area of the stage.
b. Write expressions to represent the side lengths of Fuchsia's stage.

Fuchsia looks at the stage and says, "That's too small!" So Rockstar Platforms goes back to the drawing board and designs another square stage with an area of $4x^2 + 20x + 25$ square feet.

2. What are the side lengths of Fuchsia's new stage?

The company calls Fuchsia back out to look at the stage. "I guess it will do," she says, "but I would have preferred a stage with an area of $5x^2 + 12x + 4$ square feet." Rockstar Platforms' foreman explains, "But then your demand for a square stage would not be met."

3. Explain why Fuchsia's demand for a square stage would not be met if the stage had an area of $5x^2 + 12x + 4$ square feet.

"Leave it like it is," concedes Fuchsia. "Now, I need you to put up a video screen." Fuchsia wants a large rectangular video screen set up behind her so her fans can see her from far away. Rockstar Platforms' foreman has a plan for a video screen with an area of $2x^2 + 11x + 14$. "Boss, that's not going to work," his assistant cautions. "It's going to be too long for the stage. "It'll work," the foreman insists.

4. Who is correct, the foreman or his assistant? Justify your response.

After finishing Fuchsia's stage, Rockstar Platforms is contracted to set up a stage for Mich.i.el. Mich.i.el wants a rectangular stage with an area of $2x^2 + 7x + 5$ square feet, but he makes one very specific request. In order to fit all of his backup dancers, the length of the stage must be $x + 4$ feet.

5. How wide will Rockstar Platforms have to make the stage to meet Mich.i.el's request?

Two nights later, Mich.i.el and Fuchsia run into each other at a party and start arguing over who had the bigger stage.

6. How many times larger was Fuchsia's stage than Mich.i.el's stage? Write an expression in simplest form that represents the ratio of the area of Fuchsia's stage to the area of Mich.i.el's stage.

Factoring and Simplifying Rational Expressions

ROCK STAR DEMANDS

Scoring Guide	Exemplary	Proficient	Emerging	Incomplete
	The solution demonstrates the following characteristics:			
Mathematics Knowledge and Thinking (Items 1b, 2, 5, 6)	• Clear understanding of and accuracy in factoring polynomials and simplifying rational expressions	• Adequate understanding of and accuracy in factoring polynomials and simplifying rational expressions	• Partial understanding of factoring polynomials and simplifying rational expressions that results in some incorrect answers	• Little or no understanding of factoring polynomials and simplifying rational expressions
Problem Solving (Items 1b, 2, 5, 6)	• Appropriate and efficient strategy that results in a correct answer	• Strategy that may include unnecessary steps but results in a correct answer	• Strategy that results in some incorrect answers	• No clear strategy when solving problems
Mathematical Modeling/ Representations (Item 1a)	• Clear and accurate diagram of the stage	• Correct diagram of the stage	• Partially accurate diagram of the stage	• Incomplete or inaccurate diagram of the stage
Reasoning and Communication (Items 3, 4)	• Ease and accuracy describing the relationship between a mathematical expression and a real-world constraint • Precise use of appropriate math terms and language to identify and explain whether the foreman or his assistant is correct	• Little difficulty describing the relationship between a mathematical expression and a real-world constraint • Correct identification of whether the foreman or his assistant is correct with an adequate explanation	• Partially correct description of the relationship between a mathematical expression and a real-world constraint • Misleading or confusing explanation of whether the foreman or his assistant is correct	• Little or no understanding of how a mathematical expression might relate to a real-world constraint • Incomplete or inaccurate explanation of whether the foreman or his assistant is correct

Quadratic Functions

5

Unit Overview

In this unit you will study a variety of ways to solve quadratic functions and systems of equations and apply your learning to analyzing real world problems.

Key Terms

As you study this unit, add these and other terms to your math notebook. Include in your notes your prior knowledge of each word, as well as your experiences in using the word in different mathematical examples. If needed, ask for help in pronouncing new words and add information on pronunciation to your math notebook. It is important that you learn new terms and use them correctly in your class discussions and in your problem solutions.

Math Terms

- quadratic function
- standard form of a quadratic function
- parabola
- vertex of a parabola
- maximum
- minimum
- parent function
- axis of symmetry
- translation
- vertical stretch
- vertical shrink
- transformation
- reflection
- factored form
- zeros of a function
- roots
- completing the square
- discriminant
- imaginary numbers
- imaginary unit
- complex numbers
- piecewise-defined function
- nonlinear system of equations

ESSENTIAL QUESTIONS

How are quadratic functions used to model, analyze, and interpret mathematical relationships?

Why is it advantageous to know a variety of ways to solve and graph quadratic functions?

EMBEDDED ASSESSMENTS

This unit has three embedded assessments, following Activities 30, 33, and 35. They will allow you to demonstrate your understanding of graphing, identifying, and modeling quadratic functions, solving quadratic equations and solving nonlinear systems of equations.

Getting Ready

Write your answers on notebook paper. Show your work.

1. Determine each product.
   a. $(x - 2)(3x + 5)$
   b. $2y(y + 6)(y - 1)$
2. Factor each polynomial.
   a. $2x^2 + 14x$
   b. $3x^2 - 75$
   c. $x^2 + 7x + 10$
3. If $f(x) = 3x - 5$, find each value.
   a. $f(4)$
   b. $f(-2)$
4. Solve the equation. $4x - 5 = 19$
5. Solve the inequality.
   $\frac{1}{3}x + 9 > 13$
6. Explain how to graph $2x + y = 4$.

The following graph compares calories burned when running and walking at constant rates of 10 mi/h and 2 mi/h, respectively.

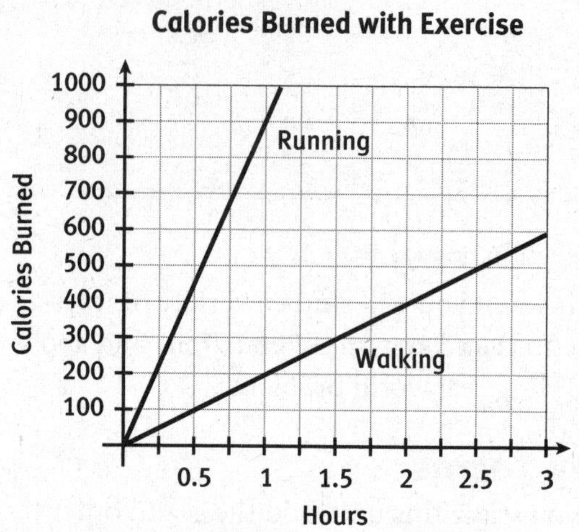

7. What does the ordered pair (1.5, 300) represent on this graph?
8. How many calories would be burned after four hours when running and after four hours when walking?

Introduction to Quadratic Functions

Touchlines

Lesson 29-1 Modeling with a Quadratic Function

My Notes

Learning Targets:

- Model a real-world situation with a quadratic function.
- Identify quadratic functions.
- Write a quadratic function in standard form.

SUGGESTED LEARNING STRATEGIES: Create Representations, Interactive Word Wall, Marking the Text, Look for a Pattern, Discussion Groups

Coach Wentworth coaches girls' soccer and teaches algebra. Soccer season is starting, and she needs to mark the field by chalking the touchlines and goal lines for the soccer field. Coach Wentworth can mark 320 yards for the total length of all the touchlines and goal lines combined. She would like to mark the field with the largest possible area.

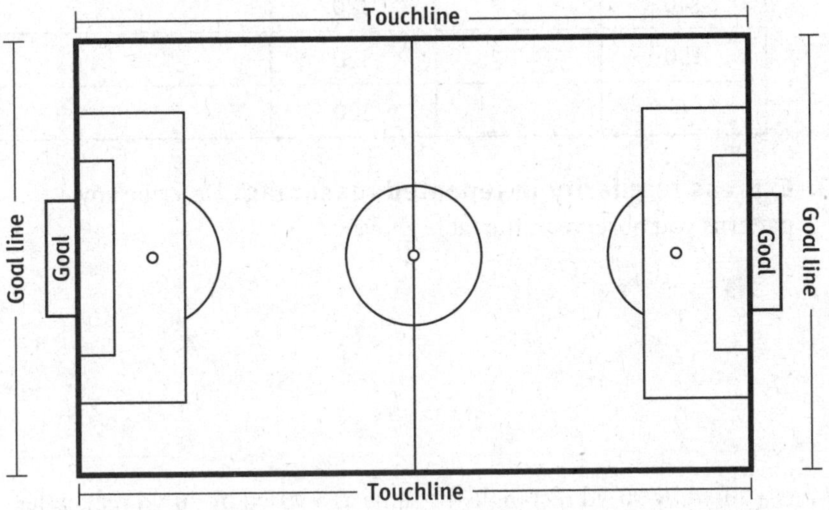

FIFA regulations require that all soccer fields be rectangular in shape.

1. How is the perimeter of a rectangle determined? How is the area of a rectangle determined?

CONNECT TO SPORTS

FIFA stands for *Fédération Internationale de Football Association* (International Federation of Association Football) and is the international governing body of soccer.

My Notes

2. Complete the table below for rectangles with the given side lengths. The first row has been completed for you.

Length (yards)	Width (yards)	Perimeter (yards)	Area (square yards)
10	150	320	1500
20		320	
40		320	
60		320	
80		320	
100		320	
120		320	
140		320	
150		320	
l		320	

3. Express regularity in repeated reasoning. Describe any patterns you observe in the table above.

4. Is a 70-yd by 90-yd rectangle the same as a 90-yd by 70-yd rectangle? Explain your reasoning.

5. Graph the data from the table in Item 2 as ordered pairs.

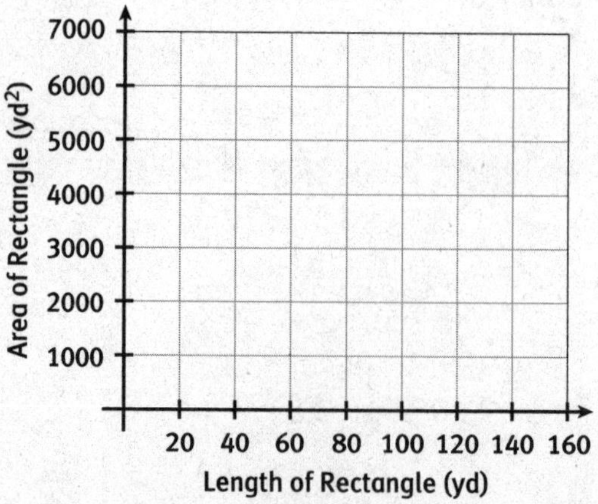

My Notes

6. Use the table and the graph to explain why the data in Items 2 and 5 are not linear.

7. Describe any patterns you see in the graph above.

8. What appears to be the largest area from the data in Items 2 and 5?

9. Write a function $A(l)$ that represents the area of a rectangle whose length is l and whose perimeter is 320.

The function $A(l)$ is called a ***quadratic function*** because the greatest degree of any term is 2 (an x^2 term). The ***standard form of a quadratic function*** is $y = ax^2 + bx + c$ or $f(x) = ax^2 + bx + c$, where a, b, and c are real numbers and $a \neq 0$.

10. Write the function $A(l)$ in standard form. What are the values of a, b, and c?

My Notes

Check Your Understanding

11. For the function $f(x) = x^2 + 2x + 3$, create a table of values for $x = -3, -2, -1, 0, 1$. Then sketch a graph of the quadratic function on grid paper.

12. Barry needs to find the area of a rectangular room with a width that is 2 feet longer than the length. Write an expression for the area of the rectangle in terms of the length.

13. Critique the reasoning of others. Sally states that the equation $g(x) = x^3 + 10x^2 - 3x$ represents a quadratic function. Explain why Sally is incorrect.

14. Write the quadratic function $f(x) = (3 - x)^2$ in standard form.

LESSON 29-1 PRACTICE

15. Create tables to graph $y = 3x$ and $y = 3x^2$ on grid paper. Explain the differences between the graphs.

16. Pierre uses the function $r(t) = t + 2$ to model his rate r in mi/h t minutes after leaving school. Complete the table and use the data points to graph the function.

t	$r(t)$
0	
1	
2	
3	
4	

17. Pierre uses the function $d(t) = t(t + 2)$ to model his distance d in miles from home. Add another column to your table in Item 16 to represent $d(t)$. Sketch a graph of $d(t)$ on the same coordinate plane as the graph of $r(t)$.

18. What types of functions are represented in Items 16 and 17?

19. Write each quadratic function in standard form.

a. $g(x) = x(x - 2) + 4$ **b.** $f(t) = 3 - 2t^2 + t$

20. Attend to precision. Determine whether each function is a quadratic function. Justify your responses.

a. $S(r) = 2\pi r^2 + 20\pi r$ **b.** $f(a) = \dfrac{a^2 + 4a - 3}{2}$

c. $f(x) = 4x^2 - 3x + 2$ **d.** $g(x) = 3x^{-2} + 2x - 1$

e. $f(x) = 4^2x - 3$ **f.** $h(x) = \dfrac{4}{x^2} - 3x + 2$

My Notes

Learning Targets:

- Graph a quadratic function.
- Interpret key features of the graph of a quadratic function.

SUGGESTED LEARNING STRATEGIES: Interactive Word Wall, Create Representations, Construct an Argument, Marking the Text, Discussion Groups

1. Use a graphing calculator to graph $A(l)$ from Item 9 in Lesson 29-1. Sketch the graph on the grid below.

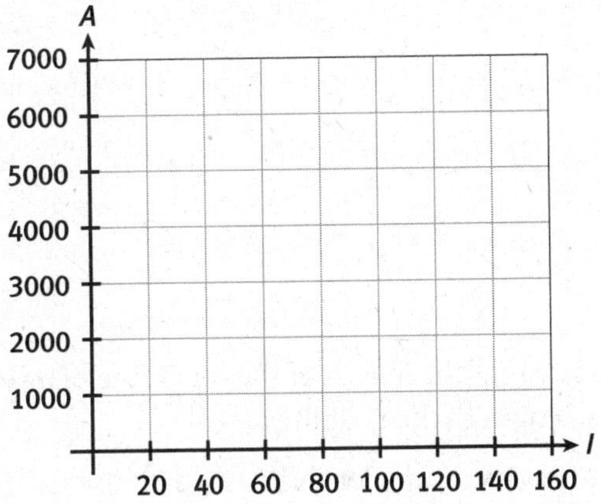

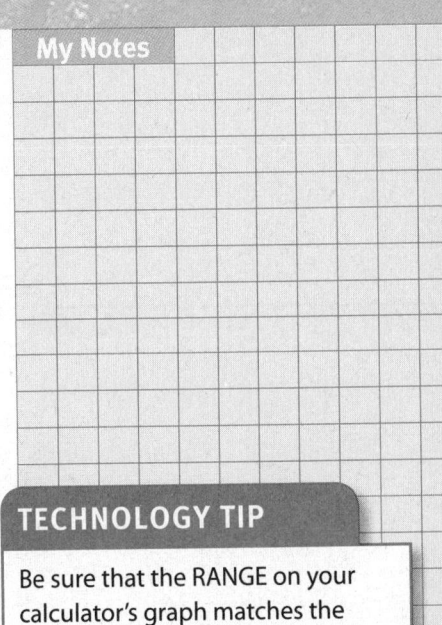

TECHNOLOGY TIP

Be sure that the RANGE on your calculator's graph matches the range shown in the grid.

The graph of a quadratic function is a curve called a ***parabola***. A parabola has a point at which a maximum or minimum value of the function occurs. That point is called the ***vertex of a parabola***. The *y*-value of the vertex is the ***maximum*** or ***minimum*** of the function.

2. Identify the vertex of the graph of $A(l)$ in Item 1. Does the vertex represent a maximum or a minimum of the function?

CONNECT TO AP

AP Calculus students find these same maximum or minimum values of functions in *optimization* problems.

3. Examine the graph of $A(l)$. For what values of the length is the area increasing? For what values of the length is the area decreasing?

My Notes

4. Describe the point where the area changes from increasing to decreasing.

5. Use the table, the graph, and/or the function to determine the reasonable domain and range of the function $A(l)$. Describe each using words and an inequality.

FIFA regulations state that the length of the touchline of a soccer field must be greater than the length of the goal line.

6. **Reason abstractly.** Can Coach Wentworth use the rectangle that represents the largest area of $A(l)$ for her soccer field? Explain why or why not.

My Notes

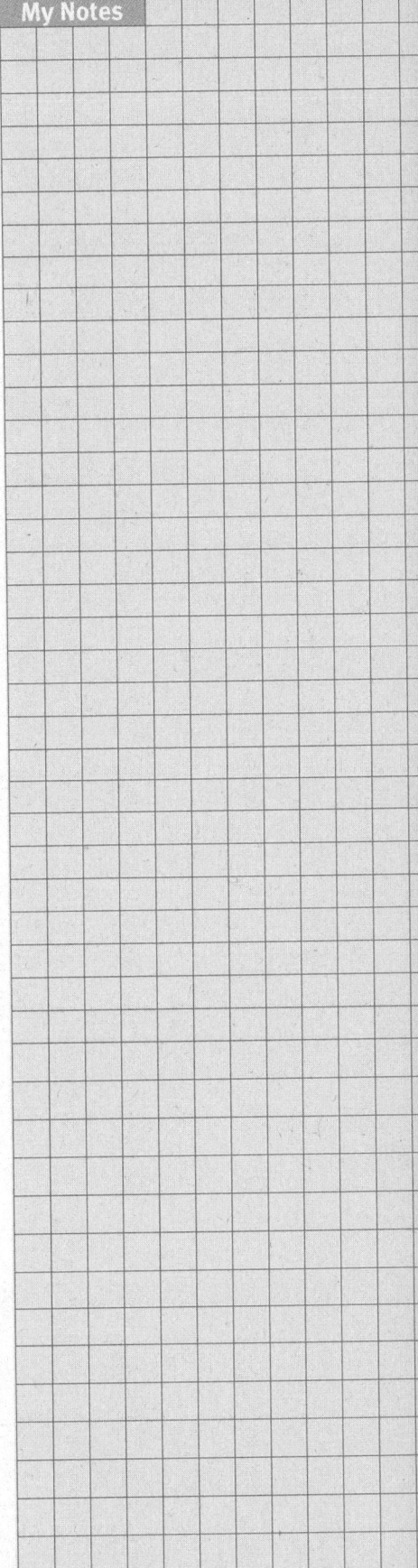

FIFA regulations also state that the length of the touchlines of a soccer field must be at least 100 yds, but no more than 130 yds. The goal lines must be at least 50 yds, but no more than 100 yds.

7. **Construct viable arguments.** Determine the dimensions of the FIFA regulation soccer field with the largest area and a 320-yd perimeter. Support your reasoning with multiple representations.

8. Consider the quadratic function $f(x) = x^2 - 2x - 3$.
   a. Write the function in factored form by factoring the polynomial $x^2 - 2x - 3$.

   b. To find the x-intercepts of $f(x)$, use the factored form of $f(x)$ and solve the equation $f(x) = 0$.

   c. A parabola is symmetric over the vertical line that contains the vertex. How do you think the x-coordinate of the vertex relates to the x-coordinates of the x-intercepts? Use the symmetry of a parabola to support your answer.

   d. Write the vertex of the quadratic function.

My Notes

Check Your Understanding

9. Complete the table for the quadratic function $f(x) = -x^2 - 4x - 3$. Then graph the function.

x	$f(x)$
−4	
−3	
−2	
−1	
0	

10. Identify the maximum or minimum value of the quadratic function in Item 9.

11. Consider the quadratic function $y = -x^2 + 2x - 3$.

a. Create a table of values for the function for domain values 0, 1, 2, 3, and 4.

b. Sketch a graph of the function. Identify and label the vertex. What is the maximum value of the function?

c. Use inequalities to write the domain, range, and values of x for which y is decreasing.

LESSON 29-2 PRACTICE

12. Write the quadratic function $g(x) = x(x - 2) + 4$ in standard form.

For Items 13–16, use the quadratic function $f(x) = x^2 + 6x + 5$.

13. Create a table of values and graph $f(x)$.

14. Use your graph to identify the maximum or minimum value of $f(x)$.

15. Write the domain and range of $f(x)$ using inequalities.

16. Determine the values of x for which $f(x)$ is increasing.

17. Make use of structure. Sketch a graph of a quadratic function with a maximum. Now sketch another graph of a quadratic function with a minimum. Explain the difference between the increasing and decreasing behavior of the two functions.

ACTIVITY 29 PRACTICE

Write your answers on notebook paper. Show your work.

Lesson 29-1

1. The base of a rectangular window frame must be 1 foot longer than the height. Which of the following is an equation for the area of the window in terms of the height?
   A. $A(h) = h + 1$
   B. $A(h) = (h + 1)h$
   C. $A(h) = h^2 - 1$
   D. $A(h) = h^2 + h + 1$

2. Which of the following is an equation for the area of an isosceles right triangle, in terms of the base?

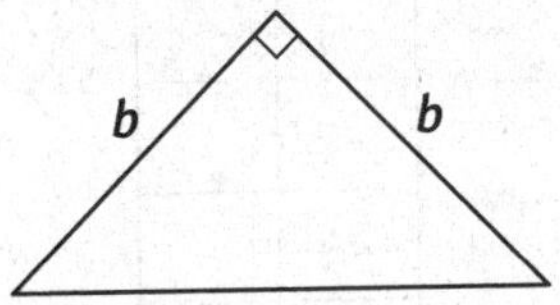

   A. $A(b) = b^2$
   B. $A(b) = \frac{1}{2}b^2$
   C. $A(b) = \frac{\sqrt{b^2}}{2}$
   D. $A(b) = \frac{1}{2}b$

3. Ben is creating a triangle that has a base that is twice the length of the height.
   a. Write an expression for the base of the triangle in terms of the height.
   b. Write a function for the area, $A(h)$, of the triangle in terms of the height.
   c. Complete the table and then graph the function.

h	$A(h)$
1	
2	
3	
4	
5	

Use the following information for Items 4–8.

Jenna is in charge of designing the screen for a new smart phone. The design specs call for a rectangular screen that has an outside perimeter of 12 inches.

4. Complete the table for the screen measurements.

Width, w	Length, l	Area, $A(w)$
1	5	5
2		
3		
4		
5		
w		

5. Write a function $A(w)$ for the area of the screen in terms of the width.
6. Graph the function $A(w)$. Label each axis.
7. Determine the domain and range of the function.
8. Determine the maximum area of the screen that Jenna can design. What are the dimensions of this screen?
9. Samantha's teacher writes the function $f(x) = 2x^3 - 2x(3 - x + x^2) + 3$.
   a. Barry tells Samantha that the function cannot be quadratic because it contains the term $2x^3$. What should Barry do to the function before making this assumption?
   b. Is the function a quadratic function? Explain.

10. Identify whether each function is quadratic.

a. $y = 2x - 3^2$

b. $y = 3x^2 - 2x$

c. $y = 2 - \frac{3}{x^2} + x$

11. State whether the data in each table are linear. Explain why or why not.

x	y
0	5
1	2
2	−1
3	−4
4	−7

x	y
0	5
1	2
2	1
3	2
4	5

Lesson 29-2

12. Write each quadratic function in standard form.

a. $y = 3x - 5 + x^2$

b. $y = 6 - 5x^2$

c. $y = -0.5x + \frac{3}{4}x^2 - \pi$

13. For each function, complete the table of values. Graph the function and identify the maximum or minimum of the function.

a. $y = x^2 + 4x - 1$

x	y
−3	
−2	
−1	
0	
1	

b. $y = -x^2 + 8x - 13$

x	y
2	
3	
4	
5	
6	

c. $y = x^2 - 1$

x	y
0	
−1	
−2	
−3	
−4	

14. Use your graphing calculator to graph several functions $y = ax^2 + b$ using positive and negative values of a and b. Do the signs of a and b appear to affect whether the function has a minimum or maximum value? Make a conjecture.

MATHEMATICAL PRACTICES
Construct Viable Arguments and Critique the Reasoning of Others

15. As part of her math homework, Kylie is graphing a quadratic function. After plotting several points, she notices that the dependent values are increasing as the independent values increase. She reasons that the function will eventually reach a maximum value and then begin to decrease. Is Kylie correct? Why or why not?

Graphing Quadratic Functions

Transformers

Lesson 30-1 Translations of the Quadratic Parent Function

My Notes

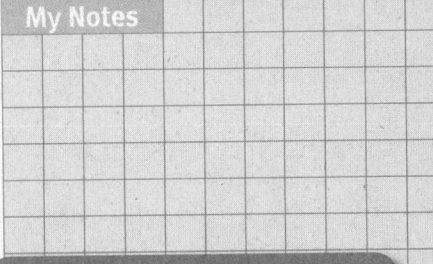

Learning Targets:

- Graph translations of the quadratic parent function.
- Identify and distinguish among transformations.

SUGGESTED LEARNING STRATEGIES: Look for a Pattern, Create Representations, Sharing and Responding, Quickwrite, Think-Pair-Share

The function $y = x^2$ or $f(x) = x^2$ is the quadratic ***parent function***.

1. Complete the table for $y = x^2$. Then graph the function.

x	y
−3	
−2	
−1	
0	
1	
2	
3	

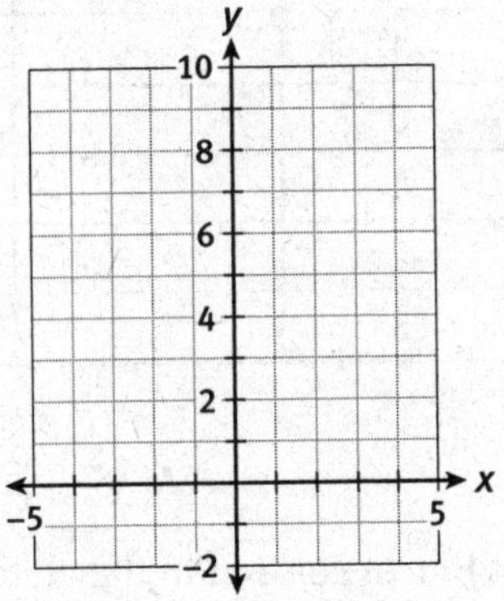

MATH TERMS

A **parent function** is the most basic function of a particular category or type. For example, the linear parent function is $y = x$ or $f(x) = x$.

2. Use words such as *vertex*, *maximum*, *minimum*, *increasing*, and *decreasing* to describe the graph of $y = x^2$.

3. The line that passes through the vertex of a parabola and divides the parabola into two symmetrical parts is the parabola's ***axis of symmetry***.

a. Draw the axis of symmetry for the graph of $y = x^2$ as a dashed line on the graph in Item 1.

b. Write the equation for the axis of symmetry you drew in Item 1.

My Notes

4. Complete the second and third columns of the table below. (Leave the fourth column blank for now.) Use your results to explain why the function $y = x^2$ is not linear.

x	$y = x^2$	Difference Between Consecutive y-Values ("First Differences")	"Second Differences"
−3			
−2			
−1			
0			
1			
2			
3			

5. The fourth column in the table in Item 4 is titled "Second Differences." Complete the fourth column by finding the change in consecutive values in the third column. What do you notice about the values?

6. a. Complete the table for $f(x) = x^2$ and $g(x) = x^2 + 3$. Then graph each function on the same coordinate grid.

x	$f(x) = x^2$	$g(x) = x^2 + 3$
−3		
−2		
−1		
0		
1		
2		
3		

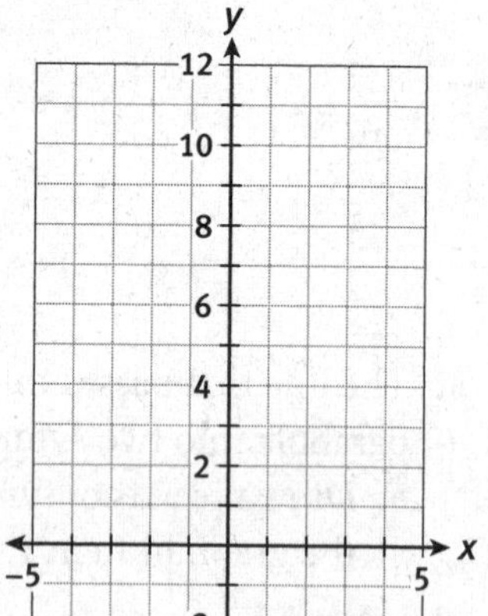

b. Identify the vertex, domain, range, and axis of symmetry for each function.

	$f(x) = x^2$	$g(x) = x^2 + 3$
Vertex		
Domain		
Range		
Axis of symmetry		

My Notes

7. a. Complete the table for $f(x) = x^2$ and $h(x) = x^2 - 4$. Then graph each function on the same coordinate grid.

x	$f(x) = x^2$	$h(x) = x^2 - 4$
−3		
−2		
−1		
0		
1		
2		
3		

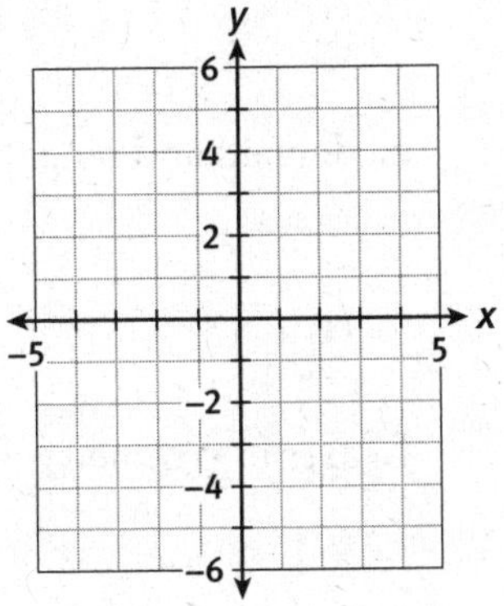

b. Identify the vertex, domain, range, and axis of symmetry for each function.

	$f(x) = x^2$	$h(x) = x^2 - 4$
Vertex		
Domain		
Range		
Axis of symmetry		

8. Compare the functions $g(x)$ and $h(x)$ in Items 6 and 7 to the parent function $f(x) = x^2$. Describe any patterns you notice in the following.

a. equations

b. tables

c. graphs

My Notes

d. vertices

e. domain and range

f. axes of symmetry

The changes to the parent function in Items 6 and 7 are examples of vertical translations. A ***translation*** of a graph is a change that shifts the graph horizontally, vertically, or both. A translation does not change the shape of the graph.

9. Make use of structure. How does the value of k in the equation $g(x) = x^2 + k$ change the graph of the parent function $f(x) = x^2$?

Check Your Understanding

For Items 10 and 11, predict the translations of the graph of $f(x) = x^2$ for each function. Confirm your predictions by graphing each equation on the same coordinate grid as the graph of the parent function.

10. $g(x) = x^2 - 2$

11. $h(x) = x^2 + 4$

12. The graphs of two functions are shown below, along with the graph of the parent function. Determine the value of c for each function.

a. $g(x) = x^2 + c$

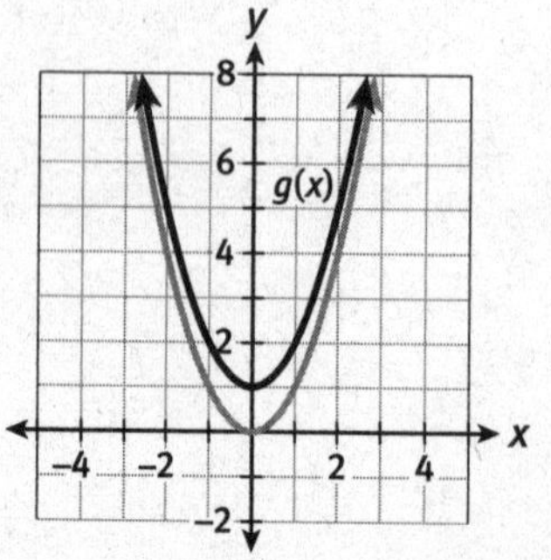

b. $h(x) = x^2 + c$

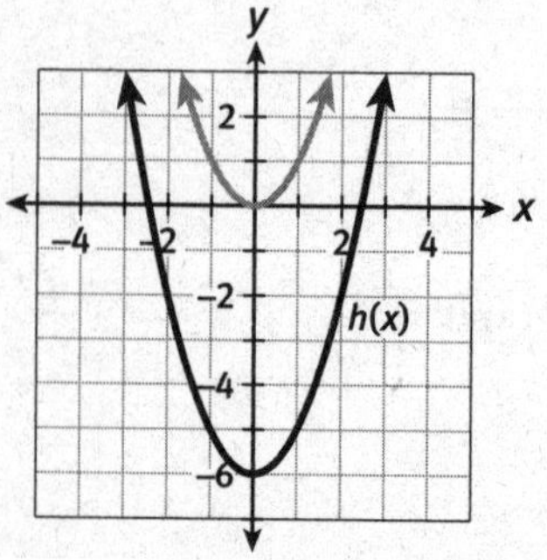

My Notes

13. a. Complete the table for $f(x) = x^2$ and $k(x) = (x + 3)^2$. Then graph both functions on the same coordinate grid.

x	$f(x) = x^2$	$k(x) = (x + 3)^2$
−3		
−2		
−1		
0		
1		
2		
3		

b. Identify the vertex, domain, range, and axis of symmetry for each function.

	$f(x) = x^2$	$k(x) = (x + 3)^2$
Vertex		
Domain		
Range		
Axis of symmetry		

14. a. Complete the table for $f(x) = x^2$ and $p(x) = (x - 4)^2$. Then graph both functions on the same coordinate grid.

x	$f(x) = x^2$	$p(x) = (x - 4)^2$
−1		
0		
1		
2		
3		
4		
5		

My Notes

b. Identify the vertex, domain, range, and axis of symmetry for each function.

	$f(x) = x^2$	$p(x) = (x - 4)^2$
Vertex		
Domain		
Range		
Axis of symmetry		

15. Compare the functions $k(x)$ and $p(x)$ in Items 13 and 14 to the parent function $f(x) = x^2$. Describe any patterns you notice in the following.

a. equations

b. graphs

c. vertices

d. domain and range

e. axes of symmetry

My Notes

The changes to the parent function in Items 13 and 14 are examples of horizontal translations.

16. Express regularity in repeated reasoning. How does the value of h in the equation $k(x) = (x + h)^2$ change the graph of the parent function $k(x) = x^2$?

Check Your Understanding

For Items 17 and 18, predict the translations of the graph of $f(x) = x^2$ for each equation. Confirm your predictions by graphing each equation on the same coordinate grid as the graph of the parent function.

17. $k(x) = (x - 2)^2$

18. $p(x) = (x + 1)^2$

19. The graphs of two functions are shown below, along with the graph of the parent function. Determine the value of h for each function. Then write each function in standard form.

a. $k(x) = (x + h)^2$

b. $p(x) = (x + h)^2$

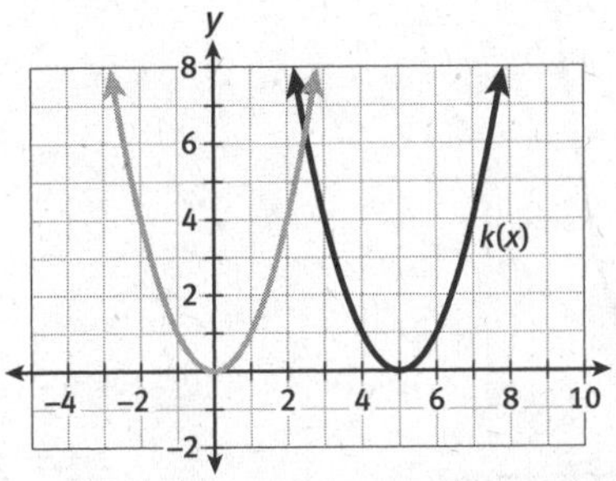

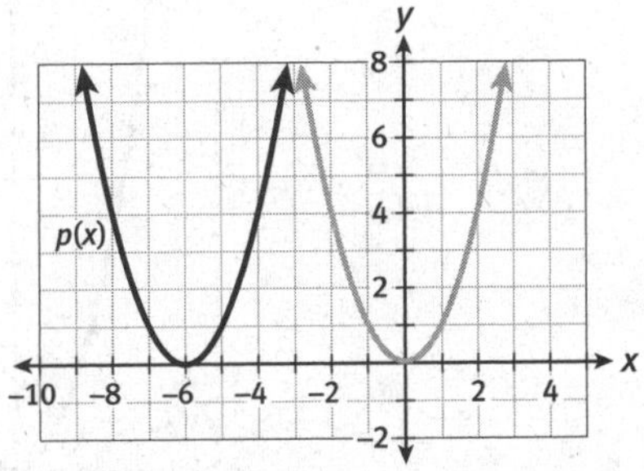

LESSON 30-1 PRACTICE

20. The graph of $g(x)$ is a vertical translation 2 units up from the graph of $f(x) = x^2$. Which of the following equations describes $g(x)$?

A. $g(x) = x^2 + 2$
B. $g(x) = x^2 - 2$
C. $g(x) = (x + 2)^2$
D. $g(x) = (x - 2)^2$

21. For each function, use translations to graph the function on the same coordinate grid as the parent function $f(x) = x^2$. Then identify the vertex, domain, range, and axis of symmetry.

a. $g(x) = x^2 + 1$
b. $h(x) = x^2 - 1$
c. $k(x) = (x + 1)^2$
d. $n(x) = (x - 1)^2$

22. Attend to precision. Write an ordered pair that represents the vertex of the graph of each quadratic function.

a. $f(x) = x^2 + c$
b. $f(x) = (x - k)^2$

My Notes

Learning Targets:

- Graph vertical stretches and shrinks of the quadratic parent function.
- Identify and distinguish among transformations.

SUGGESTED LEARNING STRATEGIES: Create Representations, Look for a Pattern, Sharing and Responding, Think-Pair-Share, Discussion Groups

1. Complete the table for $f(x) = x^2$ and $g(x) = 2x^2$.

x	$f(x) = x^2$	$g(x) = 2x^2$
−2		
−1		
0		
1		
2		

a. The graph of $f(x) = x^2$ is shown below. Graph $g(x) = 2x^2$ on the same coordinate grid.

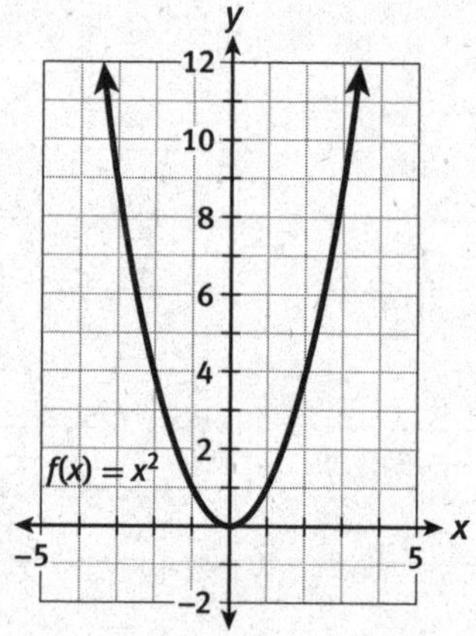

b. Identify the vertex, domain, range, and axis of symmetry for each function.

	$f(x) = x^2$	$g(x) = 2x^2$
Vertex		
Domain		
Range		
Axis of symmetry		

My Notes

2. Complete the table for $f(x) = x^2$ and $h(x) = \frac{1}{2}x^2$.

x	$f(x) = x^2$	$h(x) = \frac{1}{2}x^2$
−2		
−1		
0		
1		
2		

a. The graph of $f(x) = x^2$ is shown below. Graph $h(x) = \frac{1}{2}x^2$ on the same coordinate grid.

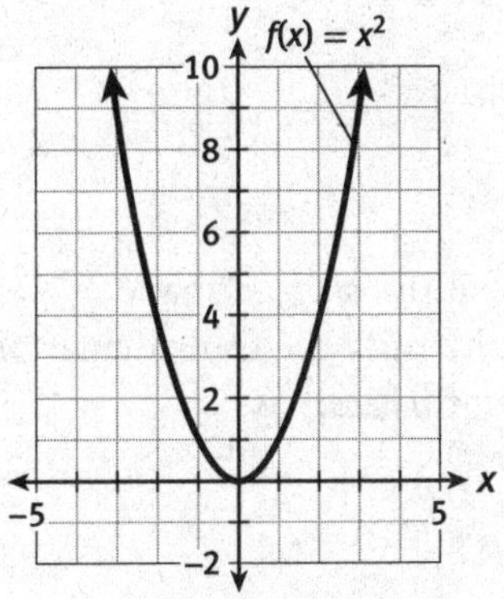

b. Identify the vertex, domain, range, and axis of symmetry for each function.

	$f(x) = x^2$	$h(x) = \frac{1}{2}x^2$
Vertex		
Domain		
Range		
Axis of symmetry		

3. Compare the functions $g(x)$ and $h(x)$ in Items 1 and 2 to the parent function $f(x) = x^2$. Describe any patterns you notice in the following.

a. equations

DISCUSSION GROUP TIPS

As you listen to the group discussion, take notes to aid comprehension and to help you describe your own ideas to others in your group. Ask questions to clarify ideas and to gain further understanding of key concepts.

My Notes

b. tables

c. graphs

d. vertex, domain and range, and axes of symmetry

The change to the parent function in Item 1 is a ***vertical stretch*** by a factor of 2 and the change in Item 2 is a ***vertical shrink*** by a factor of $\frac{1}{2}$. A vertical stretch or shrink changes the shape of the graph.

4. Reason quantitatively. For $a > 0$, how does the value of a in the equation $g(x) = ax^2$ change the graph of the parent function $f(x) = x^2$?

MATH TIP

Create an organized summary of the single transformations of a quadratic function. Include a graph of the parent function, graph of the transformation, the equation, and a verbal description of the transformation.

5. A change in the position, size, or shape of a parent graph is a ***transformation***. Identify the transformations that have been introduced so far in this activity.

Vertical stretching and shrinking also applies to linear functions. Below is the graph of the parent linear function $f(x) = x$.

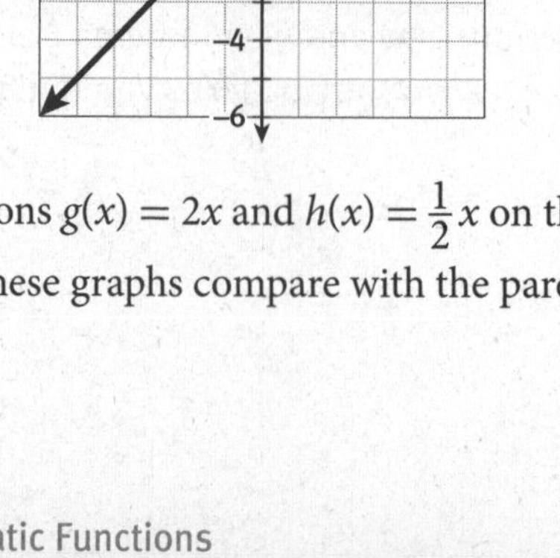

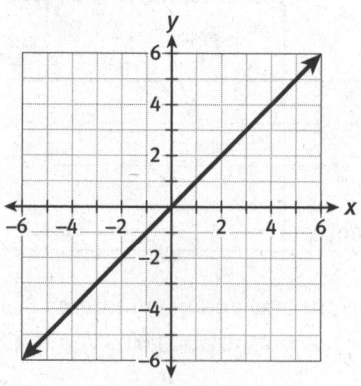

6. Graph the functions $g(x) = 2x$ and $h(x) = \frac{1}{2}x$ on the coordinate grid above. How do these graphs compare with the parent linear function?

My Notes

Check Your Understanding

For Items 7 and 8, predict the change from the graph of $f(x) = x^2$ for each function. Confirm your predictions by graphing each function on the same coordinate grid as the graph of the parent function.

7. $g(x) = 3x^2$

8. $h(x) = \frac{1}{4}x^2$

Write the equation for each transformation of the graph of $f(x) = x^2$.

9. a vertical stretch by a factor of 5

10. a vertical shrink by a factor of $\frac{1}{5}$

LESSON 30-2 PRACTICE

11. The graph of $g(x)$ is a vertical stretch of the graph of $f(x) = x^2$ by a factor of 7. Which of the following equations describes $g(x)$?

A. $g(x) = x^2 + 7$

B. $g(x) = 7x^2$

C. $g(x) = (x + 7)^2$

D. $g(x) = \frac{1}{7}x^2$

12. For each function, use transformations to graph the function on the same coordinate grid as the parent function $f(x) = x^2$.

a. $g(x) = 4x^2$

b. $h(x) = \frac{1}{4}x^2$

c. $k(x) = 0.5x^2$

d. $p(x) = \frac{3}{4}x^2$

13. Reason abstractly. Chen begins with a quadratic data set, which contains the vertex (0, 0). He multiplies every value in the range by 3 to create a new data set. Which of the following statements are true?

A. The function that represents the new data set must be $y = 3x^2$.

B. The graphs of the original data set and the new data set will have the same vertex.

C. The graph of the new data set will be a vertical stretch of the graph of the original data set.

D. The graph of the new data set will be translated up 3 units from the graph of the original data set.

14. How are the graphs of linear functions $f(x) = x$, $k(x) = \frac{3}{4}x$, and $t(x) = \frac{4}{3}x$ the same? How are they different?

My Notes

Learning Targets:

- Graph reflections of the quadratic parent function.
- Identify and distinguish among transformations.
- Compare functions represented in different ways.

SUGGESTED LEARNING STRATEGIES: Look for a Pattern, Create Representations, Predict and Confirm, Sharing and Responding, Think-Pair-Share

1. The quadratic parent function $f(x) = x^2$ is graphed below. Graph $g(x) = -x^2$ on the same coordinate grid.

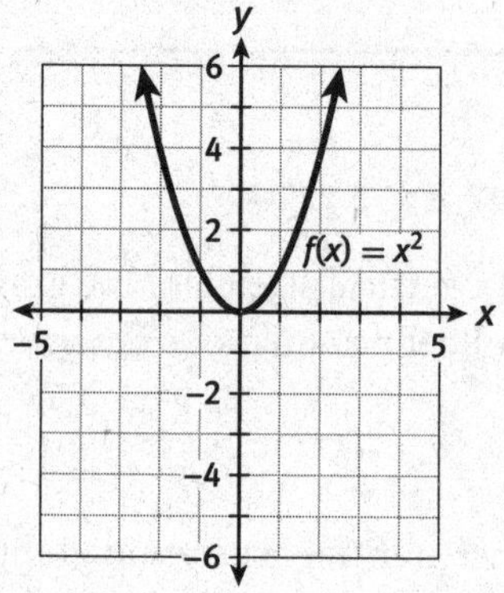

2. Compare $g(x)$ and its graph to the parent function $f(x) = x^2$.

The change to the parent function in Item 1 is a ***reflection*** over the x-axis. A *reflection* of a graph is the mirror image of the graph over a line. A reflection preserves the shape of the graph.

3. How does the sign of a in the function $g(x) = ax^2$ affect the graph?

A graph may represent more than one transformation of the graph of the parent function. The order in which multiple transformations are performed is determined by the order of operations as indicated in the equation.

My Notes

4. How is the graph of $f(x) = x^2$ transformed to produce the graph of $g(x) = 2x^2 - 5$?

5. Use the transformations you described in Item 4 to graph the function $g(x) = 2x^2 - 5$.

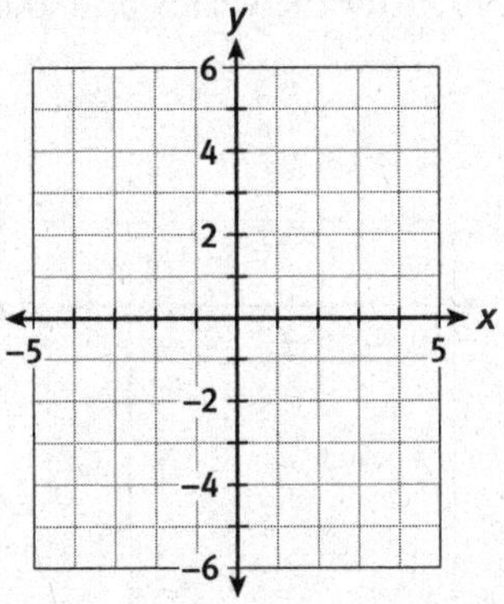

6. How would you transform of the graph of $f(x) = x^2$ to produce the graph of the function $h(x) = -\frac{1}{2}x^2 + 4$?

7. Use the transformations you described in Item 6 to graph the function $h(x) = -\frac{1}{2}x^2 + 4$. Identify the vertex and axis of symmetry of the parabola.

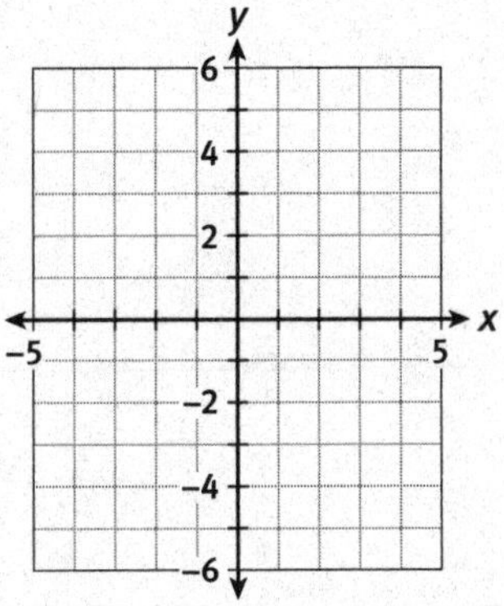

My Notes

8. How would you transform of the graph of $f(x) = x^2$ to produce the graph of the function $k(x) = -\frac{1}{2}(x+4)^2$?

9. Use the transformations you described in Item 8 to graph the function $k(x) = -\frac{1}{2}(x+4)^2$. Identify the vertex and axis of symmetry of the parabola.

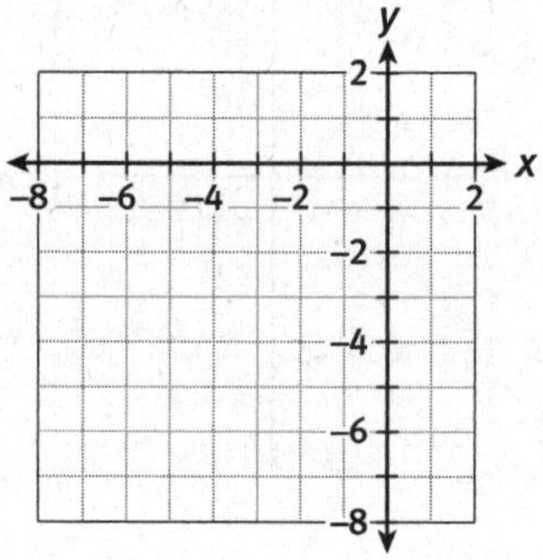

10. Reason abstractly. The graph of $g(x) = a(x - h)^2 + k$ represents multiple transformations of the graph of the parent function, $f(x) = x^2$. Describe how each value transforms the graph of $g(x)$ from the graph of $f(x)$.

a. k

b. h

c. $|a|$

d. the sign of a

My Notes

11. Examine the function $r(x) = 2(x - 4)^2 + 3$.

a. Describe the transformations from the graph of $f(x) = x^2$ to the graph of $r(x) = 2(x - 4)^2 + 3$.

b. Use the transformations you described in Part (a) to graph the function.

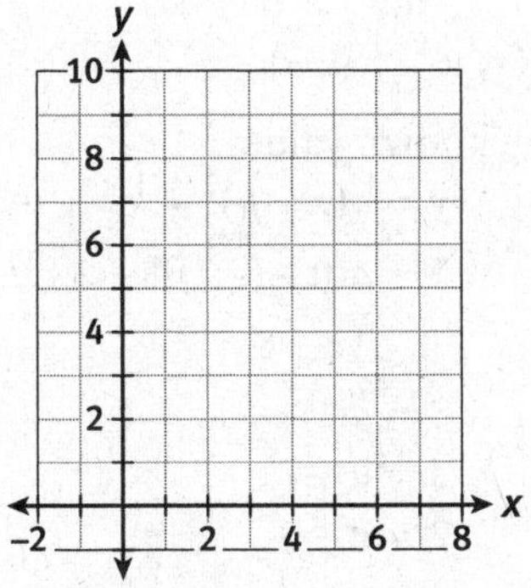

c. Identify the vertex and axis of symmetry of the parabola.

12. The vertex of the graph of $f(x) = x^2$ is (0, 0).

a. Describe how transformations can be used to determine the vertex of the graph of $g(x) = a(x - h)^2 + k$ without graphing.

b. Identify the vertex of the graph of $y = 3(x - 4)^2 + 7$.

Check Your Understanding

13. Without graphing, determine the vertex of the graph of each function.

a. $y = x^2 + 6$

b. $y = 3x^2 - 8$

c. $y = (x - 4)^2$

d. $y = (x - 2)^2 + 1$

e. $y = 2(x + 2)^2 - 5$

My Notes

Example A

For the quadratic function shown in the graph, write the equation in standard form.

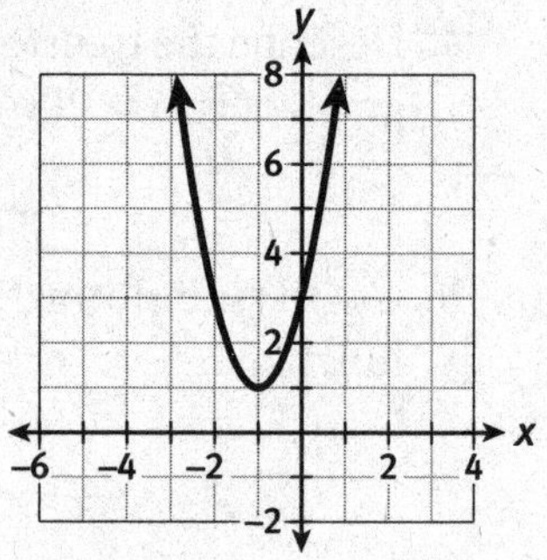

Step 1: Identify the vertex.

$(h, k) = (-1, 1)$

Step 2: Choose another point on the graph.

$(x, y) = (0, 3)$

Step 3: Substitute the known values.

$$y = a(x-h)^2 + k$$
$$3 = a(0-(-1))^2 + 1$$

Step 4: Solve for *a*.

$$3 = a(1)^2 + 1$$
$$3 = a + 1$$
$$2 = a$$

Step 5: Write the equation of the function by substituting the values of *a*, *h*, and *k*.

$$y = a(x-h)^2 + k$$
$$y = 2(x+1)^2 + 1$$

Step 6: Multiply and combine like terms to write the equation in standard form.

$$y = 2(x+1)^2 + 1$$
$$y = 2(x^2 + 2x + 1) + 1$$
$$y = 2x^2 + 4x + 2 + 1$$
$$y = 2x^2 + 4x + 3$$

Solution: $y = 2x^2 + 4x + 3$

Try These A

For each quadratic function, write its equation in standard form.

a.

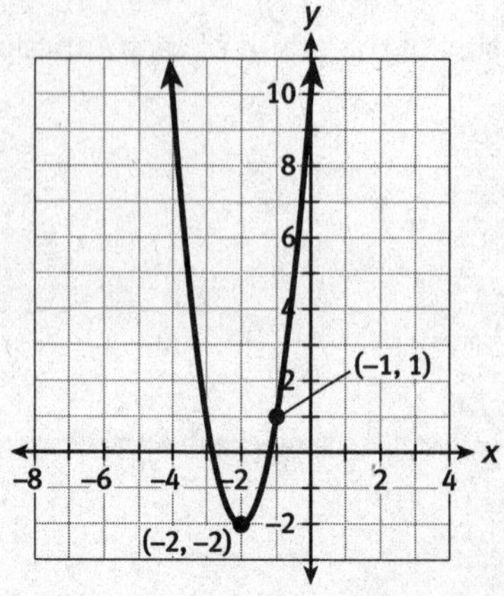

b.

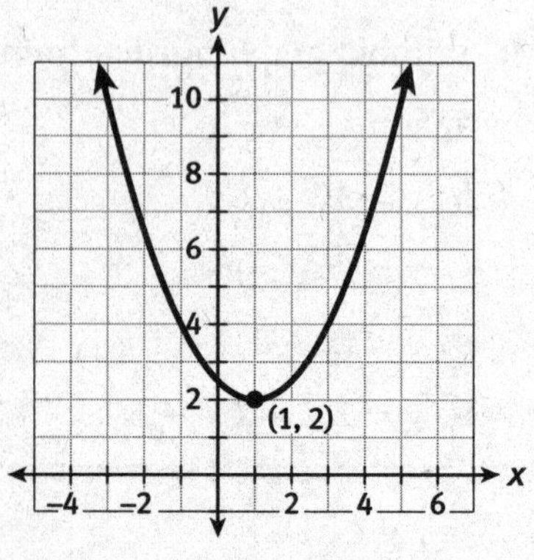

My Notes

The vertex of a parabola represents the *minimum* value of the quadratic function if the parabola opens upward. The vertex of a parabola represents the *maximum* value of the quadratic function if the parabola opens downward.

14. Make sense of problems. The equation and graph below represent two different quadratic functions.

Function 1: $y = x^2 - 3$

Function 2:

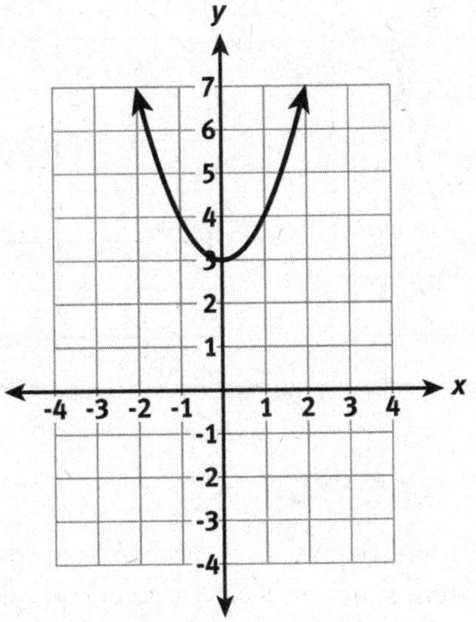

Which function has the greater minimum value? Justify your response.

15. The verbal description and the table below represent two different quadratic functions.

Function 1: The graph of function 1 is the graph of the parent function stretched vertically by a factor of 4 and translated 1 unit up.

Function 2:

x	y
−5	4
−4	1
−3	0
−2	1
−1	4

Which function has the greater minimum value? Justify your response.

DISCUSSION GROUP TIPS

If you do not understand something in your group discussions, ask for help or raise your hand for help. Describe your questions as clearly as possible, using synonyms or other words when you do not know the precise words to use.

My Notes

Check Your Understanding

16. Which of the following represents a quadratic function whose minimum value is less than the minimum value of the function $y = 3x^2 + 4$?

A. $y = 2(x + 1)^2 + 6$

B. The function whose graph is translated right 1 unit and up 4 units from the graph of the parent function

C.

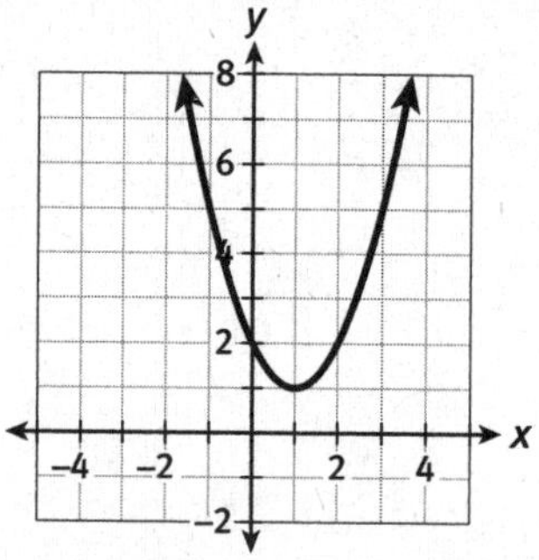

D.

x	y
−3	9
−2	6
−1	5
0	6
1	9

LESSON 30-3 PRACTICE

17. For each function, identify how its graph has been transformed from the graph of the parent function $f(x) = x^2$. Then graph each function.

a. $g(x) = \frac{1}{2}x^2 + 3$ **b.** $h(x) = -x^2 + 4$

18. Write the equation of the function whose graph has been transformed from the graph of the parent function as described.

a. Translated up 9 units and right 2 units

b. Stretched vertically by a factor of 5 and translated down 12 units

19. Write an equation for each function.

a.

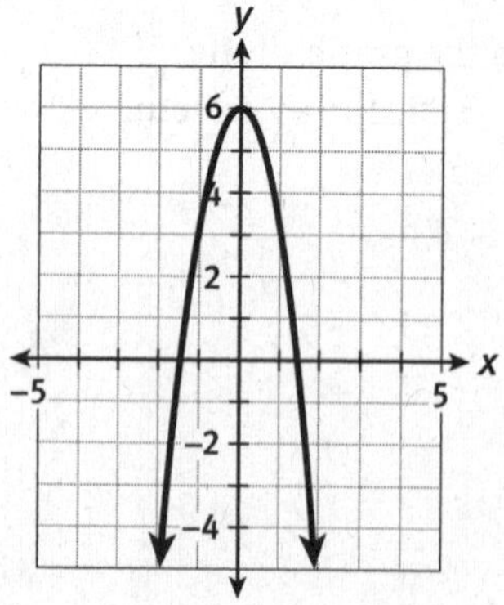

b.

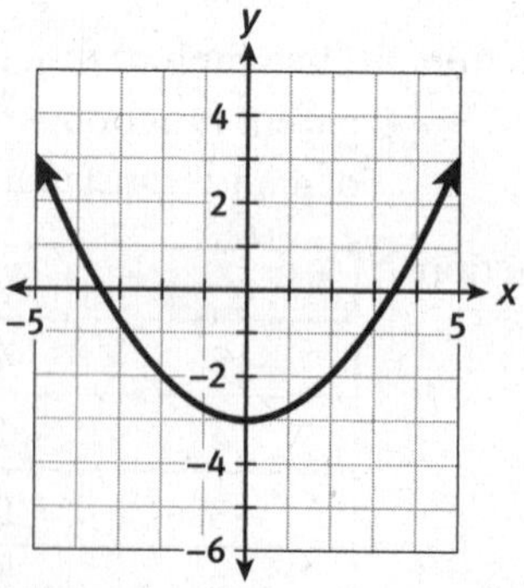

20. **Critique the reasoning of others.** Kumara graphs a quadratic function, and the vertex of the graph is (2, 3). Phillip states that his function $y = -x^2 + 7$ has a greater maximum value than Kumara's. Is Phillip correct? Justify your response.

ACTIVITY 30 PRACTICE

Write your answers on notebook paper. Show your work.

Lesson 30-1

For each quadratic function, describe the transformation of its graph from the graph of the parent function $f(x) = x^2$.

1. $g(x) = x^2 + 7$
2. $y = x^2 - \frac{3}{5}$
3. $h(x) = (x + 4)^2$
4. $y = \left(x - \frac{1}{2}\right)^2$
5. The graph of which of the following has a vertex at the point (0, 8)?
   A. $y = x^2 + 8$
   B. $y = x^2 - 8$
   C. $y = (x + 8)^2$
   D. $y = (x - 8)^2$
6. Identify the range of the function $y = x^2 - 3$.
7. Determine the equation of the axis of symmetry of the graph of $y = x^2 - 3$.
8. Identify the range of the function $y = (x - 11)^2$.
9. Determine the equation of the axis of symmetry of the graph of $y = (x - 11)^2$.
10. Write the equation of each quadratic function.
   a.

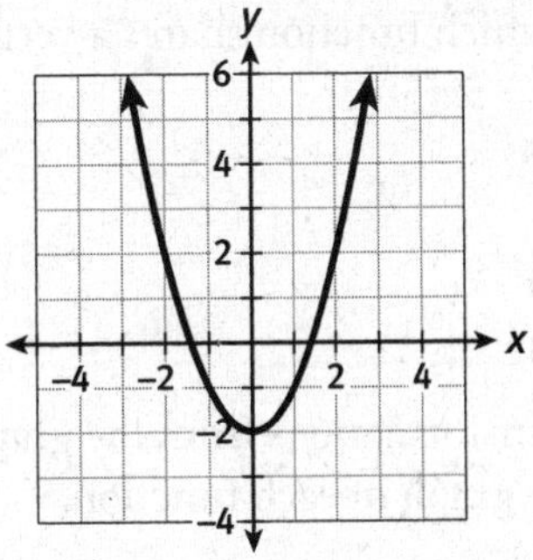

   b. The quadratic function whose graph is translated 10 units to the left from the graph of the parent function $f(x) = x^2$

   c.

x	y
−2	4.5
−1	1.5
0	0.5
1	1.5
2	4.5

Lesson 30-2

11. The graph of which function shares a vertex with the graph of $y = 5x^2 + 3$?

A. $y = 3x^2 + 5$
B. $y = x^2 + 3$
C. $y = (x + 3)^2$
D. $y = (x - 5)^2$

12. Identify the transformations from the graph of $f(x) = x^2$ to the graph of each function.

a. $y = 12x^2$
b. $y = \frac{2}{5}x^2$

13. Write the equation of the function whose graph has been transformed from the graph of the parent function as described.

a. a vertical stretch by a factor of 6
b. a vertical shrink by a factor of $\frac{1}{2}$

14. Write the equation of the quadratic function shown in the graph.

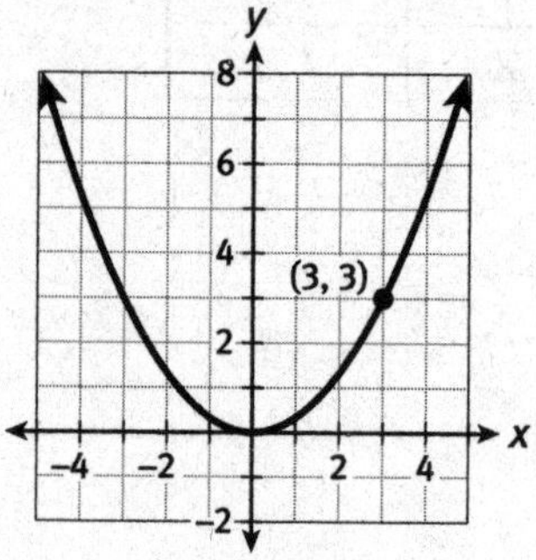

Lesson 30-3

15. Identify the transformations from the graph of the parent function $f(x) = x^2$ to the graph of each function. Then graph the function.

a. $y = 2x^2 + 1$
b. $y = -(x + 3)^2$
c. $y = (x + 1)^2 + 4$
d. $y = -\frac{1}{2}(x - 1)^2 + 1$

16. Write the equation of the quadratic function whose graph has the given vertex and passes through the given point.

a. vertex: $(0, -1)$; point $(1, 0)$
b. vertex: $(2, 0)$; point $(0, 4)$
c. vertex: $(-1, 5)$; point $(2, -4)$

MATHEMATICAL PRACTICES
Construct Viable Arguments and Critique the Reasoning of Others

17. Explain why the domains of the functions $y = x^2$ and $y = x^2 + c$, for any real number c, are the same. Then write a statement about the ranges of the functions $y = x^2$ and $y = (x + c)^2$.

In 1680, Isaac Newton, scientist, astronomer, and mathematician, used a comet visible from Earth to prove that some comets follow a parabolic path through space as they travel around the sun. This and other discoveries like it help scientists to predict past and future positions of comets.

1. Assume the path of a comet is given by the function $y = -x^2 + 4$.
   a. Graph the path of the comet. Explain how you graphed it.
   b. Identify the vertex of the function.
   c. Identify the maximum or minimum value of the function.
   d. Identify the domain and range.
   e. Write the equation for the axis of symmetry.

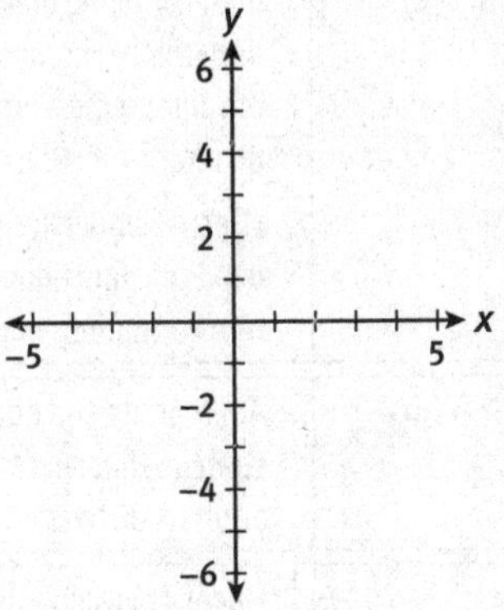

2. Identify the table that represents a parabolic comet path. Explain and justify your choice.

A.

x	y
−2	−1
−1	−4
0	−5
1	−4
2	−1

B.

x	y
−2	4
−1	−3
0	−5
1	−3
2	4

3. The graph at right shows a portion of the path of a comet represented by a function in the form $y = a(x - h)^2 + k$. Determine the values of a, h, and k and write the equation of the function.

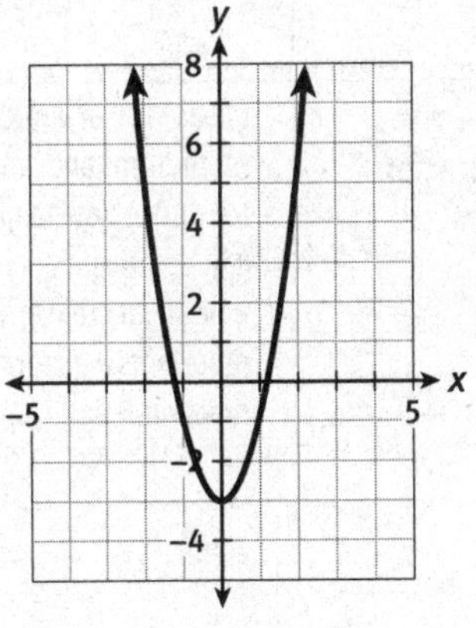

4. The equation and graph below represent two different quadratic functions for the parabolic paths of comets.

Function 1:

$y = -3x^2 + 4$

Function 2:

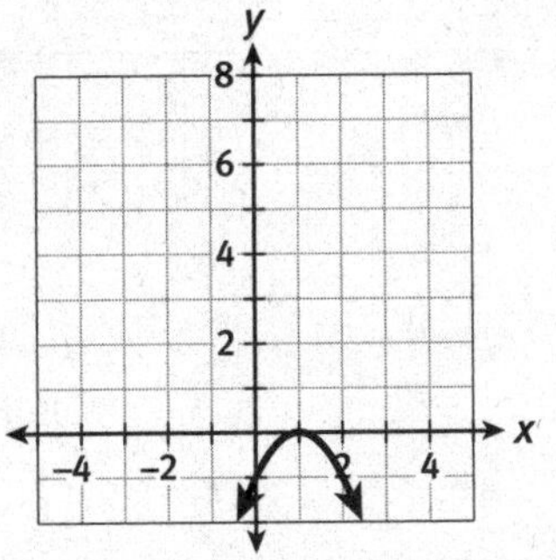

Identify the maximum value of each function. Which function has the greater maximum value?

Scoring Guide	Exemplary	Proficient	Emerging	Incomplete
	The solution demonstrates the following characteristics:			
Mathematics Knowledge and Thinking (Items 2, 3)	• Clear and accurate understanding of how to determine whether a table of values represents a quadratic function • Effective understanding of quadratic functions as transformations of $y = x^2$	• Largely correct understanding of how to determine whether a table of values represents a quadratic function • Adequate understanding of quadratic functions as transformations of $y = x^2$	• Difficulty determining whether a table of values represents a quadratic function • Partial understanding of quadratic functions as transformations of $y = x^2$	• Inaccurate or incomplete understanding of how to determine whether a table of values represents a quadratic function • Little or no understanding of quadratic functions as transformations of $y = x^2$
Problem Solving (Items 2, 4)	• Appropriate and efficient strategy that results in a correct answer	• Strategy that may include unnecessary steps but results in a correct answer	• Strategy that results in a partially incorrect answer	• No clear strategy when solving problems
Mathematical Modeling / Representations (Items 1–4)	• Fluency in using tables, equations, and graphs to represent quadratic functions • Effective understanding of how to graph a quadratic function and identify key features from the graph	• Little difficulty using tables, equations, and graphs to represent quadratic functions • Largely correct understanding of how to graph a quadratic function and identify key features from the graph	• Some difficulty using tables, equations, and/or graphs to represent quadratic functions • Partial understanding of how to graph a quadratic function and/or identify key features from the graph	• Significant difficulty using tables, equations, and/or graphs to represent quadratic functions • Inaccurate or incomplete understanding of how to graph a quadratic function and/or identify key features from the graph
Reasoning and Communication (Items 1a, 2)	• Precise use of appropriate math terms and language to explain how to graph a quadratic function and whether a table of values represents a quadratic function	• Adequate explanation of how to graph a quadratic function and whether a table of values represents a quadratic function	• Misleading or confusing explanation of how to graph a quadratic function and/or whether a table of values represents a quadratic function	• Incomplete or inaccurate explanation of how to graph a quadratic function and/or whether a table of values represents a quadratic function

Trebuchet Trials

Lesson 31-1 Solving by Graphing or Factoring

My Notes

Learning Targets:

- Use a graph to solve a quadratic equation.
- Use factoring to solve a quadratic equation.
- Describe the connection between the zeros of a quadratic function and the x-intercepts of the function's graph.

SUGGESTED LEARNING STRATEGIES: Visualization, Summarizing, Paraphrasing, Think-Pair-Share, Quickwrite

Carter, Alisha, and Joseph are building a *trebuchet* for an engineering competition. A trebuchet is a medieval siege weapon that uses gravity to launch an object through the air. When the counterweight at one end of the throwing arm drops, the other end rises and a projectile is launched through the air. The path the projectile takes through the air is modeled by a parabola.

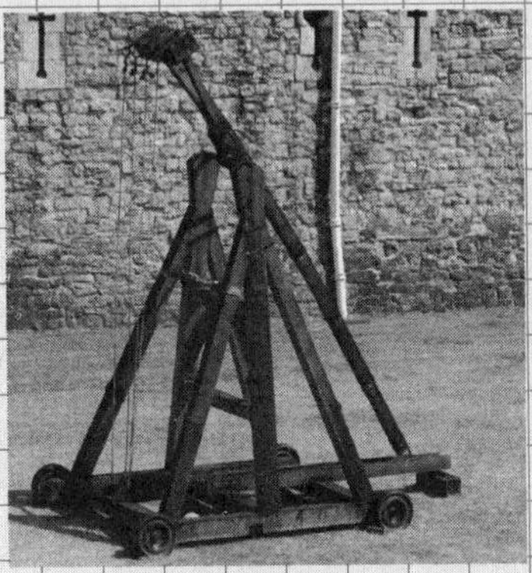

To win the competition, the team must build their trebuchet according to the competition specifications to launch a small projectile as far as possible. After conducting experiments that varied the projectile's mass and launch angle, the team discovered that the ball they were launching followed the path given by the quadratic equation $y = -\frac{1}{8}x^2 + 2x$.

CONNECT TO PHYSICS

The throwing arm of a trebuchet is an example of a class I lever.

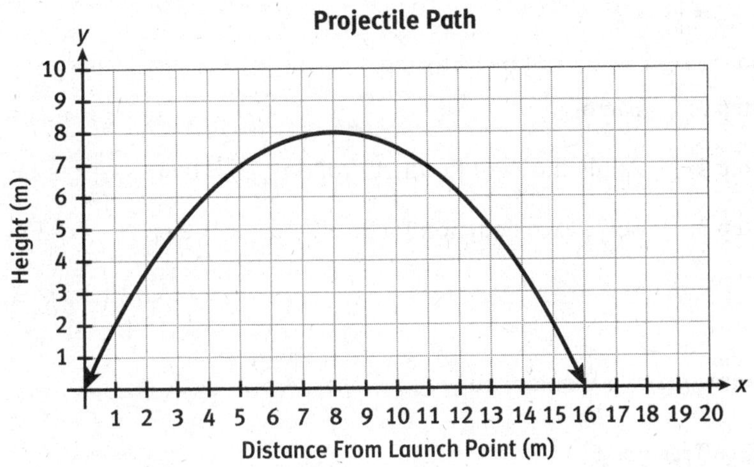

1. **Make sense of problems.** How far does the ball land from the launching point?

2. What is the maximum height of the ball?

3. What are the x-coordinates of the points where the ball is on the ground?

My Notes

To determine how far the ball lands from the launching point, you can solve the equation $-\frac{1}{8}x^2 + 2x = 0$, because the height y equals 0.

4. Verify that $x = 0$ and $x = 16$ are solutions to this equation.

5. Without the graph, could you have determined these solutions? Explain.

MATH TERMS

A polynomial is in **factored form** when it is expressed as the product of one or more polynomials.

The ***factored form*** of a polynomial equation provides an effective way to determine the values of x that make the equation equal 0.

Zero Product Property

If $ab = 0$, then either $a = 0$ or $b = 0$.

Example A

Solve $-\frac{1}{8}x^2 + 2x = 0$ by factoring.

Step 1: Factor. $x\left(-\frac{1}{8}x + 2\right) = 0$

Step 2: Apply the Zero Product Property. $x = 0$ or $-\frac{1}{8}x + 2 = 0$

Step 3: Solve each equation for x. $x = 0$ or $-\frac{1}{8}x = -2$

$$(-8)\left(-\frac{1}{8}\right)x = -2(-8)$$

$$x = 16$$

Solution: $x = 0$ or $x = 16$

Try These A

Solve each quadratic equation by factoring.

a. $x^2 - 5x - 14 = 0$

b. $3x^2 - 6x = 0$

c. $x^2 + 3x = 18$

My Notes

6. How do the solutions to the projectile path equation $-\frac{1}{8}x^2 + 2x = 0$ in Example A relate to the equation's graph?

The graph of the function $y = x^2 - 6x + 5$ is shown below.

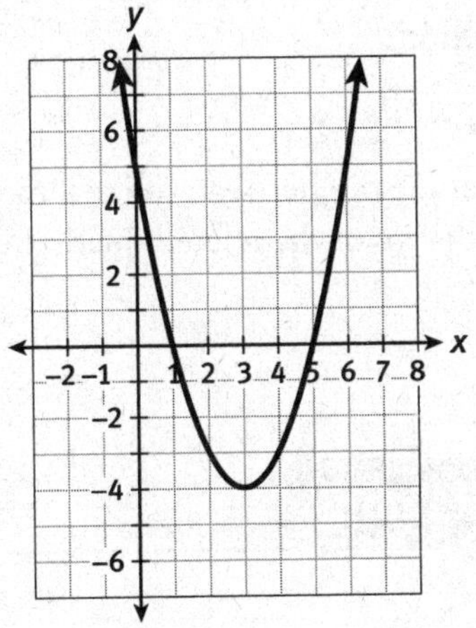

7. Identify the x-intercepts of the graph.

8. What is the x-coordinate of the vertex?

9. Describe the x-coordinate of the vertex with respect to the two x-intercepts.

10. a. Factor the quadratic expression $x^2 - 6x + 5$.

b. Each of the factors you found in part a is a linear expression. Solve the related linear equation for each factor by setting it equal to 0.

c. Solve the related quadratic equation for the expression in Part a. That is, solve $x^2 - 6x + 5 = 0$.

11. a. How do the linear factors of $x^2 - 6x + 5$ you found in Item 10a relate to the x-intercepts of the graph of the function $y = x^2 - 6x + 5$ shown above?

b. Without graphing, how could you determine the x-intercepts of the graph of the quadratic function $y = x^2 + x - 12$?

My Notes

The x-coordinates of the x-intercepts of a quadratic function $y = ax^2 + bx + c$ are the ***zeros of the function***. The solutions of a quadratic equation $ax^2 + bx + c = 0$ are the **roots** of the equation.

MATH TIP

The relationship described in Item 12b is between the *real* roots of a quadratic equation and the zeros of the related quadratic function. Later in this unit you will see quadratic equations that have non-real roots. The same relationship does not exist between non-real roots and zeros of the related function.

12. The quadratic function $y = ax^2 + bx + c$ is related to the equation $ax^2 + bx + c = 0$ by letting y equal zero.

a. Why do you think the x-coordinates of the x-intercepts are called the zeros of the function?

b. Describe the relationship between the real *roots* of a quadratic equation and the *zeros* of the related quadratic function.

Check Your Understanding

Solve by factoring.

13. $x^2 - 81 = 0$

14. $\frac{1}{2}x^2 - 2x = 0$

Identify the zeros of the quadratic function. How are the linear factors of the quadratic expression related to the zeros?

15. $y = x^2 + 8x + 7$

16. $y = x^2 - 3x + 2$

LESSON 31-1 PRACTICE

17. Make use of structure. Use the graph of the quadratic function $y = 2x^2 + 6x$ shown to determine the roots of the quadratic equation $0 = 2x^2 + 6x$.

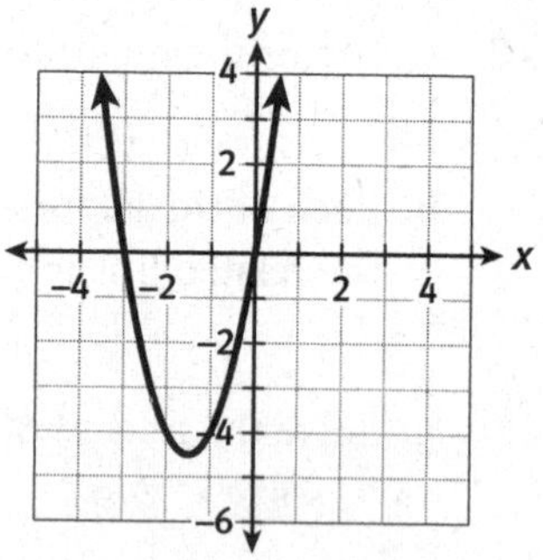

Solve by factoring.

18. $x^2 - 2x + 1 = 0$

19. $2x^2 - 7x - 4 = 0$

20. $x^2 + 5x = 0$

21. Write a quadratic equation whose roots are 3 and -6.

22. A whale jumps vertically from a pool at Ocean World. The function $y = -16x^2 + 32x$ models the height of a whale in feet above the surface of the water after x seconds.

a. What is the maximum height of the whale above the surface of the water?

b. How long is the whale out of the water? Justify your answer.

23. Construct viable arguments. Is it possible for two different quadratic functions to share the same zeros? Use a graph to justify your response.

My Notes

Learning Targets:

- Identify the axis of symmetry of the graph of a quadratic function.
- Identify the vertex of the graph of a quadratic function.

SUGGESTED LEARNING STRATEGIES: Close Reading, Marking the Text, Think-Pair-Share, Predict and Confirm, Discussion Groups

The *axis of symmetry* of the parabola determined by the function $y = ax^2 + bx + c$ is the vertical line that passes through the vertex.

The equation for the axis of symmetry is $x = -\frac{b}{2a}$.

The vertex is on the axis of symmetry. Therefore, the *x*-coordinate of the vertex is $-\frac{b}{2a}$.

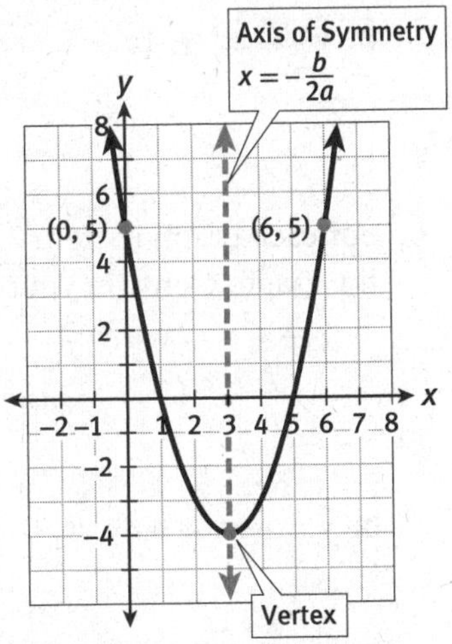

MATH TIP

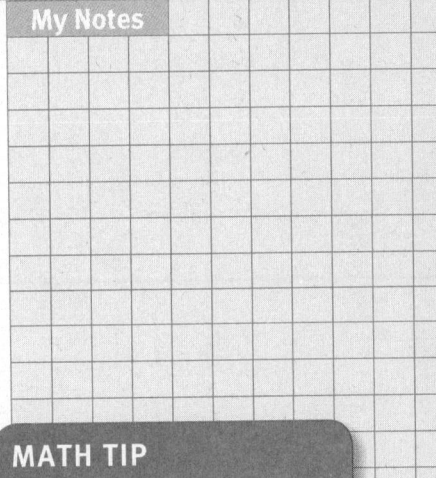

Substitute the *x*-coordinate of the vertex into the quadratic equation to find the *y*-coordinate of the vertex.

Each point on a quadratic graph will have a mirror image point with the same *y*-coordinate that is equidistant from the axis of symmetry. For example, the point (0, 5) is reflected over the axis of symmetry to the point (6, 5) on the graph.

1. Give the coordinates of two additional points that are reflections over the axis of symmetry in the graph above.

2. For each function, identify the zeros graphically. Confirm your answer by setting the function equal to 0 and solving by factoring.

a. $y = x^2 + 2x - 8$ **b.** $y = -x^2 + 4x + 5$

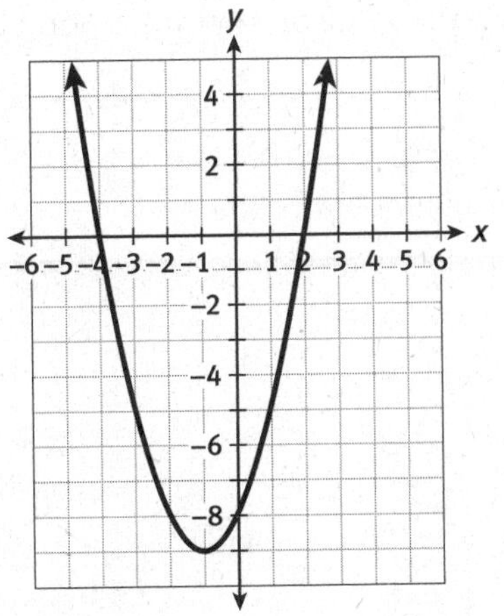

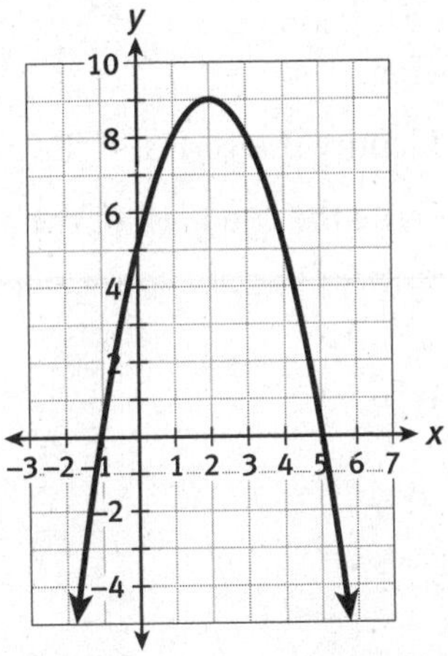

My Notes

3. For each graph in Item 2, determine the x-coordinate of the vertex by finding the x-coordinate exactly in the middle of the two zeros. Confirm your answer by calculating the value of $-\frac{b}{2a}$.
   a. $y = x^2 + 2x - 8$

   b. $y = -x^2 + 4x + 5$

4. For each graph in Item 2, determine the y-coordinate of the vertex from the graph. Confirm your answer by evaluating the function at $x = -\frac{b}{2a}$.
   a. $y = x^2 + 2x - 8$

   b. $y = -x^2 + 4x + 5$

5. **Reason quantitatively.** For each graph in Item 2, determine the maximum or minimum value of the function.
   a. $y = x^2 + 2x - 8$

   b. $y = -x^2 + 4x + 5$

Check Your Understanding

For the function $y = -x^2 + x + 6$, use the value of $-\frac{b}{2a}$ to respond to the following.

6. Identify the vertex of the graph.
7. Write the equation of the axis of symmetry.

My Notes

LESSON 31-2 PRACTICE

Use the value of $-\frac{b}{2a}$ to determine the vertex and to write the equation for the axis of symmetry for the graph of each of the following quadratic functions.

8. $y = x^2 - 10x$

9. $y = x^2 - 4x - 32$

10. $y = x^2 + x - 12$

11. $y = 6x - x^2$

12. Describe the graph of a quadratic function that has its vertex and a zero at the same point.

13. Model with mathematics. An architect is designing a tunnel and is considering using the function $y = -0.12x^2 + 2.4x$ to determine the shape of the tunnel's entrance, as shown in the figure. In this model, y is the height of the entrance in feet and x is the distance in feet from one end of the entrance.

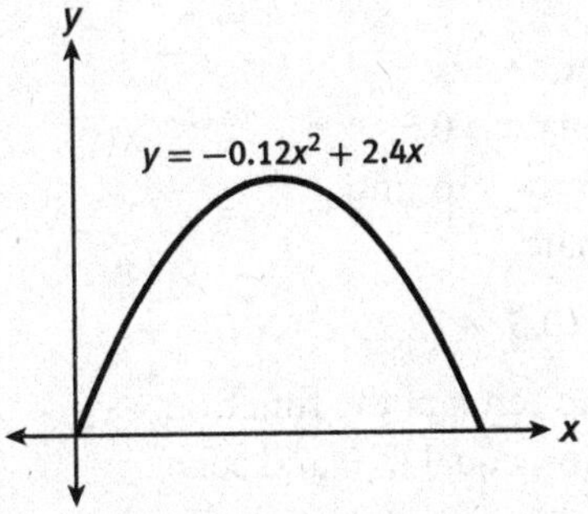

a. How wide is the tunnel's entrance at its base?
b. What is the vertex? What does it represent?
c. Could a truck that is 14 feet tall pass through the tunnel? Explain.

My Notes

Learning Targets:

- Use the axis of symmetry, the vertex, and the zeros to graph a quadratic function.
- Interpret the graph of a quadratic function.

SUGGESTED LEARNING STRATEGIES: Create Representations, Close Reading, Note Taking, Identify a Subtask

If a quadratic function can be written in factored form, you can graph it by finding the vertex and the zeros.

Example A

Graph the quadratic function $y = x^2 - x - 12$.

Step 1: Determine the axis of symmetry using $x = -\frac{b}{2a}$.

The axis of symmetry is $x = 0.5$.

$y = x^2 - x - 12$

$a = 1, b = -1$

$x = -\frac{(-1)}{2(1)} = \frac{1}{2} = 0.5$

Step 2: Determine the vertex.
The x-coordinate is 0.5.
Substitute 0.5 for x to find the y-coordinate.

The vertex is $(0.5, -12.25)$.

$y = (0.5)^2 - (0.5) - 12$
$= 0.25 - 0.5 - 12$
$= -12.25$

Step 3: Determine the zeros of the function.
Set the function equal to 0 and solve.

The zeros are -3 and 4.

$x^2 - x - 12 = 0$
$(x - 4)(x + 3) = 0$
$x - 4 = 0$ or $x + 3 = 0$
$x = 4$ or $x = -3$

Step 4: Graph the function.

Plot the vertex $(0.5, -12.25)$ and the points where the function crosses the x-axis $(-3, 0)$ and $(4, 0)$. Connect the points with a smooth parabolic curve.

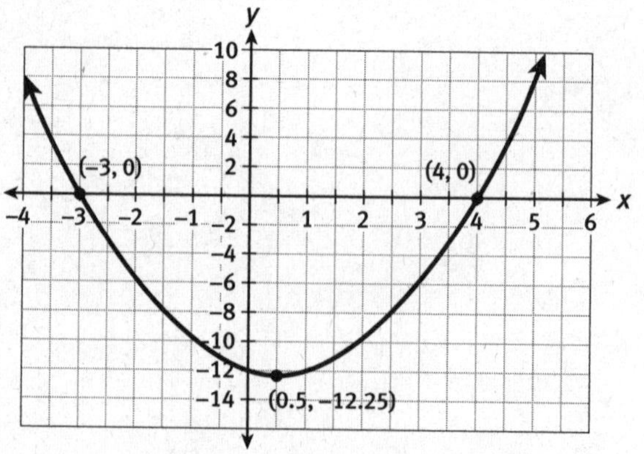

My Notes

Try These A

a. Check the graph in Example A by plotting two more points on the graph. First choose an x-value, and then find the y-value by evaluating the function. Plot this point. Then plot the reflection of the point over the axis of symmetry to get another point. Verify that both points are on the graph.

Graph the quadratic functions by finding the vertex and the zeros. Check your graphs.

b. $y = x^2 - 2x - 8$
c. $y = 4x - x^2$
d. $y = x^2 + 4x - 5$

Joseph, Carter, and Alisha tested a new trebuchet designed to launch the projectile even further. They also refined their model to reflect a more accurate launch height of 1 m. The new projectile path is given by the function $y = -\frac{1}{19}(x^2 - 18x - 19)$.

1. Graph the projectile path on the coordinate axes below.

Distance Projectile Path

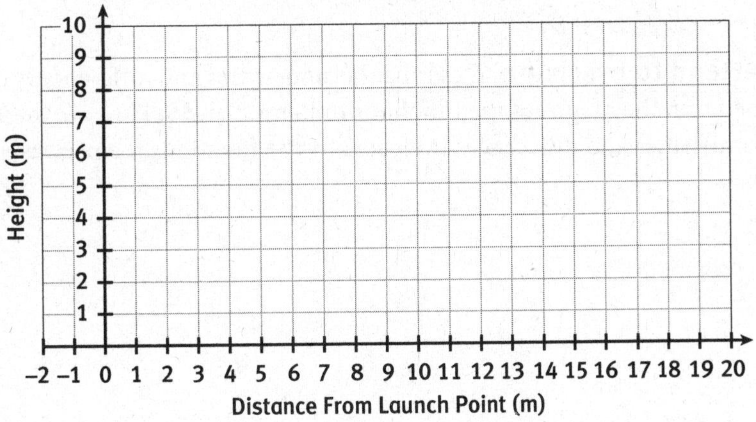

CONNECT TO AP

In AP Calculus, you will represent projectile motion with special types of equations called parametric equations.

My Notes

2. **Make sense of problems.** Last year's winning trebuchet launched a projectile a horizontal distance of 19.5 m. How does the team's trebuchet compare to last year's winner?

Check Your Understanding

Graph each quadratic function.

3. $y = x^2 - 11x + 30$
4. $y = x^2 - 9$

LESSON 31-3 PRACTICE

Graph the quadratic functions. Label the vertex, axis of symmetry, and zeros on each graph.

5. $y = x^2 + 5x$
6. $y = -x^2 + 2x$
7. $y = -x^2 + \frac{1}{4}$
8. $y = x^2 - 3x - 4$
9. $y = -x^2 + 3x + 4$
10. **Attend to precision.** Describe the similarities and differences between the graphs of the functions in Items 8 and 9. How are these similarities and differences indicated by the functions themselves?

ACTIVITY 31 PRACTICE

Write your answers on notebook paper.
Show your work.

Lesson 31-1

Use the graphs to determine the zeros of the quadratic functions.

1. $y = 0.5x^2$

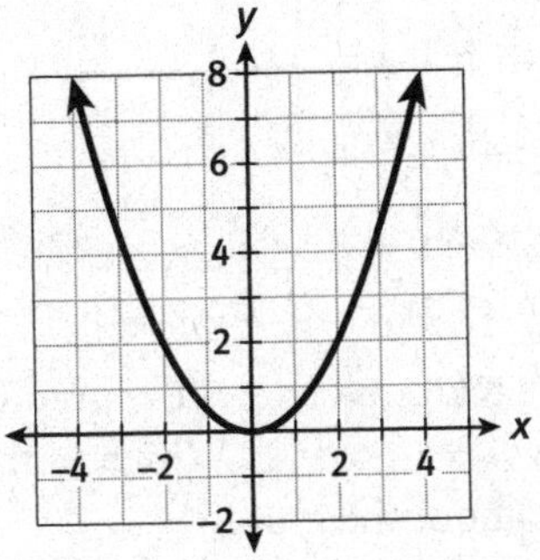

2. $y = x^2 - 4$

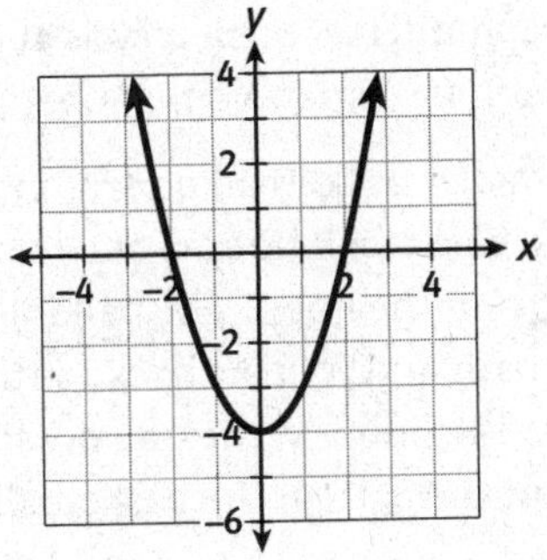

3. $y = -x^2 + 4x - 3$

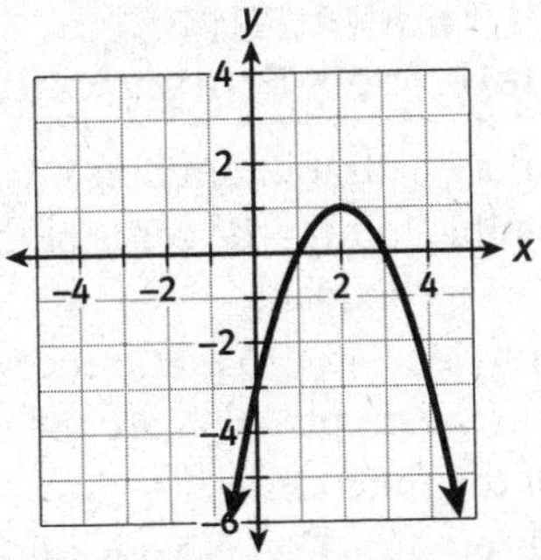

4. Identify the zeros of the quadratic function $y = (x - a)(x - b)$.

5. Which of the following functions has zeros at $x = 2$ and $x = -3$?
A. $y = x^2 + x - 6$ **B.** $y = -x^2 + 5x + 6$
C. $y = 2x^2 + x - 3$ **D.** $y = x^2 + 2x - 3$

6. What are the solutions to the equation $x^2 - 6x = -5$?
A. $-1, 5$ **B.** $2, 3$
C. $1, 5$ **D.** $-2, 3$

Solve by factoring.

7. $0 = x^2 - 25$

8. $0 = x^2 + 17x$

9. $0 = x^2 - 5x - 66$

10. $0 = x^2 + 99x - 100$

11. $0 = -x^2 - 4x + 32$

12. $0 = x^2 + 10x + 25$

13. $0 = x^2 + 8x + 7$

14. A fountain at a city park shoots a stream of water vertically from the ground. The function $y = -8x^2 + 16x$ models the height of the stream of water in feet after x seconds.
a. What is the maximum height of the stream of water?
b. At what time does the stream of water reach its maximum height?
c. For how many seconds does the stream of water appear above ground?
d. Identify a reasonable domain and range for this function.

15. Write a quadratic equation whose roots are -4 and 8. How do the roots relate to the zeros and factors of the associated quadratic function?

16. Write a quadratic function whose zeros are 2 and 9. How do the zeros relate to the factors of the quadratic function?

Lesson 31-2

17. If a quadratic function has zeros at $x = -4$ and $x = 6$, what is the x-coordinate of the vertex?

18. What is the vertex of the quadratic function $y = x^2 + 8x + 15$?

19. Write the equation for the axis of symmetry of the graph of the quadratic function $y = 2x^2 + 4x - 1$.

Use the following information for Items 20–24.

A diver in Acapulco jumps from a cliff. His height y, in meters, as a function of x, his distance from the cliff base in meters, is given by the quadratic function $y = 100 - x^2$, for $x \geq 0$.

20. Graph the function representing the cliff diver's height.
  a. Identify the vertex of the graph
  b. Identify the x-intercepts of the graph.

21. Determine the solutions of the equation $100 - x^2 = 0$.

22. How do the solutions to the equation in Item 21 relate to the x-intercepts of the graph in Item 20?

23. How high is the cliff from which the diver jumps?

24. How far from the base of the cliff does the diver hit the water?

25. Which of these functions has a graph with the axis of symmetry $x = -2$?
  A. $y = x^2 + 4x - 2$
  B. $y = x^2 - 4x + 2$
  C. $y = 2x^2 + 2x - 3$
  D. $y = 2x^2 - 2x + 3$

Lesson 31-3

26. Lisa correctly graphed a quadratic function and found that its vertex was in Quadrant I. Which function could she have graphed?
  A. $y = x^2 + 4x + 2$
  B. $y = x^2 - 4x + 2$
  C. $y = x^2 - 4x + 6$
  D. $y = x^2 + 4x + 6$

27. Which of these is the equation of the parabola graphed below?

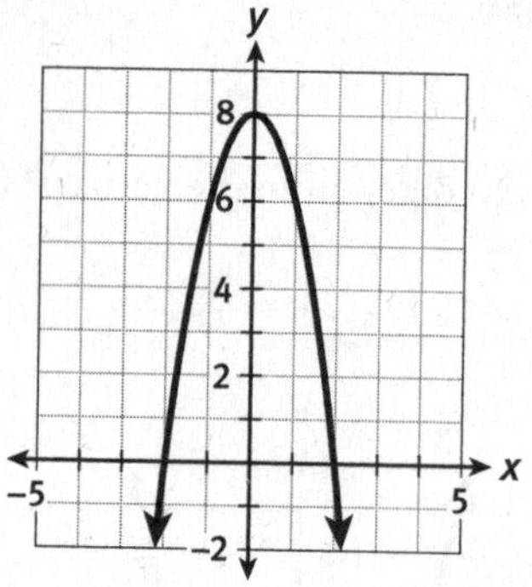

  A. $y = 2x^2 - 2x + 8$ **B.** $y = -2x^2 + 8$
  C. $y = -x^2 - 8x - 15$ **D.** $y = -x^2 - 8$

28. DeShawn's textbook shows the graph of the function $y = x^2 + x - 6$. Which of these is a true statement about the graph?
  A. The axis of symmetry is the y-axis.
  B. The vertex is $(0, -6)$.
  C. The graph intersects the x-axis at $(-3, 0)$.
  D. The graph is a parabola that opens downward.

29. Kim throws her basketball up from the ground toward the basketball hoop from a distance of 20 feet away from the hoop. The ball follows a parabolic path and returns back to the gym floor 5 feet from the hoop. Write one possible equation to represent the path of the basketball. Explain your answer.

MATHEMATICAL PRACTICES
Use Appropriate Tools Strategically

30. Consider the quadratic function $y = x^2 - x - 3$.
  a. Is it possible to find the zeros of the function by factoring? Explain.
  b. Use your calculator to graph the function. Based on the graph, what are the approximate zeros of the function?
  c. Use the zero function of your calculator to find more accurate approximations for the zeros. Round to the nearest tenth.

Keeping it Quadratic

Lesson 32-1 The Square Root Method

My Notes

Learning Targets:

- Solve quadratic equations by the square root method.
- Provide examples of quadratic equations having a given number of real solutions.

SUGGESTED LEARNING STRATEGIES: Guess and Check, Simplify the Problem, Think-Pair-Share, Create Representations

Nguyen is trying to build a square deck around his new hot tub. To decide how large a deck he should build, he needs to determine the side length, x, of different sized decks given the possible area of each deck. He knows that the area of a square is equal to the length of a side squared. Using this information, Nguyen writes the following equations to represent each of the decks he is considering.

1. Solve each equation. Be prepared to discuss your solution methods with your classmates.

a. $x^2 = 49$

b. $x^2 = 100$

c. $x^2 = 15$

d. $2x^2 = 18$

e. $x^2 - 4 = 0$

f. $x^2 + 2 = 0$

g. $x^2 + 3 = 3$

MATH TIP

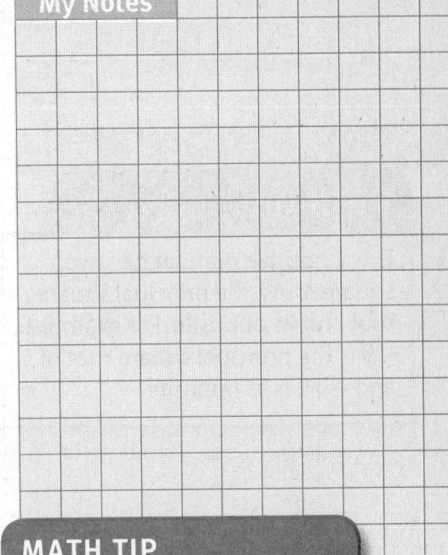

A solution of an equation makes the equation true. For example, $x + 5 = 7$ has the solution $x = 2$ because $2 + 5 = 7$.

2. Refer to the equations in Item 1 and their solutions.

a. What do the equations have in common?

b. What types of numbers are represented by the solutions of these equations?

c. How many solutions do the equations have?

d. **Reason quantitatively.** Which solutions are reasonable for side lengths of the squares? Explain.

My Notes

To solve a quadratic equation of the form $ax^2 + c = 0$, isolate the x^2-term and then take the square root of both sides.

MATH TIP

Every positive number has two square roots, the principal square root and its opposite. For example, $\sqrt{5}$ is the principal square root of 5 and $-\sqrt{5}$ is its opposite.

Example A

Solve $3x^2 - 6 = 0$ using square roots.

Step 1: Add 6 to both sides. $3x^2 - 6 = 0$

$3x^2 = 6$

Step 2: Divide both sides by 3. $\frac{3x^2}{3} = \frac{6}{3}$

$x^2 = 2$

Step 3: Take the square root of both sides. $\sqrt{x^2} = \pm\sqrt{2}$

$x = +\sqrt{2}$ or $-\sqrt{2}$

Solution: $x = +\sqrt{2}$ or $x = -\sqrt{2}$

READING MATH

The $\pm$ symbol is read "plus or minus."

Try These A

Solve each equation using square roots.

a. $x^2 - 10 = 1$

b. $\frac{x^2}{4} = 1$

c. $4x^2 - 6 = 14$

3. Quadratic equations can have 0, 1, or 2 real solutions. Fill in the table below with equations from the first page that represent the possible numbers of solutions.

MATH TIP

The square root of a negative number is not a real number, so equations of the form $x^2 = c$ where $c < 0$ have no real solutions.

Number of Solutions	Result When x^2 is Isolated	Example(s)
Two	x^2 = positive number	
One	$x^2 = 0$	
No real solutions	x^2 = negative number	

My Notes

4. **Reason abstractly.** A square frame has a 2-in. border along two sides as shown in the diagram. The total area is 66 in^2. Answer the questions to help you write an equation to find the area of the unshaded square.

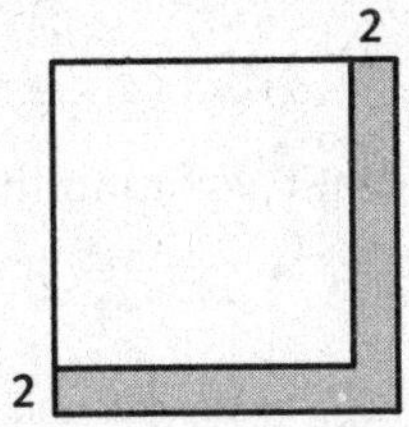

a. Label the sides of the unshaded square x.

b. Fill in the boxes to write an equation for the total area in terms of x.

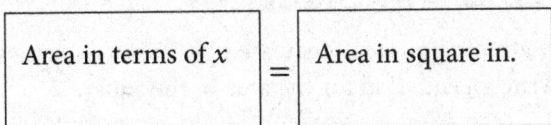

Area in terms of x	=	Area in square in.

You can solve quadratic equations like the one you wrote in Item 4 by isolating the variable.

Example B

Solve $(x+2)^2 = 66$ using square roots. Approximate the solutions to the nearest hundredth.

Step 1: Take the square root of both sides.

$$(x+2)^2 = 66$$
$$\sqrt{(x+2)^2} = \pm\sqrt{66}$$
$$x+2 = \pm\sqrt{66}$$

Step 2: Subtract 2 from both sides.

$$x = -2 \pm \sqrt{66}$$
$$x = -2+\sqrt{66} \text{ or } x = -2-\sqrt{66}$$

Step 3: Use a calculator to approximate the solutions.

Solution: $x \approx -10.12$ or $x \approx 6.12$

5. Are both solutions to this equation valid in the context of Item 4? Explain your response.

Try These B

Solve each equation using square roots.

a. $(x-5)^2 = 121$ **b.** $(2x-1)^2 = 6$ **c.** $x^2 - 12x + 36 = 2$

My Notes

Check Your Understanding

Solve each equation using square roots.

6. $x^2 + 12 = 13$ **7.** $(x - 4)^2 = 1$ **8.** $3x^2 - 6 = 15$

9. Give an example of a quadratic equation that has
a. one real solution.
b. no real solutions.
c. two real solutions.

LESSON 32-1 PRACTICE

10. If the length of a square is decreased by 1 unit, the area will be 8 square units. Write an equation for the area of the square.

11. Calculate the side length of the square in Item 10.

Solve each equation.

12. $x^2 - 22 = 0$ **13.** $(x + 5)^2 - 4 = 0$

14. $x^2 - 4x + 4 = 0$ **15.** $(x + 1)^2 = 12$

16. Model with mathematics. Alaysha has a square picture with an area of 100 square inches, including the frame. The width of the frame is x inches.

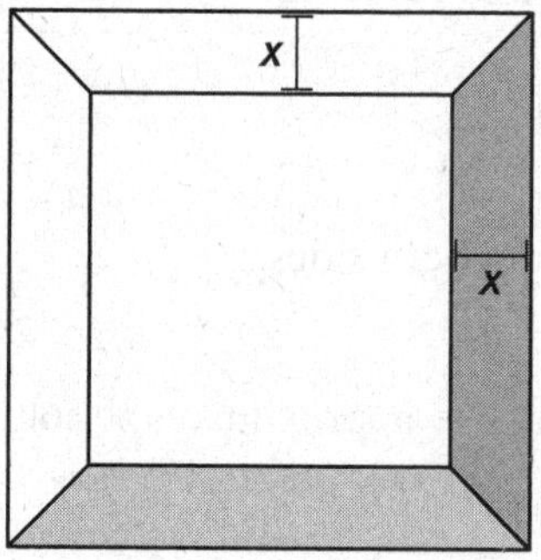

a. Write an equation in terms of x for the area A of the picture inside the frame.
b. If the area of the picture inside the frame is 64 square inches, what are the possible values for x?
c. If the area of the picture inside the frame is 64 square inches, how wide is the picture frame? Justify your response.

My Notes

Learning Targets:

- Solve quadratic equations by completing the square.
- Complete the square to analyze a quadratic function.

SUGGESTED LEARNING STRATEGIES: Note Taking, Graphic Organizer, Identify a Subtask

As shown in Example B in Lesson 32-1, quadratic equations are more easily solved with square roots when the side with the variable is a perfect square. When a quadratic equation is written in the form $x^2 + bx + c = 0$, you can complete the square to transform the equation into one that can be solved using square roots. ***Completing the square*** is the process of adding a term to the variable side of a quadratic equation to transform it into a perfect square trinomial.

Example A

Solve $x^2 + 10x - 6 = 0$ by completing the square.

Step 1: Isolate the variable terms. Add 6 to both sides.
$x^2 + 10x - 6 = 0$
$x^2 + 10x = 6$

Step 2: Transform the left side into a perfect square trinomial.
Divide the coefficient of the x-term by 2. $10 \div 2 = 5$
Square the 5 to determine the constant. $5^2 = 25$
Complete the square by adding 25 to both sides of the equation.
$x^2 + 10x + \square = 6 + \square$
$x^2 + 10x + \boxed{25} = 6 + \boxed{25}$
$x^2 + 10x + 25 = 31$

Step 3: Solve the equation.
Write the trinomial in factored form. $(x + 5)(x + 5) = 31$
Write the left side as a square of a binomial. $(x + 5)^2 = 31$
Take the square root of both sides. $\sqrt{(x+5)^2} = \pm\sqrt{31}$
Solve for x. $(x + 5) = \pm\sqrt{31}$
Leave the solutions in $\pm$ form. $x = -5 \pm \sqrt{31}$

Solution: $x = -5 \pm \sqrt{31}$

MATH TIP

Use a graphic organizer to help you complete the square.

	x	5
x	x^2	$5x$
5	$5x$	25

$x^2 + 5x + 5x + 25 = (x + 5)(x + 5)$
$x^2 + 10x + 25 = (x + 5)^2$

Try These A

Make use of structure. Solve each quadratic equation by completing the square.

a. $x^2 - 8x + 3 = 11$

b. $x^2 + 7 = 2x + 8$

My Notes

CONNECT TO PHYSICS

Quadratic functions can describe the *trajectory*, or path, of a moving object, such as a ball that has been thrown.

Completing the square is useful to help analyze specific features of quadratic functions, such as the maximum or minimum value and the possible number of zeros.

When a quadratic equation is written in the form $y = a(x - h)^2 + k$, you can determine whether the function has a maximum or minimum value based on a and what that value is based on k. This information can also help you determine the number of x-intercepts.

Example B

Analyze the quadratic function $y = x^2 - 6x + 13$ by completing the square.

Step 1: Complete the square on the right side of the equation. $y = x^2 - 6x + 13$

Isolate the variable terms. $y - 13 = x^2 - 6x$

Transform the right side into a perfect square trinomial. $y - 13 + 9 = x^2 - 6x + 9$

Factor and simplify. $y - 4 = (x - 3)^2$

Write the equation in the form $y = a(x - h)^2 + k$. $y = (x - 3)^2 + 4$

Step 2: Identify the direction of opening and whether the vertex represents a maximum or minimum.

Since the value of a is positive, the parabola opens upward and the vertex represents a minimum. Opens upward; Vertex is a minimum.

Step 3: Determine the maximum or minimum value. This parabola is the graph of the parent function $y = x^2$ translated up 4 units. So the parent function's minimum value of 0 is increased to 4. Minimum value: 4

Step 4: Determine the number of x-intercepts. Since the minimum value of the function is $y = 4$ and the parabola opens upward, the function will never have a y-value less than 4. The graph will never intersect the x-axis, so there are no x-intercepts. no x-intercepts

Step 5: Verify by graphing.

Solution: The graph of the quadratic function opens upward, the minimum value is 4, and there are no x-intercepts.

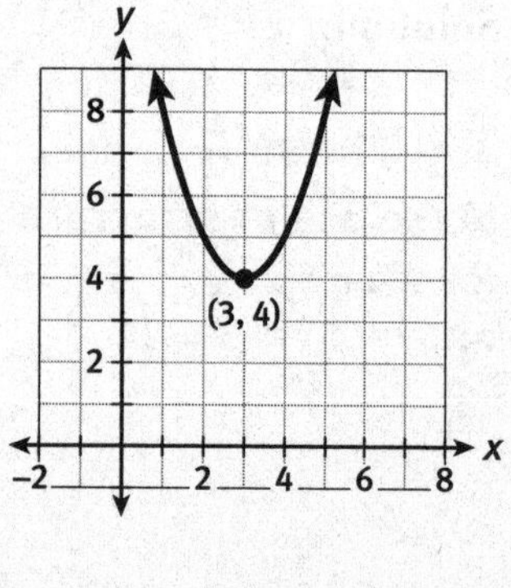

My Notes

Try These B

Write each of the following quadratic functions in the form $y = a(x - h)^2 + k$. Identify the direction of opening, vertex, maximum or minimum value, and number of x-intercepts.

a. $y = x^2 - 4x + 9$

b. $y = -x^2 - 6x - 8$

c. $y = x^2 + 8x + 15$

Check Your Understanding

Solve by completing the square.

1. $x^2 + 2x + 3 = 0$

2. $x^2 + 6x + 4 = 0$

LESSON 32-2 PRACTICE

Solve by completing the square.

3. $2 = x^2 - 10x$

4. $4x = x^2 - 4x - 32$

5. $-2x^2 + 4 = -x^2 + x - 7$

6. $x + 1 = 6x - x^2$

Complete the square to determine the vertex and maximum or minimum value. Determine the number of x-intercepts.

7. $y = x^2 - 2x + 2$

8. $y = -x^2 + 8x - 6$

9. **Make sense of problems.** In a model railroad, the track is supported by an arch that is represented by $y = -x^2 + 10x - 16$, where y represents the height of the arch in inches and x represents the distance in inches from a cliff. Complete the square to answer the following questions.

a. How far is the center of the arch from the cliff?

b. What is the maximum height of the arch?

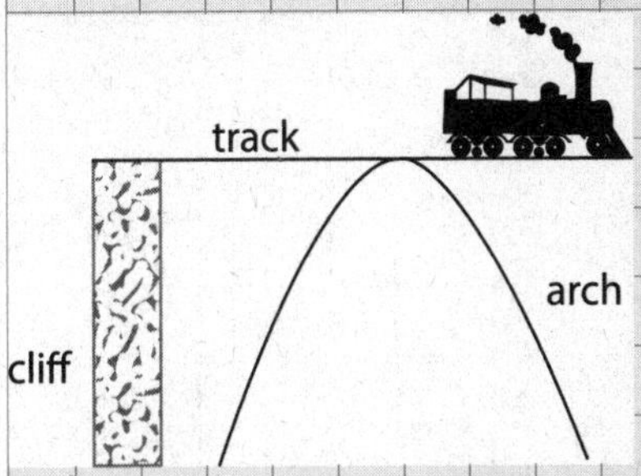

10. The bubbler is the part of a drinking fountain that produces a stream of water. The water in a drinking fountain follows a path given by $y = -x^2 + 6x + 4.5$, where y is the height of the water in centimeters above the basin, and x is the distance of the water from the bubbler. What is the maximum height of the water above the basin?

My Notes

Learning Targets:

- Derive the quadratic formula.
- Solve quadratic equations using the quadratic formula.

SUGGESTED LEARNING STRATEGIES: Close Reading, Note Taking, Identify a Subtask

Generalizing a solution method into a formula provides an efficient way to perform complicated procedures. You can complete the square on the general form of a quadratic equation $ax^2 + bx + c = 0$ to find a formula for solving all quadratic equations.

$$ax^2 + bx + c = 0$$

$$x^2 + \frac{b}{a}x = -\frac{c}{a}$$

____

$$x^2 + \frac{b}{a}x + \left(\frac{b}{2a}\right)^2 = -\frac{c}{a} + \left(\frac{b}{2a}\right)^2$$

____

$$x + \frac{b}{2a} = \pm\sqrt{\frac{b^2 - 4ac}{4a^2}}$$

____

$$x = \frac{-b \pm \sqrt{b^2 - 4ac}}{2a}$$

Quadratic Formula

When $a \neq 0$, the solutions of $ax^2 + bx + c = 0$ are

$$x = \frac{-b \pm \sqrt{b^2 - 4ac}}{2a}.$$

My Notes

To apply the quadratic formula, make sure the equation is in standard form $ax^2 + bx + c = 0$. Identify the values of a, b, and c in the equation and then substitute these values into the quadratic formula. If the expression under the radical sign is not a perfect square, write the solutions in simplest radical form or use a calculator to approximate the solutions.

Example A

Solve $x^2 + 3 = 6x$ using the quadratic formula.

Step 1: Write the equation in standard form.

$$x^2 + 3 = 6x$$
$$x^2 - 6x + 3 = 0$$

Step 2: Identify a, b, and c.

$$a = 1, b = -6, c = 3$$

Step 3: Substitute these values into the quadratic formula.

$$x = \frac{-(-6) \pm \sqrt{(-6)^2 - 4(1)(3)}}{2(1)}$$

Step 4: Simplify using the order of operations.

$$x = \frac{6 \pm \sqrt{36 - 12}}{2} = \frac{6 \pm \sqrt{24}}{2}$$

Step 5: Write as two solutions.

$$x = \frac{6 + \sqrt{24}}{2} \text{ or } x = \frac{6 - \sqrt{24}}{2}$$

Solution: Use a calculator to approximate the two solutions.
$x \approx 5.45$ or $x \approx 0.55$

If you do not have a calculator, write your solution in simplest radical form. To write the solution in simplest form, simplify the radicand and then divide out any common factors.

$$x = \frac{6 \pm \sqrt{24}}{2} = \frac{6 \pm \sqrt{4 \cdot 6}}{2} = \frac{6 \pm 2\sqrt{6}}{2} = \frac{2(3 \pm \sqrt{6})}{2} = 3 \pm \sqrt{6}$$

Try These A

Solve using the quadratic formula.

a. $3x^2 = 4x + 3$

b. $x^2 + 4x = -2$

My Notes

Check Your Understanding

Solve using the quadratic formula.

1. $3x^2 - 5x + 1 = 0$
2. $x^2 + 6 = -8x + 12$

LESSON 32-3 PRACTICE

Solve using the quadratic formula.

3. $x^2 + 5x - 1 = 0$
4. $-2x^2 - x + 4 = 0$
5. $4x^2 - 5x - 2 = 1$
6. $x^2 + 3x = -x + 1$
7. $3x^2 = -6x + 4$
8. A baseball player tosses a ball straight up into the air. The function $y = -16x^2 + 30x + 5$ models the motion of the ball, where x is the time in seconds and y is the height of the ball, in feet.
   a. Write an equation you can solve to find out when the ball is at a height of 15 feet.
   b. Use the quadratic formula to solve the equation. Round to the nearest tenth.
   c. How many solutions did you find for Part (b)? Explain why this makes sense.
9. **Critique the reasoning of others.** José and Marta each solved $x^2 + 4x = -3$ using two different methods. Who is correct and what is the error in the other student's work?

José

$$x^2 + 4x = -3$$
$$x^2 + 4x - 3 = 0$$
$$a = 1, b = 4, c = -3$$
$$x = \frac{-4 \pm \sqrt{4^2 - 4(1)(-3)}}{2(1)}$$
$$= \frac{-4 \pm \sqrt{16 + 12}}{2}$$
$$= \frac{-4 \pm \sqrt{28}}{2}$$
$$= \frac{-4 \pm 2\sqrt{7}}{2} = -2 \pm \sqrt{7}$$

Marta

$$x^2 + 4x = -3$$
$$x^2 + 4x + \square = -3 + \square$$
$$x^2 + 4x + 4 = -3 + 4$$
$$(x + 2)^2 = 1$$
$$x + 2 = \pm\sqrt{1}$$
$$x + 2 = 1 \text{ or } x + 2 = -1$$
$$x = -1 \text{ or } x = -3$$

My Notes

Learning Targets:

- Choose a method to solve a quadratic equation.
- Use the discriminant to determine the number of real solutions of a quadratic equation.

SUGGESTED LEARNING STRATEGIES: Think-Pair-Share, Graphic Organizer, Look for a Pattern, Create a Plan, Quickwrite

There are several methods for solving a quadratic equation. They include factoring, using square roots, completing the square, and using the quadratic formula. Each of these techniques has different advantages and disadvantages. Learning how and why to use each method is an important skill.

1. Solve each equation below using a different method. State the method used.

a. $x^2 + 5x - 24 = 0$

b. $x^2 - 6x + 2 = 0$

c. $2x^2 + 3x - 5 = 0$

d. $x^2 - 100 = 0$

2. How did you decide which method to use for each equation in Item 1?

My Notes

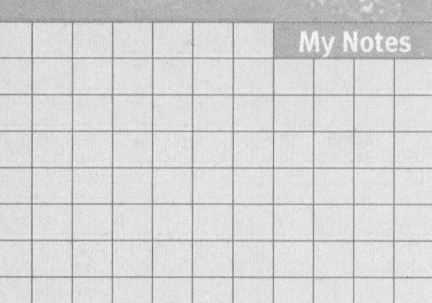

The expression $\sqrt{b^2 - 4ac}$ in the quadratic formula helps you understand the nature of the quadratic equation. The ***discriminant***, $b^2 - 4ac$, of a quadratic equation gives information about the number of real solutions, as well as the number of x-intercepts of the related quadratic function.

3. Solve each equation using any appropriate solution method. Then complete the rest of the table.

Equation	Discriminant	Solutions	Number of Real Solutions	Number of x-Intercepts	Graph of Related Quadratic Function
$x^2 + 2x - 8 = 0$					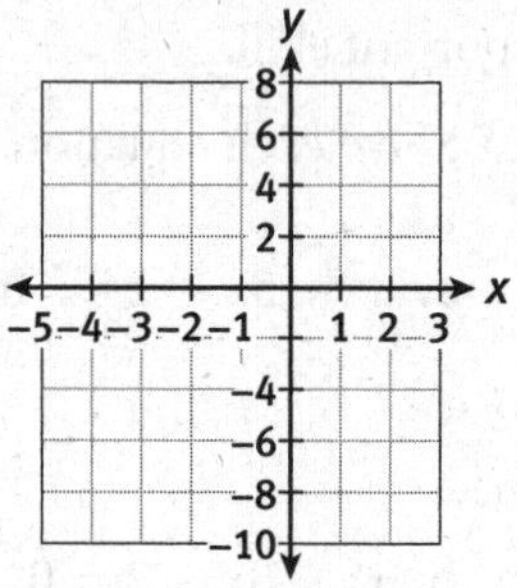
$x^2 + 2x + 1 = 0$					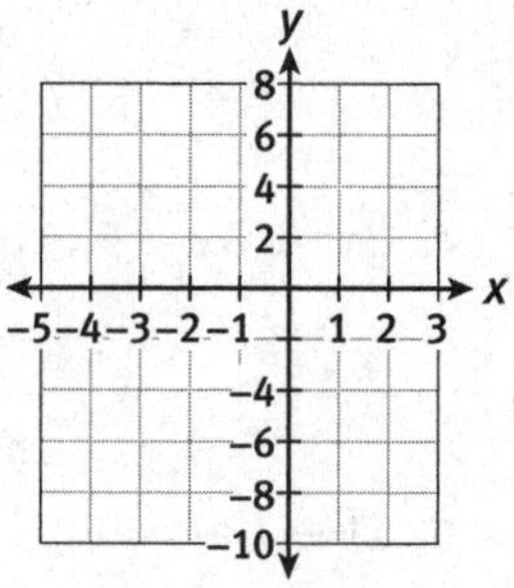
$x^2 + 2x + 5 = 0$					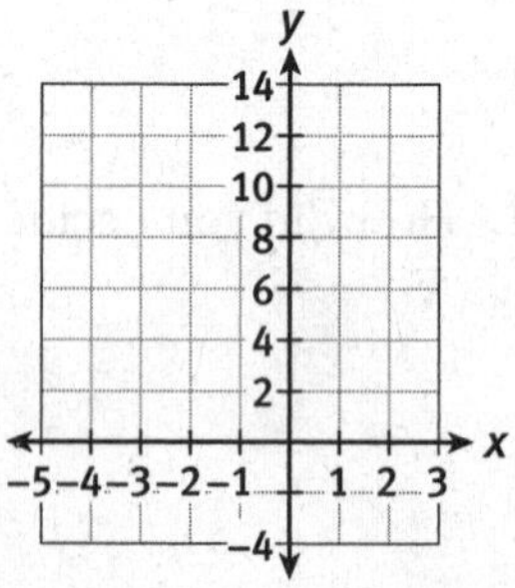

My Notes

4. Express regularity in repeated reasoning. Complete each statement below using the information from the table in Item 3.

- If $b^2 - 4ac > 0$, the equation has ______ real solution(s) and the graph of the related function has ______ x-intercept(s).
- If $b^2 - 4ac = 0$, the equation has ______ real solution(s) and the graph of the related function has ______ x-intercept(s).
- If $b^2 - 4ac < 0$, the equation has ______ real solution(s) and the graph of the related function has ______ x-intercept(s).

Check Your Understanding

Use the discriminant to determine the number of real solutions.

5. $4x^2 + 2x - 12 = 0$

6. $x^2 - 7x + 14 = 0$

7. $x^2 - 10x + 25 = 0$

LESSON 32-4 PRACTICE

For each equation, use the discriminant to determine the number of real solutions. Then solve the equation.

8. $x^2 - 1 = 0$

9. $x^2 - 4x + 4 = 0$

10. $-4x^2 + 3x = -2$

11. $x^2 - 2 = 12x$

12. $x^2 + 5x - 1 = 0$

13. $-x^2 - 2x - 10 = 0$

14. Model with mathematics. Lin launches a model rocket that follows a path given by the function $y = -0.4t^2 + 3t + 0.5$, where y is the height in meters and t is the time in seconds.

a. Explain how you can write an equation and then use the discriminant to determine whether Lin's rocket ever reaches a height of 5 meters.

b. If Lin's rocket reaches a height of 5 meters, at approximately what time(s) does it do so? If not, what is the rocket's maximum height?

My Notes

Learning Targets:

- Use the imaginary unit i to write complex numbers.
- Solve a quadratic equation that has complex solutions.

SUGGESTED LEARNING STRATEGIES: Think-Pair-Share, Close Reading, Note Taking, Construct an Argument, Identify a Subtask

When solving quadratic equations, there are always one, two, or no real solutions. Graphically, the number of x-intercepts is helpful for determining the number of real solutions.

- When there is one real solution, the graph of the related quadratic function touches the x-axis once, and the vertex of the parabola is on the x-axis.
- When there are two real solutions, the graph crosses the x-axis twice.
- When there are no real solutions, the graph never crosses the x-axis.

1. Graph the function $y = x^2 - 6x + 13$. Use the graph to determine the number of real solutions to the equation $x^2 - 6x + 13 = 0$.

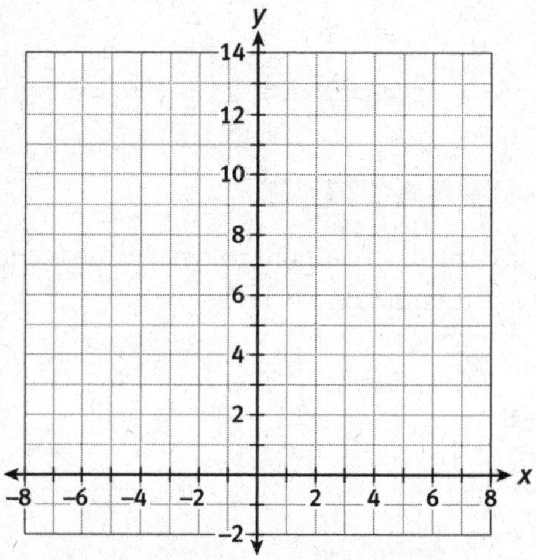

2. Construct viable arguments. What does the number of real solutions to the equation in Item 1 indicate about the value of the discriminant of the equation? Explain.

When the value of the discriminant is less than zero, there are no *real* solutions. This is different from stating there are *no* solutions. In cases where the discriminant is negative, there are two solutions that are *not* real numbers. ***Imaginary numbers*** offer a way to determine these non-real solutions. The ***imaginary unit***, i, equals $\sqrt{-1}$. Imaginary numbers are used to represent square roots of negative numbers, such as $\sqrt{-4}$.

My Notes

Example A

Simplify $\sqrt{-4}$.

Step 1: Write the radical as a product involving $\sqrt{-1}$. $\sqrt{-4} = \sqrt{-1} \cdot \sqrt{4}$

Step 2: Replace $\sqrt{-1}$ with the imaginary unit, *i*. $= i\sqrt{4}$

Step 3: Simplify the radical. The principal square root of 4 is 2. $= 2i$

Solution: $\sqrt{-4} = 2i$

Try These A

Simplify.

a. $\sqrt{-16}$ **b.** $-\sqrt{-9}$ **c.** $\sqrt{-8}$

Problems involving imaginary numbers can also result in ***complex numbers***, $a + bi$, where a and b are real numbers. In this form, a is the real part and b is the imaginary part.

Example B

Solve $x^2 - 6x + 12 = 0$.

Step 1: Identify *a*, *b*, and *c*. $a = 1, b = -6, c = 12$

Step 2: Substitute these values into the quadratic formula. $x = \dfrac{-(-6) \pm \sqrt{(-6)^2 - 4(1)(12)}}{2(1)}$

Step 3: Simplify using the order of operations. $x = \dfrac{6 \pm \sqrt{36 - 48}}{2} = \dfrac{6 \pm \sqrt{-12}}{2}$

Step 4: Simplify using the imaginary unit, *i*. $x = \dfrac{6 \pm \sqrt{-1}\sqrt{12}}{2} = \dfrac{6 \pm i\sqrt{12}}{2}$

Step 5: Simplify the radical and the fraction. $x = \dfrac{6 \pm 2i\sqrt{3}}{2} = 3 \pm i\sqrt{3}$

Solution: $x = 3 \pm i\sqrt{3}$

Try These B

Solve each equation.

a. $x^2 + 100 = 0$ **b.** $x^2 - 4x = -11$

WRITING MATH

Notice that the solution to Example A is written $2i$ and not $i2$. However, when a value includes a radical and i, i is written in front of the radical, as in $i\sqrt{3}$.

My Notes

Check Your Understanding

3. Simplify $\sqrt{-27}$.

4. Solve $x^2 + 4x + 6 = 0$.

LESSON 32-5 PRACTICE

Simplify.

5. $-\sqrt{-11}$

6. $\sqrt{-42}$

7. $\pm\sqrt{-81}$

Solve.

8. $5x^2 - 2x + 3 = 0$

9. $-x^2 - 6 = 0$

10. $(x - 1)^2 + 3 = 0$

11. **Make use of structure.** Consider the quadratic function $y = x^2 + 2x + c$, where c is a real number.

a. Write and simplify an expression for the discriminant.

b. Explain how you can use your result from Part (a) to write and solve an inequality that tells you when the function will have two zeros that involve imaginary numbers.

c. Use your results to describe the zeros of the function $y = x^2 + 2x + 3$.

ACTIVITY 32 PRACTICE

Write your answers on notebook paper. Show your work.

Lesson 32-1

Solve each equation using square roots.

1. $x^2 + 7 = 43$
2. $(x - 5)^2 + 2 = 11$
3. $x^2 - 8x + 16 = 3$
4. Antonio drops a rock from a cliff that is 400 feet high. The function $y = -16t^2 + 400$ gives the height of the rock in feet after t seconds. Write and solve an equation to determine how long it takes the rock to land at the base of the cliff. (*Hint*: At the base of the cliff, the height y is 0.)
5. Maya wants to use square roots to solve the equation $x^2 - 6x + 9 = k$, where k is a positive real number. Which of these is the best representation of the solution?
   A. $x = 3 \pm \sqrt{k}$ **B.** $x = -3 \pm \sqrt{k}$
   C. $x = \pm\sqrt{k + 3}$ **D.** $x = \pm\sqrt{k - 3}$

Lesson 32-2

6. Given the equation $x^2 - 8x = 3$, what number should be added to both sides to complete the square?
   A. -4 **B.** 8
   C. 16 **D.** 64

Write each of the following equations in the form $y = a(x - h)^2 + k$. Then identify the direction of opening, vertex, maximum or minimum value, and x-intercepts.

7. $y = x^2 - 4x + 11$
8. $y = -x^2 - 6x - 8$
9. $y = x^2 + 2x - 8$
10. A golfer stands on a platform 16 feet above a driving range. Once the golf ball is hit, the function $y = -16t^2 + 64t + 16$ represents the height of the ball in feet after t seconds.
    a. Write an equation you can solve to determine the number of seconds it takes for the ball to land on the driving range.
    b. Solve the equation by completing the square. Leave your answer in radical form.
    c. Use a calculator to find the number of seconds, to the nearest tenth, that it takes the ball to land on the driving range.
11. Which of the following is a true statement about the graph of the quadratic function $y = x^2 - 2x + 3$?
    A. The vertex of the graph is $(-1, 2)$.
    B. The graph intersects the x-axis at $x = 1$.
    C. The graph is a parabola that opens upward.
    D. There is exactly one x-intercept.

Solve by completing the square.

12. $x^2 - 4x = 12$
13. $x^2 + 10x + 21 = 0$
14. $2x^2 - 4x - 4 = 0$
15. $x^2 + 6x = -10$
16. A climbing structure at a playground is represented by the function $y = -x^2 + 4x + 1$, where y is the height of the structure in feet and x is the distance in feet from a wall. What is the maximum height of the structure?
    A. 1 foot **B.** 2 feet
    C. 4 feet **D.** 5 feet

Lesson 32-3

Solve using the quadratic formula.

17. $4x^2 - 4x = 3$

18. $5x^2 - 9x - 2 = 0$

19. $x^2 = 2x + 4$

20. A football player kicks a ball. The function $y = -16t^2 + 32t + 3$ models the motion of the ball, where t is the time in seconds and y is the height of the ball in feet.

a. Write an equation you can solve to find out when the ball is at a height of 11 feet.

b. Use the quadratic formula to solve the equation. Round to the nearest tenth.

21. Kyla was asked to solve the equation $2x^2 + 6x - 1 = 0$. Her work is shown below. Is her solution correct? If not, describe the error and give the correct solution.

$2x^2 + 6x - 1 = 0$

$a = 2, b = 6, c = -1$

$$x = \frac{-6 \pm \sqrt{6^2 - 4(2)(-1)}}{2(2)}$$

$$= \frac{-6 \pm \sqrt{36 - 8}}{4}$$

$$= \frac{-6 \pm \sqrt{28}}{4}$$

$$= \frac{-6 \pm 2\sqrt{7}}{4}$$

$$= \frac{-3 \pm \sqrt{7}}{2}$$

Lesson 32-4

Use the discriminant to determine the number of real solutions.

22. $x^2 + 3x + 5 = 0$

23. $4x^2 - 4x + 1 = 0$

24. The discriminant of a quadratic equation is -1. Which of the following must be a true statement about the graph of the related quadratic function?

A. The graph intersects the x-axis in exactly two points.

B. The graph lies entirely above the x-axis.

C. The graph intersects the x-axis at $x = -1$.

D. The graph has no x-intercepts.

25. A dolphin jumps straight up from the water. The quadratic function $y = -16t^2 + 20t$ models the motion of the dolphin, where t is the time in seconds and y is the height of the dolphin, in feet. Use the discriminant to explain why the dolphin does not reach a height of 7 feet.

Lesson 32-5

Simplify.

26. $\pm\sqrt{-2}$

27. $-\sqrt{-25}$

28. $\sqrt{-8}$

29. $-\sqrt{-121}$

30. $12 - \sqrt{-144}$

31. $\pm\sqrt{-32}$

Solve.

32. $2x^2 - 5x + 5 = 0$

33. $x^2 + x + 3 = 0$

34. $-3x^2 - 3x - 1 = 0$

35. $-x^2 - x - 2 = 0$

MATHEMATICAL PRACTICES
Construct Viable Arguments and Critique the Reasoning of Others

36. For what values of p does the quadratic function $y = x^2 + 4x + p$ have two real zeros? Justify your answer.

Applying Quadratic Equations

Rockets in Flight

Lesson 33-1 Fitting Data with a Quadratic Function

My Notes

Learning Targets:

- Write a quadratic function to fit data.
- Use a quadratic model to solve problems.

SUGGESTED LEARNING STRATEGIES: Create Representations, Look for a Pattern, Summarizing, Predict and Confirm, Discussion Groups

Cooper is a model rocket fan. Cooper's model rockets have single engines and, when launched, can rise as high as 1000 ft depending upon the engine size. After the engine is ignited, it burns for 3–5 seconds, and the rocket accelerates upward. The rocket has a parachute that will open as the rocket begins to fall back to Earth.

Cooper wanted to investigate the flight of a rocket from the time the engine burns out until the rocket lands. He set a device in a rocket, named *Spirit*, to begin collecting data the moment the engine shut off. Unfortunately, the parachute failed to open. When the rocket began to descend, it was in *free fall*.

The table shows the data that was collected.

The *Spirit*	
Time Since the Engine Burned Out (s)	Height (ft)
0	512
1	560
2	576
3	560
4	512
5	432
6	320

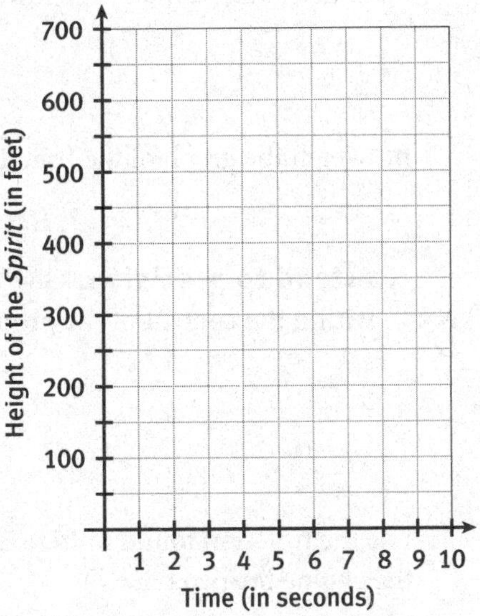

1. a. Use the table to determine whether the height of the *Spirit* can be modeled by a linear function.

b. Graph the data for the height of the *Spirit* as a function of time on the grid.

CONNECT TO SCIENCE

A *free falling* object is an object that is falling under the sole influence of gravity. The approximate value of acceleration due to gravity for an object in free fall on Earth is 32 ft/s^2 or 9.8 m/s^2.

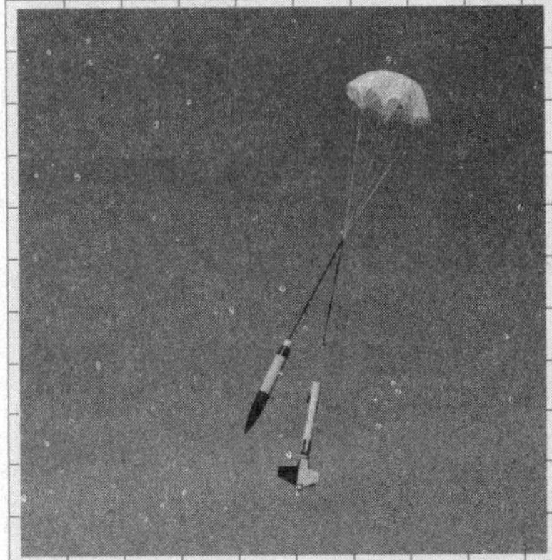

My Notes

2. Use the table and graph from Item 1.

   a. How high was the *Spirit* when the engine burned out?

   b. How long did it take the rocket to reach its maximum height after the engine cut out?

   c. Estimate the time the rocket was in free fall before it reached the earth.

CONNECT TO TECHNOLOGY

For Item 3a, enter the data from the table above Item 1 into a graphing calculator. Use the calculator's quadratic regression feature to find a representative function.

3. Use the table and graph from Item 1.

   a. **Use appropriate tools strategically.** Use a graphing calculator to determine a quadratic function $h(t)$ for the data.

   b. Sketch the graph of the function on the grid in Item 1.

   c. **Attend to precision.** Give a reasonable domain and range for $h(t)$ within the context of the problem. Be sure to include units.

4. Use the function found in Item 3 to verify the height of the *Spirit* when the engine burned out.

5. **Construct viable arguments.** Use the graph of $h(t)$ to approximate the time interval in which the *Spirit* was in free fall. Explain how you determined your answer.

My Notes

Check Your Understanding

Use the data in the table for Items 6–8.

x	0	1	2	3
$f(x)$	5	4.5	3	0.5

6. Graph the data in the table. Do the data appear to be linear? Explain.

7. Use a graphing calculator to write a function to model $f(x)$.

8. Use the function to determine the value of $f(x)$ when $x = 1.5$.

LESSON 33-1 PRACTICE

9. **Model with mathematics.** Cooper wanted to track another one of his rockets, the *Eagle*, so that he could investigate its time and height while in flight. He installed a device into the nose of the *Eagle* to measure the time and height of the rocket as it fell back to Earth. The device started measuring when the parachute opened. The data for one flight of the *Eagle* is shown in the table below.

The *Eagle*

Time Since Parachute Opened (s)	0	1	2	3	4	5	6	7	8	9
Height(ft)	625	618	597	562	513	450	373	282	177	58

a. Graph the data from the table.
b. Use a graphing calculator to determine a quadratic function $h(t)$ that models the data.

10. Use the function from Item 9b to find the time when the rocket's height is 450 ft. Verify that your result is the same as the data in the table.

11. After the parachute opened, how long did it take for the rocket to hit Earth?

My Notes

Learning Targets:

- Solve quadratic equations.
- Interpret the solutions of a quadratic equation in a real-world context.

SUGGESTED LEARNING STRATEGIES: Create Representations, Predict and Confirm, Think-Pair-Share, Discussion Groups

1. The total time that the *Spirit* was in the air after the engine burned out is determined by finding the values of t that make $h(t) = 0$.

a. Rewrite the equation identified in Item 3a in Lesson 33-1 and set it equal to 0.

b. Completely factor the equation.

c. Make use of structure. Identify and use the appropriate property to find the time that the *Spirit* took to strike Earth after the engine burned out.

2. Draw a horizontal line on the graph in Item 1 in Lesson 33-1 to indicate a height of 544 ft above Earth. Estimate the approximate time(s) that the *Spirit* was 544 ft above Earth.

My Notes

3. The time(s) that the *Spirit* was 544 ft above Earth can be determined exactly by finding the values of t that make $h(t) = 544$.

a. Rewrite the equation from Item 3a in Lesson 33-1 and set it equal to 544.

b. Is the method of factoring effective for solving this equation? Justify your response.

c. Is the quadratic formula effective for solving this problem? Justify your response.

d. Determine the time(s) that the rocket was 544 ft above Earth. Round your answer to the nearest thousandth of a second. Verify that this solution is reasonable compared to the estimated times from the graph in Item 2.

CONNECT TO AP

In AP Calculus, calculations are approximated to the nearest thousandth.

e. Attend to precision. Explain why it is more appropriate in this context to round to the thousandths place rather than to use the exact answer or an approximation to the nearest whole number.

4. Cooper could not see the *Spirit* when it was higher than 528 ft above Earth.

a. Calculate the values of t for which $h(t) = 528$.

b. Reason quantitatively. Write an inequality to represent the values of t for which the rocket was not within Cooper's sight.

My Notes

Check Your Understanding

The path of a rocket is modeled by $h(t) = -16t^2 + 45t + 220$, where h is the height in feet and t is the time in seconds.

5. Determine the time t when the rocket was on the ground. Round to the nearest thousandth.

6. Identify the times, t, when the height $h(t)$ was greater than 220 ft.

LESSON 33-2 PRACTICE

Solve the quadratic equations. Round to the nearest thousandth.

7. $-16t^2 + 100t + 316 = 0$

8. $-16t^2 + 100t + 316 = 100$

Make sense of problems. For Items 9–12, use the function $h(t) = -16t^2 + 8t + 30$, which represents the height of a diver above the surface of a swimming pool t seconds after she dives.

9. The diver begins her dive on a platform. What is the height of the platform above the surface of the swimming pool? How do you know?

10. At what time does the diver reach her maximum height? What is the maximum height?

11. How long does it take the diver to reach the water? Round to the nearest thousandth.

12. Determine the times at which the diver is at a height greater than 20 ft. Explain how you arrived at your solution.

ACTIVITY 33 PRACTICE

**Write your answers on notebook paper.
Show your work.**

Lesson 33-1

A model rocket is launched from the ground. Its height at different times after the launch is recorded in the table below. Use the table for Items 1–7.

Time Since Launch of Rocket (s)	Height of Rocket (ft)
1	144
2	256
3	336
4	384
5	400
6	384
7	336

1. Are the data in the table linear? Justify your response.
2. Use a graphing calculator to determine a quadratic function $h(t)$ for the data.
3. Identify a reasonable domain for the function within this context.
4. How high is the rocket after 8 seconds?
5. After how many seconds does the rocket come back to the ground?
6. What is the maximum height the rocket reached?
7. a. At what time(s) t will the height of the rocket be equal to 300 ft?
   b. How many times did you find in Part (a)? Explain why this makes sense in the context of the problem.
8. Which of the following functions best models the data in the table below?

x	$f(x)$
1	29
2	66
3	101
4	134
5	165
6	194

A. $f(x) = x + 28$
B. $f(x) = -x^2 + 40x - 10$
C. $f(x) = -16x^2 + 10x + 100$
D. $f(x) = -|x + 3| + 42$

9. As part of a fireworks display, a pyrotechnics team launches a fireworks shell from a platform and collects the following data about the shell's height.

Time Since Launch of Shell (s)	Height of Shell (ft)
1	68
2	100
3	100
4	68
5	4

Which of the following is a true statement about this situation?
A. The launch platform is 68 ft above the ground.
B. The maximum height of the shell is 100 ft.
C. The shell hits the ground after 6 seconds.
D. The shell starts to descend 2.5 seconds after it is launched.

Lesson 33-2

A soccer player passes the ball to a teammate, and the teammate kicks the ball. The function $h(t) = -16t^2 + 14t + 4$ represents the height of the ball, in feet, t seconds after it is kicked. Use this information for Items 10–16.

10. What is the height of the ball at the moment it is kicked? Justify your answer.

11. Graph the function.

12. Calculate the time at which the ball reaches its maximum height.

13. What is the maximum height of the ball?

14. Assuming no one touches the ball after it is kicked, determine the time when the ball falls to the ground.

15. Identify a reasonable domain and range for the function.

16. Determine the times when the ball is higher than 6 ft. Explain how you arrived at your solution.

Casey is standing on the roof of a building. She tosses a ball into the air so that her friend Leon, who is standing on the sidewalk, can catch it. The function $y = -16x^2 + 32x + 80$ models the height of the ball in feet, where x is the time in seconds. Use this information for Items 17–20.

17. Leon lets the ball hit the sidewalk. Determine the total time the ball is in the air until it hits the sidewalk.

18. Is the ball ever at a height of 100 ft? Justify your answer.

19. How many times is the ball at a height of exactly 92 ft?

A. never
B. one time
C. two times
D. three times

20. Casey solves the equation shown below. What does the solution of the equation represent?

$$10 = -16x^2 + 32x + 80$$

A. The height of the ball after 10 seconds
B. The time when the ball is at a height of 10 ft
C. The time when the ball has traveled a total distance of 10 ft
D. The time it takes the ball to rise vertically 10 ft from the rooftop

The function $h(t) = -16t^2 + 50t + k$, where $k > 0$, gives the height, in feet, of a marble t seconds after it is shot into the air from a slingshot. Determine whether each statement is always, sometimes, or never true.

21. The initial height of the marble is k feet.

22. There is some value of t for which the height of the marble is 0.

23. The graph of the function is a straight line.

24. The marble reaches a height of 50 ft.

25. The marble reaches a height of 65 ft.

26. The maximum height of the marble occurs at $t = 1$ second.

MATHEMATICAL PRACTICES
Reason Abstractly and Quantitatively

27. Why do you think quadratic functions are used to model free-fall motion instead of linear functions?

Every fall the Physics Club hosts an annual egg-drop contest. The goal of the egg-drop contest is to construct an egg-protecting package capable of providing a safe landing upon falling from a fifth-floor window.

During the egg-drop contest, each contestant drops an egg about 64 ft to a target placed at the foot of a building. The area of the target is about 10 square feet. Points are given for targeting, egg survival, and time to reach the target.

Colin wanted to win the egg-drop contest, so he tested one of his models with three different ways of dropping the package. These equations represent each method.

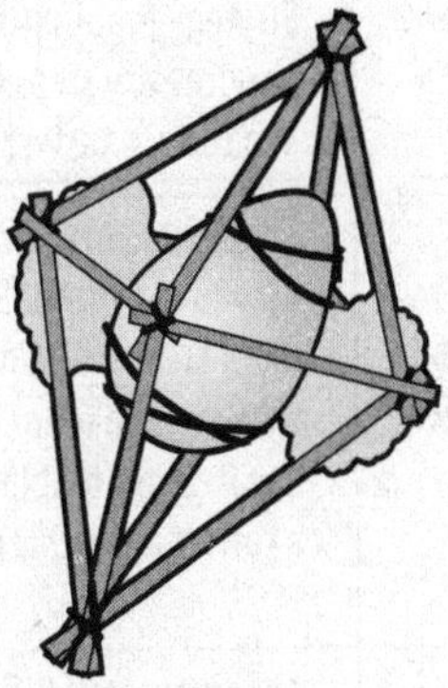

Method A	$h(t) = -16t^2 + 64$
Method B	$h(t) = -16t^2 - 8t + 64$
Method C	$h(t) = -16t^2 - 48t + 64$

1. Quadratic equations can be solved by using square roots, by factoring, or by using the quadratic formula. Solve the three equations above to find t when $h(t) = 0$. Use a different solution method for each equation. Show your work, and explain your reasoning for choosing the method you used.
2. Colin found that the egg would not break if it took longer than 1.5 seconds to hit the ground. Which method(s)—A, B, or C—will result in the egg not breaking?
3. Colin tried another method. This time he recorded the height of the egg at different times after it was dropped, as shown in the table below.

Elapsed Time Since Drop of Egg (s)	Height of Egg (ft)
0	64
0.25	62
0.5	58
0.75	52
1.00	44
1.25	34

   a. Use a graphing calculator to determine a function $h(t)$ that models the quadratic data.
   b. After how many seconds will the egg hit the ground? Show the work that justifies your answer mathematically.
   c. At what time, t, will the egg be 47 feet above the ground? Show the work that justifies your answer mathematically.

Scoring Guide	Exemplary	Proficient	Emerging	Incomplete
	The solution demonstrates the following characteristics:			
Mathematics Knowledge and Thinking (Item 1)	• Effective understanding of and accuracy in solving quadratic equations	• Adequate understanding of how to solve quadratic equations, leading to solutions that are usually correct	• Difficulty solving quadratic equations	• Inaccurate or incomplete understanding of how to solve quadratic equations
Problem Solving (Item 2)	• Appropriate and efficient strategy that results in a correct answer	• Strategy that may include unnecessary steps but results in a correct answer	• Strategy that results in some incorrect answers	• No clear strategy when solving problems
Mathematical Modeling/ Representations (Item 3)	• Clear and accurate understanding of how to use technology to model real-world data and how to use a graph to solve a real-world problem	• Some difficulty understanding how to use technology to model real-world data and/or how to use a graph to solve a real-world problem, but a correct answer is present	• Partial understanding of how to use technology to model real-world data and/or how to use a graph to solve a real-world problem that results in some incorrect answers	• Little or no understanding of how to use technology to model real-world data and/or how to use a graph to solve a real-world problem
Reasoning and Communication (Items 1, 3b, 3c)	• Precise use of appropriate math terms and language to explain a choice of solution method • Clear and accurate use of mathematical work to justify an answer	• Adequate explanation of choice of solution method • Correct use of mathematical work to justify an answer	• Misleading or confusing explanation of choice of solution method • Partially correct justification of an answer using mathematical work	• Incomplete or inaccurate explanation of choice of solution method • Incorrect or incomplete justification of an answer using mathematical work

Modeling with Functions

ACTIVITY 34

Photo App
Lesson 34-1 Constructing Models

My Notes

Learning Targets:

- Construct linear, quadratic, and exponential models for data.
- Graph and interpret linear, quadratic, and exponential functions.

SUGGESTED LEARNING STRATEGIES: Discussion Groups, Look for a Pattern, Create Representations, Sharing and Responding, Construct an Argument

Jenna, Cheyenne, and Kim all use a photo app on their smartphones to share their photos online and to track how many people follow their photo stories. The people who choose to follow the girls' photo stories can also stop following or "unsubscribe" at any time. After the first eight months, the data for each of the girl's monthly number of followers was compiled and is shown in the table below.

Months Since Account Was Opened	Jenna's Total Followers	Cheyenne's Total Followers	Kim's Total Followers
1	37	2	16
2	42	5	29
3	46	9	38
4	52	16	42
5	56	33	43
6	63	65	39
7	66	128	31
8	72	251	20

1. Describe any patterns you observe in the table. As you share your ideas with your group, be sure to use mathematical terms and academic vocabulary precisely. Make notes to help you remember the meaning of new words and how they are used to describe mathematical concepts.

My Notes

2. Reason quantitatively. Use the terms *linear*, *quadratic*, or *exponential* to identify the type of function that would best model each girl's total number of followers during the first eight months.

a. Jenna:

b. Cheyenne:

c. Kim:

3. Use the regression function of a graphing calculator to write a function that could be used to model each girl's total number of followers over the first eight months.

a. Jenna:

b. Cheyenne:

c. Kim:

4. Approximately how many followers does Jenna gain each month? Justify your response.

5. Critique the reasoning of others. Cheyenne tells the other girls she thinks her following is almost doubling each month. Is she correct? Justify your response using Cheyenne's function or the data in the table.

6. Model with mathematics. Use the functions from Item 3 to create graphs to represent each of the girl's total number of followers over the first eight months. Label at least three points on each graph.

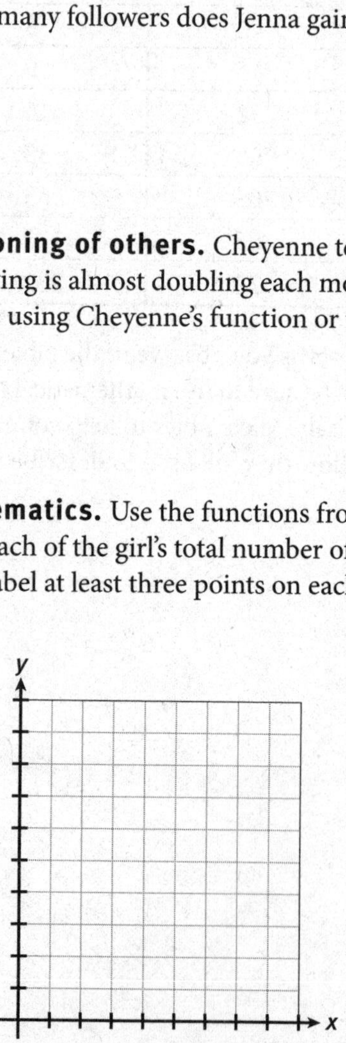

My Notes

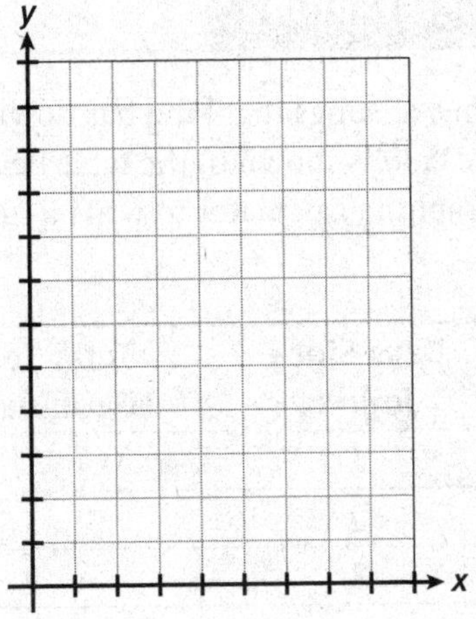

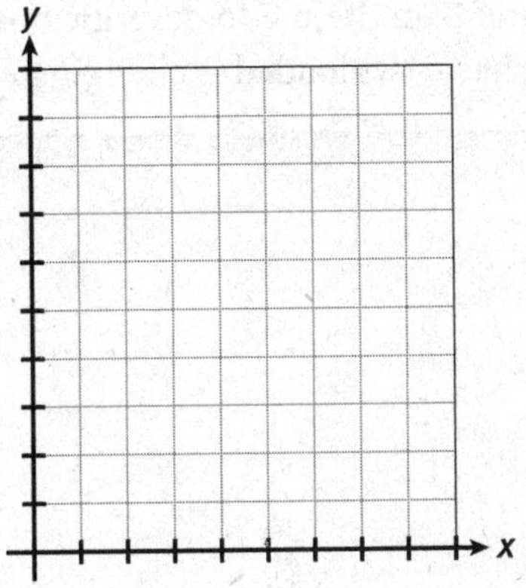

a. Describe the similarities and differences between the reasonable domains and ranges of each of the functions represented by the graphs.

b. Identify the maximum values, if they exist, of each of the functions represented by the graphs.

My Notes

Check Your Understanding

7. The total number of songs that Ping has downloaded since he joined an online music club is shown in the table below. Use the regression function of a graphing calculator to write a function that best models the data.

Days Since Joining	Total Songs Downloaded
1	1
2	3
3	9
4	28
5	81

8. Use your function from Item 7 to describe how the total number of songs that Ping has downloaded is changing each day.

My Notes

LESSON 34-1 PRACTICE

Alice and Ahmad have started an in-home pet grooming business. They each recorded the number of clients they visited each month for the first seven months. Their data are shown in the table below.

Months in Business	Number of Alice's Clients	Number of Ahmad's Clients
1	5	0
2	7	1
3	10	2
4	11	4
5	13	8
6	15	15
7	18	32

9. Examine the data in the table to determine the type of function that would best model the number of each groomer's clients.

10. Use the regression function of a graphing calculator to find a function to represent the number of Alice's clients.

11. Use the regression function of a graphing calculator to find a function to represent the number of Ahmad's clients.

12. Graph the functions from Items 10 and 11 on separate graphs.

13. Attend to precision. Describe the similarities and differences between the two functions. Compare and contrast the following features:

a. domain and range
b. maximum and minimum values
c. rate of increase per day
d. groomer with the most clients during different months

My Notes

Learning Targets:

- Identify characteristics of linear, quadratic, and exponential functions.
- Compare linear, quadratic, and exponential functions.

SUGGESTED LEARNING STRATEGIES: Visualization, Discussion Groups, Construct an Argument, Create Representations

1. Rewrite the functions you found for Jenna, Kim, and Cheyenne in Lesson 34-1.

   Jenna:

   Cheyenne:

   Kim:

2. Which of the three girls—Jenna, Kim, or Cheyenne—has a constant rate of change in her number of followers per month? Explain.

3. Which girl had the greatest number of followers initially? Justify your response using both the functions and the graphs.

4. Did any of the girls experience followers who "unsubscribed" from, or stopped following, their photo story? Explain how you know.

5. Which girl's photo story gained the most followers over the eight months? Justify your response using the functions or the graphs.

6. **Critique the reasoning of others.** Kim states that even if the number of Jenna's followers had grown twice as quickly as it did, Cheyenne's followers would still eventually outnumber Jenna's followers. Is this assumption reasonable? Justify your response.

My Notes

7. Write a new function that describes the total number of Jenna's and Kim's followers combined.

8. Determine the reasonable domain and range, as well as any maximum or minimum values, of the function you wrote in Item 7.

9. **Construct viable arguments.** Will the number of Cheyenne's followers ever exceed the total number of Jenna and Kim's followers combined? Justify your response.

Check Your Understanding

Write the letter of the description that matches the given function.

10. $f(x) = -x^2 - 4x - 3$

11. $f(x) = 4x - 3$

12. $f(x) = 3(4)^x$

A. has a constant rate of change

B. has a maximum value

C. increases very quickly at an ever increasing rate

My Notes

LESSON 34-2 PRACTICE

The graphs show the number of times two online retailers—Roberto's Books and Tyler's Time to Read—recommended the bestselling book *My Story* to their customers after *x* days. Use the graphs for Items 13–16.

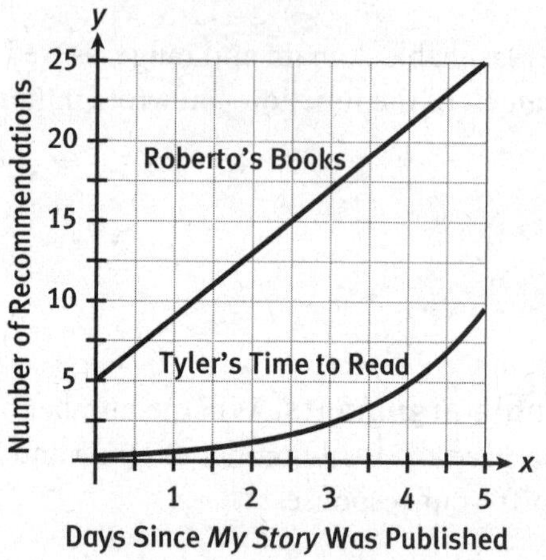

13. Who had given more recommendations of *My Story* after three days?

14. Who recommended *My Story* the same number of times each day? How many times was it recommended each day?

15. If this model continues, Roberto's Books will have recommended *My Story* approximately 1465 times after one year. Is this a reasonable amount? Justify your response.

16. Will the number of times Tyler's Time to Read recommends *My Story* ever exceed the number of times Roberto's Books recommends *My Story*? Explain your reasoning.

17. Make use of structure. The total number of times that Roberto's Books recommended the bestseller *A Fatal Memory* after *x* days is modeled by the function $f(x) = 3x + 2$. The total number of times that Penny's Place recommended the same book after *x* days is modeled by the function $g(x) = -4x^2 + 12$. Write a function to model the combined number of times that Roberto's Books and Penny's Place recommended *A Fatal Memory*.

MATH TIP

You can combine algebraic representations for two different functions by adding, subtracting, multiplying, or dividing the expressions.

My Notes

Learning Targets:

- Compare piecewise-defined, linear, quadratic, and exponential functions.
- Write a verbal description that matches a given graph.

SUGGESTED LEARNING STRATEGIES: Look for a Pattern, Create Representations, Critique Reasoning, Think-Pair-Share, Marking the Text

Rosa begins using the photo app to create a photo story and track her number of followers. The table below shows her results.

Months Since Account Was Started	Rosa's Total Followers
1	25
2	25
3	25
4	25
5	50
6	50
7	50
8	50

1. Describe any patterns you observe in the table.

2. Graph the data in the table.

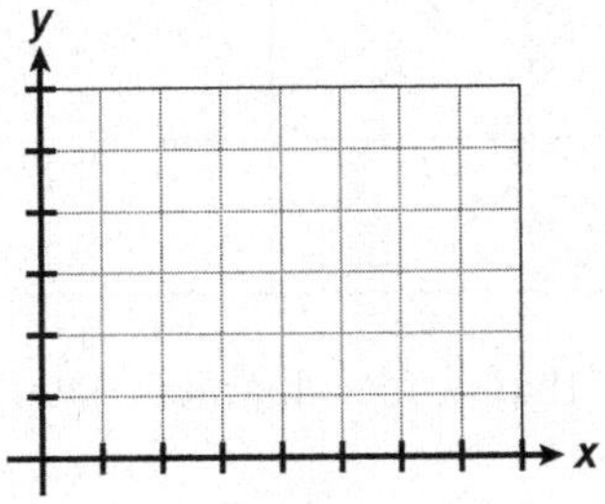

3. Write a function to model Rosa's number of followers for the first four months.

4. Write a function to model Rosa's number of followers for months five through eight.

My Notes

MATH TERMS

A **piecewise-defined function** is a function that is defined differently for different disjoint intervals in its domain. For example, for the piecewise function $f(x)=\begin{cases} x \text{ when } x<0 \\ 2 \text{ when } x\geq 0 \end{cases}$, $f(x)=x$ when $x<0$ and $f(x)=2$ when $x\geq 0$.

5. Write a ***piecewise-defined function*** to represent the number of followers Rosa has in any given month.

6. Critique the reasoning of others. Rosa says that her total number of followers is changing at a constant rate of 25 followers per month for the first four months. Is Rosa's statement correct? Explain your reasoning.

Juanita has recorded the number of followers for her latest photo story over the last seven days. She finds that the function $f(x) = -5|x - 4| + 45$ represents the number of followers after x days.

7. Complete the table for the number of followers each day.

Days Since Photo Story Was Posted	Number of Followers
1	
2	
3	
4	
5	
6	
7	

8. Graph the function that models Juanita's data.

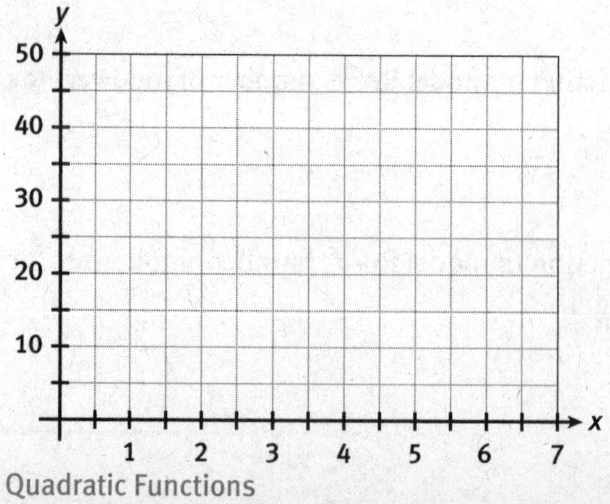

My Notes

9. Determine the reasonable domain and range, as well as the maximum and minimum values of the function within the context of the problem.

10. **Reason abstractly.** Describe the similarities and differences between the function $f(x) = -5|x - 4| + 45$ and a quadratic function.

11. The graphs below show the number of followers for two photo stories over one week. Describe how the number of followers changed over time.

a.

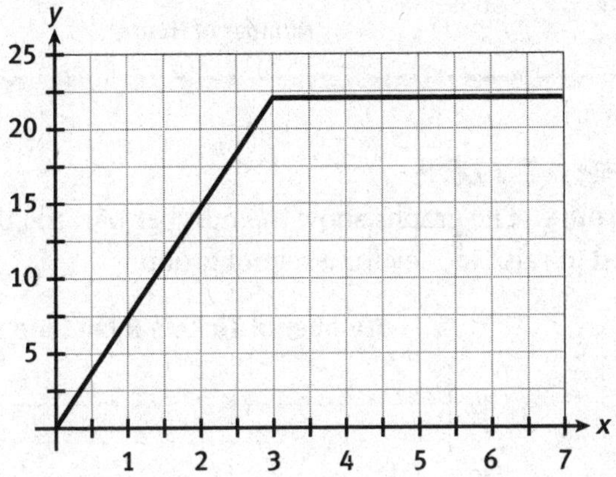

b.

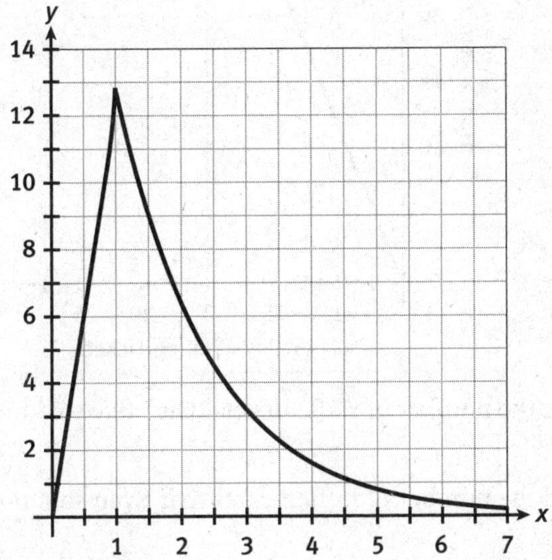

My Notes

Check Your Understanding

Use the graph for Items 12 and 13.

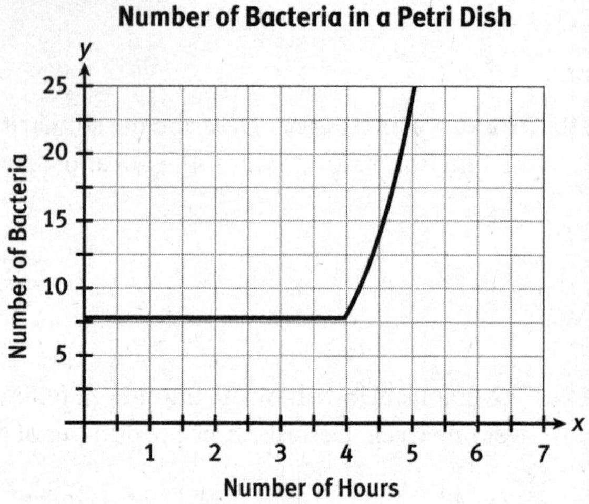

12. During which time period was the number of bacteria constant?

13. Describe the type of change in the number of bacteria for hours four through seven.

LESSON 34-3 PRACTICE

Make sense of problems. The graphs show the number of raffle tickets Mike and Ryan each sold to raise money for a school group.

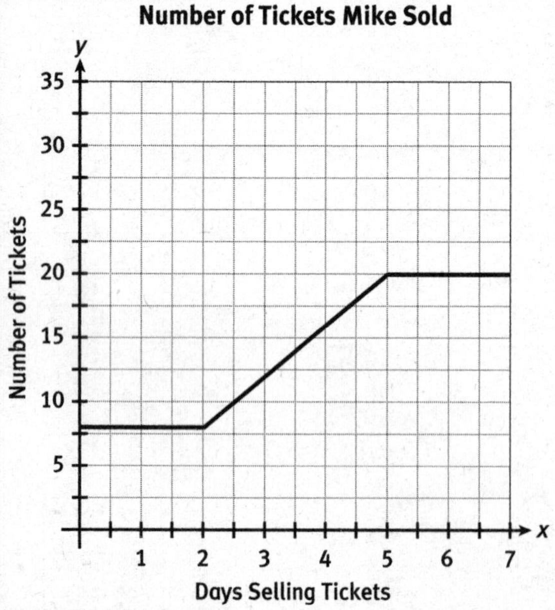

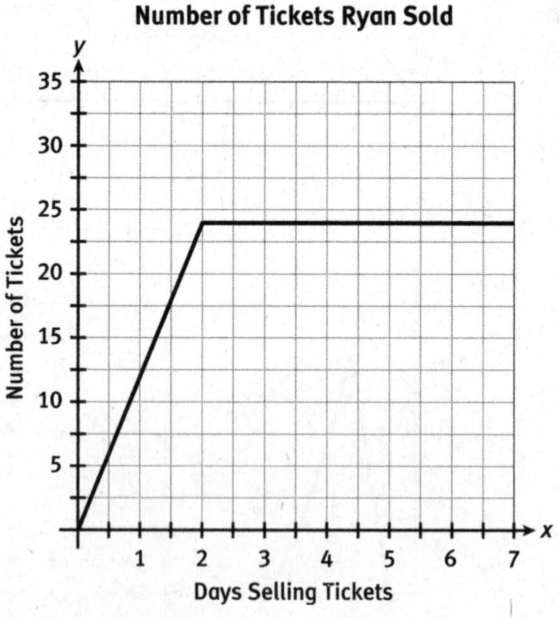

14. For which days did the number of raffle tickets that Mike sold increase at a constant rate?

15. For which days did the number of raffle tickets that Ryan sold not increase?

16. Describe any patterns you see in the number of tickets that Mike sold over the seven days.

17. Compare and contrast any patterns you observe in the number of tickets that Mike sold and the number of tickets Ryan sold over the seven days.

ACTIVITY 34 PRACTICE

Write your answers on notebook paper. Show your work.

Li and Alfonso have both opened accounts with an online music store. The data in the table below show the total number of songs each of them has downloaded since opening his account. Use the table for Items 1–12.

Days Since Account Opened	Total Number of Songs Downloaded by Li	Total Number of Songs Downloaded by Alfonso
1	1	5
2	2	15
3	5	26
4	8	34
5	16	45
6	33	56
7	65	67

Lesson 34-1

1. What type of function would best model the number of songs that Li has downloaded after x days?

A. linear
B. quadratic
C. exponential
D. absolute value

2. What type of function would best model the number of songs that Alfonso has downloaded after x days?

A. linear
B. quadratic
C. exponential
D. absolute value

3. Use the regression function of a graphing calculator to write a function that models the number of songs that Li has downloaded after x days.

4. Use the regression function of a graphing calculator to write a function that models the number of songs that Alfonso has downloaded after x days.

5. Determine the reasonable domain and range for each function.

Lesson 34-2

6. How would the number of songs downloaded by Alfonso change if the rate of change of Alfonso's downloads remained the same, but he had not downloaded any songs on day 1?

7. Describe the similarities and differences between the rates of change in the number of songs downloaded by the two boys.

8. If the models continue to represent the number of songs downloaded, who do you predict will have downloaded more songs after 30 days? Explain your reasoning.

9. If the rate of change of the number of songs downloaded by Alfonso tripled, how many songs will he have downloaded after 30 days? Is this a reasonable number?

10. How many songs will Li have downloaded after 30 days if his model continues? Is this a reasonable number?

Lesson 34-3

Caily opened an account at the same time with the same online music store. The following piecewise function represents the total number of songs she has downloaded over the first x days.

$$f(x) = \begin{cases} 10 \text{ when } 1 \le x < 15 \\ 30 \text{ when } 15 \le x < 25 \\ 45 \text{ when } 25 \le x \le 30 \end{cases}$$

Use this function for Items 11 and 12.

11. Describe the number of songs Caily has downloaded during this time.

12. Describe the difference between the number of songs downloaded by Caily and by Alfonso over the first 30 days.

13. The functions $g(x) = 10x + 2$ and $h(x) = -0.5x^2 + 4x + 30$ represent the total numbers of two different types of fish in a pond over x weeks. Write a function $k(x)$ that represents the combined number of fish during the same period of time.

14. Graph the function $k(x)$ from Item 13. What is the maximum value of the function?

MATHEMATICAL PRACTICES
Construct Viable Arguments and Critique the Reasoning of Others

15. The graph shows the number of "likes" a blogger received for a blog post t days after it was posted.

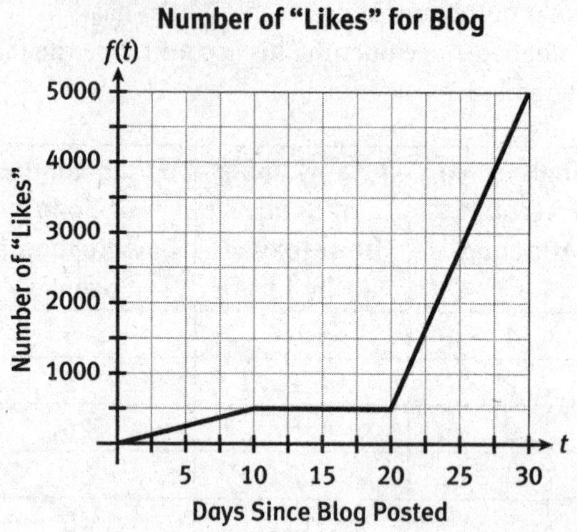

The blogger believes that there may have been something wrong with how the "likes" were recorded between days 10 and 20. Why might she believe this?

Systems of Equations

Population Explosion

Lesson 35-1 Solving a System Graphically

My Notes

Learning Targets:

- Write a function to model a real-world situation.
- Solve a system of equations by graphing.

SUGGESTED LEARNING STRATEGIES: Marking the Text, Create Representations, Sharing and Responding, Discussion Groups, Visualization

Professor Hearst is studying different types of bacteria in order to determine new ways to prevent their population overgrowth. Each bacterium in the first culture that she examines divides to produce another bacterium once each minute. In the second culture, she observes that the number of bacteria increases by 10 bacteria each second.

CONNECT TO BIOLOGY

Bacteria can be harmful or helpful to other organisms, including humans. Bacteria are commonly used for food fermenting, waste processing, and pest control.

1. Each population began with 10 bacteria. Complete the table for the population of each bacteria sample.

Elapsed Time (minutes)	Population of Sample A	Population of Sample B
0	10	10
1		
2		
3		
4		
5		

2. **Reason quantitatively.** Describe the type of function that would best model each population.

My Notes

3. Write a function $A(t)$ to model the number of bacteria present in Sample A after t minutes.

4. Write a function $B(t)$ to model the number of bacteria present in Sample B after t minutes.

5. **Use appropriate tools strategically.** Use a graphing calculator to graph $A(t)$ and $B(t)$ on the same coordinate plane.

   a. Sketch the graph below and label several points on each graph.

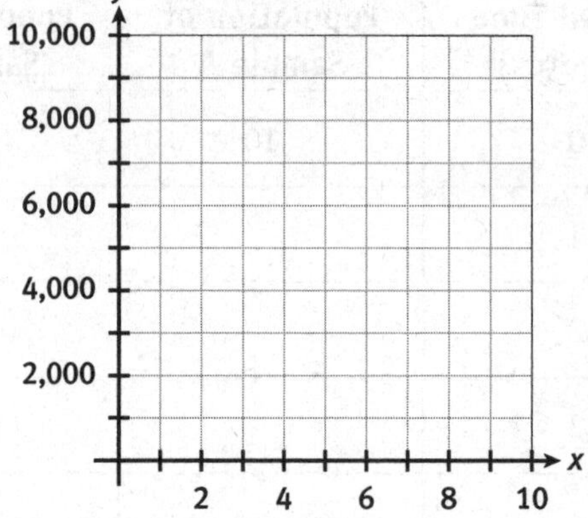

   b. Determine the points of intersection of the two graphs. Round non-integer values to the nearest tenth.

   c. **Make use of structure.** What do the points of intersection indicate about the two graphs? Explain.

   d. Interpret the meaning of the points of intersection within the context of the bacteria samples.

6. Which bacteria population contains more bacteria? Explain.

CONNECT TO TECHNOLOGY

You can use a graphing calculator to determine the point of intersection in several ways. You may choose to graph the functions and determine the point of intersection. You also may use the table feature to identify the value of x when y_1 and y_2 are equal.

My Notes

The solutions you found in Item 5b are solutions to the ***nonlinear system of equations*** $A(t) = 10(2)^t$ and $B(t) = 600t + 10$.

Just as with linear systems, you can solve nonlinear systems by graphing each equation and determining the intersection point(s).

MATH TERMS

A **nonlinear system of equations** is a system in which at least one of the equations is not linear.

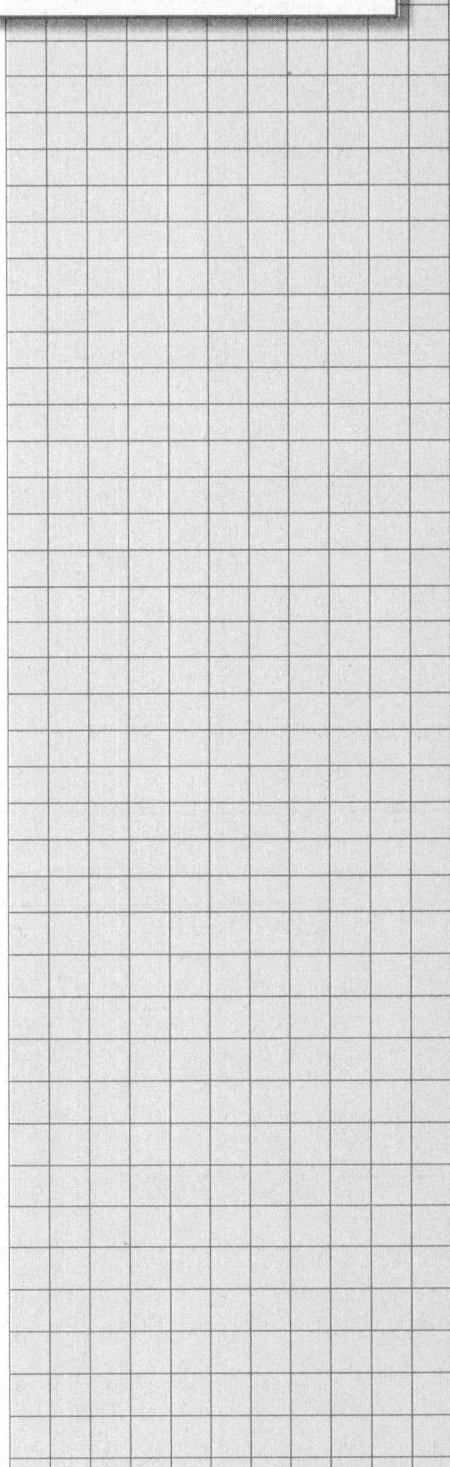

7. Solve each system of equations by graphing.

a. $y = 2x + 1$
$y = x^2 + 1$

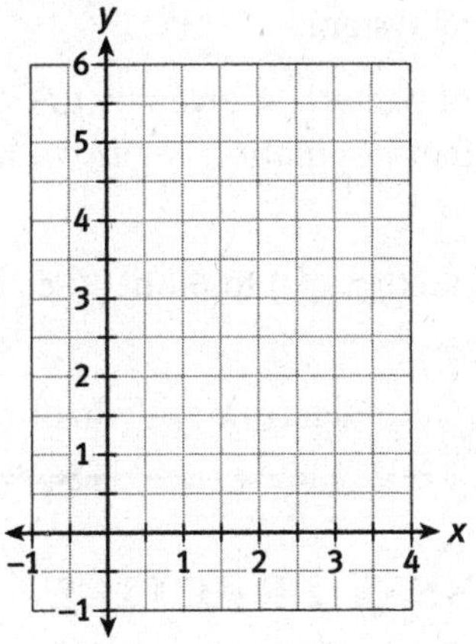

b. $y = x - 1$
$y = 3^x - 2$

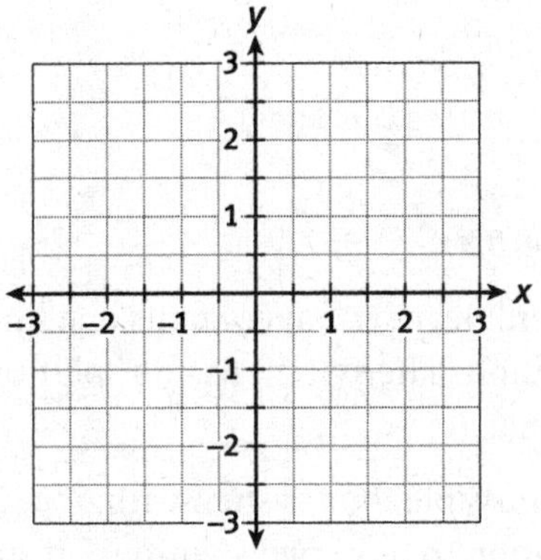

c. $y = 2(5)^x$
$y = -x^2 + 2x - 2$

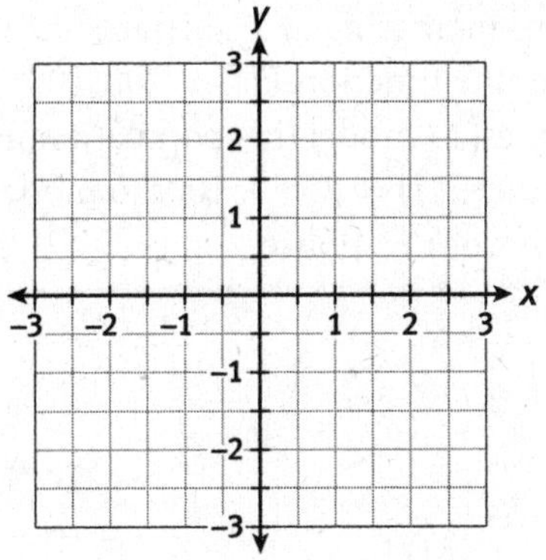

My Notes

Check Your Understanding

8. Examine the graphs in Item 7. How many solutions are possible for a nonlinear system of linear, quadratic, and/or exponential equations? Describe how this is different from the number of possible solutions for a linear system.

A population of bacteria is given by $f(t) = t^2 + 2t + 30$, where t is in minutes. Another population begins with 10 bacteria and doubles every minute.

9. Write a function $g(t)$ to model the population of the second sample of bacteria.

10. Use a graph to determine at what time the two populations are equal.

LESSON 35-1 PRACTICE

For Items 11–13, solve each system of equations by graphing.

11. $y = -2x + 4$
$y = -x^2 + 3$

12. $y = x^2 + 5x - 3$
$y = -x^2 + 5x + 1$

13. $y = -4x^2 - 1$
$y = 4(0.5)^x$

14. A nonlinear system contains one linear equation and one quadratic equation. The system has no solutions. Sketch a possible graph of this system.

15. Is it possible for a system with one linear equation and one exponential equation to have two solutions? If so, sketch a graph that could represent such a system. If not, explain why not.

16. **Critique the reasoning of others.** A population of 200 bacteria begins increasing at a constant rate of 100 bacteria per minute. Francis writes the function $P(t) = 200(100)^t$, where t represents the time in minutes, to model this population. Fred disagrees. He writes the function $P(t) = 200 + 100t$ to model this population. Who is correct? Justify your response.

My Notes

Learning Targets:

- Write a system of equations to model a real-world situation.
- Solve a system of equations algebraically.

SUGGESTED LEARNING STRATEGIES: Create a Plan, Identify a Subtask, Construct an Argument, Close Reading, Note Taking, Visualization, Think-Pair-Share, Discussion Groups

Just as with linear systems, nonlinear systems of equations can be solved algebraically.

Example A

Solve the system of equations algebraically.

$$y = -x + 3$$
$$y = x^2 - 2x - 3$$

Step 1: The first equation is solved for y. Substitute the expression for y into the second equation.
$y = x^2 - 2x - 3$
$(-x+3) = x^2 - 2x - 3$

Step 2: Solve the resulting equation using any method. In this case, solve by factoring.
$0 = x^2 - x - 6$
$0 = (x+2)(x-3)$

Step 3: Apply the Zero Product Property. $x + 2 = 0$ or $x - 3 = 0$

Step 4: Solve each equation for x. $x = -2$ or $x = 3$

Step 5: Calculate the corresponding y-values by substituting the x-values from Step 4 into one of the original equations.
When $x = -2$, $y = -(-2) + 3 = 5$
When $x = 3$, $y = -(3) + 3 = 0$

Step 6: Check the solution by substituting into the other original equation.
When $x = -2$, $y = (-2)^2 - 2(-2) - 3 = 5$
When $x = 3$, $y = (3)^2 - 2(3) - 3 = 0$

Solution: $(-2, 5)$ and $(3, 0)$

Try These A

Solve each system algebraically.

a. $y = -x + 2$
$y = x^2 - x + 2$

b. $y = 2x^2 - 7$
$y = 7x - 3$

c. $y = -x + 3$
$y = x^2 - 2x - 4$

My Notes

1. Lauren solved the following system of equations algebraically and found two solutions.

$$y = -x + 3$$
$$y = x^2 - 3x - 4$$

Will solved the system by graphing and said that there is only one solution. Who is correct? Justify your response both algebraically and graphically.

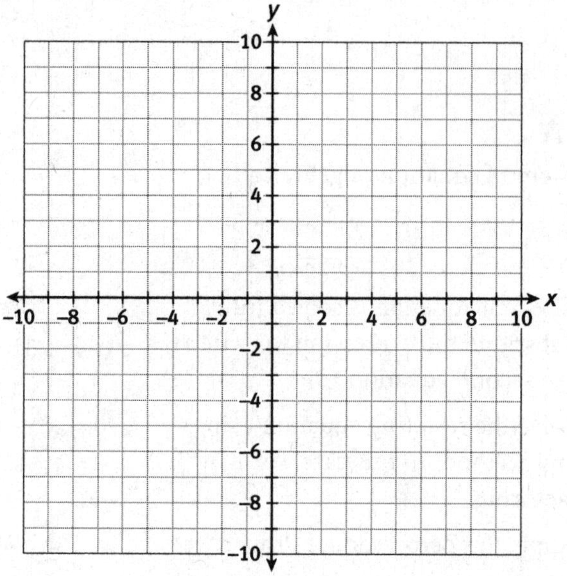

2. Model with mathematics. Deshawn drops a ball from the 520-foot-high observation deck of a tower. The height of the ball in feet after t seconds is given by $f(t) = -16t^2 + 520$. At the moment the ball is dropped, Zoe begins traveling up the tower in an elevator that starts at the ground floor. The elevator travels at a rate of 12 feet per second. At what time will Zoe and the ball pass by each other?

a. Write a function $g(t)$ to model Zoe's height above the ground after t seconds.

b. Write a system of equations using the function modeling the height of the ball and the function you wrote in Part (a).

c. Solve the system that you wrote in Part (b) algebraically. Round to the nearest hundredth, if necessary.

d. Interpret the meaning of the solution in the context of the problem. Does the solution you found in Part (c) make sense? Explain.

e. Determine the height at which Zoe and the ball pass by each other. Explain how you found your answer.

3. At the same moment that Deshawn drops the ball, Joey begins traveling down the tower in another elevator that starts at the observation deck. This elevator also travels at a rate of 12 feet per second.

a. Write a function $h(t)$ to model Joey's height above the ground after t seconds. Explain any similarities or differences between this function and the function in Item 2a.

b. Solve the system of equations algebraically.

c. Interpret the solution in the context of the problem.

d. **Construct viable arguments.** Determine whether Joey or the ball reaches the ground first. Justify your response.

My Notes

Check Your Understanding

4. How many solutions could the following system of equations have?

$$y = x + 8$$
$$y = x^2 - 10x + 36$$

5. Solve the system of equations in Item 4 using any appropriate algebraic method.

LESSON 35-2 PRACTICE

Solve each system of equations algebraically.

6. $y = 16x - 13$
$y = 4x^2 + 3$

7. $y = 5$
$y = -x^2 - x + 1$

8. $y = x$
$y = x^2 + 2x - 4$

9. Jessica has decided to solve a system of equations by graphing. Her graph is shown. Why might she prefer to solve this system algebraically?

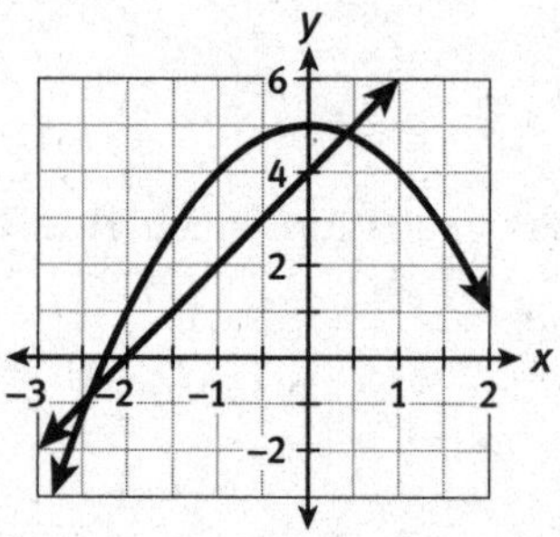

10. Make sense of problems. A competitive diver dives from a 33-foot high diving board. The height of the diver in feet after t seconds is given by $u(t) = -16t^2 + 4t + 33$. At the moment the diver begins her dive, another diver begins climbing the diving board ladder at a rate of 2 feet per second. At what height above the pool deck do the two divers pass each other?

ACTIVITY 35 PRACTICE

Write your answers on notebook paper. Show your work.

Lesson 35-1

1. Which function models the size of a neighborhood that begins with one home and doubles in size every year?
   A. $P(t) = t + 2$
   B. $P(t) = 2t + 1$
   C. $P(t) = 2(1)^t$
   D. $P(t) = (2)^t$

2. Which function models the size of a neighborhood that begins with 4 homes and increases by 6 homes every year?
   A. $P(t) = 4t + 6$
   B. $P(t) = 6t + 4$
   C. $P(t) = 6^t + 4$
   D. $P(t) = 4(6)^t$

3. When will the number of homes in Items 1 and 2 be equal?

For Items 4 and 5, sketch the graph of a system that matches the description. If no such system exists, write *not possible.*

4. The system contains a linear equation and an exponential equation. There are no solutions.

5. The system contains two exponential equations. There is one solution.

For Items 6–8, solve each system of equations by graphing.

6. $y = -x^2 + 3$
   $y = x^2 + 4$

7. $y = 0.5x^2 + x - 2$
   $y = x - 2$

8. $y = 2x^2 + 5$
   $y = -2x + 5$

9. Twin sisters Tamara and Sandra each receive \$50 as a birthday present.

   Tamara puts her money into an account that pays 3% interest annually. The amount of money in Tamara's account after x years is given by the function $t(x) = 50(1.03)^x$.

   Sandra puts her money into a checking account that does not pay interest, but she plans to deposit \$50 per year into the account. The amount of money in Sandra's account after x years is given by the function $s(x) = 50x + 50$.

   Use a graphing calculator to graph this system of equations. When will Tamara and Sandra have an equal amount of money in their accounts?

10. Josie and Jamal sold granola bars as a fund raiser, and they each started with 128 granola bars to sell.

    Josie sold 30 granola bars every day. The number of granola bars that Josie sold after x days is given by the function $y = -30x + 128$.

    Every day, Jamal sold half the number of the granola bars than he sold the day before. The number of granola bars that Jamal sold after x days is given by the function $y = 128(0.5)^x$.

    Use a graphing calculator to graph this system of equations. After how many days did Josie and Jamal have the same number of unsold granola bars?

For Items 11 and 12, write a system of equations to model the situation. Then solve the system by graphing.

11. A sample of bacteria starts with 2 bacteria and doubles every minute. Another sample starts with 4 bacteria and increases at a constant rate of 2 bacteria every minute. When will the populations be equal?

12. Jennie and James plan to save money by raking leaves. Jennie already has 1 penny. With each bag of leaves she rakes, she doubles the amount of money she has. James earns 10 cents per bag. When will Jennie have more money than James?

Lesson 35-2

For Items 13–17, solve each system of equations algebraically.

13. $y = 3x^2 - x - 2$
$y = 2x + 3$

14. $y = x^2 - 81$
$y = 18x - 161$

15. $y = x^2 + 4$
$y = 4x$

16. $y = 3x + 3$
$y = x^2 + 3x + 2$

17. $y = 3x$
$y = 2x^2$

For Items 18 and 19, write a system of equations to model the scenarios. Then solve the system of equations algebraically.

18. Simone is driving at a rate of 60 mi/h on the highway. She passes Jethro just as he begins accelerating onto the highway from a complete stop. The distance that Jethro has traveled in feet after t seconds is given by the function $f(t) = 5.5t^2$. When will Jethro catch up to Simone? (*Hint:* Use the fact that 60 mi/h is equivalent to 88 ft/s to write a function that gives the distance Simone has traveled.)

19. Jermaine is playing soccer next to his apartment building. He kicks the ball such that the height of the ball in feet after t seconds is given by the function $g(t) = -16t^2 + 48t + 2$. At the same moment that he kicks the ball, Jermaine's father begins descending in the elevator from his apartment at a rate of 5 ft/s. The apartment is 30 feet above the ground floor. At what time(s) are Jermaine's father and the soccer ball at the same height?

MATHEMATICAL PRACTICES
Construct Viable Arguments and Critique the Reasoning of Others

20. Rachel is solving systems of equations and has concluded that the quadratic formula is always an appropriate solution method when solving a nonlinear system algebraically. Do you agree with Rachel's conclusion? Use examples to support your reasoning.

Emilio loves sports and sports memorabilia. He has collected many different items over the years, but his favorite items are a signed baseball card, the catcher's mitt of his favorite catcher from 1979, and a vintage pennant from his favorite team. Emilio enjoys keeping track of how much his items are worth and has tracked their values for the last 10 years. The table below shows the values he has recorded thus far.

Years Since Item Acquired	Signed Baseball Card	Catcher's Mitt	Vintage Pennant
1	$50.00	$30.00	$6.00
2	$53.00	$37.00	$13.00
3	$56.00	$45.00	$22.00
4	$59.00	$55.00	$33.00
5	$62.00	$68.00	$46.00
6	$65.00	$83.00	$61.00
7	$68.00	$101.00	$78.00
8	$71.00	$124.00	$97.00
9	$74.00	$152.00	$118.00
10	$77.00	$186.00	$141.00

1. Identify the type of function that can be used to represent the value of each of the items shown in the table. Use the regression functions of a graphing calculator to determine functions $S(t)$, $C(t)$, and $P(t)$ that model the value of the signed baseball card, the catcher's mitt, and the vintage pennant, respectively.

2. In addition to the items in the table, Emilio also has a baseball that he caught during his favorite game of all time. The function $B(t) = -100|t-5|+1300$ can be used to model the value of the ball. Graph this function and each of the other three functions on separate coordinate planes.

3. Identify the appropriate domain for each function. Justify your responses.

4. Which item reached the greatest value during the last 10 years? What is that maximum value? Justify your response.

5. Which item's value is changing at a constant rate? Support your response using a graph or the table.

6. Which item or items have a decreasing value? For what values of t do the function(s) decrease? Justify your response.

7. Which item's value is increasing the fastest? Explain your reasoning.

8. After how many years will the signed baseball card be worth more than the catcher's mitt? Use graphs to justify your response.

9. Create a system of equations to represent the relationship between the values of the signed baseball card and the vintage pennant. Solve the system algebraically. Interpret the solution within the context of the sports memorabilia. Describe how the values of the two items compare before and after the time represented by the system's solution.

Scoring Guide	Exemplary	Proficient	Emerging	Incomplete
	The solution demonstrates the following characteristics:			
Mathematics Knowledge and Thinking (Items 1–9)	• Clear and accurate understanding of piecewise-defined, linear, quadratic, and exponential functions and the key features of their graphs • Effective understanding of how to solve systems of equations graphically and algebraically	• Largely correct understanding of piecewise-defined, linear, quadratic, and exponential functions and the key features of their graphs • Adequate understanding of how to solve systems of equations graphically and algebraically	• Partial understanding of piecewise-defined, linear, quadratic, and exponential functions and the key features of their graphs • Some difficulty solving systems of equations graphically and/or algebraically	• Inaccurate or incomplete understanding of piecewise-defined, linear, quadratic, and exponential functions and the key features of their graphs • Little or no understanding of how to solve systems of equations graphically and/or algebraically
Problem Solving (Items 4, 8)	• Appropriate and efficient strategy that results in a correct answer	• Strategy that may include unnecessary steps but results in a correct answer	• Strategy that results in a partially incorrect answer	• No clear strategy when solving problems
Mathematical Modeling / Representations (Items 1–9)	• Clear and accurate understanding of how to find, graph, interpret, and compare regression functions that model real-world data • Effective understanding of how to write and interpret the solution of a system of equations that represents a real-world scenario	• Largely correct understanding of how to find, graph, interpret, and compare regression functions that model real-world data • Adequate understanding of how to write and interpret the solution of a system of equations that represents a real-world scenario	• Partial understanding of how to find, graph, interpret, and/or compare regression functions that model real-world data • Some difficulty writing and/or interpreting the solution of a system of equations that represents a real-world scenario	• Inaccurate or incomplete understanding of how to find, graph, interpret, and/or compare regression functions that model real-world data • Little or no understanding of how to write and/or interpret the solution of a system of equations that represents a real-world scenario
Reasoning and Communication (Items 3–9)	• Precise use of appropriate math terms and language to describe and compare characteristics of several functions	• Adequate description and comparison of characteristics of several functions	• Misleading or confusing description and comparison of characteristics of several functions	• Incomplete or inaccurate description and comparison of characteristics of several functions

Probability and Statistics

6

Unit Overview

In this unit you will investigate univariate data, using statistics and graphs to compare different distributions and to comment on similarities and differences among them. You will also use two-way tables to summarize bivariate categorical data and find a "best-fit line" to summarize bivariate numerical data. You will use technology to calculate a measure of strength and direction for relationships that are linear in form. Finally, you will learn to distinguish between correlation/association and causation.

Key Terms

As you study this unit, add these and other terms to your math notebook. Include in your notes your prior knowledge of each word, as well as your experiences in using the word in different mathematical examples. If needed, ask for help in pronouncing new words and add information on pronunciation to your math notebook. It is important that you learn new terms and use them correctly in your class discussions and in your problem solutions.

Academic Vocabulary

- cluster
- census
- associate

Math Terms

- sample
- sampling error
- measurement error
- standard deviation
- outlier
- normal distribution
- *z* score
- correlate
- correlation coefficient
- residual
- best-fit line
- residual plot
- two-way frequency table
- segmented bar graph
- row percentages

ESSENTIAL QUESTIONS

- How are dot plots, histograms, and box plots used to learn about distributions of numerical data?
- How can the scatter plot, best-fit line, and correlation coefficient be used to learn about linear relationships in bivariate numerical data?
- How can a two-way table be used to learn about associations between two categorical variables?
- When is it reasonable to interpret associations as evidence for causation?

EMBEDDED ASSESSMENTS

These assessments, following Activities 37 and 40, will give you an opportunity to demonstrate your understanding of how graphical displays and numerical summaries are used to compare distributions and of methods for summarizing and describing relationships in bivariate data.

Getting Ready

Write your answers on notebook paper.
Show your work.

1. Each scatter plot below shows a set of (x, y) coordinate pairs with an approximate linear trend. Estimate the equations of the trend lines for the graphs below.

a.

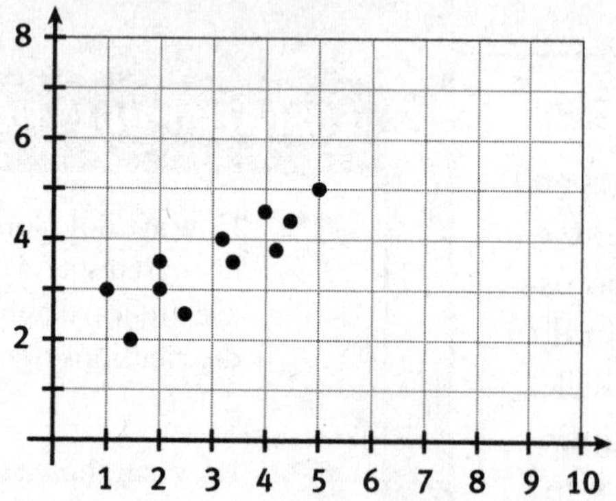

b.

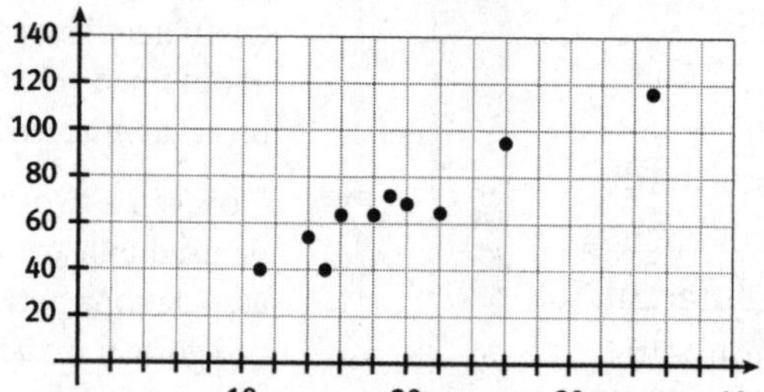

c.

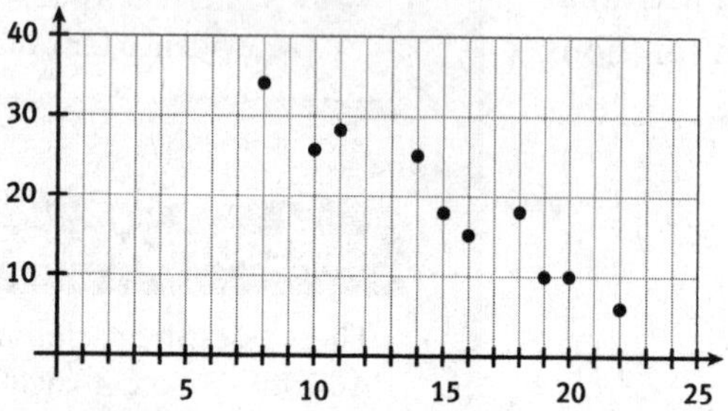

2. For each trend line below, interpret the slope of the trend line in relation to the variable quantities.
   a. $y = 75 + 0.19x$, where x is the number of miles traveled and y is the cost of an airline flight in dollars
   b. $y = 100 - 1.2x$, where x is the number of hours of TV watched per week and y is the test score on last week's test
3. An election was held at Greg's school; Greg and his friend Mary were both nominated. The table below shows the results of the voting.

Grade	Voted for Greg	Voted for Mary	Voted for Another Candidate	Total
Seventh	63	81		150
Eighth		71	13	250
Total	229		19	400

   a. Fill in the three remaining cells of the table above.
   b. What percentage of seventh graders voted for Greg? What percentage of eighth graders voted for Greg?
4. The heights (to the nearest inch) of 14 students are given below. Use these data for parts a–d.

68	66	67	70	66	68	67
69	65	67	64	66	63	65

   a. Compute the mean height of these students.
   b. Compute the median height of these students.
   c. Construct a dot plot of the heights of the students.
   d. Describe the shape of the distribution shown in the dot plot.

Measures of Center and Spread

To Text, or Not to Text

Lesson 36-1 Mean, Median, Mode, and MAD

My Notes

Learning Targets:

- Interpret differences in center and spread of data in context.
- Compare center and spread of two or more data sets.
- Determine the mean absolute deviation of a set of data.

SUGGESTED LEARNING STRATEGIES: Summarizing, Interactive Word Wall, Create Representations, Look for a Pattern, Think-Pair-Share

Zach is a high school student who enjoys texting with friends after school. Recently, Zach's parents have become concerned about the amount of time that he spends text messaging on school nights.

Zach decides to compare the amount of time he spends text messaging to that of his good friend Olivia. Both of them record the number of minutes they spend text messaging on school nights for one week.

	Sunday	Monday	Tuesday	Wednesday	Thursday
Zach	10 min	60 min	20 min	135 min	75 min
Olivia	60 min	60 min	60 min	60 min	60 min

One way to describe a set of data is by explaining how the data *cluster* around a value, or its center. The measures of center include the **mean,** the **median,** and the **mode.**

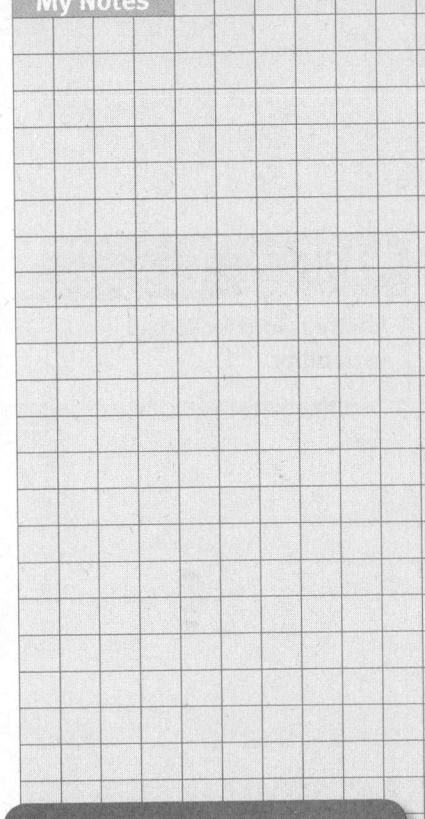

ACADEMIC VOCABULARY

To ***cluster*** means to group around.

1. Find the mean amount of time that Zach spends text messaging each night. Show how you determined your answer.

2. Find the mean amount of time that Olivia spends text messaging each night. Show how you determined your answer.

MATH TIP

The **mean** of a set of data is found by adding the values and dividing the sum of the values by the number of values.

The **median** of a set of data is found by listing the values in order and finding the value in the middle. If there is an even number of values, the median is the mean of the two values in the middle.

The **mode** of a set of data is the value that appears most often. If there are two or more values that appear most often, then each of these values is the mode. If no data value repeats, then there is no mode.

My Notes

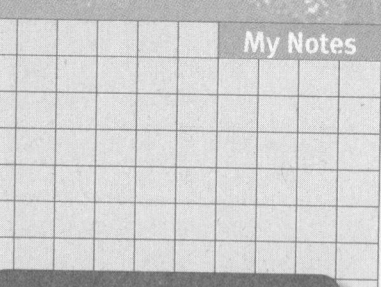

3. Reason quantitatively. Compare the amounts of time that Zach and Olivia spend text messaging. Describe similarities and differences.

MATH TIP

Another word for spread is **variability.**

Zach knows that data can be described by center and also by spread. **Spread** indicates how far apart the data values are in the set. Measures of spread include the **range** and the **mean absolute deviation.**

Zach asks his friend Trey to record the amount of time he spends text messaging on school nights. To measure spread, Zach chooses the *range.*

4. Find the mean and range of Trey's data.

	Sunday	Monday	Tuesday	Wednesday	Thursday
Zach	10 min	60 min	20 min	135 min	75 min
Trey	10 min	10 min	135 min	135 min	10 min

5. Complete the table below. How do the mean and range of Trey's data compare to those of Zach's?

	Mean	Range
Trey		
Zach		

DISCUSSION GROUP TIPS

As you share your ideas, be sure to use mathematical terms and academic vocabulary precisely. Make notes to help you remember the meaning of new words and how they are used to describe mathematical concepts.

6. Construct viable arguments. Describe how the two data sets are different. Did the mean and range help you to identify these differences? Explain.

My Notes

Because the range is based on only two values, it does not reflect any variation in the data between the greatest and least values. The range is greatly influenced by extreme values.

Another measure of spread that is not as influenced by extremes is the mean absolute deviation, which is computed using all the data values. The **mean absolute deviation** is the mean (average) of the absolute values of the deviations of the data. The **deviation** is a measure of how far a data value is from the mean.

7. To find the mean absolute deviation of Zach's data, begin by completing the table. Use the mean for Zach's data that you calculated in Item 1.

Zach	Deviation	Absolute Deviation
Time x	Time − Mean $= (x - \bar{x})$	$\|$Time − Mean$\|$ $= \|x - \bar{x}\|$
10		
60		
20		
135		
75		

WRITING MATH

The symbol $\bar{x}$ is used to represent the mean of a set of values.

8. To finish calculating the mean absolute deviation of Zach's data, find the mean of the numbers in the third column. Determine the sum of the numbers in the third column and then divide by the number of data values (the number of items in the first column).

My Notes

9. **Reason abstractly.** Why would statisticians use the mean absolute deviation rather than the mean of the deviations (in the second column)?

	Sunday	Monday	Tuesday	Wednesday	Thursday
Zach	10 min	60 min	20 min	135 min	75 min
Olivia	60 min	60 min	60 min	60 min	60 min
Trey	10 min	10 min	135 min	135 min	10 min

10. Trey's text messaging minutes are shown in the table above.
    a. Find the mean absolute deviation for the amount of time that Trey spends text messaging.

    b. Why is the mean absolute deviation for Trey's data set greater than the mean absolute deviation for Zach's data set?

11. Olivia's text messaging minutes are also given in the table above Item 10.
    a. Find the mean absolute deviation for the amount of time that Olivia spends text messaging.

    b. **Make sense of problems.** Explain why Olivia's mean absolute deviation is descriptive of her data.

My Notes

Check Your Understanding

12. During the annual food drive, Mr. Binford's homeroom collected canned goods for a month. The numbers of cans collected are given below.

Boys

12	42	69	91	97	61
15	37	104	38	82	90
51	96	19	66	8	24

Girls

20	63	18	89	67	19
66	108	96	24	16	44

Compare and contrast the results for the boys and girls using the mean and the range.

13. Remi recorded data on her car's fuel efficiency for five trips in the table below.

Trip	1	2	3	4	5
Miles per Gallon	23.7	25.5	25.2	24.8	25.4

a. Calculate the absolute deviation for each trip if the average number of miles per gallon for the trips was 24.9.

b. Find the mean absolute deviation for the five trips.

Zach would like to assure his parents that his text messaging time is not unusual for a high school student. His average text messaging time is the same as Olivia's and Trey's average times, but the variability differs for each set of data.

Before reporting to his parents, Zach decides to gather additional information. He considers using a ***census*** of the 1250 students in his school.

14. Conducting a census is often difficult. What difficulties might Zach encounter if he proceeds with his census?

ACADEMIC VOCABULARY

A ***census*** gathers information about every member of the population.

My Notes

MATH TERMS

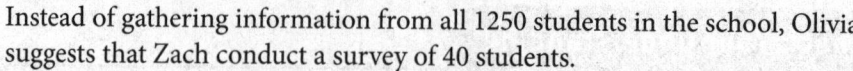

A **sample** is a portion of the population. A good sample looks like the entire population and provides useful information about the population.

Sampling error occurs because particular subgroups of the population are missing from the sample or are over-represented. Sampling error is also called **sample selection bias**.

Instead of gathering information from all 1250 students in the school, Olivia suggests that Zach conduct a survey of 40 students.

15. What is the population of Zach's survey? What is the ***sample***?

Olivia warns Zach to choose his sample wisely. A selection method that produces samples that are not representative of the total population will introduce a ***sampling error*** into the process.

16. Explain why each sampling method produces sampling error when surveying the school's population.

a. Zach asks the 30 students in his algebra class how many minutes they spend per school night text messaging with friends.

b. Zach leaves questionnaires on a table in the main hall. One of the questions asks students to indicate the number of minutes they spend per school night text messaging with friends.

My Notes

Only sampling methods that incorporate random chance into the sample selection method can hope to avoid sampling error.

17. Zach obtains a list of the names of every student in the school. Explain how Zach could use the list to randomly select a sample of 40 students.

18. Error can also occur when the method for obtaining a response is flawed. Suppose that Zach has properly selected a random sample of 40 students. Determine why each method produces ***measurement error***.

a. Zach gives each student a questionnaire. One question states: "It is very important for young people to have time to socialize with each other and today's method of communication seems to be text messaging. How many minutes do you typically spend text messaging with friends on school nights?"

b. Zach verbally surveys 40 students. Zach feels that the student responses are too low, so he reminds them that he is trying to convince his parents that 60 minutes per night is not too much time to spend text messaging.

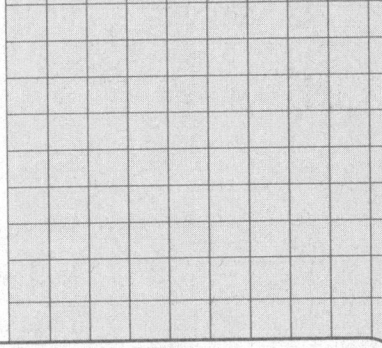

MATH TERMS

Measurement error occurs when incorrect or misleading data are collected that can be ascribed to the interviewer, the respondent, the survey instrument, or the method used for recording the data.

CONNECT TO AP

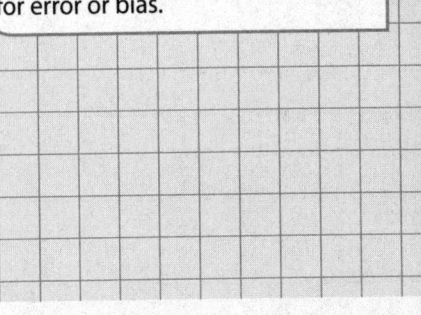

In AP Statistics, it is important to understand whether a particular sample or method for gathering information contains the potential for error or bias.

My Notes

19. Attend to precision. Gathering good data involves avoiding error in the process.

a. Write a question that Zach can use to gather data about text messaging habits and avoid measurement error.

b. Describe a method that Zach can use to collect answers from a group of 40 students and avoid sampling error.

Check Your Understanding

Classify each as a sample or a census.

20. Surveying every ninth-grade student regarding the new dress code policy for the students in all grades in the school to determine the student opinion

21. Asking every eighth-grade student whom they will vote for in the upcoming eighth-grade student election

22. Selecting every fourth student from an alphabetical list of students to gather data on absenteeism for the school

My Notes

LESSON 36-1 PRACTICE

Work with your group to answer Items 23–27. As you discuss your solutions, speak clearly and use precise mathematical language. Remember to use complete sentences and words such as *and, or, since, for example, therefore, because of* to make connections between your thoughts.

23. Vernice asked 12 classmates to record the number of hours they spent watching television during one week. The table shows the data she collected.

10	11	22	7	17	17
20	31	0	12	19	23

a. Calculate the mean and mean absolute deviation for the data.
b. What statements could Vernice make about the viewing habits of these classmates?

24. Rewrite each question to avoid measurement error.
a. Don't you agree that seniors should be dismissed early on Fridays at least once each semester?
b. Drinking beverages with sugar promotes tooth decay and obesity. How many soft drinks with sugar did you drink in the past week?

25. Scores from the same benchmark test were collected from two algebra classes, each with 30 students enrolled. One class had a mean score of 79 with a mean absolute deviation of 5, and the other had a mean score of 81 with a mean absolute deviation of 10. What can be said about the distribution of scores on this test for the two classes?

26. Mitch and a group of his friends have estimated how long it will take each of them to run 400 meters around the track. Their estimates in seconds are 115, 76, 94, 81, 78, 99, 68, and 84.
a. Calculate the mean and the range.
b. Which estimate stands out as unusual?
c. What might be a reason for such an unusual estimate?

27. Construct viable arguments. What are the advantages of taking a census of the population instead of a sample? Describe some examples when it would be worth the time and effort to conduct a census.

My Notes

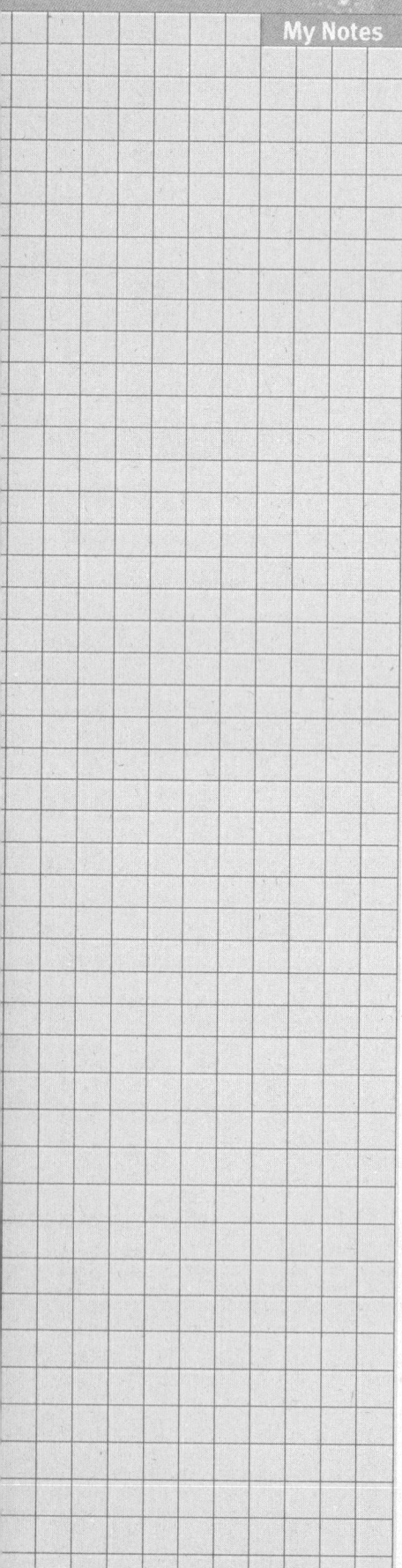

Learning Targets:

- Use summation and subscript notation.
- Calculate and interpret the standard deviation of a numerical data set.
- Select appropriate measures of spread by examining the shape of a distribution.

SUGGESTED LEARNING STRATEGIES: Summarizing, KWL Chart, Create Representations, Self Revision/Peer Revision, Think-Pair-Share

In the previous lesson, you studied one way to describe the spread, or variability, in a data set—the mean absolute deviation (MAD). The mean absolute deviation is the mean (or average) difference of the data values from the mean of a numerical data set.

Consider the following data from C|NET (www.cnet.com), a tech media website that publishes information about technology and consumer electronics. These data are taken from a review of cell phone battery lifetimes. The table below gives the talk times (in hours) for the top 10 brands of batteries.

Talk Time (hours)	
19.78	11
14.55	10.7
13.4	10.6
12.75	10.6
12	10.3

1. Calculate the mean for these 10 talk times.

2. Complete the table below by finding the absolute values of the deviations (the absolute value of the difference between each data value and the mean calculated in Item 1).

Talk Time and Deviations from the Mean			
Value	|Deviation|	Value	|Deviation|
19.78	7.212	11.00	1.568
14.55		10.70	
13.40		10.60	
12.75		10.60	
12.00		10.30	

3. Attend to precision. The mean absolute deviation for this data set is the mean of the 10 absolute deviations. Calculate the mean absolute deviation.

My Notes

Another common measure of variability is the ***standard deviation***. Before calculating the standard deviation, let's introduce some notation.

The Greek letter $\sum$ (sigma) is used to indicate a sum of several values.

For example, $\sum_{i=1}^{5} i$ means "the sum of the values of *i* as *i* goes from 1 to 5":

i ends at this value.

$$\sum_{i=1}^{5} i = 1+2+3+4+5 = 15$$

i starts at this value.

In the summation above, *i* can be replaced by any expression. Also, the values of *i* can start and end at any number:

$$\sum_{i=3}^{6}(i+2) = (3+2)+(4+2)+(5+2)+(6+2) = 26$$

4. Determine each sum.

a. $\sum_{i=1}^{4} i^2$

b. $\sum_{i=0}^{2} 2^i$

c. $\sum_{i=2}^{7} |5-i|$

The standard deviation is similar to the MAD in that it is based on deviations from the mean. The formulas below show the similarities between the MAD and the standard deviation *s*.

$$\text{MAD} = \frac{\sum_{i=1}^{n}|x_i - \bar{x}|}{n} \qquad s = \sqrt{\frac{\sum_{i=1}^{n}(x_i - \bar{x})^2}{n-1}}$$

As you can see, the calculation of the standard deviation is a little more involved than MAD. Why this additional complexity? The basic answer is that this measure has some advantages in more advanced statistical settings.

You will calculate both the MAD and standard deviation using a data set consisting of four battery talk times (in hours):

$$x_1 = 7.00,\ x_2 = 10.00,\ x_3 = 8.10,\ x_4 = 9.32$$

The mean of these 4 observations is 8.605.

MATH TERMS

The **standard deviation** is a measure of variability in a data set.

DISCUSSION GROUP TIPS

As needed, refer to the Glossary to review definitions and pronunciations of key terms. Incorporate your understanding into group discussions to confirm your knowledge and use of key mathematical language.

MATH TIP

In sigma notation, the letter *i* is called the **index of summation.**

MATH TIP

In a data set with *n* values, the individual values are referred to as $x_1, x_2, x_3, \ldots, x_n$. In the formulas for the MAD and the standard deviation, x_i, as *i* goes from 1 to *n*, refers to these values.

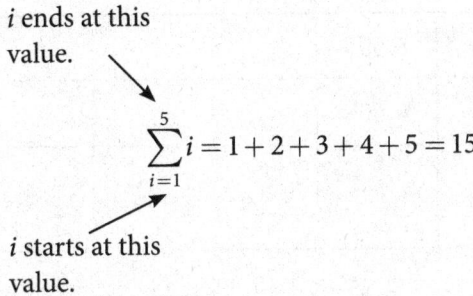

My Notes

5. Complete the table. Then compute the MAD and the standard deviation.

x_i	$\bar{x}$	$\lvert x_i - \bar{x} \rvert$	$(x_i - \bar{x})^2$
$x_1 = 7.00$	8.605		
$x_2 = 10.00$	8.605		
$x_3 = 8.10$	8.605		
$x_4 = 9.32$	8.605		
$\sum_{i=1}^{4} x_i =$		$\sum_{i=1}^{4} \lvert x_i - \bar{x} \rvert =$	$\sum_{i=1}^{4} (x_i - \bar{x})^2 =$

$$\text{mean absolute deviation} = \frac{\sum_{i=1}^{4} \lvert x_i - \bar{x} \rvert}{n} =$$

$$\text{standard deviation} = \sqrt{\frac{\sum_{i=1}^{4} (x_i - \bar{x})^2}{n-1}} =$$

6. The weights (in ounces) of three newborns are:

$$w_1 = 120; \quad w_2 = 115; \quad w_3 = 125$$

a. Compute the mean of these three weights.

b. Complete the table below and calculate both the MAD and the standard deviation.

w_i	$\bar{w}$	$\lvert w_i - \bar{w} \rvert$	$(w_i - \bar{w})^2$
120			
115			
125			
$\sum_{i=1}^{3} w_i =$		$\sum_{i=1}^{3} \lvert w_i - \bar{w} \rvert =$	$\sum_{i=1}^{3} (w_i - \bar{w})^2 =$

$$\text{mean absolute deviation} = \frac{\sum_{i=1}^{3} \lvert w_i - \bar{w} \rvert}{n} =$$

$$\text{standard deviation} = \sqrt{\frac{\sum_{i=1}^{3} (w_i - \bar{w})^2}{n-1}} =$$

You can see that calculating the standard deviation can be a lot of work. Fortunately, calculators and computers can be used to do the calculations.

My Notes

Check Your Understanding

7. Calculate the mean and standard deviation of the original data for the cell phone battery talk times (shown below) using a calculator or computer software.

Talk Times (hours)	
19.78	11.00
14.55	10.70
13.40	10.60
12.75	10.60
12.00	10.30

LESSON 36-2 PRACTICE

The table shows the speeds of the 10 fastest roller coasters in the United States. Use the table for Items 8–11.

Fastest Roller Coasters (mi/h)	
76	81
85	67
74	72
63	59
73	80

8. Find $|x - \bar{x}|$ for each data value.

Value	76	81	85	67	74	72	63	59	73	80
$\lvert x - \bar{x} \rvert$										

9. Find the mean absolute deviation using the values you calculated in Item 8.

10. Using technology, calculate the standard deviation for the data set.

11. **Make sense of problems.** A new roller coaster is being built, scheduled to be complete in the spring of next year. It is projected to reach speeds of 90 mi/h. This new value will cause the lowest speed to be removed from the list. Calculate the new standard deviation and describe the changes.

ACTIVITY 36 PRACTICE

Write your answers on notebook paper. Show your work.

1. For each question, decide:
- Is it a statistical question?
- If not, explain why not and rewrite the question so that it is a statistical question.

a. How many states have you visited?

b. Do you like chocolate ice cream?

c. Do you watch TV at night?

d. How many sports do you play?

2. Write three examples of statistical questions whose responses would show different levels of variability. Include at least one question whose responses would show "lots" of variability and at least one question whose responses would show "little" variability.

3. Describe the variability associated with the question "Do students in my school need more homework each night?" Do you think there will be a lot or a little variability? Explain.

4. Suppose the results of a survey about where your classmates were born showed great variability. What would these results tell you about the people in your class?

5. The following question was asked of your classmates today before school started:

"How many states have you visited in the last year?"

The responses to this question showed little variability. How could you change this question to gather answers that show greater variability?

For Items 6 and 7, use the data set $x_1 = 2.3$, $x_2 = 4.1$, $x_3 = 1.6$, and $x_4 = 2.0$.

6. Compute the value of $\frac{x_1 + x_2}{x_3}$.

7. Compute the value of $\sum_{i=1}^{3} x_i$.

Use the information below for Items 8 and 9.

The amount of caffeine in beverages presents an important health concern, especially for women of childbearing age. In a recent study of carbonated sodas, the numbers of milligrams of caffeine detected were as follows:

29.5, 38.2, 39.6, 29.5, 31.7, 27.4, 45.4, 48.2, 36.0, 33.8, 19.4, 18.0, 34.6

8. Calculate the mean, mean absolute deviation, and standard deviation for these data.

MATHEMATICAL PRACTICES
Reason Abstractly and Quantitatively

9. In the report of the caffeine levels, five sodas were left out because no caffeine was detected. If these sodas were given a value of 0 mg of caffeine and added to the data above, how would the mean and standard deviation change?

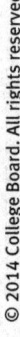

Dot and Box Plots and the Normal Distribution

Disturbing Coyotes

Lesson 37-1 Dot Plots and Box Plots

Learning Targets:

- Construct representations of univariate data in a real-world context.
- Describe characteristics of a data distribution, such as center, shape, and spread, using graphs and numerical summaries.
- Compare distributions, commenting on similarities and differences among them.

SUGGESTED LEARNING STRATEGIES: Summarizing, Paraphrasing, Look for a Pattern, Discussion Groups, Quickwrite

Professional wildlife managers and the public are concerned with the impact of human activity on wildlife. One measure studied is animals' "home range," the typical area in which an animal spends its time.

Researchers were concerned that the home ranges of some coyotes in a portion of Colorado were affected by military maneuvers involving jeeps, tanks, helicopters, and jet fighter flyovers. To evaluate these potential effects, several coyotes were collared with radio transmitters. The researchers used the transmitters to track the movement of the coyotes. Coyotes were monitored before, during, and after the military maneuvers.

Home Range Before Maneuvers (km^2)	Home Range During Maneuvers (km^2)	Home Range After Maneuvers (km^2)
3.9	7.5	3.1
5.4	32.6	5.4
5.7	3.2	4.5
4.8	2.7	4.1
5.3	7.3	8.6
5.3	9.1	8.0
5.5	18.0	6.4
11.4	6.5	7.8
3.6	2.1	1.0
3.5	5.2	3.7
6.3	4.3	10.7

A **dot plot** is an effective method for representing univariate (one-variable) data when dealing with small data sets.

1. A dot plot of the "Before" data is shown below. Make dot plots of the "During" and "After" data sets.

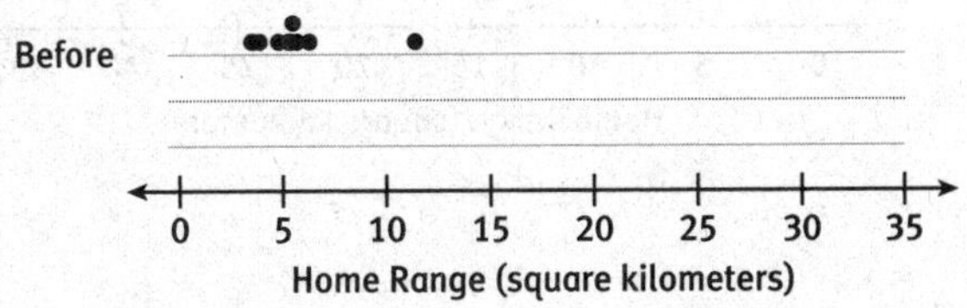

My Notes

MATH TIP

To make a dot plot, draw a number line with an appropriate scale for the data. Then mark a dot for each data value above the appropriate number on the number line.

My Notes

2. Compare and contrast the centers and spreads of the three data sets.

MATH TIP

The first quartile (Q_1) is the median of the data values to the left of the overall median, and the third quartile (Q_3) is the median of the data values to the right of the overall median. A box plot (sometimes called a box-and-whisker plot) is a graph of the five-number summary that consists of a central box from Q_1 to Q_3 that has a vertical line segment at the median. Horizontal line segments ("whiskers") extend from the box to the minimum and maximum data values.

A five-number summary provides a numerical summary of a set of data. It is used to construct a **box plot.** The five-number summary for the "Before" home ranges is shown below, together with the resulting box plot.

Home Ranges: Before Maneuvers	
Minimum:	3.5
First quartile (Q_1):	3.9
Median:	5.3
Third quartile (Q_3):	5.7
Maximum:	11.4

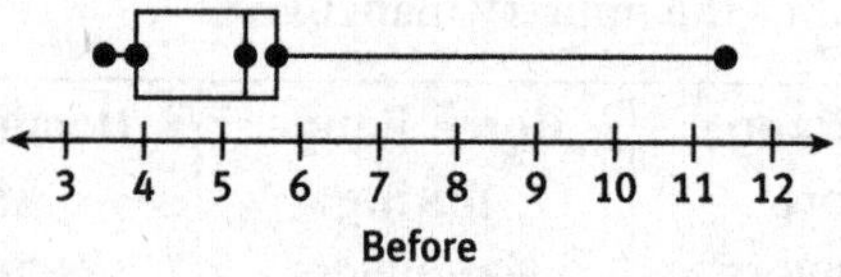

3. Create five-number summaries of the "During" and "After" data.

Home Ranges: During Maneuvers	
Minimum:	
First quartile:	
Median:	
Third quartile:	
Maximum:	

Home Ranges: After Maneuvers	
Minimum:	
First quartile:	
Median:	
Third quartile:	
Maximum:	

4. Use the summaries in Item 3 to construct box plots of the "During" and "After" data sets in the space below.

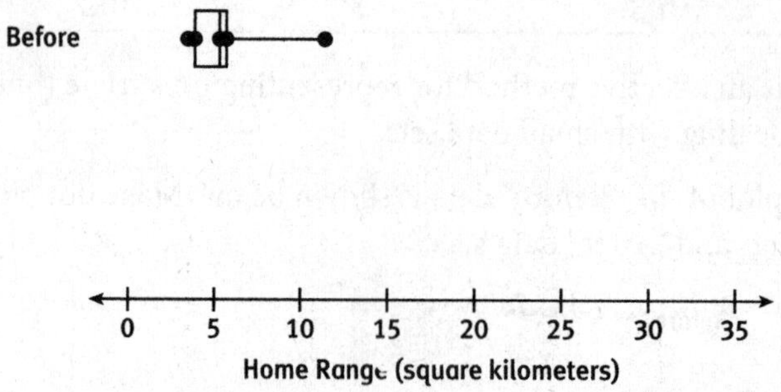

My Notes

5. Based on the box plots and five-number summaries in Items 3 and 4:

a. Which data set seems to have the least overall spread? Which data set seems to have the greatest overall spread?

b. Which data set seems to have the least spread in its "middle 50%" box? Which data set seems to have the greatest spread in its "middle 50%" box?

Remember that the initial concern before the data gathering was that the home ranges of the local coyotes might change during the military maneuvers. To investigate this concern, you will use the graphs, numerical summaries, and comparisons you have developed as a starting point for your analysis.

6. Reason abstractly and quantitatively. Based on the data and graphs, does it appear that there was a substantial change in the coyotes' home ranges during the military maneuvers? Write a few sentences specifically comparing the "Before" and "During" data sets. Use numerical values where possible.

7. Do there appear to be any substantial permanent changes to the coyotes' home ranges after the military maneuvers? Write a few sentences specifically comparing the "Before" and "After" data sets. Use numerical values where possible.

My Notes

8. The "During" data set contains two values that are far away from the rest of the data. The "Before" data set contains one such value also. Suppose you wish to call attention to the fact there are such far-away data values. Which type of plot—the dot plot or the box plot—would be your choice? Why?

Check Your Understanding

9. A teacher in a statistics class allows her students to use notes about statistical procedures on tests. She believes that a teacher-made study sheet will be more effective in helping students recall the procedures. In each of her three classes she used one of three helping strategies: (a) student-made notes with information about the procedures, (b) teacher-made information printed on paper in the form of a flowchart, and (c) teacher-made information delivered by computer access during the exam. Each of her classes has 18 students. The test scores for her students are given as percent correct.

Student Notes	89	15	39	15	31	69	39	54	31	62	46	39	54	39	15	46	23	31
Paper Help	76	24	77	71	18	29	59	77	41	77	77	47	71	82	82	82	59	65
Computer Help	100	13	73	73	33	53	60	60	27	80	80	47	73	80	80	93	60	53

a. In order to compare the results for these three groups, construct a dot plot for each of the three data sets.

b. Describe the three data sets with specific attention to center and spread.

10. For each group, create a five-number summary for these data.

	Student Notes	Paper Help	Computer Help
Minimum			
First quartile			
Median			
Third quartile			
Maximum			

11. Use the summaries in Item 10 to construct box plots of the three data sets: "Notes," "Paper," and "Computer."

12. a. For these data, what advantages do you see in using the dot plots to display the data sets?

b. For these data, what advantages do you see in using the box plots to display the data sets?

13. Comparing the test scores for these groups and using specific information from the five-number summary and/or your dot plots and box plots, answer the following in a few sentences.

a. Which group appears to have done the best on the exam?

b. Which group appears to have done the worst on the exam?

LESSON 37-1 PRACTICE

Some researchers believed that one reason students often have unhealthy sleeping habits is that they don't adequately manage their time. The researchers wanted to test whether providing information to students about time management could help. Eighteen students were divided into three groups. Students in Group 1 were taught to use a planner for time management and asked to sleep 7–8 hours daily. Students in Group 2 were taught to use a planner but not given any instruction on how many hours to sleep daily. Students in Group 3 were simply instructed to sleep as they usually do. At the conclusion of the study the participants were given a questionnaire to measure their anxiety levels. (A high score on the questionnaire indicates low levels of anxiety.)

Dot plots of the data from the questionnaire are shown below. Use the dot plots for Items 14 and 15.

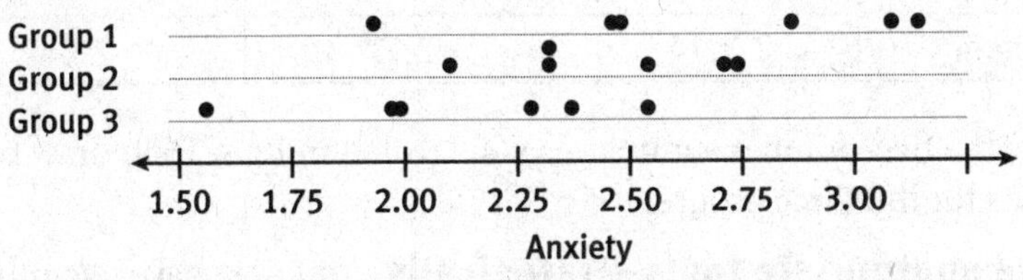

14. In a few sentences, compare the centers of these data sets. Do the three groups have approximately equal centers? If not, how do they differ?

15. In a few sentences, compare the spreads of these data sets. Do the three groups have very similar spreads? If not, how do they differ?

My Notes

A table of the data from the questionnaires is shown below. Use the table for Item 16.

Group 1: Planner & Sleep Instruction	Group 2: Planner Only	Group 3: Neither Planner nor Sleep Instruction
2.86	2.32	1.56
3.14	2.71	1.97
2.48	2.32	2.28
1.93	2.54	2.54
2.46	2.74	1.99
3.08	2.10	2.37

16. Create a five-number summary for each group.

	Group 1: Planner & Sleep Instruction	Group 2: Planner Only	Group 3: Neither Planner nor Sleep Instruction
Minimum	1.93	2.10	1.56
First quartile		2.32	
Median		2.43	2.135
Third quartile			2.37
Maximum	3.14	2.74	2.54

17. Use the five-number summaries you created in Item 16 to draw box plots for the three groups.

18. **Use appropriate tools strategically.** You now have two different graphs for these data. Which graph—dot plots or box plots—do you feel makes it easier to compare centers and spreads? Explain your reasoning.

My Notes

Learning Targets:

- Use modified box plots to summarize data in a way that shows outliers.
- Compare distributions, commenting on similarities and differences among them.

SUGGESTED LEARNING STRATEGIES: Summarizing, Paraphrasing, Think-Pair-Share, Create Representations, Quickwrite

You already are familiar with many ways to summarize data graphically and numerically. In this lesson, you will see a new type of plot called a **modified box plot.**

Below are the dot plots of the data about the coyotes from Lesson 37-1. Refer back to Lesson 37-1 for the actual data values.

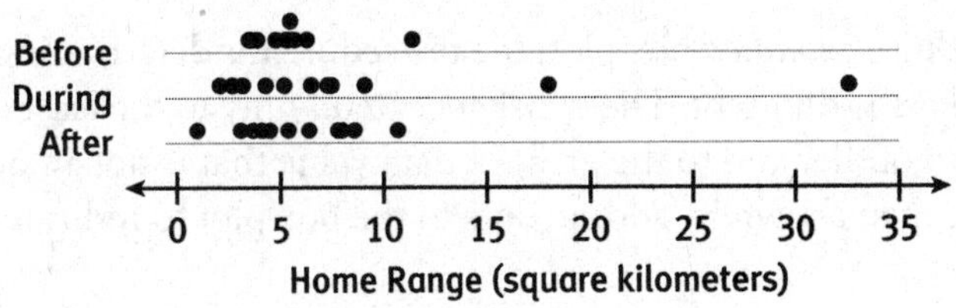

Notice that there are two unusually large values in the "During" data set and one in the "Before" data set.

Outliers are values that differ so much from the rest of a one-variable data set that attention is drawn to them. Outliers may arise for many reasons, including measurement errors or recording errors. It may also be the case that there actually are unusual values in a data set.

Outliers can also occur in bivariate (two-variable) data. It is important to consider the impact of outliers when summarizing and analyzing data.

MATH TERMS

An **outlier** is a data point that is unusual enough that it draws attention during the data analysis.

1. Which, if any, of the data values shown in the dot plots do you consider to be outliers? List each data value that you think might be an outlier and the data set from which it came.

2. Describe the method you used in Item 1 to determine whether a data value is an outlier. Is it based on distance? Is it based on the concentration of other data values? Compare your method with those of your classmates.

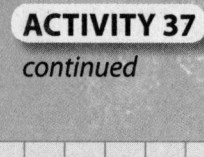

My Notes

MATH TIP

Notation Review

Q_1 First quartile

Q_3 Third quartile

$IQR = Q_3 - Q_1$

It is not surprising that people do not always agree about how different a value must be in order to be called an outlier. For consistency, a data value is considered to be an outlier if it is more than $1.5 \times (IQR)$ from the nearest quartile. Recall that *IQR* stands for **interquartile range** and is the difference between the third quartile (Q_3) and the first quartile (Q_1): $IQR = Q_3 - Q_1$.

Mathematically, this means that a data value *x* is an outlier if:

$$x > Q_3 + 1.5(IQR)$$
or
$$x < Q_1 - 1.5(IQR)$$

We will use this definition of an outlier to draw a *modified* box plot. The idea behind modifying the box plot is to create a plot that shows outliers.

When creating a *modified* box plot, the procedure for determining the length of the whiskers is different. The whiskers extend only to the least data value that is not an outlier and to the greatest data value that is not an outlier. Outliers are then shown by adding dots to the box plot to indicate their locations.

Let's walk through the steps together using the data from the "Before" data set. Here are the data, arranged in order, as well as the five-number summary and box plot of the data from Lesson 37-1:

3.5	3.6	3.9	4.8	5.3	5.3	5.4	5.5	5.7	6.3	11.4

Home Ranges: Before Maneuvers	
Minimum:	3.5
First quartile:	3.9
Median:	5.3
Third quartile:	5.7
Maximum:	11.4

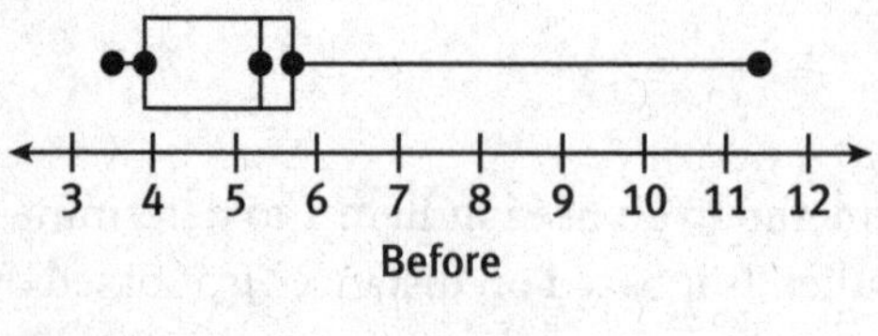

The maximum value, 11.4, is much greater than the other data values. In fact, it is more than 5 units away from the next-greatest value of 6.3. The remaining data (from the minimum value of 3.5 to 6.3) have a range of only 2.8.

Now let's modify the box plot to show outliers.

My Notes

3. For the "Before" data set, how great or how small must a data value be to be identified as an outlier? Perform the calculations below to find the upper and lower boundary values that separate outliers from the rest of the data.

$$Q_3 + 1.5(IQR) =$$

$$Q_1 - 1.5(IQR) =$$

4. Are any data values less than the lower boundary for outliers identified in Item 3?

5. Are any data values greater than the upper boundary for outliers identified in Item 3?

Next, identify the least and greatest values in the data set that are not outliers. These values will determine the endpoints of the whiskers in the modified box plot.

6. What is the least data value that is not an outlier?

7. What is the greatest data value that is not an outlier?

My Notes

Now you have everything you need to draw the modified box plot. The modified box plot is constructed as follows:

Step 1. Draw the box as usual.

Step 2. Extend the whiskers to the least and greatest data values that are not outliers.

Step 3. Place dots above the scale to indicate the outliers.

8. Follow the directions above to draw the modified box plot for the "Before" data:

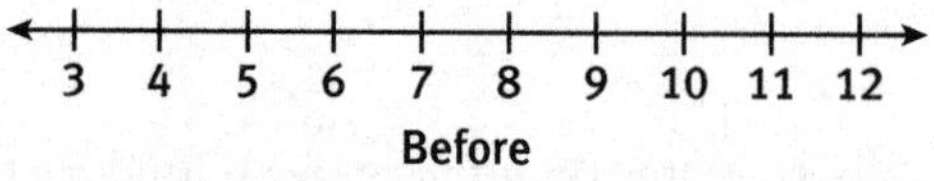

Check Your Understanding

9. A data set has a third quartile of 64 and a first quartile of 29. What are the upper and lower boundary values that separate outliers from the rest of the data set?

10. Make sense of problems. If the third quartile of the data set in Item 9 were increased by 10, how would this change the upper boundary for outliers? Explain your answer.

11. Critique the reasoning of others. In a survey of 21 teenage girls about their text message usage for one month, the five-number summary is minimum = 0, $Q_1 = 1$, median = 31, $Q_3 = 56$, and maximum = 1305. In analyzing these data, Michelle determined that there are no outliers. Do you agree or disagree? Explain.

My Notes

LESSON 37-2 PRACTICE

12. Describe the effects an outlier can have on a set of data.

13. Construct viable arguments. In a data set, the third quartile is 36 and the first quartile is 12. Would a value of 52 be considered an outlier? Why or why not?

14. Lisa recorded the heights of her classmates. Her data are shown in the table below.

Heights of Classmates in Inches				
61	58	62	60	57
67	68	61	64	70
72	64	63	59	69

Calculate the upper and lower boundaries for outliers for the heights of Lisa's classmates.

15. List two values that would be considered outliers in the data set in Item 14. Include one value that is less than the lower boundary for outliers and one that is greater than the upper boundary for outliers.

My Notes

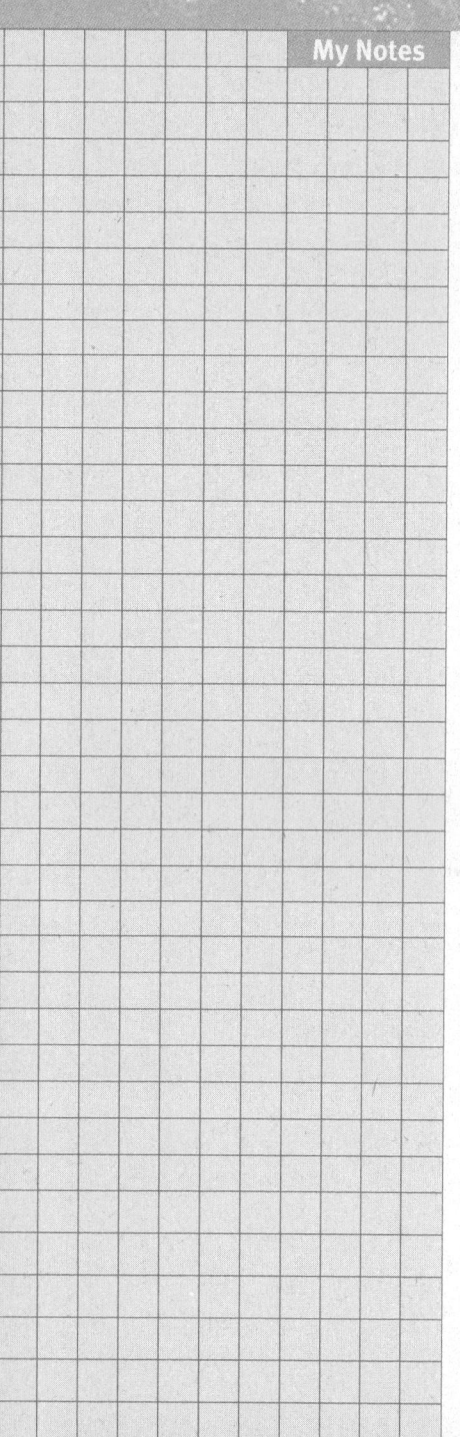

Learning Targets:

- Use the mean and standard deviation to fit a normal distribution.
- Develop an understanding of the normal distribution.
- Use technology to estimate the percentages under the normal curve.

SUGGESTED LEARNING STRATEGIES: Visualization, Think-Pair-Share, Create Representations, Look for a Pattern, Quickwrite

There is a relationship between sleep quality and health. Many studies have concluded that good-quality rest helps to relieve health issues such as high blood pressure, depression, weight gain or loss, and fatigue.

1. How many hours did you spend sleeping in the last 24 hours?

2. Gather all of your classmates' answers to Item 1. In the *My Notes* section of this page, create a dot plot of the number of hours that students in your class spent sleeping.

3. Use the dot plot to describe the data. Identify the mean, standard deviation, maximum, and minimum.

4. What do you consider to be a normal amount of time spent sleeping in a 24-hour period?

Data can be distributed in different ways.

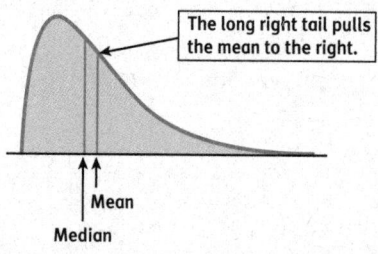

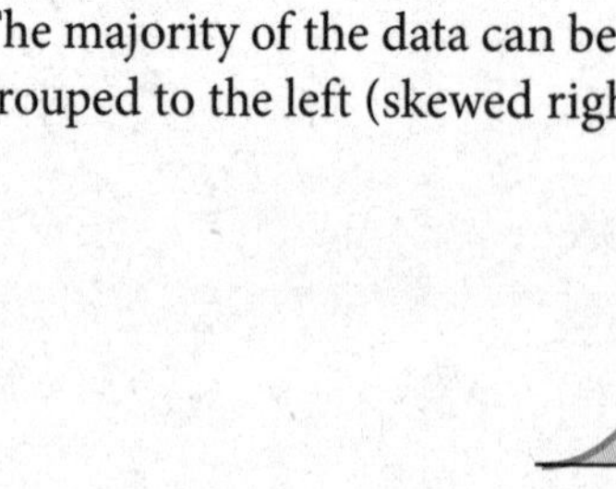

The majority of the data can be grouped to the left (skewed right).

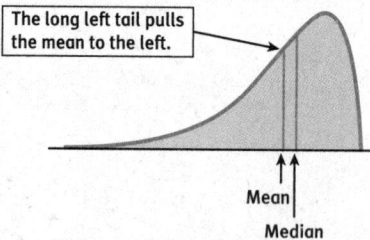

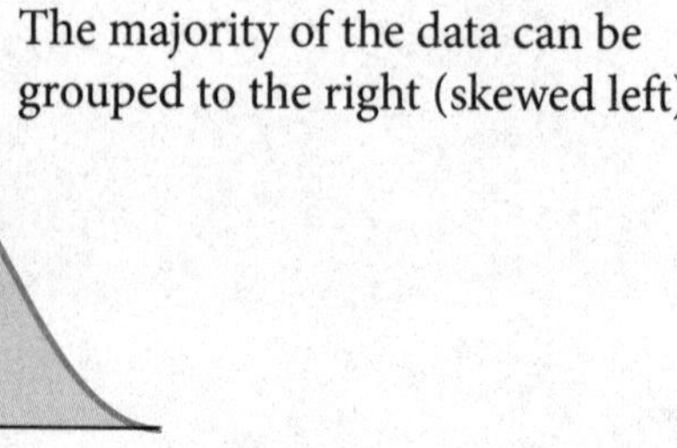

The majority of the data can be grouped to the right (skewed left).

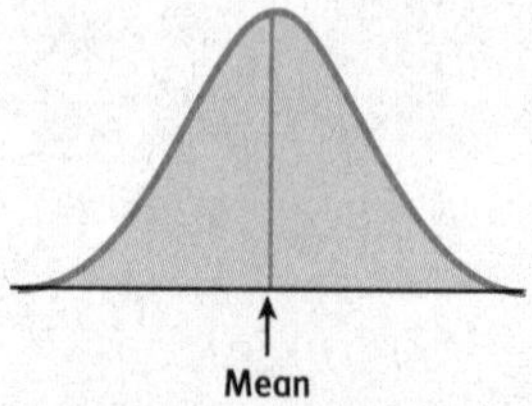

The majority of the data can be grouped symmetrically around the center.

This last type of distribution is called a ***normal distribution***.

MATH TERMS

In a **normal distribution**, the data values are grouped symmetrically about the mean, with most of the data values occurring near the mean. Because of its shape, a normal distribution is sometimes called a bell curve.

My Notes

Many real-world data are approximately normally distributed—for instance, height, weight, grades, blood pressure, and time people spend sleeping. Because so many data sets can be modeled by a normal distribution, a table of values is used to analyze and make decisions about normally distributed data. For this to be accomplished, the data values are converted to "scores" that can be easily compared. This is called *standardizing*. A standard score means that the data have a mean of 0 and a standard deviation of 1.

5. Below is a list of the hours slept by the students in Mr. Trent's class. Calculate the mean number of hours these students slept.
2, 4, 5, 6, 7, 7, 7, 8, 9, 10

6. Find the probability that a student in Mr. Trent's class slept more than 7 hours.

Below is a graph of the hours slept by Mr. Trent's students. Dashed vertical lines have been drawn at one standard deviation above and below the mean.

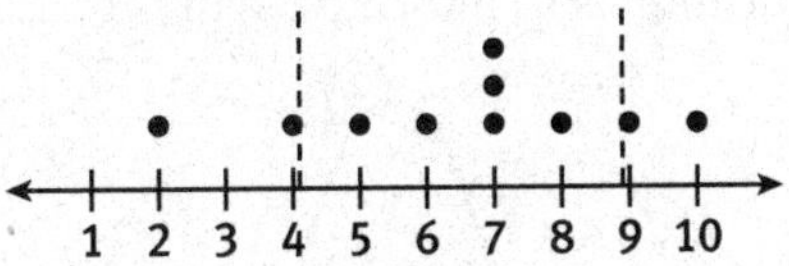

7. Calculate the percent of students whose time spent sleeping is within one standard deviation above or below the mean.

The first step in standardizing the data values from Mr. Trent's class is to calculate each value's deviation from the mean.

8. For each data value, calculate the deviation from the mean, $(x - \bar{x})$, and fill in the second column of the table. Leave the third column blank for now.

Hours Slept and Deviation from the Mean

Hours	$(x - \bar{x})$	$\frac{x - \bar{x}}{s}$
2		
4		
5		
6		
7		
7		
7		
8		
9		
10		

My Notes

9. Calculate the sum of the deviations from the mean.

10. Explain the numerical value that you calculated in Item 9.

The next step is to divide by the standard deviation.

11. The standard deviation s for Mr. Trent's class is 2.4. Use this to complete the last column of the table in Item 8. Round values to the nearest thousandth, if necessary.

MATH TERMS

A ***z* score** is a standard score that indicates by how many standard deviations a data value is above or below the mean.

The numbers in the last column of the table represent the standardized scores for each data value. A standardized score is called a ***z* score:** $z = \frac{x - \bar{x}}{s}$. The set of z scores is a data set with mean 0 and standard deviation 1. This represents a standardized version of the original data set.

12. Specific z scores of ± 1 are indicated with vertical lines on the graph below. Draw vertical lines to indicate z scores of ± 2 and ± 3.

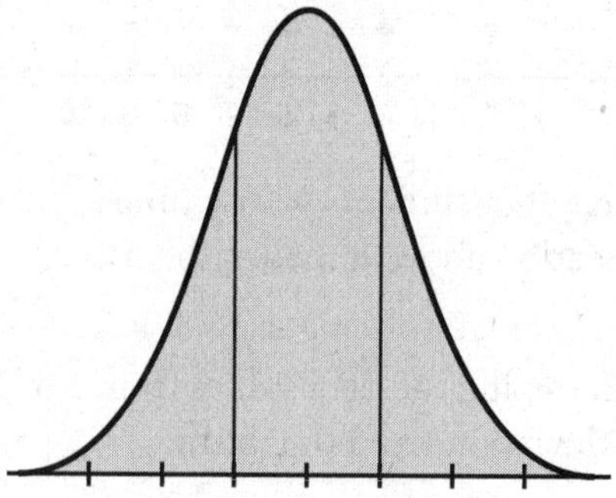

The graph above shows a standardized normal distribution, which can be used to find probabilities. For example, the probability that a data value lies between −1 and +1 standard deviation from the mean is about 68%.

Use a normal distribution and a calculator to estimate the probability that a randomly selected student from Mr. Trent's class slept between 7 and 10 hours. The steps for one type of graphing calculator are:

Step 1. Press 2nd VARS.

Step 2. Select normalcdf(.

Step 3. Determine the lower boundary and enter that number, 7, followed by a comma.

Step 4. Determine the upper boundary and enter that number, 10, followed by a comma.

Step 5. Enter the mean, 6.5, followed by a comma.

Step 6. Enter the standard deviation, 2.4, followed by).

Step 7. Press Enter.

My Notes

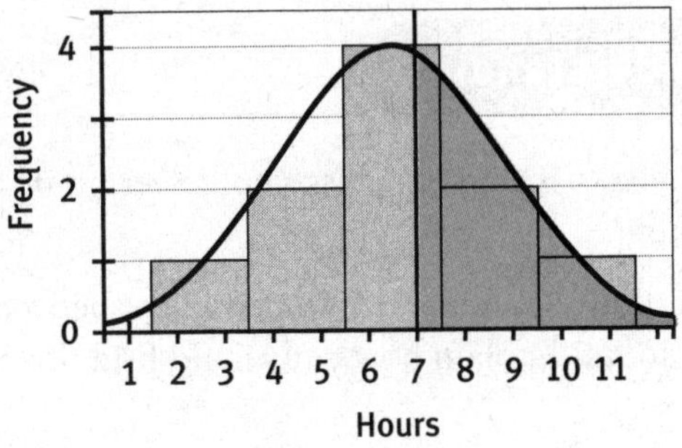

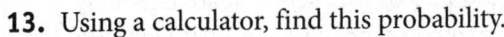

13. Using a calculator, find this probability.

14. Compare your answer to Item 13 with your answer to Item 6.

15. Make sense of problems. Use a normal distribution and a graphing calculator to estimate the probability that a student in Mr. Trent's class slept fewer than 6 hours. Does the answer that the calculator gives make sense when you look at the data? Explain.

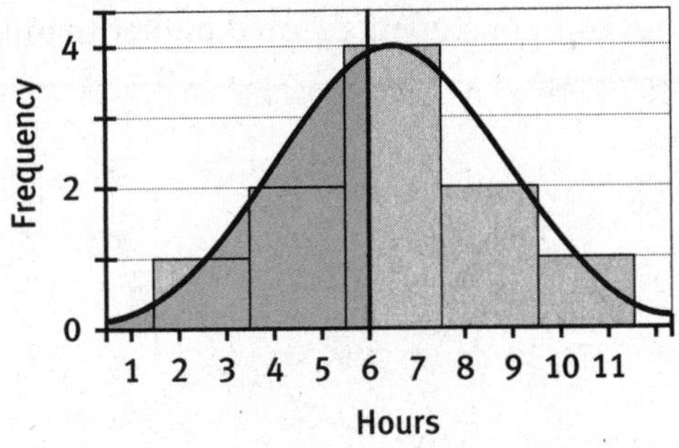

16. Estimate the probability that a student in Mr. Trent's class slept fewer hours than you did in the last 24 hours.

My Notes

Check Your Understanding

17. How many students from your class slept fewer hours than you did in the last 24 hours?

18. What percent of the students in your class slept between 6 and 8 hours in the last 24 hours? Explain how you would find this value and then calculate it.

19. You have discussed several examples of data sets that are normally distributed. Give an example of a real-world data set that would not be normally distributed and explain why.

20. The weights of 1.69-oz bags of M&Ms are normally distributed with a mean of 1.69 oz and a standard deviation of 0.05 oz. What is the probability that a bag selected at random weighs between 1.76 oz and 2 oz?

21. What is the probability that a bag of M&Ms selected at random weighs less than 1.62 oz?

22. Ian scored a 78 on last week's algebra test. The scores for the class were normally distributed with an average of 72 and a standard deviation of 5. What proportion of students scored higher than Ian?

My Notes

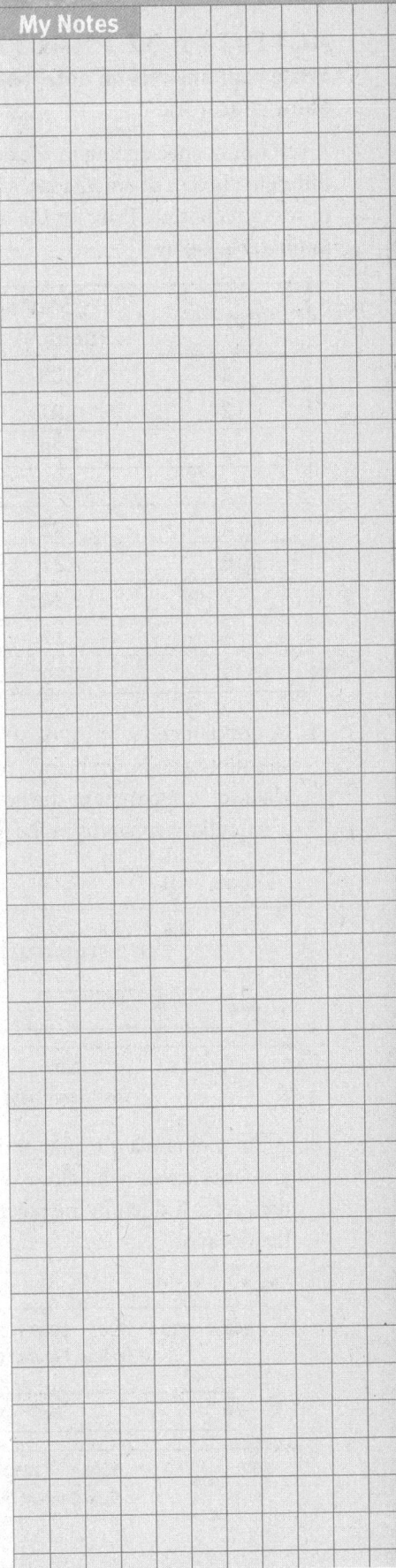

LESSON 37-3 PRACTICE

Model with mathematics. The times between eruptions of Old Faithful, a geyser at Yellowstone National Park, vary from 44 to 122 minutes. The average time between eruptions is 91 minutes. The table below lists the times between eruptions for January 1, 2011.

Time in Minutes	
85	85
85	92
100	88
85	90
99	100
91	101
86	96

23. Draw a dot plot and calculate the mean and standard deviation.

24. Determine the interval that represents 1 standard deviation on either side of the mean. Calculate the proportion of data values that lie in this interval, and show this on your dot plot.

25. Use a normal distribution to estimate the proportion of data values that lie between 85 and 92 minutes.

26. Compare your answers to Items 24 and 25.

27. Write a statement that explains how a normal distribution is related to the data on eruptions of Old Faithful.

ACTIVITY 37 PRACTICE

Write your answers on notebook paper. Show your work.

A restaurant specializing in Mexican food offers nine different choices of enchiladas. They vary in cost and in sodium content. Data for the nine options are given in the table below.

Option	Cost/Serving (dollars)	Sodium Content (mg)
1	3.03	780
2	1.07	1570
3	1.28	1500
4	1.53	1370
5	1.05	1700
6	1.27	1330
7	2.34	440
8	2.47	520
9	2.09	660

1. A dot plot and a box plot of the Cost/Serving amounts are shown below. What feature of the data set is apparent in the box plot but not particularly apparent in the dot plot?

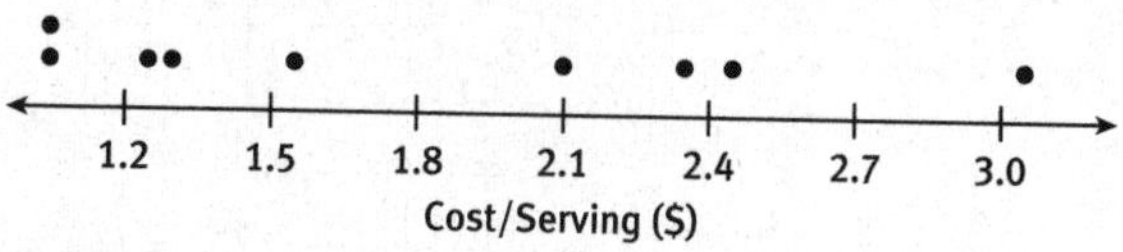

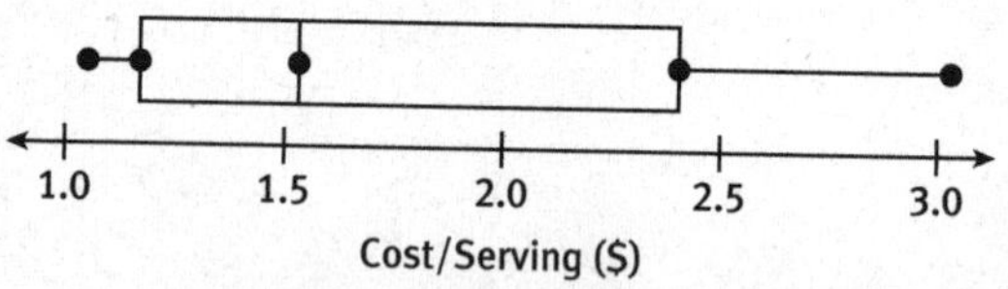

2. A dot plot and a box plot of the Sodium Content amounts are shown below. What feature of the data set is hidden in the box plot but apparent in the dot plot?

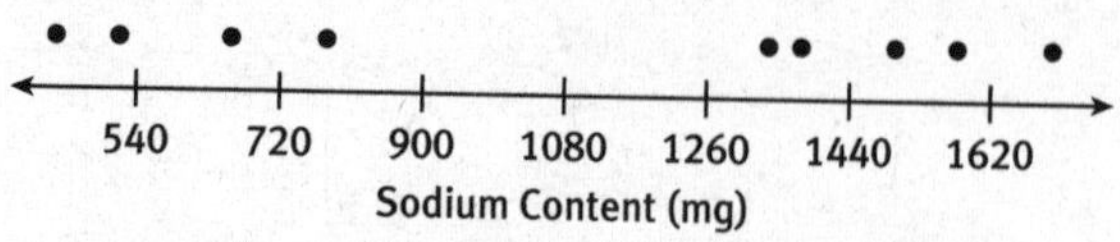

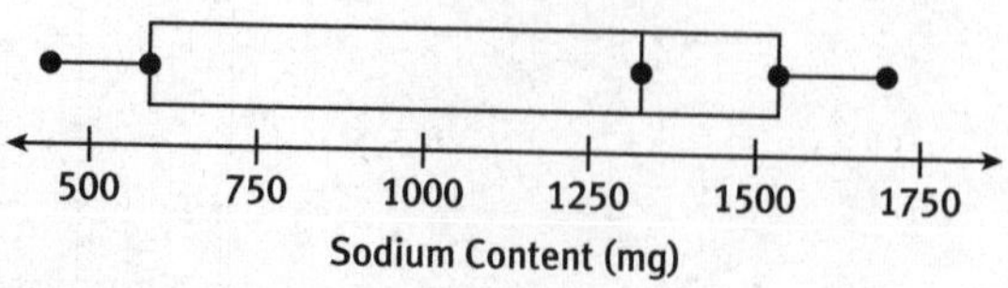

3. In a study of raptors in the western United States, 110 Cooper's hawks were trapped and their weights (in grams) recorded. A dot plot of these weights is shown below. What interesting feature do you notice about this data set? What do you think might be the reason for this interesting feature?

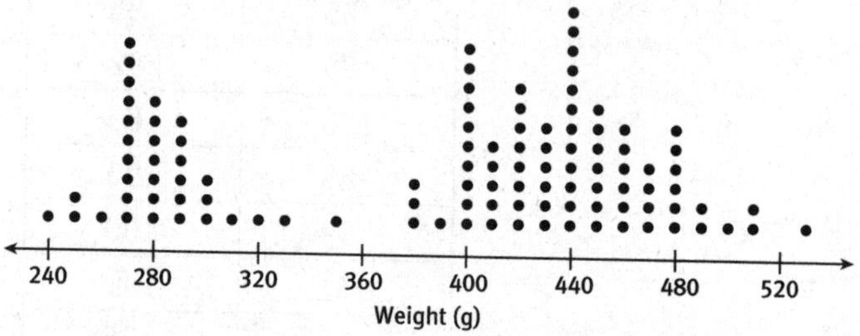

For winter sports enthusiasts, the thickness of ice is a significant safety issue. The Minnesota Department of Natural Resources recommends that ice thickness be at least 4 inches for walking or skating on the ice, and at least 5 inches for operating a snowmobile or all-terrain vehicle on the ice. Ice thicknesses (in inches) were measured at 10 randomly selected locations on the surface of a lake. The thicknesses were as follows:

5.8, 6.4, 6.9, 7.2, 5.1, 4.9, 4.3, 5.8, 7.0, 6.8

4. Construct a dot plot of the ice thicknesses.

5. On the basis of your dot plot, do you think it is safe to play hockey on this lake? Explain why or why not.

6. On the basis of your dot plot, do you think it is safe to operate a snowmobile on this lake? Explain why or why not.

7. Calculate the mean ice thickness for the locations in this sample.

8. Calculate the standard deviation of the ice thicknesses.

9. If the mean of the thicknesses were greater and the standard deviation were the same, would you be more worried or less worried about operating a snowmobile on the ice on this lake? Explain.

10. If the mean of the thicknesses were the same and the standard deviation were greater, would you be more worried or less worried about walking or skating on the ice on this lake? Explain.

Distributors of soft drinks are aware that end-aisle displays in stores are effective for increasing sales. A distributor is testing new designs of displays, where the image on the display is varied. Each image pictures one or two smiling people holding an open container of the soft drink. The distributor would like to know which images increase sales the most. The three different images are one man, one woman, and a pair of individuals (one man and one woman). Each image was used in 11 stores for 1 month, and the percent increases in sales compared to the same month of the previous year were recorded. Data from the 33 stores are shown in the table below.

Percent Increase in Sales

Image: One Man	Image: One Woman	Image: Man and Woman
4.79	5.71	8.18
5.71	6.29	9.14
5.74	7.44	9.70
5.54	6.03	9.25
4.43	5.54	7.40
6.42	5.90	9.25
6.07	5.23	8.42
3.97	7.96	8.12
5.84	4.75	9.07
5.55	4.68	7.84
6.76	5.90	8.09
5.62	5.71	9.18

11. For each of these three data sets, calculate the five-number summary and the upper and lower boundaries for outliers.

	Man	Woman	Both
Minimum			
First quartile			
Median			
Third quartile			
Maximum			
Lower outlier boundary			
Upper outlier boundary			

12. Use the information from the table in Item 11 to sketch modified box plots of these three data sets. Be sure to indicate any outliers.

13. In a few sentences, describe the similarities and differences among the three data sets.

14. Which image would you recommend that the distributor use and why?

A regional symphony orchestra needs money to repair their theater, which was seriously damaged by flooding. They have tested three different methods of asking for donations: mail, phone, and direct appeal at social gatherings. Each method was used with 11 potential donors, and the amounts donated for each method are shown below. Use the table for Items 15–17.

Contributions ($)

Mail	Phone	Direct
1000	1700	900
1500	1800	1000
1200	1900	1200
1800	1750	1500
1600	2000	1200
1100	1700	1550
1000	1800	1000
1250	1850	1100
1400	1500	1250
1300	900	1250
1400	1400	1350

15. Complete the table below.

Data Summary Table ($)

	Mail	Phone	Direct
Minimum			
First quartile	1100	1500	1000
Median	1300	1750	1200
Third quartile	1500	1850	1350
Maximum			
Lower outlier boundary			
Upper outlier boundary			

16. Construct modified box plots for the different methods.

17. Based on the data and your box plots, which method would you recommend and why?

18. Sometimes it is not clear whether a box plot is a modified box plot or a standard box plot. If you were looking at a box plot and outliers were not visible, what characteristics of the plot would lead you to believe it was standard rather than modified?

The following values represent the number of states visited by students in a class:

3, 12, 17, 2, 21, 14, 14, 8, 45, 29

Use these data for Items 19–22.

19. Find the interquartile range and any outliers for the data set.

20. If you found an outlier in Item 19, what does this number represent? Does it make sense that this number would be an outlier in this context? Explain your answer.

21. Create a modified box plot for the data.

22. Two new students joined the class, both of whom have visited only two states each. What effects, if any, does this have on the upper and lower boundaries for outliers?

MATHEMATICAL PRACTICES
Use Appropriate Tools Strategically

23. In Item 12 of Lesson 37-1, you were asked about advantages of using box plots and dot plots to describe and compare distributions of scores. Do you think the advantages you found would exist not only for these data, but for numerical data in general? Explain.

Comparing Univariate Distributions

SPLITTING THE BILL

In restaurant settings, groups of diners are faced with the problem of how to pay the bill. Three common methods are (a) each pays his or her own (PYO), (b) the bill is split evenly (SE), and (c) someone else pays the entire bill (FREE).

In a recent experiment, diners were told which method would be used to see if this would have an effect on what people ordered. Twelve diners were told that they would each pay their own share of the bill. The cost of the food and drink ordered by each individual was determined. This was repeated with a different group of 12 people who were told the bill would be split evenly and a third group of 12 diners who were told that the experimenter would pay for everyone. The data from the first two groups (PYO and SE) are summarized in the table and the box plots below.

Cost of Food and Drink (dollars)			
Statistics	Pay Your Own (PYO)	Split Evenly (SE)	Free Meal (FREE)
Minimum:	12.00	22.00	
First quartile:	31.00	40.00	
Median:	39.50	46.50	
Third quartile:	45.75	61.50	
Maximum:	59.00	81.00	
IQR:	14.75	21.50	
Mean:	37.29	50.92	
Standard deviation:	12.54	14.33	

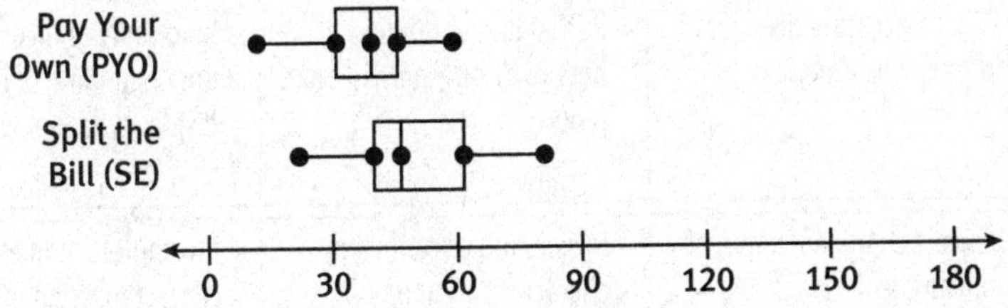

1. Using only the box plots, comment on similarities and differences in the two data sets.
2. Based on the summary statistics in the table, comment on the center and spread of the data sets for those in the PYO group and those in the SE group.

The cost data for the FREE group are shown in the table below.

168	123	101	94	81	75
69	61	59	57	51	49

SPLITTING THE BILL

3. Complete the FREE column in the table from the previous page.

4. On the scale below, sketch the box plot for the FREE group.

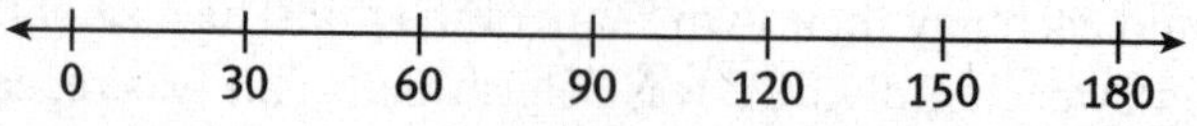

5. If you wanted to describe variability in the FREE group, would you use the *IQR* or the standard deviation? Explain your choice.

Scoring Guide	Exemplary	Proficient	Emerging	Incomplete
	The solution demonstrates the following characteristics:			
Mathematics Knowledge and Thinking (Items 1–5)	• Clear and accurate understanding of univariate data distributions, including center, shape, and spread	• Adequate understanding of univariate data distributions, including center, shape, and spread	• Partial understanding of univariate data distributions, center, shape, and spread	• Little or no understanding of univariate data distributions, center, shape, or spread
Problem Solving (Item 4)	• An appropriate and efficient strategy that results in a correct answer	• A strategy that may include unnecessary steps but results in a correct answer	• A strategy that results in some incorrect answers	• No clear strategy when solving problems
Mathematical Modeling / Representations (Items 1–5)	• Clear and accurate understanding of representations of univariate data • Clear and accurate understanding of how to display data in box plots	• Adequate understanding of representations of univariate data • Mostly accurate displays of data in box plots	• Partial understanding of representations of univariate data • Partial understanding of how to display data in box plots	• Little or no understanding of representations of univariate data • Inaccurate or incomplete understanding of how to display data in box plots
Reasoning and Communication (Items 1, 2, 5)	• Precise use of appropriate math terms and language to characterize univariate data distributions, including center, spread, and variability	• Correct characterization of univariate data distributions, including center, spread, and variability	• Misleading or confusing characterization of univariate data distributions	• Incomplete or inaccurate characterization of univariate data distributions

Correlation

What's the Relationship?

Lesson 38-1 Scatter Plots

My Notes

Learning Targets:

- Describe a linear relationship between two numerical variables in terms of direction and strength.
- Use the correlation coefficient to describe the strength and direction of a linear relationship between two numerical variables.

SUGGESTED LEARNING STRATEGIES: Graphic Organizer, Think-Pair-Share, Create Representations, Predict and Confirm, Quickwrite

Scatter plots are used to visualize the relationship between two numerical variables. When you look at a scatter plot, determine whether there appears to be a relationship (pattern) between the two variables.

For example, consider the following three scatter plots.

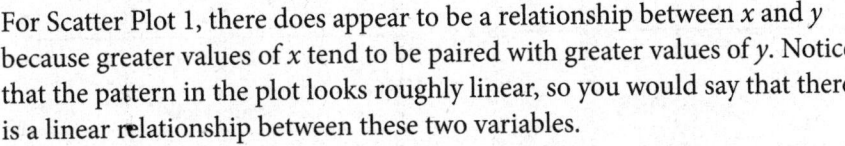

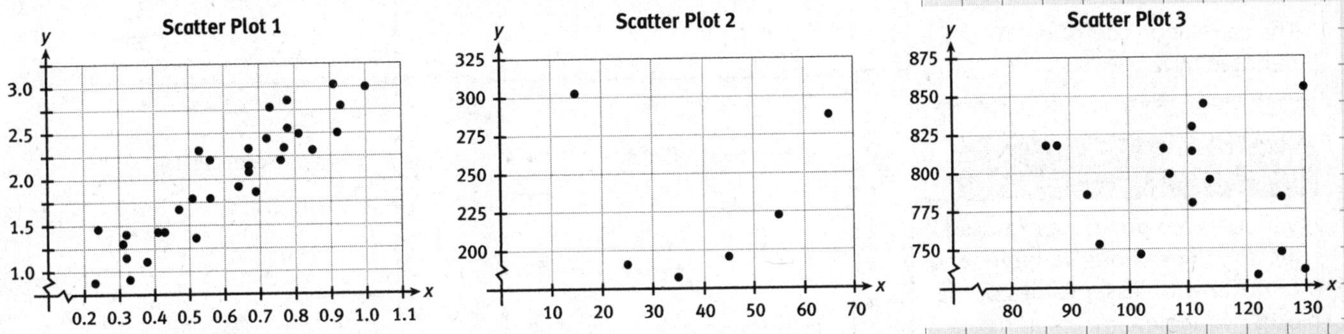

For Scatter Plot 1, there does appear to be a relationship between x and y because greater values of x tend to be paired with greater values of y. Notice that the pattern in the plot looks roughly linear, so you would say that there is a linear relationship between these two variables.

Two numerical variables are related if they tend to vary together in a predictable way.

1. For Scatter Plot 2 above, does there appear to be a relationship between x and y? If so, describe the pattern.

2. For Scatter Plot 3 above, does there appear to be a relationship between x and y? If so, describe the pattern.

My Notes

MATH TERMS

Two numerical variables are **correlated** if one variable tends to increase (or decrease) as the other variable increases.

MATH TERMS

The **correlation coefficient** is a measure of the strength and direction of a linear relationship.

The correlation coefficient is denoted by *r*.

MATH TIP

Variables are **positively related** if lesser values of one variable tend to occur with lesser values of the other variable.

Variables are **negatively related** if lesser values of one variable tend to occur with greater values of the other variable.

DISCUSSION GROUP TIP

As you read and define new terms, discuss their meanings with other group members and make connections to prior learning.

Just as the mean and standard deviation are used to describe center and variability in a data set, there is a summary statistic to describe the strength (how close the points are to a line) and direction (positive or negative) of a linear relationship. This statistic is called the ***correlation coefficient*** and is denoted by *r*.

3. Given below are seven scatter plots and seven verbal descriptions of relationships. Match each scatter plot with the appropriate description. (Each scatter plot goes with one and only one description.)

Scatter Plot 1

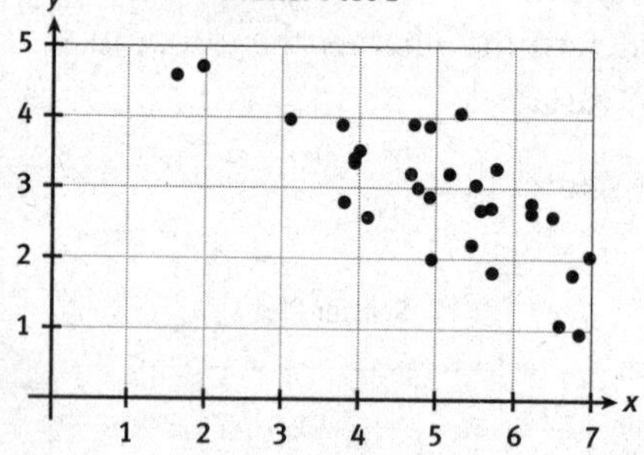

Scatter Plot 2

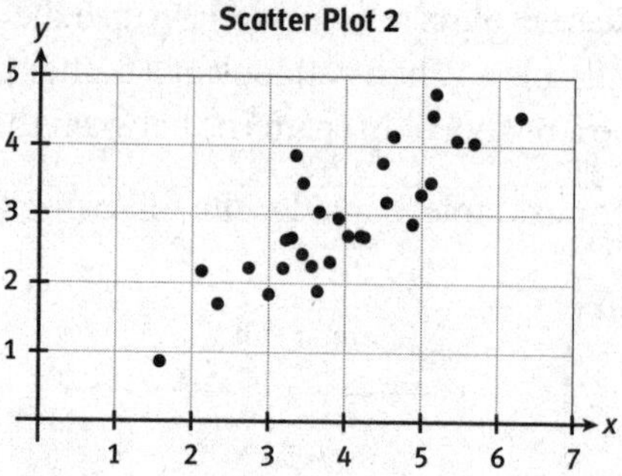

Scatter Plot 3

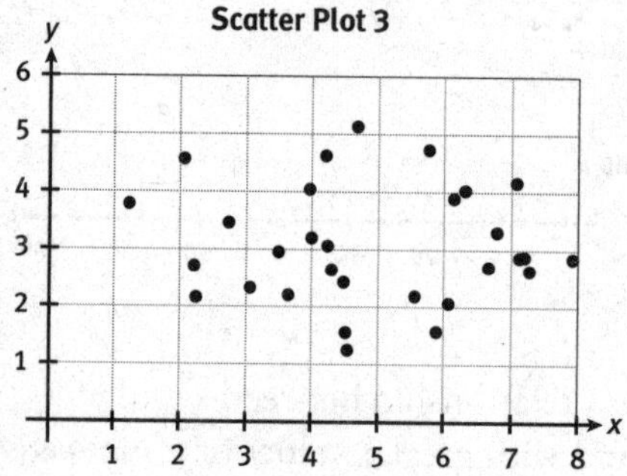

Scatter Plot 4

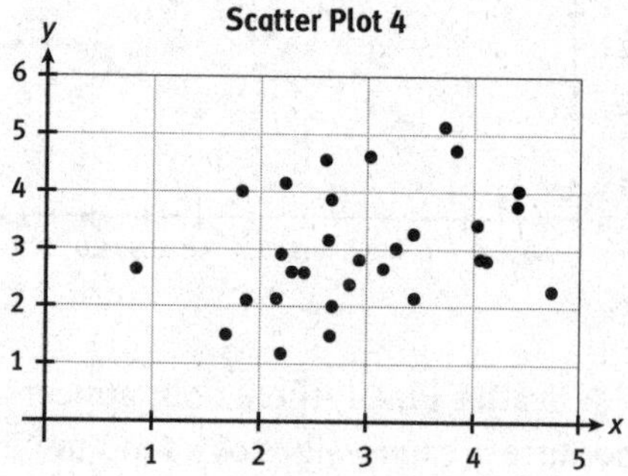

Scatter Plot 5

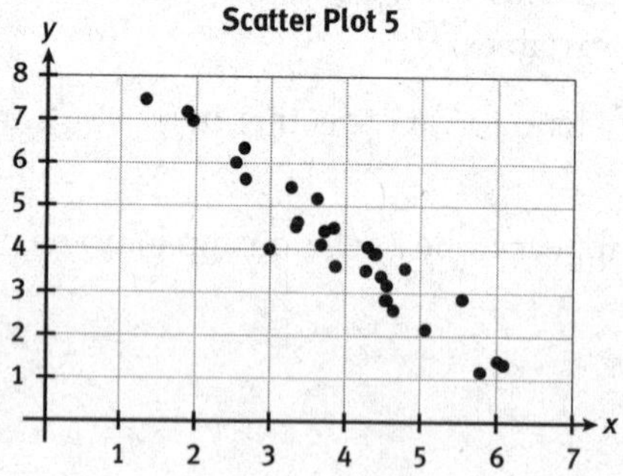

Scatter Plot 6

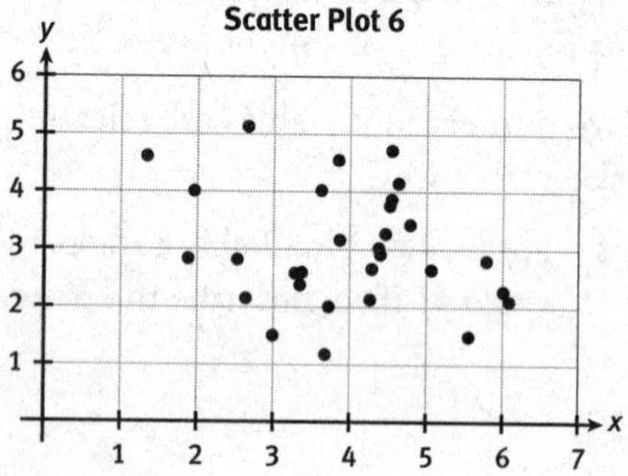

Scatter Plot 7

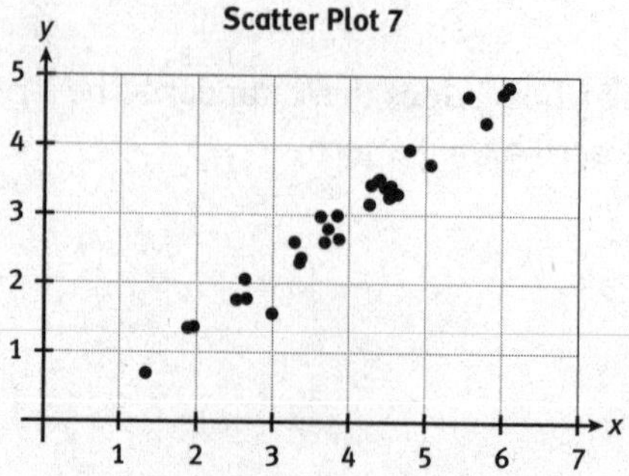

My Notes

A. Very strong positive linear relationship ($r = 0.981$) ______

B. Relatively strong positive linear relationship ($r = 0.828$) ______

C. Relatively weak positive linear relationship ($r = 0.310$) ______

D. Very slight or no linear relationship ($r = 0.043$) ______

E. Relatively weak negative linear relationship ($r = -0.238$) ______

F. Relatively strong negative linear relationship ($r = 0.772$) ______

G. Very strong negative linear relationship ($r = -0.95$) ______

4. What feature(s) of the scatter plots did you consider when deciding whether a relationship was positive or negative?

5. What feature(s) of the scatter plots did you consider when deciding whether a relationship was relatively weak, relatively strong, or very strong?

6. **Make sense of problems.** Examine the values of r for each relationship in Item 3. How does the value of r relate to the scatter plots? What makes r increase or decrease?

Here is a summary of important characteristics of r:

- The value of r quantifies the strength of a linear relationship.
- The sign of r describes the direction of the relationship: positive or negative.
- r ranges in value between -1 (perfect negative linear relationship) and $+1$ (perfect positive linear relationship).

My Notes

Check Your Understanding

The following table displays costs to travel, round-trip, to various cities from Cedar Rapids, Iowa. The costs are calculated assuming a June 1 departure and a 3-day stay. Driving costs were calculated based on $0.20 per mile.

Travel Cost (dollars)			
Destination	**Train**	**Plane**	**Car**
New York City	268	391	204
Chicago	74	453	49
Atlanta	483	703	168
Washington, D.C.	254	577	186
New Orleans	338	342	189
Denver	221	384	160
Albuquerque	354	486	222
Seattle	510	647	367
San Francisco	290	435	385
Los Angeles	390	299	362
Kansas City	184	523	64

Scatter plots of cost versus distance for each of the three travel methods are shown below.

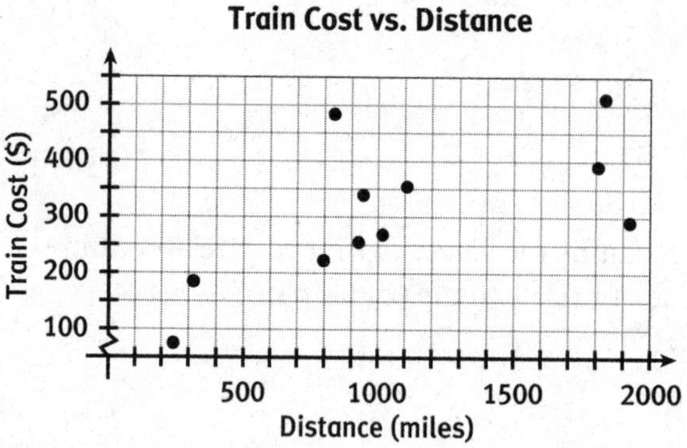

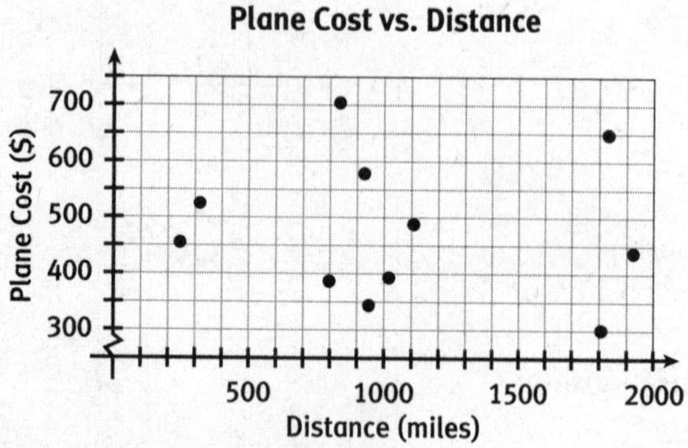

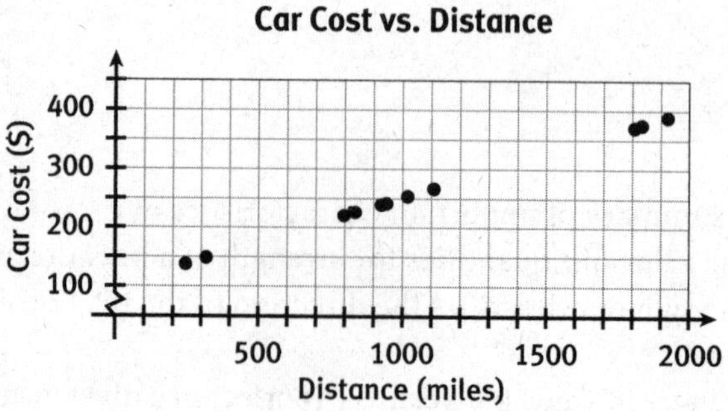

My Notes

7. **Reason abstractly.** How would you describe the relationship between cost and distance for each method of transportation? Be sure to indicate whether you think the relationship is linear and to comment on the strength and direction of the relationship.
   a. Train
   b. Plane
   c. Car

LESSON 38-1 PRACTICE

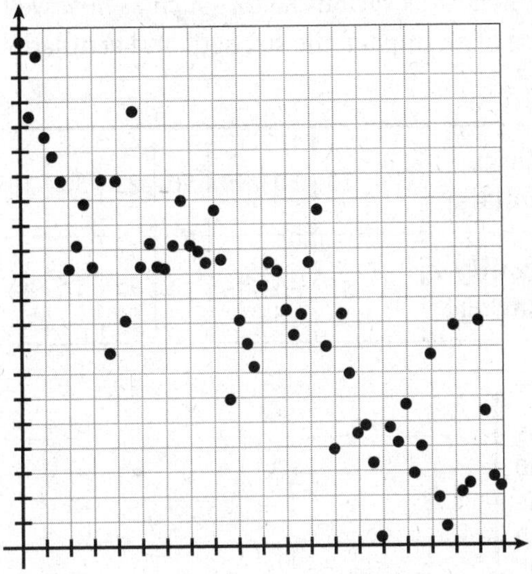

8. Describe the relationship shown in the scatter plot above.
9. **Reason abstractly.** In your own words, describe the similarities and differences between a scatter plot that shows a strong positive relationship and a scatter plot that shows a weak positive relationship.
10. What type of relationship would you expect to see between height and age? Explain your answer.
11. Describe two real-world quantities that would have a strong negative relationship.
12. Describe two real-world quantities that would have no correlation.
13. For positive linear relationships, as the value of r increases, is the linear relationship getting stronger or weaker?

My Notes

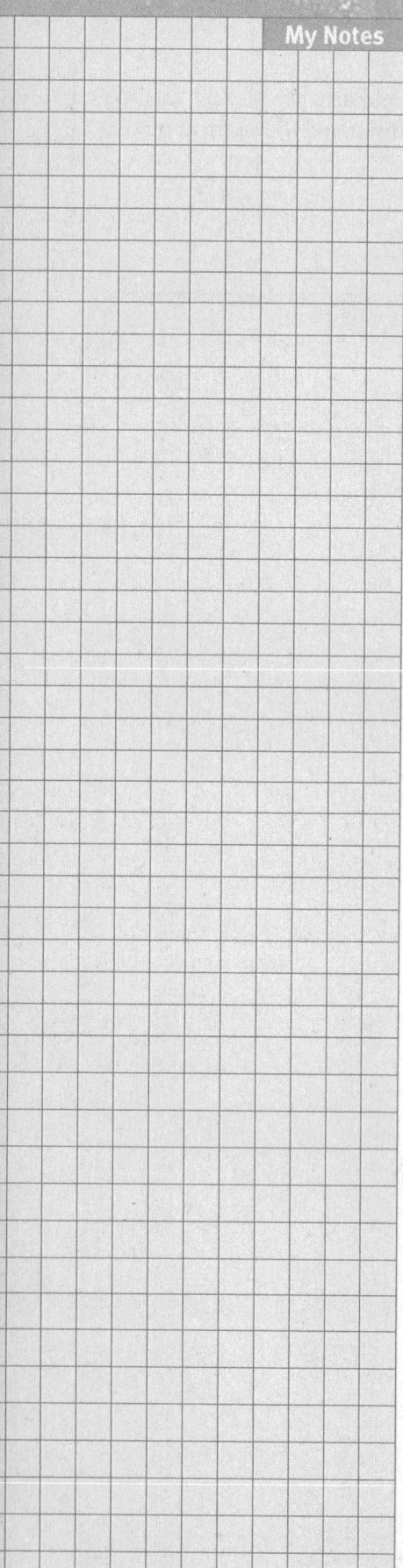

Learning Targets:

- Calculate correlation.
- Distinguish between correlation and causation.

SUGGESTED LEARNING STRATEGIES: Think-Pair-Share, Vocabulary Organizer, Quickwrite

The calculation of *r* gives additional information that helps to describe the data.

This data set shows price (in dollars) and quality ratings for 12 different brands of bike helmets. The quality rating is a number from 0 (worst) to 100 (best) that measures various factors such as how well the helmet absorbed the force of an impact, the strength and ventilation of the helmet, and its ease of use.

Bicycle Helmets													
	Price (dollars)	35	20	30	40	50	23	30	18	40	28	20	25
	Quality Rating	65	61	60	55	54	47	47	43	42	41	40	32

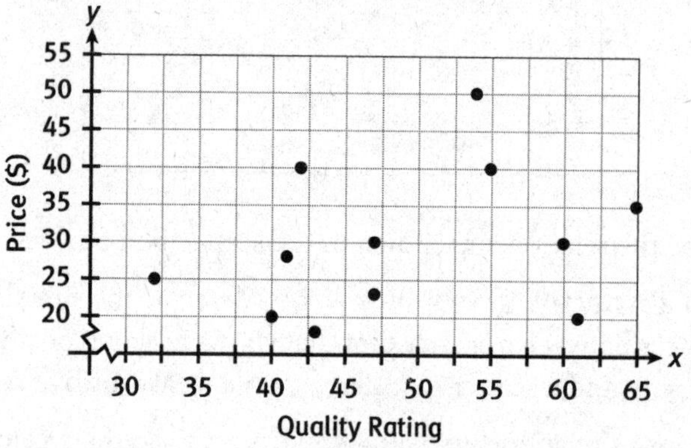

1. **a.** How would you describe the relationship between price and quality rating?

   b. Make a prediction of what you think the correlation coefficient might be.

2. Using a graphing calculator, enter the prices as one list and the quality ratings as another list. What is the value of the correlation coefficient for these two variables?

My Notes

3. **Reason abstractly.** How would you interpret the value of the correlation coefficient in the context of this problem?

At this point you have used scatter plots to visually represent the relationship between two numerical variables and you have used a numerical measure to describe the strength and direction of a linear relationship. When a relationship is uncovered by statistics, the next task is to explain its meaning.

Sometimes the interpretation of a relationship may not be obvious. For example, across European countries there is a positive linear relationship between the number of storks and the number of newborn babies. Do storks bring babies? Are storks attracted by babies? Are both babies and storks brought by the tooth fairy? Do parents with newborns have warmer houses, and therefore their chimneys attract storks looking for warm places to nest?

Humans want to make sense of their world, and sometimes leap too quickly from seeing a correlation to inferring *causation* (a cause-and-effect relationship between two variables). This tendency should be resisted! There are many reasons why two variables might be related other than cause and effect.

Here are some common examples where a correlation should not be interpreted as a cause-and-effect relationship:

- The number of fire engines responding to a fire is positively correlated with the total damage. (Should fewer fire engines—perhaps 0—be sent to fires to reduce damage?)
- The number of people drowning at beaches is positively correlated with ice cream sales. (Is ice cream dangerous?)
- Shoe size is strongly correlated with reading ability. (Should parents start their children off with size 12?)
- The number of doctors per 1000 people is positively correlated with the rate of serious disease. (Are doctors spreading disease?)

4. **Make sense of problems.** For each of the correlations above, what do you think is the correct explanation for the correlation?

Be sure to use common sense when determining causation.

My Notes

Check Your Understanding

5. For each of the following pairs of variables, indicate whether you would expect a positive correlation, a negative correlation, or a correlation close to 0. Explain your choice.

a. Weight of a car and gas mileage
b. Size and selling price of a house
c. Height and weight
d. Height and number of siblings

The table below gives data on age and number of cell phone calls made in a typical day for each person in a random sample of 10 people. Use the table for Items 6–8.

Age (years)	Number of Cell Phone Calls
55	6
33	5
60	1
38	2
55	4
19	15
30	10
33	3
37	3
52	5

6. Sketch a scatter plot of these data using the grid below.

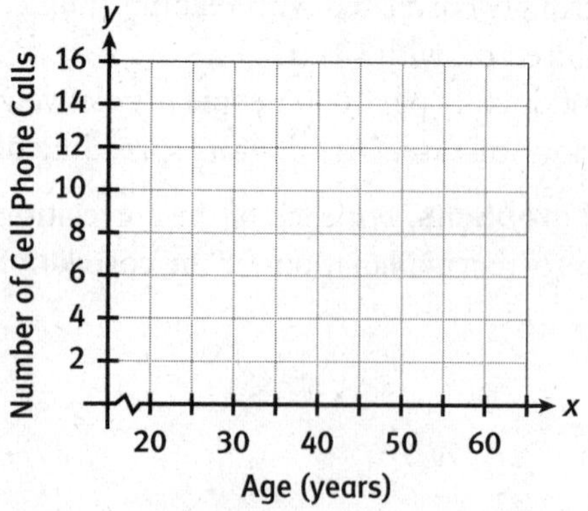

My Notes

7. Describe the direction and strength of the relationship between these two variables.

8. Calculate the value of the correlation coefficient. Do the sign and the magnitude of the correlation coefficient agree with your answer in Item 7? Explain.

9. Suppose that you shot an arrow into the air and kept track of how high it was every 1.0 second. If you made a scatter plot of the data (time, height), the resulting pattern of points would be in the shape of a parabola. Do you feel the correlation coefficient should be used to describe the strength of the relationship between time and height? Why or why not?

LESSON 38-2 PRACTICE

Model with mathematics. Consider tablet computers with 9- to 10-inch screens.

10. For tablet computers, do you think there is a relationship between price and battery life? If so, do you think the relationship is positive or negative?

11. For tablet computers, do you think there is a relationship between price and weight? If so, do you think the relationship is positive or negative?

My Notes

Data for tablet computers with 9- to 10-inch screens are shown in the table and scatter plots below.

9- to 10-inch Tablets		
Price (dollars)	Battery Life (hours)	Weight (pounds)
730	11.6	1.3
570	8.4	1.0
600	11.6	1.3
600	9.3	1.2
600	9.1	1.3
800	8.9	1.3
600	10.5	1.6
850	11.5	1.6
500	11.0	1.6
470	9.0	1.5
500	8.6	1.4
500	8.4	1.6
480	7.4	1.7
500	8.6	1.7
570	7.7	1.7
780	8.1	1.7
580	9.5	2.1

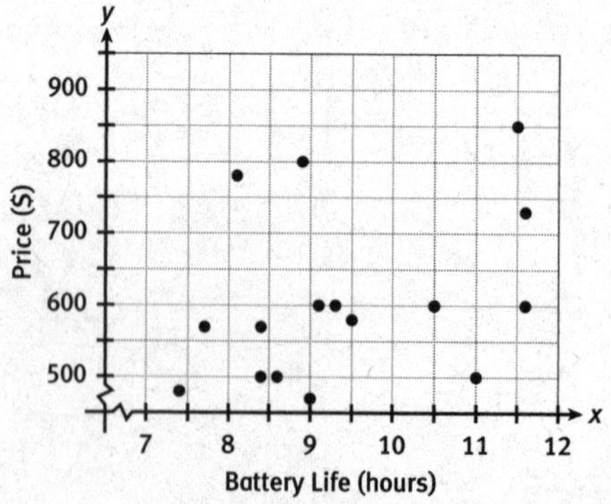

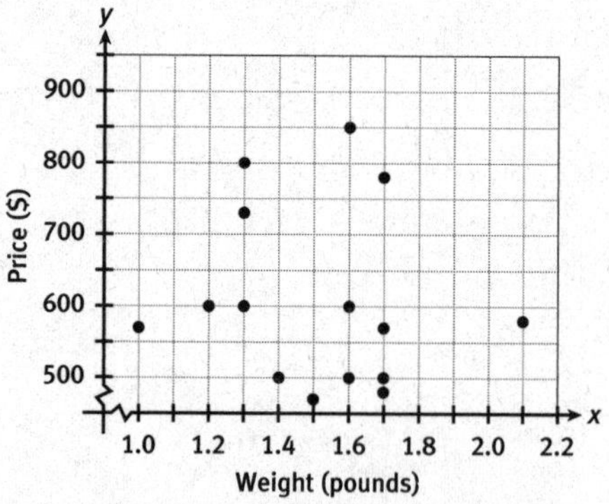

12. Calculate the correlation coefficient for Price and Battery Life.

13. Calculate the correlation coefficient for Price and Weight.

14. Are the values of the correlation coefficients consistent with your predictions in Items 10 and 11? Explain.

ACTIVITY 38 PRACTICE

Write your answers on notebook paper.
Show your work.

1. Describe as precisely as you can how the appearance of a scatter plot showing a positive linear relationship between two quantitative variables differs from the appearance of a scatter plot showing a negative relationship between two quantitative variables.

2. Describe as precisely as you can how the appearance of a scatter plot showing a strong linear relationship between two quantitative variables differs from the appearance of a scatter plot showing a weak linear relationship between two quantitative variables.

The basking shark is the second-largest fish (after the whale shark) swimming in the oceans today. In a study of these creatures, their length and average swimming speed were measured from a safe distance. The results are shown in the table below. Use the table for Items 3–5.

Body Length (meters)	Average Speed (meters/sec)
4.0	0.89
4.5	0.83
4.0	0.76
6.5	0.94
5.5	0.94

3. Sketch a scatter plot of these data.

4. Calculate the correlation coefficient for these data using your available technology.

5. How would you describe this relationship in terms of strength and direction? Support your description with specific references to the scatter plot and/or the correlation coefficient.

One danger of premature human birth is low birth weight. It is thought that low birth weight results in small hippocampus volume, which might be cause for concern because the hippocampus is important in later brain functioning. The scatter plot below displays data from a study of the relationship between hippocampus volume and birth weight in premature infants. The correlation coefficient for these data is $r = 0.51$.

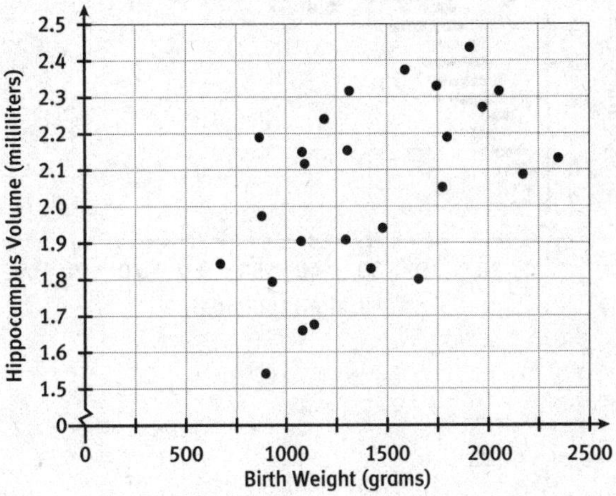

6. Describe the strength and direction of this relationship.

7. Does this relationship appear to be reasonably described as linear? Explain.

When young children are prepared for surgery, a tracheal tube is inserted to allow the unconscious child to breathe. It is very important to get the correct insertion depth. Researchers investigated the relationship between best insertion depth and the weight of the child in a large sample of children, and a scatter plot of their data is shown. The correlation coefficient is $r = 0.878$.

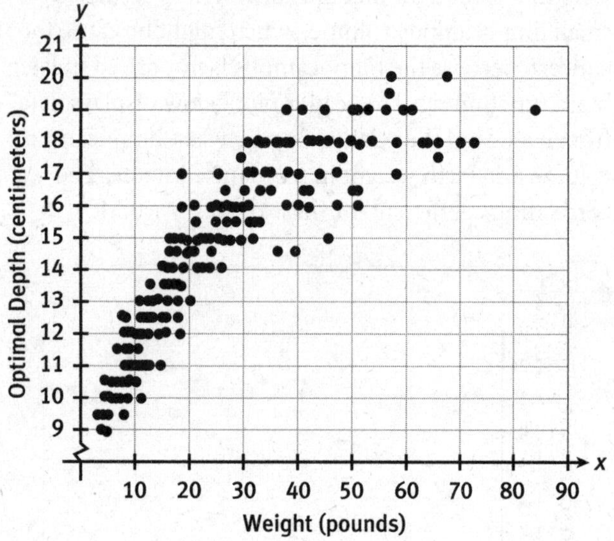

8. Describe the strength and direction of this relationship.

MATHEMATICAL PRACTICES
Reason Abstractly and Quantitatively

9. Does this relationship appear to be reasonably described as linear? Why or why not?

The Best-Fit Line

Regressing Linearly

Lesson 39-1 Line of Best Fit

My Notes

Learning Targets:

- Describe the linear relationship between two numerical variables using the best-fit line.
- Use the equation of the best-fit line to make predictions and compare the predictions to actual values.

SUGGESTED LEARNING STRATEGIES: Look for a Pattern, Interactive Word Wall, Predict and Confirm, Graphic Organizer, Discussion Groups

In recent activities, you created scatter plots as a way to graphically summarize bivariate numerical data. In addition, you learned how to use the correlation coefficient as a numerical summary of the strength and direction of a linear relationship.

In this activity, you will see a way to summarize bivariate numerical data called the "best-fit line." You will use technology to determine the slope and y-intercept of the best-fit line for a data set.

1. The scatter plots below show linear relationships of different strengths and directions. For each scatter plot, use your judgment to draw a line that you feel best represents the linear relationship.

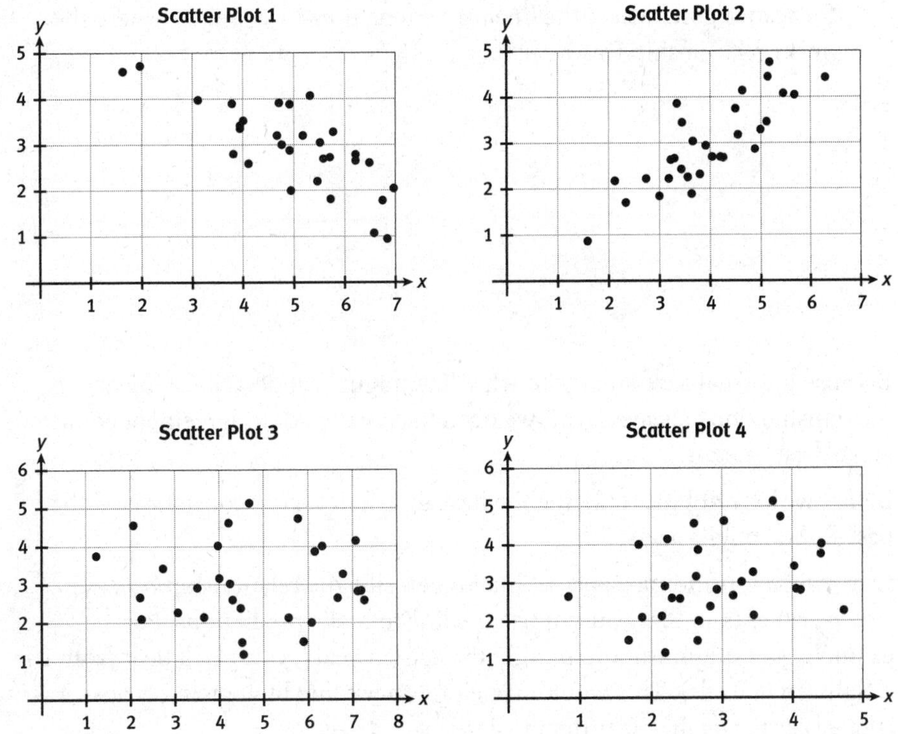

My Notes

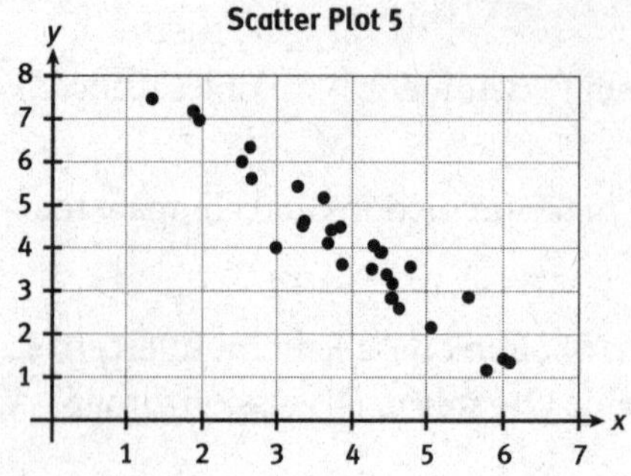

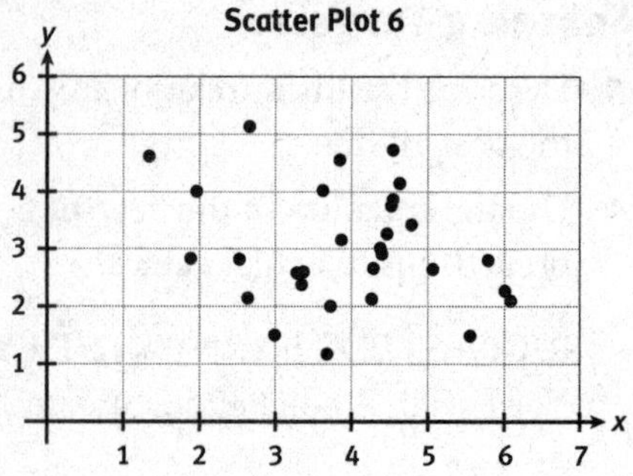

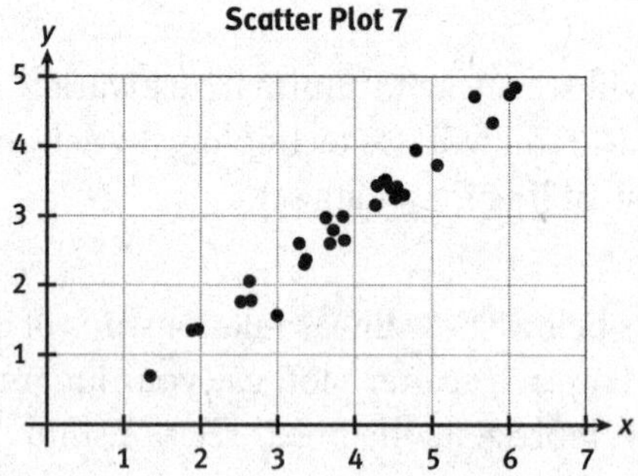

2. Compare the lines you drew with the lines drawn by another student in your class. Did you draw identical lines? Were your lines more similar for scatter plots where the linear relationship was strong or where the linear relationship was weak?

Because informal assessments of what line might best describe a linear relationship don't always agree, we need to come to some agreement about what "best" means.

Before we look at how to define the best-fit line, let's first consider how the best-fit line might be used.

One reason for finding a best-fit line to describe the relationship between two variables is so that you can use the line to make predictions. For example, you might want to predict the age (in years) of a black bear from its weight (in pounds). This would be helpful to wildlife biologists, because it is a lot easier to weigh a bear than to ask a bear its age!

My Notes

Suppose you know that for adult black bears, the relationship between age and weight can be approximately described by the line

$$y = -3.69 + 0.115x$$

where y = age in years and x = weight in pounds. You can use this equation to predict the age of a bear that weighs 100 pounds:

$$\text{predicted age} = -3.69 + 0.115(100) = -3.69 + 11.5 = 7.81 \text{ years}$$

3. Using the equation $y = -3.69 + 0.115x$, what is the predicted age of a bear that weighs 115 pounds?

The line $y = -3.69 + 0.115x$ is the best-fit line for the following data. These data are from a study in which nine black bears of known age were weighed.

Bear	Weight (x)	Age (y)
1	88.2	6.5
2	88.2	7.5
3	92.6	5.5
4	110.3	8.0
5	112.5	10.5
6	112.5	9.5
7	119.1	10.5
8	121.3	9.0
9	130.1	11.5

4. Construct a scatter plot for the bear data.

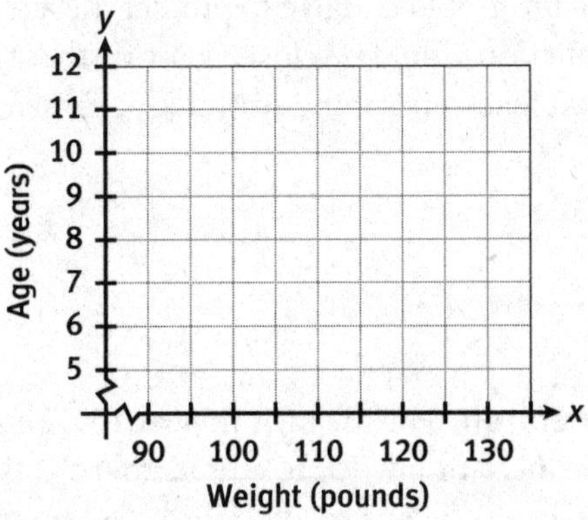

5. Add the best-fit line to your scatter plot.

(*Hint*: Find two points on the line by picking two x values and using the equation of the best-fit line to find the corresponding predicted ages. Then plot these two (x, predicted age) pairs on the scatter plot and draw the line that goes through those two points.)

My Notes

Below is a scatter plot of the bear data, the best-fit line, and a line that is not the best-fit line.

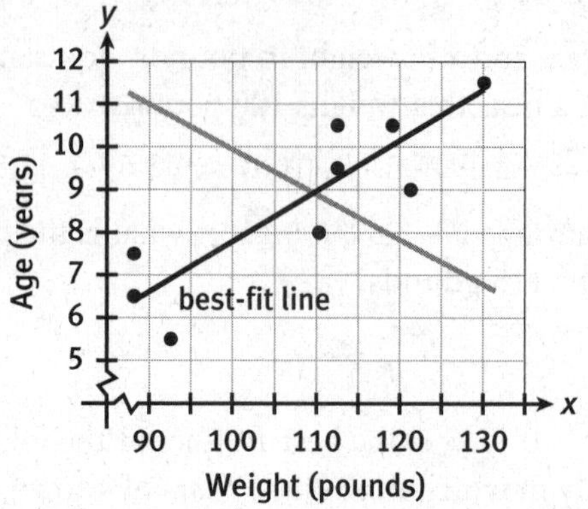

6. Why is the best-fit line a better description of the relationship between age and weight than the other line graphed above?

7. One bear in the data set (Bear 3) was 5.5 years old and weighed 92.6 pounds. If you used the best-fit line ($y = -3.69 + 0.115x$) to predict the age of this bear based on its weight, how far off would you be from the bear's actual age?

8. If you used the line graphed above to predict the age of this bear, do you think your prediction would be closer to or further from the bear's actual age? What feature(s) of the scatter plot shown above supports your answer?

9. **Attend to precision.** For the bear that was 5.5 years old and weighed 92.6 pounds, the best-fit line led to a predicted age that was greater than the bear's actual age. Will age predictions based on the best-fit line be greater than the actual age for *all* of the bears in the data set? If so, explain why. If not, give an example of a bear in the data set for which the predicted age is less than the bear's actual age.

My Notes

Check Your Understanding

10. The scatter plot below shows two lines, Line 1 and Line 2. One of these lines is the best-fit line. Which one is it?

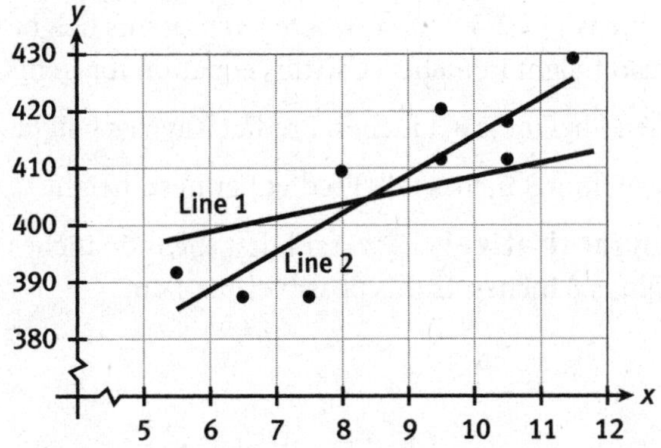

11. Suppose that for students taking a statistics class, the best-fit line for a data set where y is a student's test score (out of 100 points) and x is the number of hours spent studying for the test is $y = 43 + 12x$.

a. What is the predicted test score for a student who studied for one hour?

b. What is the predicted test score for a student who studied for three hours?

My Notes

LESSON 39-1 PRACTICE

Mr. Trent examined some data on head height and a person's actual height and found that a person's height is about 7.5 times his or her head height. "Head height" refers to the distance from the top of the head to the bottom of the chin. Using the data he gathered, Mr. Trent found that the equation of the best-fit line is $y = 2.5 + 7.5x$, where y represents height in inches and x represents head height in inches. Use this equation for Items 12–14.

12. Xavier's head height is 8.3 inches. Predict Xavier's height.

13. Tamisha is 5 foot 3 inches tall. Predict her head height.

14. Reason quantitatively. Tori said that she is 64 inches tall and her head height is 8 inches. Is this possible? Explain.

My Notes

Learning Targets:

- Use technology to determine the equation of the best-fit line.
- Describe the linear relationship between two numerical variables using the best-fit line.
- Use residuals to investigate whether a given line is an appropriate model of the relationship between numerical variables.

SUGGESTED LEARNING STRATEGIES: Predict and Confirm, Look for a Pattern

Below is a scatter plot of the bear data with the best-fit line and the points in the scatter plot labeled according to which bear the data point represents.

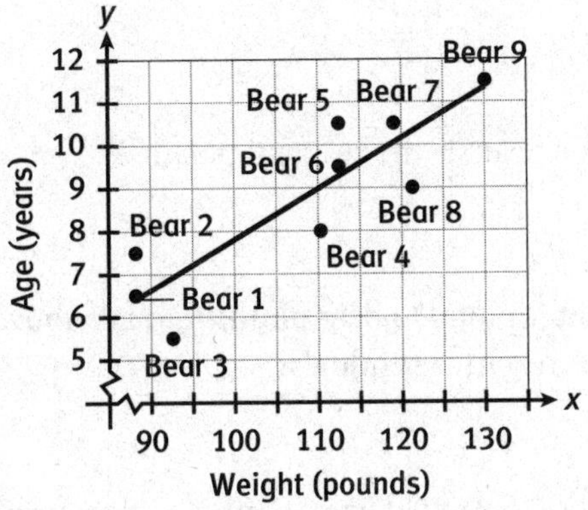

1. For which bears does the best-fit line predict an age that is less than the bear's actual age?

2. Look at the points in the scatter plot that correspond to the bears whose predicted ages are less than their actual ages. What do the points all have in common relative to the best-fit line?

My Notes

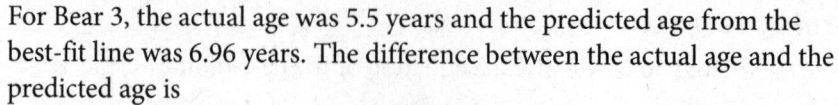

For Bear 3, the actual age was 5.5 years and the predicted age from the best-fit line was 6.96 years. The difference between the actual age and the predicted age is

$$5.5 - 6.96 = -1.46$$

3. Look at the scatter plot and locate the point corresponding to Bear 3. What does 1.46 represent in terms of the scatter plot?

MATH TERMS

A **residual** is a difference between an actual *y*-value and a predicted *y*-value.

Residual = actual *y* − predicted *y*

The difference between an actual *y*-value and a predicted *y*-value is called a ***residual***. A residual is positive when the actual *y*-value is greater than the predicted *y*-value.

4. When is a residual negative?

5. For which of the bears is the residual positive?

6. Look at the scatter plot. Do data points that fall above the best-fit line have positive or negative residuals?

The table below shows the actual ages, predicted ages using the best-fit line, and residuals for the nine bears.

Bear	Age (years)	Predicted Age (years)	Residual
1	6.5	6.45	0.05
2	7.5	6.45	1.05
3	5.5	6.96	−1.46
4	8.0	9.00	−1.00
5	10.5	9.25	1.25
6	9.5	9.25	0.25
7	10.5	10.01	0.49
8	9.0	10.26	−1.26
9	11.5	11.27	0.23

7. Make sense of problems. What is the sum of all nine residuals? Does this value surprise you? Explain why or why not.

Note: The sum of the residuals for the best-fit line is equal to zero. Here, because of rounding in the calculation of the slope and *y*-intercept of the best-fit line and rounding in calculating the predicted values, the sum of the residuals is not exactly 0.

My Notes

8. The scatter plot below shows the best-fit line and another line. If you ignore the sign of the residuals, which line has greater residuals overall? (*Hint*: Look at the distances of points to each of the two lines.)

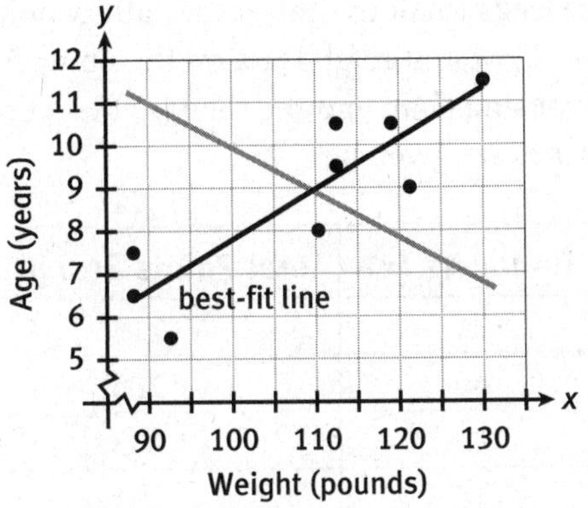

A line is a good description of a bivariate data set if the residuals tend to be small overall. To measure the overall "goodness" of a line, you might think about adding all of the residuals. The problem with this is that some residuals are positive and some are negative, and so you can get a sum that is zero (or close to zero) even for lines that are not good descriptions of the data. So, instead of judging the "goodness" of a line by looking at the sum of the residuals, we look at the **sum of the squared residuals** (**SSR**, for short). The squared residuals are all positive, so positive and negative values don't offset one another.

MATH TIP

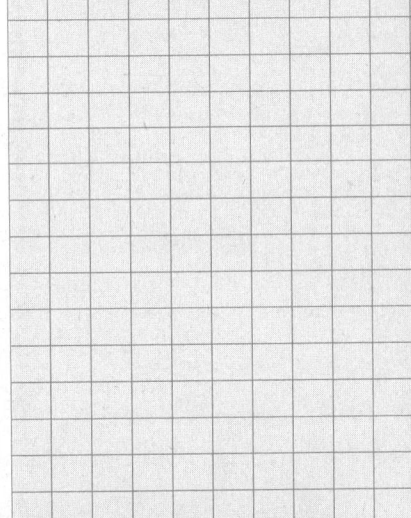

SSR stands for the sum of the squared residuals.

9. Look again at the scatter plot above that shows the bear data and the two different lines. Which line do you think has the lesser SSR? Explain your reasoning.

The ***best-fit line*** for a particular data set is the line that has the least sum of squared residuals (least SSR). In the scatter plot with the two lines, not only does the best-fit line have an SSR less than that of the other line shown, it has an SSR less than that of *any* other line.

MATH TERMS

The **best-fit line** is the line for which the sum of squared residuals (SSR) is less than that of any other line.

Calculating the equation of the best-fit line by hand is very time-consuming, especially if there are a lot of values in the data set. Because of this, you will use a graphing calculator or computer software to do the calculations.

10. Enter the bear data and use technology to verify that the equation of the best-fit line is $y = -3.69 + 0.115x$.

My Notes

Check Your Understanding

The men's basketball coach at Grinnell College employs a style of basketball known as "system ball." The idea behind system ball is that forcing turnovers on defense leads to more shots, especially 3-point shots, on offense, and thus a higher point total. Data on the number of turnovers committed by the opposing team and the total points scored by Grinnell for a sample of seven games are given below.

Turnovers (x)	Total Points Scored (y)
36	115
45	126
26	103
18	106
25	117
31	128
22	96

11. Construct a scatter plot for this data set.

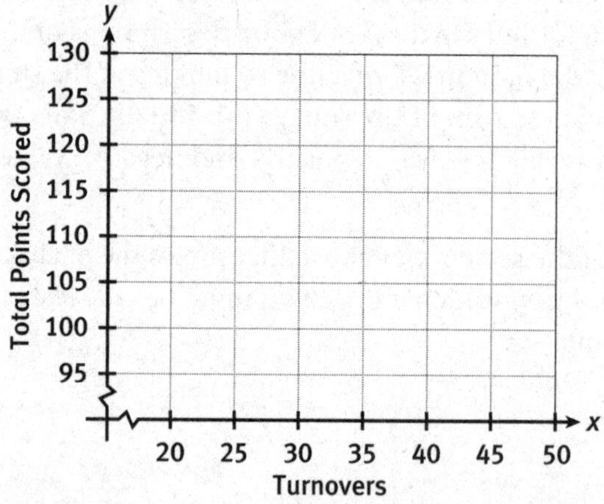

12. Based on the scatter plot, how would you describe the relationship between x and y?

13. Use technology to find the equation of the best-fit line.

14. Use the best-fit line to predict the total points scored in a game with 30 turnovers.

My Notes

LESSON 39-2 PRACTICE

The table below shows the historical minimum wage (in dollars per hour) for the State of New York. Use the table for Items 15–18.

Year (x)	Wage (y)
1962	1.15
1968	1.60
1974	2.00
1978	2.65
1981	3.35
1990	3.80
1991	4.25
2000	5.15
2005	6.00
2012	7.25

15. Construct a scatter plot of the data.

16. Does there appear to be a relationship between the year and the minimum wage? If so, describe the relationship.

17. **Use appropriate tools strategically.** Find the equation for the best-fit line and the correlation coefficient.

18. **Construct viable arguments.** Do the equation you found and the correlation coefficient support your answer in Item 16? Explain.

My Notes

Learning Targets:

- Interpret the slope of the best-fit line in the context of the data.
- Distinguish between scatter plots that show a linear relationship and those where the relationship is not linear.

SUGGESTED LEARNING STRATEGIES: Quickwrite, Look for a Pattern, Visualization, Create Representations, Guess and Check

Once you have determined the equation of the best-fit line, it is possible to interpret the slope of the line. For the bear data, the slope of the best-fit line is 0.115. This means that for each additional pound of weight, the predicted age increases by 0.115 years.

MATH TIP

The slope of the best-fit line can be interpreted as the change in the value of the predicted *y* variable that is associated with a change of 1 in the value of the *x* variable.

1. The best-fit line for a data set where y is the time to complete a task (in seconds) and x is the room temperature (in degrees Fahrenheit) is $y = 128 + 2x$. The slope of this line is 2. Interpret this value in the context of this problem.

2. What does it mean when the slope of the best-fit line is negative?

My Notes

3. The best-fit line for a data set where y is the fuel efficiency of a car (in miles per gallon) and x is the weight of the car (in pounds) is $y = 40 - 0.005x$. The slope of this line is -0.005. Interpret this value in the context of this problem.

It is sometimes also possible to interpret the y-intercept of the best-fit line, but this is not always a sensible thing to do. The y-intercept of the line is the point whose x-coordinate is 0. But it doesn't make sense to predict the age of a bear whose weight is 0 or the fuel efficiency of a car that weighs 0 pounds. So, in most cases, you won't want to interpret the y-intercept.

When is it NOT okay to use the best-fit line to make predictions?

There are three situations when it doesn't make sense to use the best-fit line to make predictions.

You should not use the best-fit line

- to predict a value that is far outside the range of values in the data set used to find the best-fit line.
- when the linear relationship between x and y is very weak, which means that the residuals are very large overall.
- when the relationship between x and y is not linear and would be better described by a curve.

4. Make sense of problems. Why do you think it is not a good idea to predict a value that is far outside the range of the data used to find the best-fit line? For example, why would it not be a good idea to predict the fuel efficiency of a 500-pound car if the data set had only cars that weighed between 2000 and 3500 pounds?

My Notes

5. Even though the best-fit line has an SSR that is less than the SSR of any other line, the residuals might still be large. That is, the points in the data set might still tend to fall far from the line. If this is the case, do you think predictions based on the line will tend to be close to actual y-values? Explain your reasoning.

Sometimes there is a relationship between two numerical variables, but the relationship is not linear. For example, consider the scatter plot below. This scatter plot was constructed using data on y, the time to finish a marathon (in minutes), and x, the age (in years), for six women.

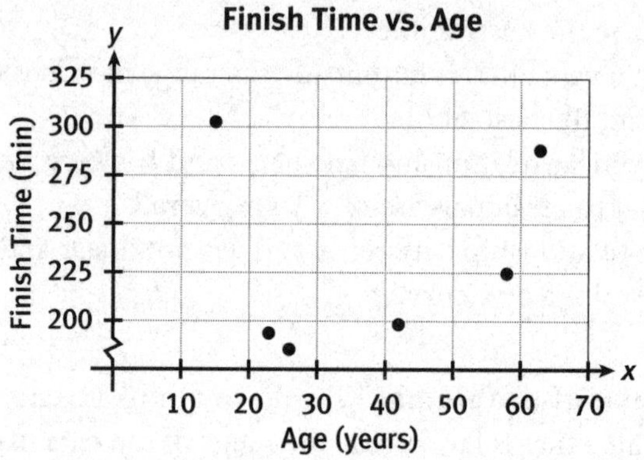

This scatter plot shows a nonlinear relationship between finish time and age.

My Notes

6. Below are scatter plots that show the best-fit line and the best-fit quadratic curve. To predict finish time based on age, would you recommend using the best-fit line or the best-fit quadratic curve? Explain why you made this choice.

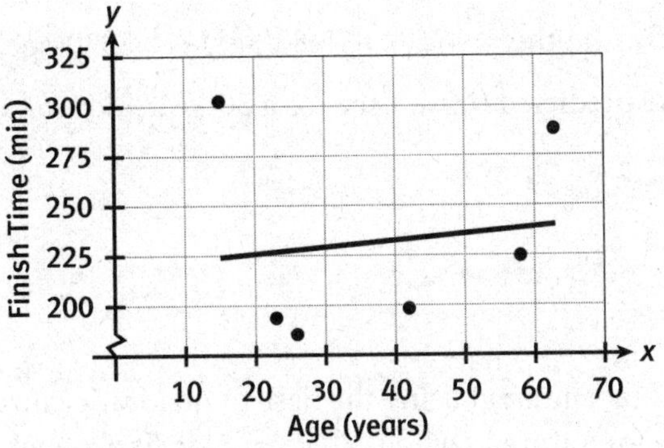

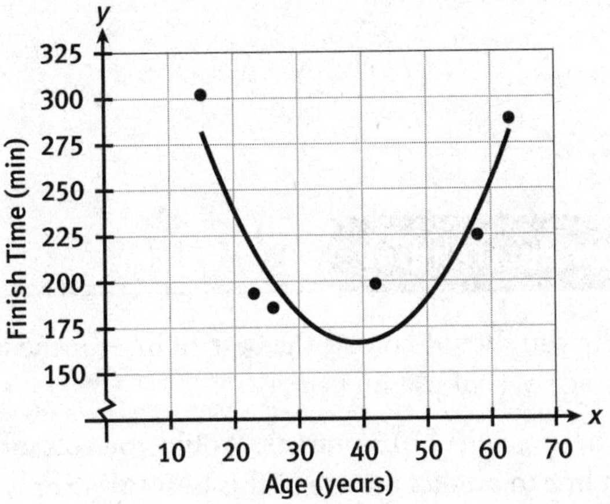

My Notes

7. What do you think it means to say a quadratic curve is the best-fit quadratic curve?

8. The equation of the best-fit quadratic curve for the marathon data is

$$\text{finish time} = 473.3 - 15.80(\text{age}) + 0.2030(\text{age})^2$$

What is the predicted finish time for a 30-year-old woman?

9. Would you recommend using the best-fit quadratic curve to predict the finish time for a woman who is 80 years old? Explain why or why not.

Check Your Understanding

10. Explain why you should not use the best-fit line for the bear data to predict the age of a 40-pound bear.

11. Suppose a bear weighs 110.5 pounds. Would you recommend using the best-fit line to predict the age of this bear? Why or why not?

12. The Chestnut High School tennis team recorded the amount of time that each player's last tennis match lasted as well as the number of calories the player burned during the match. Dakota plays number 1 singles and her match lasted 3 hours; she burned 1900 calories. Briana is number 2 singles; she beat her opponent with no problem and her match lasted only 45 minutes. She burned 475 calories. Maria beat her opponent in 1.5 hours and she burned 950 calories. Determine the equation of the best-fit line and interpret the slope in the context of the problem.

My Notes

LESSON 39-3 PRACTICE

Use the following information for Items 13–16.

At a recent football game you noticed that students tend to be near the same height as their parent of the same gender. You surveyed several students and their parents and recorded the data below.

Student's Height (in.)	Parent's Height (in.)
61	56
62	57
63	62
65	66
65	60
66	66
67	68
68	63
68	69
70	71
72	68
71	73
73	67
74	71

13. **Model with mathematics.** Draw a scatter plot and describe the relationship between the students' heights and their parents' heights.
14. Using a graphing calculator, determine the equation of the best-fit line and calculate the correlation coefficient.
15. Use the best-fit line to predict the height of your classmate Tyler's father. Tyler is 70 inches tall.
16. **Attend to precision.** Tyler's father came to pick him up and you asked his height. He is 6 feet tall. What is his residual and where does the data point lie in relation to the best-fit line?

My Notes

Learning Targets:

- Create a residual plot given a set of data and the equation of the best-fit line.
- Use residuals to investigate whether a line is an appropriate description of the relationship between numerical variables.

SUGGESTED LEARNING STRATEGIES: Create Representations, Look for a Pattern, Quickwrite

In the marathon data scatter plot from the previous lesson, it is obvious that the relationship between finish time and age is not linear. But sometimes, the nonlinearity of the relationship between two variables isn't always this obvious. One way to decide whether a curve describes a relationship better than a line is to look at a residual plot.

A ***residual plot*** is a scatter plot of the (*x*, residual) pairs.

MATH TERMS

A **residual plot** is a scatter plot of the (*x*, residual) pairs.

To see how to make a residual plot, let's return to the bear data. For the bear data set, the data and the residuals for the best-fit line are shown in the table below.

Bear	Weight (pounds)	Age (years)	Predicted Age (years)	Residual
1	88.2	6.5	6.45	0.05
2	88.2	7.5	6.45	1.05
3	92.6	5.5	6.96	−1.46
4	110.3	8.0	9.00	−1.00
5	112.5	10.5	9.25	1.25
6	112.5	9.5	9.25	0.25
7	119.1	10.5	10.01	0.49
8	121.3	9.0	10.26	−1.26
9	130.1	11.5	11.27	0.23

The first (*x*, residual) pair is (88.2, 0.05). This point has been graphed below.

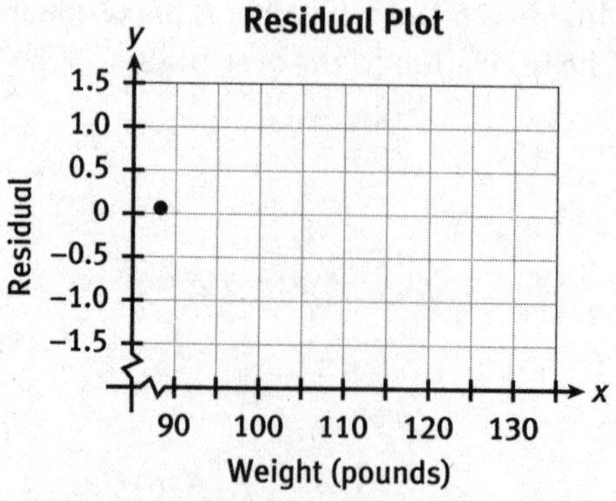

1. Complete the residual plot by adding the other eight (*x*, residual) points to the plot.

Notice that there is no pattern in the residual plot for the bear data best-fit line. The points appear to be scattered at random in this plot.

Now look at a residual plot for the best-fit line for the marathon data set.

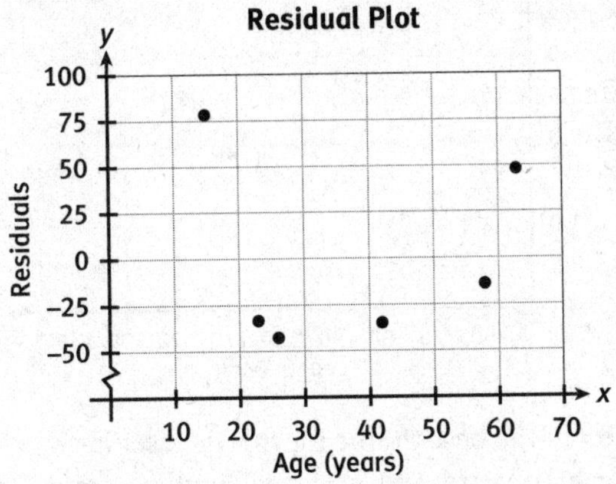

Here there is a very strong curved pattern. It is this pattern in the residual plot that confirms that a line is **not** the best way to describe the relationship between finish time and age.

Now let's look at another example where biologists were interested in predicting the ages of lobsters. The data below are the shell lengths and ages for 12 lobsters.

Shell Length (mm)	Age (years)
63	1.00
83	1.42
109	1.82
111	1.80
118	2.17
143	3.70
90	1.40
125	2.51
136	2.92
75	1.10
142	3.17
148	3.75

My Notes

MATH TIP

A strong pattern in the residual plot for the best-fit line indicates that a line is not the best way to describe the relationship.

A scatter plot of the data is shown below.

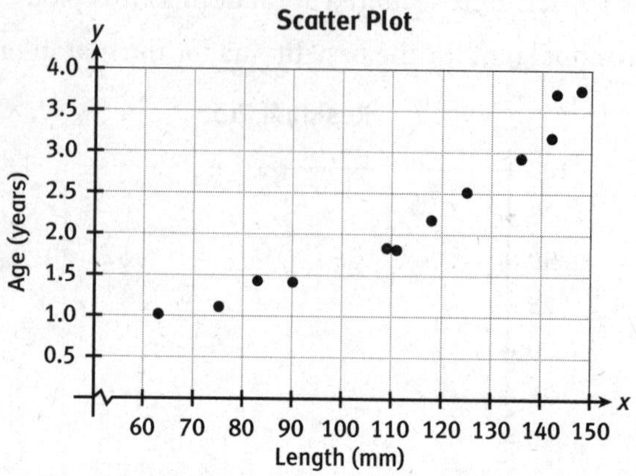

2. The equation of the best-fit line for this data set is $y = -1.41 + 0.0325x$, where y = age (in years) and x = shell length (in mm). Use this equation to complete the following table. Round values to the nearest hundredth if necessary.

Shell Length (mm)	Age (years)	Predicted Age (years)	Residual
63	1.00	0.64	0.36
83	1.42		0.13
109	1.82		−0.31
111	1.80	2.20	−0.40
118	2.17	2.43	
143	3.70	3.24	
90	1.40	1.52	−0.12
125	2.51	2.65	−0.14
136	2.92	3.01	−0.09
75	1.10		
142	3.17		
148	3.75		

A residual plot (scatter plot of the $(x, \text{residual})$ pairs) is shown here:

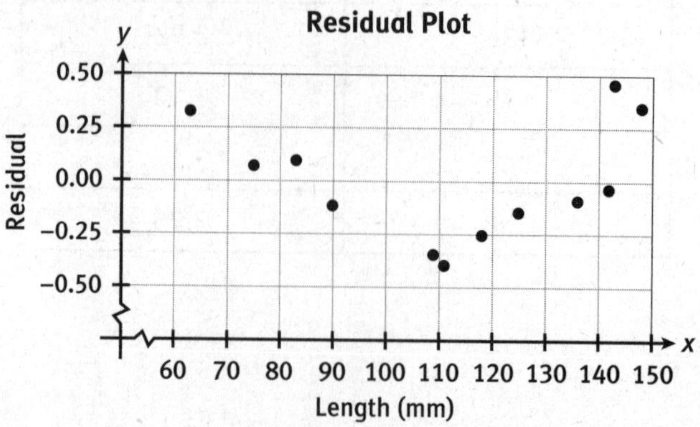

My Notes

3. Describe the pattern in the residual plot.

4. **Critique the reasoning of others.** The biologists who collected these data decided to use an exponential curve to describe the relationship between age and shell length. Do you think this was a reasonable choice? Explain why or why not.

Check Your Understanding

No tortilla chip lover likes soggy chips, so chip makers want to know what makes them crispy. The data below are from an experiment to see how the moisture content of tortilla chips is related to the frying time.

Here, x = frying time (in seconds) and y = moisture content (in percent).

Frying Time (x)	Moisture Content (y)
5	16.3
10	9.7
15	8.1
20	4.2
25	3.4
30	2.9
45	1.9
60	1.3

5. Construct a scatter plot for this data set.

6. Find the equation of the best-fit line.

My Notes

7. Compute the predicted values and the residuals.

x	y	Predicted y	Residual
5	16.3		
10	9.7		
15	8.1		
20	4.2		
25	3.4		
30	2.9		
45	1.9		
60	1.3		

8. Construct a residual plot.

9. Construct viable arguments. Based on the scatter plot and the residual plot, would you recommend using the best-fit line to describe the relationship between x and y? Explain.

My Notes

LESSON 39-4 PRACTICE

10. Why is it important to look at the scatter plot and the residual plot before deciding whether it is appropriate to describe the relationship between two numerical variables using the best-fit line?

11. A recent study of birds of prey resulted in data on $x =$ wing length and $y =$ total weight for 16 different species. The scatter plot for these data is shown below. The correlation coefficient is $r = 0.897$. Describe the relationship between these variables in terms of strength, direction, and shape.

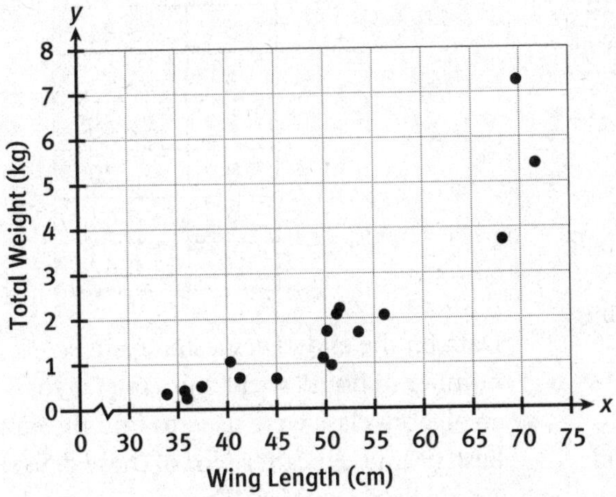

12. The residual plot for the data in Item 11 is shown below. Does the residual plot support your description of the relationship between these two variables? Explain, referring to specific characteristics of the residual plot.

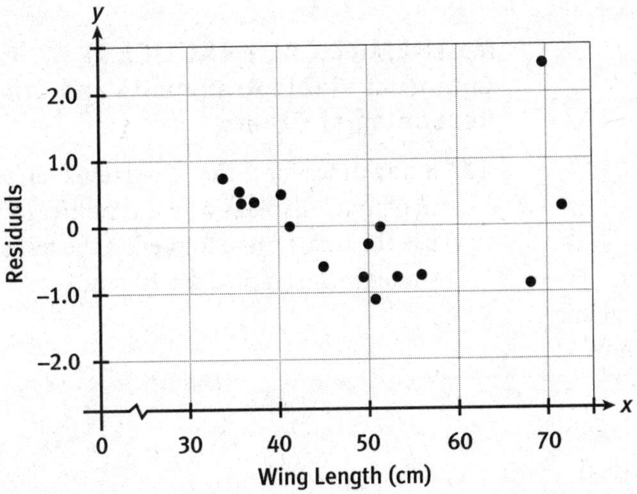

ACTIVITY 39 PRACTICE

Write your answers on notebook paper.
Show your work.

A veterinarian is studying the relationship between the weight of one-year-old golden retrievers in pounds (y) and the amount of dog food the dog is fed each day in pounds (x). A random sample of 10 one-year-old golden retrievers yielded the data in the table below. Use the table for Items 1–8.

Dog	1	2	3	4	5	6	7	8	9	10
x	0.5	1.0	0.6	1.0	1.3	1.5	0.5	0.7	0.8	0.6
y	42	67	47	55	62	71	36	42	50	43

1. Construct a scatter plot of the dog food data.

The equation of the best-fit line is $y = 24.6 + 31.7x$; the correlation coefficient is $r = 0.92$.

2. Interpret the value of the slope of the best-fit line.

3. Does it make sense to interpret the meaning of the y-intercept of the best-fit line?

4. One dog in this data set is fed 0.8 pound of food per day. If you used the best-fit line to predict the weight of this dog, how far off would you be from the dog's actual weight?

5. A box plot of the residuals is shown below. One of the residuals is an outlier. To which dog does this residual belong?

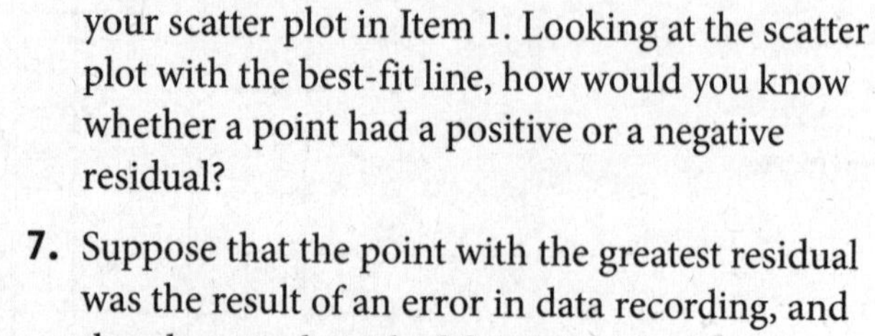

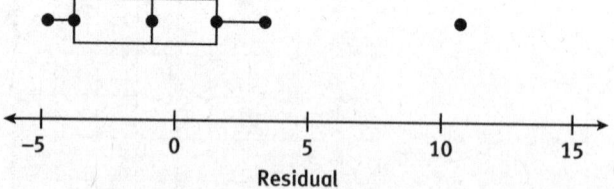

6. Suppose that you graphed the best-fit line on your scatter plot in Item 1. Looking at the scatter plot with the best-fit line, how would you know whether a point had a positive or a negative residual?

7. Suppose that the point with the greatest residual was the result of an error in data recording, and that the actual residual for that point is 2.0. Would the correlation increase or decrease? Explain your reasoning in a few sentences.

8. Below are the 10 residuals for the best-fit line. Construct a residual plot. Are there any patterns in the residual plot that indicate the relationship between x and y is not linear?

Dog	Residual
1	1.55
2	10.7
3	3.38
4	−1.3
5	−3.81
6	−1.15
7	−4.45
8	−4.79
9	0.04
10	−0.62

Data on the end of semester exam scores (y) and the number of hours spent studying (x) for 34 students in an algebra class were used to find the equation of the best-fit line. A scatter plot of these data showed a strong linear pattern. The equation of the best-fit line was $y = 42 + 11x$. Use this information for Items 9–11.

9. Interpret the value of the slope of the best-fit line.

10. Interpret the meaning of the y-intercept of the best-fit line.

MATHEMATICAL PRACTICES
Construct Viable Arguments and Critique the Reasoning of Others

11. Study times for these 34 students ranged from 0 to 6 hours. Explain why it is not reasonable to use the best-fit line to predict the test score of a student who studied for 10 hours.

Bivariate Data

Categorically Speaking

Lesson 40-1 Bivariate Categorical Data

My Notes

Learning Targets:

- Summarize bivariate categorical data in a two-way frequency table.
- Interpret frequencies and relative frequencies in two-way tables.

SUGGESTED LEARNING STRATEGIES: Summarizing, Paraphrasing, Create Representations

In previous activities, you analyzed bivariate numerical data. You measured the strength of association between two numerical variables and learned how to predict the value of one variable given the value of another, using the best-fit line. In this activity, you will work with bivariate categorical data.

The first example you will consider involves a famous data set from a tragic historical event: survival data from the luxury ship SS *Titanic*. The *Titanic* set sail on her maiden voyage on April 10, 1912. On a moonless night, April 14, she struck an iceberg and sank, and many people died. Hundreds of books and scholarly studies have tried to answer the question of how a ship so well constructed could have come to such a sad end, and government inquiries resulted in recommendations for future ship construction and safety regulations. This accident has been the topic of many popular movies, plays, and televisions specials.

Table 1 is taken from the United States Senate report dated May 28, 1912: Investigation into Loss of S. S. "Titanic". The format of the table has been preserved for historical accuracy.

Table 1: Original Senate Data

	On board.			Saved.			Lost.			Percent saved.
	Women and children.	Men.	Total.	Women and children.	Men.	Total.	Women and children.	Men.	Total.	
Passengers										
First class	156	173	329	145	54	199	11	119	130	60
Second class . . .	128	157	285	104	15	119	24	142	166	42
Third class	224	486	710	105	69	174	119	417	536	25
Total passengers	508	816	1,324	354	138	492	154	678	832	. . .
Crew	23	876	899	20	194	214	3	682	685	24
Total	531	1,692	2,223	374	332	706	157	1,360	1,517	32

There are quite a few categorical variables summarized in this table: Person Type (men, women/children); Ticket Class (first, second, third); Role (passengers, crew); and Fate (saved, lost). You will use numbers from the Senate report table to form your own tables.

My Notes

1. Use the Senate report table to answer the following questions:
   a. How many first-class passengers were there?

   b. How many of the first-class passengers were men?

   c. How many male first-class passengers were saved?

2. **Reason quantitatively.** Compare the number of male second-class passengers saved with the number of male third-class passengers saved.

3. Was the percentage of second-class passengers saved greater than or less than the percentage of third-class passengers saved?

4. Based on your answers to Items 2 and 3, does it seem like the second-class or the third-class passengers were at greater risk of death?

MATH TERMS

A **two-way frequency table** summarizes the distribution of values for bivariate categorical data.

To analyze these data we can consider two variables at a time. Bivariate categorical data are often summarized and arranged in a ***two-way frequency table***. A two-way frequency table has rows and columns corresponding to the different possible values of the two categorical variables. For example, suppose you were interested in the variables Ticket Class and Fate. There are three values for Ticket Class and two values for Fate. The data can be summarized using a table with three rows and two columns. You can label the rows with the values for Ticket Class and the columns with the values for Fate.

5. Use the information from the Senate report table to complete the two-way table Ticket Class vs. Fate.

Ticket Class vs. Fate

	Saved	Lost
First class	199	
Second class		
Third class		

Row and column totals are usually also included in a two-way table, as shown below. These totals are called *marginal* totals. The marginal totals describe the univariate distribution for each of the variables. For example, there were a total of 329 passengers with first-class tickets, and a total of 492 passengers were saved. The total number of passengers is shown in the bottom right column, the *grand total.*

Ticket Class vs. Fate Frequencies

	Saved	Lost	Total
First class	199	130	329
Second class	119	166	285
Third class	174	536	710
Total	492	832	1324

Cell counts (199) · *Marginal totals* (492, 832 and 329, 285, 710) · *Grand total* (1324)

It is common to convert these frequencies into *relative frequencies* by dividing by the grand total. These values are expressed in decimal form. As an example, consider the second-class passengers who were saved. The table in Item 5 shows that there were 119 second-class passengers who were saved. The grand total was 1324. Dividing 119 by 1324 gives 0.090 as the relative frequency.

6. Calculate the remaining relative frequencies and complete the two-way table below.

Ticket Class vs. Fate

	Saved	Lost
First class		
Second class	0.090	
Third class		

My Notes

MATH TIP

Frequency refers to the number of times a particular value occurs in a data set.

Relative frequency is the proportion of the time that a particular value occurs in a data set.

Marginal total is the sum of frequencies or relative frequencies in a row or column of a two-way table.

My Notes

Check Your Understanding

In recent years, there has been a great deal of controversy over the use of Native American-related team names and mascots for sports teams in the United States. *Sports Illustrated* magazine commissioned a survey of Native Americans living on reservations, Native Americans living off reservations, and non-Native-American sports fans to determine their feelings. One question asked about the "tomahawk chop" chant used in home games of the Atlanta Braves baseball team. Results are shown in the frequency table below.

7. Complete the frequency table below.

Person vs. Attitude Toward "Tomahawk Chop"
Frequency Table

	Non-NA Fans	NA on Res	NA off Res	Total
Like it	208	24	44	
Don't care	379		66	545
Is objectionable	156	85	24	265
Total	743	209		

8. Use the frequencies from the table above to complete the table below.

Person vs. Attitude Toward "Tomahawk Chop"
Relative Frequency Table

	Non-NA Fans	NA on Res	NA off Res	Total
Like it	0.192	0.022		0.254
Don't care		0.092	0.061	0.502
Is objectionable	0.144	0.078	0.022	
Total	0.684	0.192	0.123	

9. What is the relative frequency of non-Native-American fans who find the "tomahawk chop" objectionable?

10. What is the marginal frequency of Native Americans living on a reservation?

11. What is the marginal relative frequency of those who find the "tomahawk chop" objectionable?

My Notes

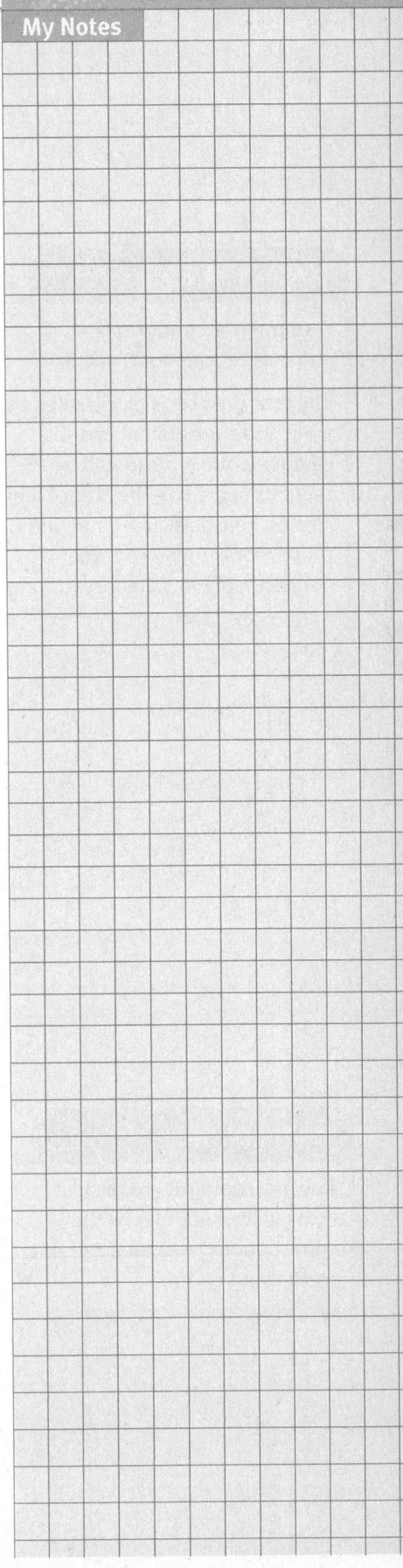

LESSON 40-1 PRACTICE

12. Complete the following table on United States usage of multimedia devices in minutes per day.

United States Usage of Multimedia in Minutes Per Day

Year	Web Browsing	Mobile Applications	Television	Total
2010	70	66	162	
2011	72	94	168	
2012	70	127	168	
Total				

13. Give the frequency of mobile application usage in 2012.

14. Determine the relative frequency of total time spent in 2010 for all usage of multimedia.

15. **Critique the reasoning of others.** Jayson states that people use multimedia devices of some type for 30% of the day. Do you agree with this statement? Why or why not?

My Notes

Learning Targets:

- Interpret frequencies and relative frequencies in two-way tables.
- Recognize and describe patterns of association in two-way tables.

SUGGESTED LEARNING STRATEGIES: Think-Pair-Share, Create Representations, Look for a Pattern

For most people, a visual presentation of data leads to faster understanding of the relationships between pairs of variables. A ***segmented bar graph*** is an effective way to present bivariate categorical data so that these relationships can be easily seen.

To see how a segmented bar graph is constructed, consider data from Major League Baseball.

Major League players are categorized as relief pitchers (RP), starting pitchers (SP), catchers (C), infielders (In), and outfielders (Out). The frequency table below summarizes data on position and team for the 75 players on three different teams (Cubs, Reds, and Pirates).

Frequencies
Positions for Three Different Baseball Teams

	RP	SP	C	In	Out	Total
Cubs	6	6	2	7	4	25
Reds	7	5	2	6	5	25
Pirates	7	4	2	8	4	25
Total	20	15	6	21	13	75

A segmented bar graph can be constructed by first calculating percentages within a row of data. Because we are interested in the percentages in each of the position categories for the different teams, we treat each individual row as a "whole." For each row, compute the percentages by first dividing each number in that row by the corresponding row total and then multiplying by 100.

The ***row percentages*** for the positions on the Cubs team are shown in the table below. Notice that the percentages in the Cubs row add to 100% because we are considering each row separately.

1. Use the frequencies in the table above to complete the following table.

Row Percentages
Positions for Three Different Baseball Teams

	RP	SP	C	In	Out	Total
Cubs	24%	24%	8%	28%	16%	100%
Reds						100%
Pirates						100%

MATH TERMS

A **segmented bar graph** summarizes categorical data.

The total data set is represented by a bar, and the different possible categories are represented by sections of the bar. The area of the section for a particular category is proportional to the relative frequency of that category.

MATH TERMS

Row percentages are calculated by dividing a number by the corresponding row total and then multiplying by 100.

My Notes

The segmented bar graph below shows the percentages of different positions (Outfield, Infield, etc.) for each of the three teams.

Positions for Three Different Baseball Teams

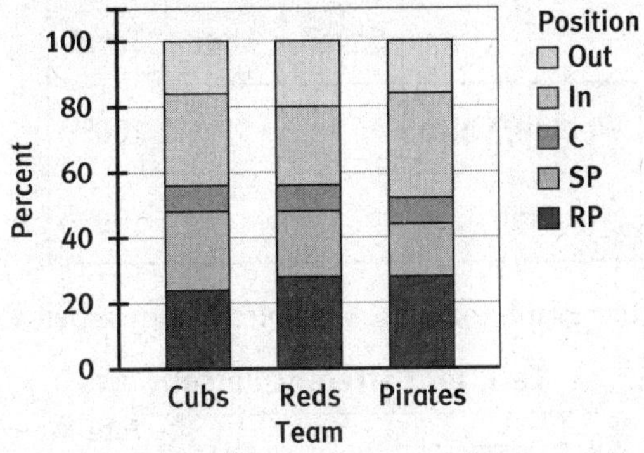

2. **Reason quantitatively.** Does it appear from the segmented bar graph that the distributions of positions are very similar for the three teams, or are they noticeably different? How are the distributions similar or different?

ACADEMIC VOCABULARY

To ***associate*** means to connect or to link.

Two categorical variables are said to be ***associated*** if knowing the value of one of the variables gives you information about the value of the other variable. With categorical data we will ask whether the distribution of values for one variable is similar to or different from the distribution of values for the other variable. A segmented bar graph presents these distributions of values graphically. For the baseball example, the distributions of the position categories (distribution across the columns in each row) are very similar for all three teams (rows). We would say that the categorical variables of position and team are **not** associated. That is, knowing the percentage of pitchers on a team does not provide information about which team it is.

The table below is a two-way frequency table for the variables Person Type and Fate on the *Titanic.*

Person Type vs. Fate

	Saved	Lost	Total
Woman/Child	354	154	508
Man	138	678	816
Total	492	832	1324

My Notes

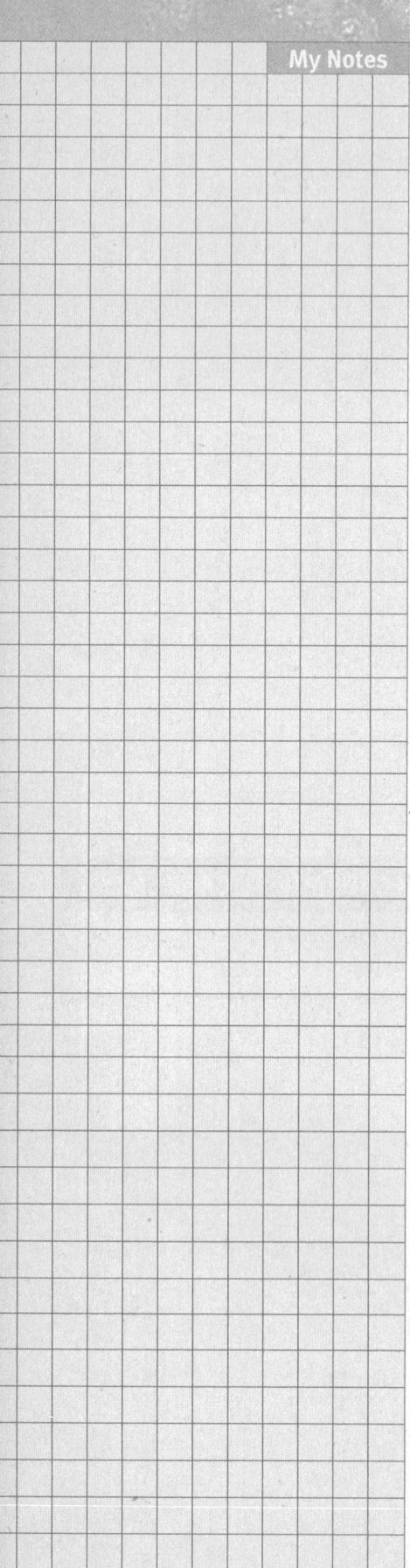

3. Calculate the percentages needed to construct a segmented bar graph by completing the table below.

Person Type vs. Fate

	Saved	Lost	Total
Woman/Child			100%
Man			100%

A segmented bar graph for these variables is shown below.

Fate for Different Persons

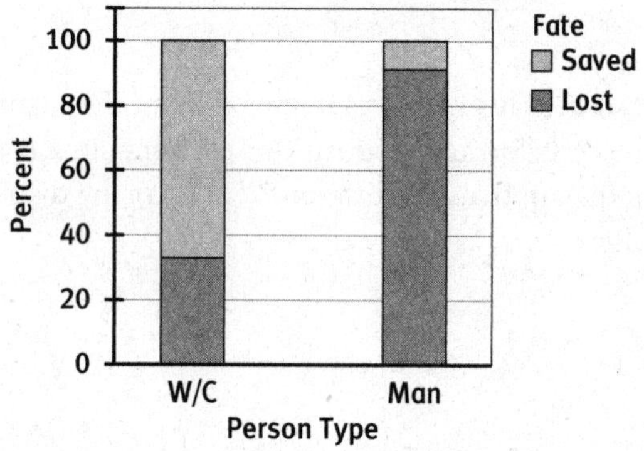

4. About what percent of the area of the bar for women and children (W/C) is ▒? Does this correspond to the percentage of women and children who were saved?

5. About what percentage of the area of the bar for men is ▓? Does this correspond to the percentage of men who were lost?

Compare this segmented bar graph with the segmented bar graph for the baseball players. The segmented bars representing baseball positions were very similar for the three teams, indicating that the position and team are not associated. The segmented bars above are quite different in terms of the percentages of Lost and Saved. There is an association between Fate and Person Type because knowing the person type (for example, women and children) does provide useful information about fate.

My Notes

The table below is a two-way frequency table for the two variables Role and Person Type for the *Titanic* data.

Role vs. Person Type

	Passenger	Crew	Total
Woman/Child	508	23	531
Man	816	876	1692
Total	1324	899	2223

To construct a segmented bar graph, one variable is placed on the horizontal axis, and the vertical axis is labeled and scaled with percentages from 0% to 100%.

The two-way frequency table above shows that 508 of the 531 women and children (96%) were passengers and 23 of 531 (4%) were crew members. The vertical bar for women and children is "segmented" to reflect this distribution of percentages.

Partial Segmented Bar Graph of Role vs. Person Type

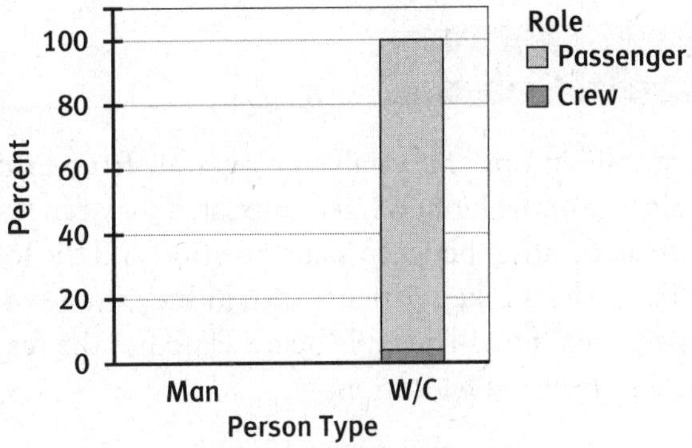

6. Complete the table below, rounding percentages to the nearest whole number.

Role vs. Person Type

	Passenger	Crew	Total
Woman/Child	96%	4%	100%
Man			100%

My Notes

7. Add the missing bar to the segmented bar graph below.

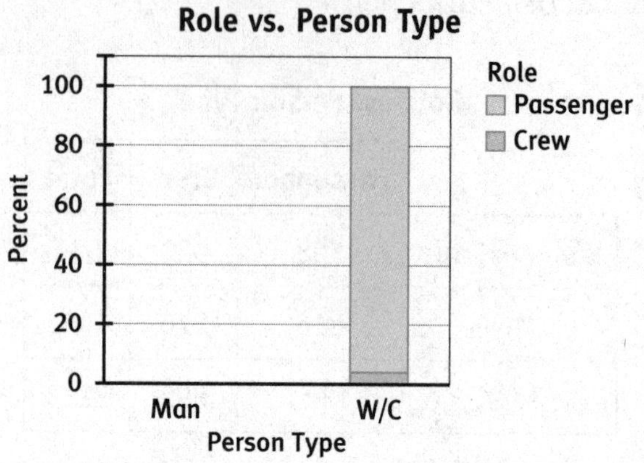

8. **Make sense of problems.** Comment on the association, if any, between these variables.

Check Your Understanding

Kidney stones are solid lumps of crystals that separate from urine and build up on the inner surface of the kidney. If left untreated they can lead to kidney failure. Is there an association between sleep position and the location of kidney stones? Researchers asked patients with kidney stones to identify their preferred sleep positions. The table below classifies the responses by sleep position and kidney stone location.

Frequencies of Kidney Stone Locations and Preferred Sleep Positions

	Left Kidney	Right Kidney	Total
Right Side Down	31	13	44
No Preference	8	9	17
Left Side Down	9	40	49
Total	48	62	110

My Notes

9. From the data summarized in the table above, find the row percentages that would be used to construct a segmented bar graph. Round the percentages to the nearest whole number and use them to complete the following table.

Kidney Stone Locations and Preferred Sleep Positions

	Left Side	Right Side	Total
Right Side Down	70%		100%
No Preference			100%
Left Side Down		82%	100%

A segmented bar graph for the kidney stone data is shown below.

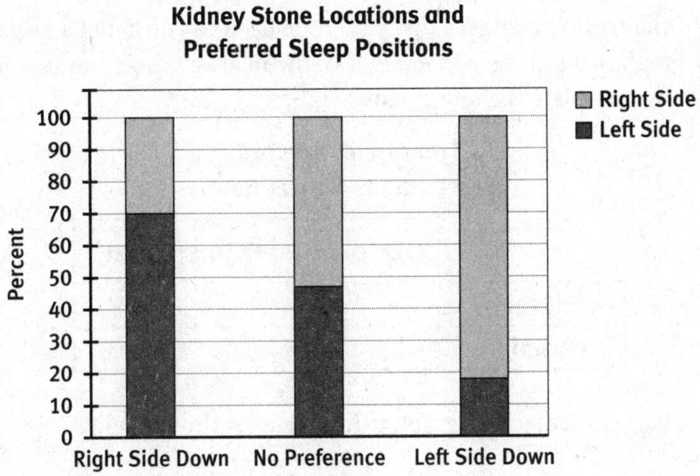

10. Make sense of problems. Does there appear to be an association between sleep position and kidney stone location? What feature(s) of the segmented bar graph support your answer?

11. How is sleep position related to the location of kidney stones?

My Notes

LESSON 40-2 PRACTICE

Does Vitamin C help prevent colds? In a study of 279 French skiers, 140 of them were given a placebo (a sham treatment with no active ingredients) and 139 of them were given Vitamin C. They were followed for one week and whether or not they caught a cold was recorded. The data from this study are shown below.

Treatment vs. Cold
Frequency Table

	Cold	No Cold	Total
Placebo	31	109	140
Vitamin C	17	122	139
Total	48	231	279

12. Find the row percentages that would be used to construct a segmented bar graph. Round the percentages to the nearest whole number and use them to complete the table below.

Treatment vs. Cold
Percentages Across Rows

	Cold	No Cold	Total
Placebo			100%
Vitamin C			100%

13. Sketch a segmented bar graph using the axes shown below.

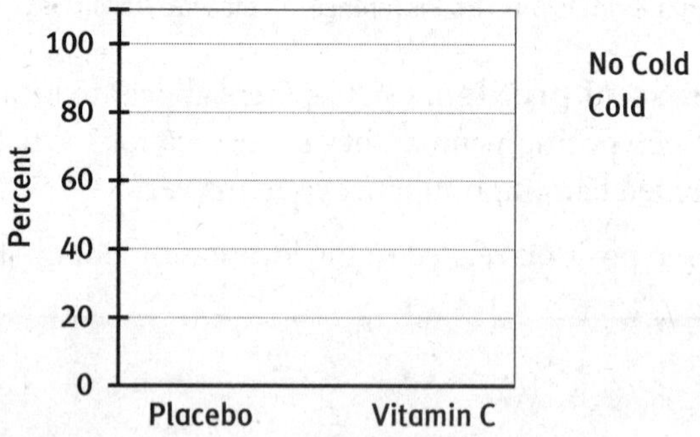

14. Make use of structure. It appears that there is an association between these two variables. Does this association suggest that Vitamin C might help prevent colds? What features of the data and the segmented bar graph support your answer?

ACTIVITY 40 PRACTICE

Write your answers on notebook paper. Show your work.

As part of the United States Census, data are collected on the number of persons in each household. The census data for four decades are summarized below.

Households by Size: Selected Years, 1970 to 2000

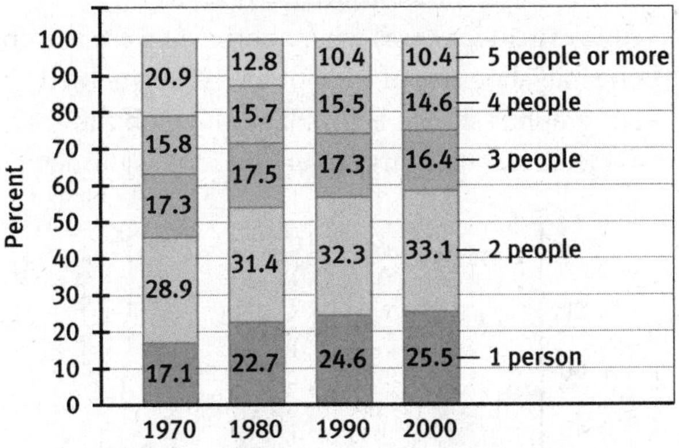

Source: U.S. Census Bureau, Current Population Survey, March Supplements: 1970 to 2000.

1. Which size household increased the most on a percentage basis between 1970 and 2000?
2. Which size household decreased the most on a percentage basis between 1970 and 2000?

Myopia (nearsightedness) is a condition in which a person's eye is slightly longer than it should be, resulting in blurry images of objects far away. Hyperopia (farsightedness) is a condition in which a person's eye is slightly shorter than it should be, resulting in blurry images of near objects. There is some evidence that nighttime light exposure during sleep before the age of two years may be associated with myopia. Data from a survey of parents of children aged 2 to 16 seen at an eye clinic are summarized in the table below.

Vision Condition vs. Light History Frequencies

	Darkness	Night Light	Room Light	Total
Hyperopia	23	17	16	56
Vision OK	66	50	29	145
Myopia	10	34	55	99
Total	99	101	100	300

3. Use the information given in the table above to complete the table of relative frequencies below.

Vision Condition vs. Light History Relative Frequencies

	Darkness	Night Light	Room Light	Total
Hyperopia	0.077	0.057	0.053	
Vision OK	0.220		0.097	0.483
Myopia		0.113	0.183	0.330
Total	0.330	0.337		

4. A segmented bar graph of Vision Condition vs. Light History is shown below. In a few sentences, describe the association between Vision Condition and Light History. (You may assume that room light has more light than a night light.)

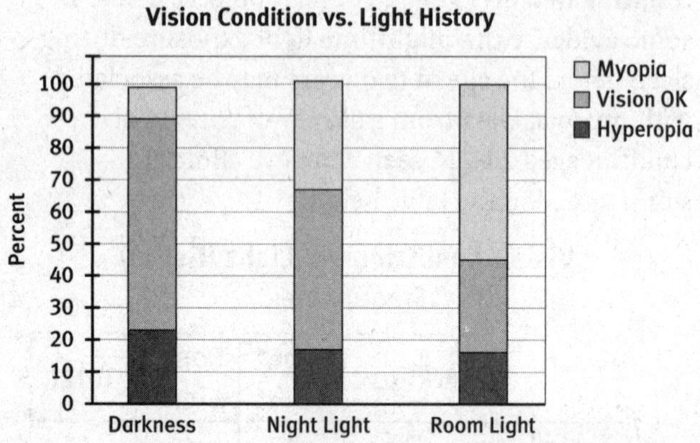

5. The segmented bar graph below shows the relationship between Ticket Class and Fate for those on the *Titanic*. Comment on the association between these two variables.

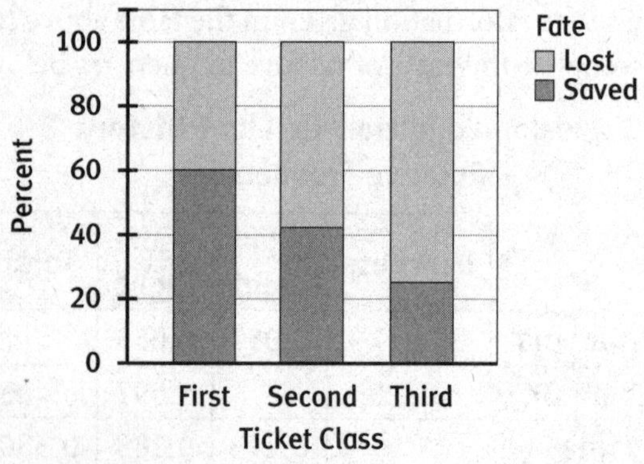

MATHEMATICAL PRACTICES
Make Sense of Problems and Persevere in Solving Them

6. In a study of right-handed men and women, data were gathered on gender and foot asymmetry. An individual was classified as having a left foot more than half a shoe size larger than the right foot (L > R), having a left foot more than half a shoe size smaller than the right foot (L < R) or having the same shoe size for both feet (L = R, which includes cases where both feet were within one half shoe size of each other). A segmented bar graph is shown below. Comment on the association between gender and foot asymmetry.

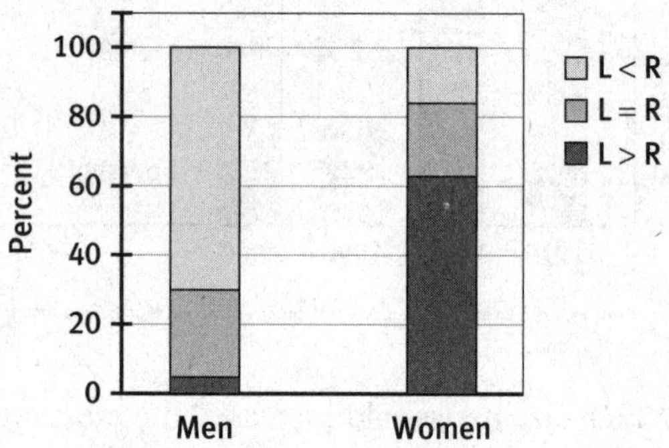

A vital tooth is one that has a functioning nerve. A nonvital tooth is one where the nerve has died. The chances of restoring a nonvital tooth to health are slim. A team of orthodontists wondered if there was a relationship between vital and nonvital tooth temperatures. They measured vital and nonvital tooth temperatures of 32 patients.

A scatter plot of the data and the best-fit line are shown below.

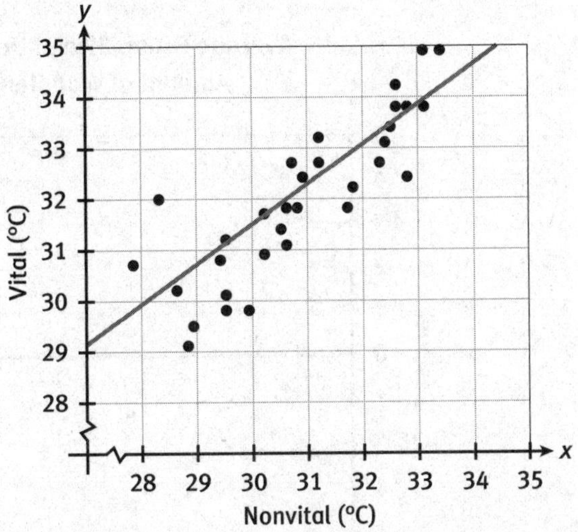

The best-fit line for predicting the vital tooth temperature from the nonvital tooth temperature is given by the equation $y = 6.0 + 0.84x$, where y is the vital tooth temperature and x is the nonvital tooth temperature. The correlation coefficient is $r = 0.856$.

1. Describe the direction of the relationship between the vital and nonvital tooth temperatures. What feature(s) of the graph support your description?

2. Describe the strength of the relationship between the vital and nonvital tooth temperatures. What feature(s) of the graph support your description?

3. One patient has a nonvital tooth temperature of 28.3°C and a vital tooth temperature of 32°C. If you used the best-fit line to predict the vital tooth temperature, how far off would you be from the actual temperature? Is the residual positive or negative?

4. The slope of the best-fit line is 0.84. Interpret this value in the context of this problem.

5. Is it sensible to interpret the y-intercept of the best-fit line for these data? Why or why not?

The social behavior of deer can be affected by their environment. Biologists observed the average group size of deer and the amount of woodland in the surrounding area. They found that as the percentage of woodland increased, deer were seen in smaller groups. Their data and a best-fit line are shown below. A residual plot is also shown. The equation of the best-fit line is $y = 7.8 - 0.069x$, where y is the average group size and x is the percentage of the environment that is woodland. The correlation coefficient is $r = -0.80$.

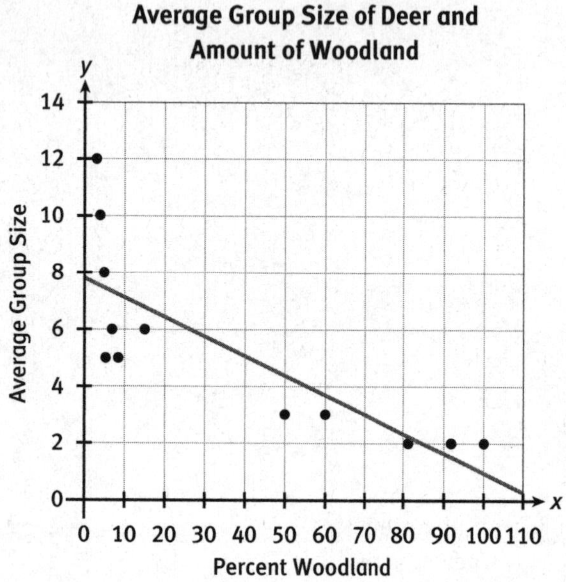

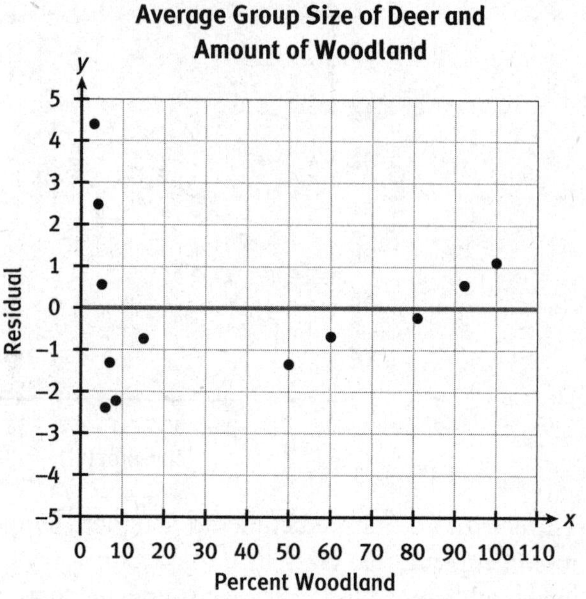

6. Comment on the direction and strength of the relationship between average group size and the percentage of woodland. Is the best-fit line an appropriate way to describe this relationship? Why or why not?

A marketing consultant for a travel agency surveyed 400 travelers about their reasons for traveling and their most important travel concerns. The data are summarized in the table below.

	Most Important Travel Concern		
Reason for Travel	**Hotel Room**	**On-Time Arrival**	**Rental Car Cost**
Business	25	150	25
Leisure	50	50	100

7. Calculate the row percentages and sketch a segmented bar graph for business travelers and leisure travelers.

8. About what percent of the travelers in this sample are most concerned about the rental car cost?

9. What is the biggest difference between the business and leisure travelers? Why do you think this is the case?

10. Is there an association between the reason for travel and the most important travel concern? What feature(s) of the segmented bar graph support your answer?

Scoring Guide	Exemplary	Proficient	Emerging	Incomplete
	The solution demonstrates the following characteristics:			
Mathematics Knowledge and Thinking (Items 1–10)	• Clear and accurate understanding of relationships and associations in bivariate data • Clear and accurate understanding of lines of best fit, correlation coefficients, and residuals	• Recognition of relationships and associations in bivariate data • Adequate understanding of lines of best fit, correlation coefficients, and residuals	• Partial recognition of relationships and associations in bivariate data • Partial understanding of lines of best fit, correlation coefficients, and residuals	• Little or no understanding of relationships and associations in bivariate data • Little or no understanding of lines of best fit, correlation coefficients, and residuals
Problem Solving (Item 3)	• Clear and accurate interpretation of bivariate data to make a prediction	• Interpreting bivariate data to make a prediction	• Difficulty making an accurate prediction from bivariate data	• Inaccurate interpretation of bivariate data
Mathematical Modeling / Representations (Items 1–7, 10)	• Clear and accurate understanding of scatter plots and lines of best fit • Clear and accurate understanding of two-way tables, relative frequency, and segmented bar graphs	• A functional understanding of scatter plots and lines of best fit • A functional understanding of two-way tables, relative frequency, and segmented bar graphs	• Partial understanding of scatter plots and lines of best fit • Partial understanding of two-way tables, relative frequency, and segmented bar graphs	• Little or no understanding of scatter plots and lines of best fit • Inaccurate understanding of two-way tables, relative frequency, and segmented bar graphs
Reasoning and Communication (Items 1, 2, 4–6, 9, 10)	• Precise use of appropriate math terms and language to characterize relationships and associations in bivariate data	• Correct characterization of relationships and associations in bivariate data	• Misleading or confusing characterization of relationships and associations in bivariate data	• Incomplete or inaccurate characterization of relationships and associations in bivariate data

Symbols

$<$	is less than
$>$	is greater than
$\leq$	is less than or equal to
$\geq$	is greater than or equal to
$=$	is equal to
$\neq$	is not equal to
$\approx$	is approximately equal to
$\lvert a \rvert$	absolute value: $\lvert 3 \rvert = 3$; $\lvert -3 \rvert = 3$
$\sqrt{\ }$	square root
%	percent
$\perp$	perpendicular
$\parallel$	parallel
(x, y)	ordered pair
$\overset{\frown}{AB}$	arc AB
$\overleftrightarrow{AB}$	line AB
$\overrightarrow{AB}$	ray AB
$\overline{AB}$	line segment AB
$\angle A$	angle A
$m\angle A$	measure of angle A
$\triangle ABC$	triangle ABC
π	pi; $\pi \approx 3.14$; $\pi \approx \frac{22}{7}$

Resources

Formulas

Perimeter	
P	= sum of the lengths of the sides
Rectangle	$P = 2l + 2w$
Square	$P = 4s$
Circumference	$C = 2\pi r$

Area	
Circle	$A = \pi r^2$
Parallelogram	$A = bh$
Rectangle	$A = lw$
Square	$A = s^2$
Triangle	$A = \frac{1}{2}bh$
Trapezoid	$A = \frac{1}{2}h(b_1 + b_2)$

Surface Area	
Cube	$SA = 6e^2$
Rectangular Prism	$SA = 2lw + 2lh + 2wh$
Cylinder	$SA = 2\pi r^2 + 2\pi rh$
Cone	$SA = \pi r^2 + \pi rl$
Regular Pyramid	$SA = B + \frac{1}{2}pl$
Sphere	$SA = 4\pi r^2$

Volume	
Cylinder	$V = Bh, B = \pi r^2$
Rectangular Prism	$V = lwh$
Triangular Prism	$V = Bh, B = \frac{1}{2}bh$
Pyramid	$V = \frac{1}{3}Bh$
Cone	$V = \frac{1}{3}\pi r^2 h$
Sphere	$V = \frac{4}{3}\pi r^3$

Linear function	
Slope	$m = \frac{y_2 - y_1}{x_2 - x_1}$
Slope-intercept form	$y = mx + b$
Point-slope form	$y - y_1 = m(x - x_1)$
Standard form	$Ax + By = C$

Quadratic Equations	
Standard Form	$ax^2 + bx + c = 0$
Quadratic Formula	$x = \frac{-b \pm \sqrt{b^2 - 4ac}}{2a}$

Other Formulas	
Pythagorean Theorem	$a^2 + b^2 = c^2$, where c is the hypotenuse of a right triangle
Distance	$d = \sqrt{(x_2 - x_1)^2 + (y_2 - y_1)^2}$
Direct variation	$y = kx$
Inverse variation	$y = \frac{k}{x}$

Temperature	
Celsius	$C = \frac{5}{9}(F - 32)$
Fahrenheit	$F = \frac{9}{5}C + 32$

Properties of Real Numbers

Reflexive Property of Equality	For all real numbers a, $a = a$.
Symmetric Property of Equality	For all real numbers a and b, if $a = b$, then $b = a$.
Transitive Property of Equality	For all real numbers a, b, and c, if $a = b$ and $b = c$, then $a = c$.
Substitution Property of Equality	For all real numbers a and b, if $a = b$, then a may be replaced by b.
Additive Identity	For all real numbers a, $a + 0 = 0 + a = a$.
Multiplicative Identity	For all real numbers a, $a \cdot 1 = 1 \cdot a = a$.
Commutative Property of Addition	For all real numbers a and b, $a + b = b + a$.
Commutative Property of Multiplication	For all real numbers a and b, $a \cdot b = b \cdot a$.
Associative Property of Addition	For all real numbers a, b, and c, $(a + b) + c = a + (b + c)$.
Associative Property of Multiplication	For all real numbers a, b, and c, $(a \cdot b) \cdot c = a \cdot (b \cdot c)$.
Distributive Property of Multiplication over Addition	For all real numbers a, b, and c, $a(b + c) = a \cdot b + a \cdot c$.
Additive Inverse	For all real numbers a, there is exactly one real number $-a$ such that $a + (-a) = 0$ and $(-a) + a = 0$.
Multiplicative Inverse	For all real numbers a and b where $a \neq 0$, $b \neq 0$, there is exactly one number $\frac{b}{a}$ such that $\frac{b}{a} \cdot \frac{a}{b} = 1$ and $\frac{a}{b} \cdot \frac{b}{a} = 1$.
Multiplication Property of Zero	For all real numbers a, $a \cdot 0 = 0$ and $0 \cdot a = 0$.
Addition Property of Equality	For all real numbers a, b, and c, if $a = b$, then $a + c = b + c$.
Subtraction Property of Equality	For all real numbers a, b, and c, if $a = b$, then $a - c = b - c$.
Multiplication Property of Equality	For all real numbers a, b, and c, if $a = b$, then $a \cdot c = b \cdot c$.
Division Property of Equality	For all real numbers a, b, and c, $c \neq 0$ if $a = b$, then $\frac{a}{c} = \frac{b}{c}$.
Zero Product Property of Equality	For all real numbers a and b, if $a \cdot b = 0$ then $a = 0$ or $b = 0$ or both a and b equal 0.
Addition Property of Inequality*	For all real numbers a, b, and c, if $a > b$, then $a + c > b + c$.
Subtraction Property of Inequality*	For all real numbers a, b, and c, if $a > b$, then $a - c > b - c$.
Multiplication Property of Inequality *	For all real numbers a, b, and c, $c > 0$, if $a > b$, then $a \cdot c > b \cdot c$. For all real numbers a, b, and c, $c < 0$, if $a > b$, then $a \cdot c < b \cdot c$.
Division Property of Inequality*	For all real numbers a, b, and c, $c > 0$ if $a > b$, then $\frac{a}{c} > \frac{b}{c}$. For all real numbers a, b, and c, $c < 0$ if $a > b$, then $\frac{a}{c} < \frac{b}{c}$.

*These properties are also true for $<$, $\leq$, $\geq$.

Properties of Exponents

For any numbers a and b and all integers m and n,

$a^m \cdot a^n = a^{m+n}$

$(a^m)^n = a^{mn}$

$(ab)^m = a^m b^m$

$\frac{a^m}{a^n} = a^{m-n}, a \neq 0$

$\left(\frac{a}{b}\right)^m = \frac{a^m}{b^m}, b \neq 0$

$a^{-n} = \frac{1}{a^n}, a \neq 0$ and $\frac{1}{a^{-n}} = a^n, a \neq 0$

$a^0 = 1, a \neq 0$

Properties of Radicals

In the expression $\sqrt[n]{a}$,

a is the radicand, $\sqrt{\ }$ is the radical symbol, and n is the root index.

$\sqrt[n]{a} = b$, if $b^n = a$ $\quad$ b is the nth root of a.

$a\sqrt{b} \pm c\sqrt{b} = (a \pm c)\sqrt{b}$, where $b \geq 0$.

$(a\sqrt{b})(c\sqrt{d}) = ac\sqrt{bd}$, where $b \geq 0, d \geq 0$.

$\frac{a\sqrt{b}}{c\sqrt{d}} = \frac{a}{c}\sqrt{\frac{b}{d}}$, where $b \geq 0, c \neq 0, d > 0$.

Table of Measures

Customary	Metric
Distance/Length	
1 foot (ft) = 12 inches (in.) 1 yard (yd) = 3 feet (ft) = 36 inches (in.) 1 mile (mi) = 5280 feet (ft)	1 centimeter (cm) = 10 millimeters (mm) 1 meter (m) = 100 centimeters (cm) 1 kilometer (km) = 1000 meters (m)
Volume	
1 cup (c) = 8 fluid ounces (fl oz) 1 pint (pt) = 2 cups (c) 1 quart (qt) = 2 pints (pt) 1 gallon (gal) = 4 quarts (qt)	1 liter (L) = 1000 milliliters (mL)
Weight/Mass	
1 pound (lb) = 16 ounces (oz)	1 gram (g) = 1000 milligrams (mg) 1 kilogram (kg) = 1000 grams (g)
Time	
1 minute (min) = 60 seconds (sec) 1 hour (hr) = 60 minutes (min) 1 day (d) = 24 hours (hr) 1 week (wk) = 7 days (d)	1 year (yr) = 365 days (d) 1 year (yr) = 52 weeks (wk) 1 year (yr) = 12 months (mo)

SpringBoard Learning Strategies

READING STRATEGIES

STRATEGY	DEFINITION	PURPOSE
Activating Prior Knowledge	Recalling what is known about a concept and using that information to make a connection to a new concept	Helps students establish connections between what they already know and how that knowledge is related to new learning
Chunking the Activity	Grouping a set of items/questions for specific purposes	Provides an opportunity to relate concepts and assess student understanding before moving on to a new concept or grouping
Close Reading	Reading text word for word, sentence by sentence, and line by line to make a detailed analysis of meaning	Assists in developing a comprehensive understanding of the text
Graphic Organizer	Arranging information into maps and charts	Builds comprehension and facilitates discussion by representing information in visual form
Interactive Word Wall	Visually displaying vocabulary words to serve as a classroom reference of words and groups of words as they are introduced, used, and mastered over the course of a year	Provides a visual reference for new concepts, aids understanding for reading and writing, and builds word knowledge and awareness
KWL Chart (Know, Want to Know, Learn)	Activating prior knowledge by identifying what students know, determining what they want to learn, and having them reflect on what they learned	Assists in organizing information and reflecting on learning to build content knowledge and increase comprehension
Marking the Text	Highlighting, underlining, and /or annotating text to focus on key information to help understand the text or solve the problem	Helps the reader identify important information in the text and make notes about the interpretation of tasks required and concepts to apply to reach a solution
Predict and Confirm	Making conjectures about what results will develop in an activity; confirming or modifying the conjectures based on outcomes	Stimulates thinking by making, checking, and correcting predictions based on evidence from the outcome
Levels of Questions	Developing literal, interpretive, and universal questions about the text while reading the text	Focuses reading, helps in gaining insight into the text by seeking answers, and prepares one for group and class discussions
Paraphrasing	Restating in your own words the essential information in a text or problem description	Assists with comprehension, recall of information, and problem solving
Role Play	Assuming the role of a character in a scenario	Helps interpret and visualize information in a problem
Shared Reading	Reading the text aloud (usually by the teacher) as students follow along silently, or reading a text aloud by the teacher and students	Helps auditory learners do decode, interpret, and analyze challenging text
Summarizing	Giving a brief statement of the main points in a text	Assists with comprehension and provides practice with identifying and restating key information
Think Aloud	Talking through a difficult text or problem by describing what the text means	Helps in comprehending the text, understanding the components of a problem, and thinking about possible paths to a solution
Visualization	Picturing (mentally and/or literally) what is read in the text	Increases reading comprehension and promotes active engagement with the text
Vocabulary Organizer	Using a graphic organizer to keep an ongoing record of vocabulary words with definitions, pictures, notes, and connections between words	Supports a systematic process of learning vocabulary

SpringBoard Learning Strategies

COLLABORATIVE STRATEGIES

STRATEGY	DEFINITION	PURPOSE
Critique Reasoning	Through collaborative discussion, respond to the arguments of others; question the use of mathematical terminology, assumptions, and conjectures to improve understanding and to justify and communicate conclusions	Helps students learn from each other as they make connections between mathematical concepts and learn to verbalize their understanding and support their arguments with reasoning and data that make sense to peers
Debriefing	Discussing the understanding of a concept to lead to consensus on its meaning	Helps clarify misconceptions and deepen understanding of content
Discussion Groups	Working within groups to discuss content, to create problem solutions, and to explain and justify a solution	Aids understanding through the sharing of ideas, interpretation of concepts, and analysis of problem scenarios
Group Presentation	Presenting information as a collaborative group	Allows opportunities to present collaborative solutions and to share responsibility for delivering information to an audience
Jigsaw	Reading different texts or passages, students become "experts" and then move to a new group to share their information; after sharing, students go back to the original group to share new knowledge	Provides opportunities to summarize and present information to others in a way that facilitates understanding of a text or passage (or multiple texts or passages) without having each student read all texts
Sharing and Responding	Communicating with another person or a small group of peers who respond to a piece of writing or proposed problem solution	Gives students the opportunity to discuss their work with peers, to make suggestions for improvement to the work of others, and/or to receive appropriate and relevant feedback on their own work
Think-Pair-Share	Thinking through a problem alone, pairing with a partner to share ideas, and concluding by sharing results with the class	Enables the development of initial ideas that are then tested with a partner in preparation for revising ideas and sharing them with a larger group

WRITING STRATEGIES

Drafting	Writing a text in an initial form	Assists in getting first thoughts in written form and ready for revising and refining
Note Taking	Creating a record of information while reading a text or listening to a speaker	Helps in organizing ideas and processing information
Prewriting	Brainstorming, either alone or in groups, and refining thoughts and organizing ideas prior to writing	Provides a tool for beginning the writing process and determining the focus of the writing
Quickwrite	Writing for a short, specific amount of time about a designated topic	Helps generate ideas in a short time
RAFT (Role of Writer, Audience, Format, and Topic)	Writing a text by consciously choosing a viewpoint (role of the writer), identifying an audience, choosing a format for the writing, and choosing a topic	Provides a framework for communicating in writing and helps focus the writer's ideas for specific points of communication
Self Revision / Peer Revision	Working alone or with a partner to examine a piece of writing for accuracy and clarity	Provides an opportunity to review work and to edit it for clarity of the ideas presented as well as accuracy of grammar, punctuation, and spelling

SpringBoard Learning Strategies

PROBLEM-SOLVING STRATEGIES

Construct an Argument	Use mathematical reasoning to present assumptions about mathematical situations, support conjectures with mathematically relevant and accurate data, and provide a logical progression of ideas leading to a conclusion that makes sense	Helps develop the process of evaluating mathematical information, developing reasoning skills, and enhancing communication skills in supporting conjectures and conclusions
Create a Plan	Analyzing the tasks in a problem and creating a process for completing the tasks by finding information needed for the tasks, interpreting data, choosing how to solve a problem, communicating the results, and verifying accuracy	Assists in breaking tasks into smaller parts and identifying the steps needed to complete the entire task
Create Representations	Creating pictures, tables, graphs, lists, equations, models, and /or verbal expressions to interpret text or data	Helps organize information using multiple ways to present data and to answer a question or show a problem solution
Guess and Check	Guessing the solution to a problem, and then checking that the guess fits the information in the problem and is an accurate solution	Allows exploration of different ways to solve a problem; guess and check may be used when other strategies for solving are not obvious
Identify a Subtask	Breaking a problem into smaller pieces whose outcomes lead to a solution	Helps to organize the pieces of a complex problem and reach a complete solution
Look for a Pattern	Observing information or creating visual representations to find a trend	Helps to identify patterns that may be used to make predictions
Simplify the Problem	Using "friendlier" numbers to solve a problem	Provides insight into the problem or the strategies needed to solve the problem
Work Backward	Tracing a possible answer back through the solution process to the starting point	Provides another way to check possible answers for accuracy
Use Manipulatives	Using objects to examine relationships between the information given	Provides a visual representation of data that supports comprehension of information in a problem

Glossary
Glosario

A

absolute value (p. 49) The distance from a number, n, to zero on a number line, written $|n|$.
valor absoluto (pág. 49) Distancia entre un número y el cero en una recta numérica, escrita como $|n|$.

absolute value equation (p. 49) An equation involving the absolute value of a variable expression.
ecuación con valor absoluto (pág. 49) Ecuación que involucra el valor absoluto de una expresión con variables.

absolute value inequality (p. 54) An inequality involving the absolute value of a variable expression.
desigualdad con valor absoluto (pág. 54) Desigualdad que involucra el valor absoluto de una expresión con variables.

arthithmetic sequence (p. 160) A sequence in which the difference of consecutive terms is constant.
progresión aritmética (pág. 160) Sucesión en la que la diferencia entre términos consecutivos es constante.

axis of symmetry of a parabola (p. 433) A line that passes through the vertex, dividing the parabola into two symmetrical halves.
eje de simetría de una parábola (pág. 433) Recta que pasa a través del vértice, dividiendo la parábola en dos mitades simétricas.

B

base (p. 287) The factor in an exponential expression that is being raised to a power.
basa (pág. 287) El factor en una expresión exponencial que esta siendo elevado a una potencia.

binomial (p. 385) A a sum or difference of two monomials.
binomio (pág. 385) Una suma o diferencia de dos monomios.

boundary line (p. 242) A line that divides the coordinate plane into two regions, called half-planes.
frontera (pág. 242) Recta que divide el plano de coordenadas en dos regiones, llamadas semiplanos.

C

causation (p. 197) The relation between two events, where the second event is a consequence of the first.
causalidad (pág. 197) La relación entre dos eventos, donde el segundo evento es una consecuencia del primer evento.

census (p. 527) A study that gathers information about every member of the population.
censo (pág. 527) Estudio que reúne información acerca de cada miembro de la población.

coefficient (p. 356) A number by which a variable is multiplied. For example, in the term $6x$, 6 is the coefficient.
coeficiente (pág. 356) Número por el cual se multiplica una variable. Por ejemplo, en el término $6x$, 6 es el coeficiente.

coincident lines (p. 264) Lines that occupy the same space or location in the plane and pass through the same set of points.
líneas coincidentes (pág. 264) Líneas que ocupan el mismo espacio o ubicación en el plano y pasan por el mismo conjunto de puntos.

common difference (p. 160) The difference between consecutive terms in an arithmetic sequence.
diferencia común (pág. 160) La diferencia entre los terminos consecutivos de una progresión aritmética.

common ratio (p. 314) The ratio, typically denoted by the letter r, between consecutive terms in a geometric sequence.
proporción común (pág. 314) La proporcion, normalmente denotado por la letra r, entre los terminos consecutivos de una progresión geométrica.

complex number (p. 481) A number of the form $a + bi$ where a and b are real and i is not real.
número complejo (pág. 481) Un número con la forma $a + bi$ donde a y b son número reales e i no es real.

compound inequality (p. 44) Two inequalities joined by the word *and* or by the word *or*.
desigualdad compuesta (pág. 44) Dos desigualdades unidas por la palabra *y* o por la palabra *o*.

compound interest (p. 341) Interest calculated on the total principal plus the interest earned or owed during the previous time period.
interés compuesto (pág. 341) Interés calculado sobre el capital total más el interés devengado o adeudado durante el período anterior.

conjunction (p. 45) Two statements joined by the word *and*.
conjunción (pág. 45) Dos enunciados unidos por la palabra *y*.

consecutive (p. 5) Refers to items that follow each other in order.
consecutivo (pág. 5) Se refiere a los elementos que se suceden en el orden.

constant term (p. 356) A term in an expression that does not change in value because it does not contain a variable. For example, the constant term in the expression $3n + 6$ is 6.
término constante (pág. 356) Término cuyo valor no cambia en una expresión, pues no contiene variables. Por ejemplo, el término constante en la expresión $3n + 6$ es 6.

correlation (p. 197) A collection of data points has a *positive correlation* if it has the property that y tends to increase as x increases. It has a *negative correlation* if y tends to decrease as x increases. A correlation is also known as an **association.**
correlación (pág. 197) Un conjunto de datos tiene una *correlación positiva* si tiene la propiedad de que y tiende a aumentar a medida que aumenta x. Tiene una *correlación negativa* si y tiende a disminuir a medida que aumenta x. Una correlación se conoce también como **asociación.**

cube root (p. 301) The cube root of a number, n, is the number which when used as a factor three times results in the product n.
raíz cúbica (pág. 301) La raíz cúbica de un número n, es el número que usado tres veces como factor da un producto de n.

D

degree of a polynomial (p. 357) The largest degree of any term in the polynomial.
grado de un polinomio (pág. 357) El grado mayor entre todos los términos del polinomio.

degree of a term (p. 356) The sum of the exponents of the variables contained in the term.
grado de un termino (pág. 356) Suma de los exponentes de las variables contenidas en el término.

dependent variable (p. 81) The variable whose value is determined by the input or value of the independent variable.
variable dependiente (pág. 81) Variable cuyo valor queda determinado por la entrada o valor de la variable independiente.

difference of two squares (p. 376) A polynomial of the form $a^2 - b^2$, which is the product of binomials of the form $(a + b)(a - b)$.
diferencia de cuadrados (pág. 376) Polinomio de la forma $a^2 - b^2$, que es el producto de binomios de la forma $(a + b)(a - b)$.

descending order (p. 357) Terms are arranged in order from largest to smallest.
orden descendente (pág. 357) Los términos se ordenan de mayor a menor.

discrete data (p. 82) A set of data with a finite number of data values; a graph of discrete data appears as individual points on a number line or coordinate plane.
datos discretos (pág. 82) Conjunto de datos con un número finito de valores; una gráfica de datos discretos aparece como puntos individuales en una recta numérica o plano de coordenadas.

discriminant (p. 478) The value $b^2 - 4ac$ for a quadratic equation written in the standard form $ax^2 + bx + c = 0$; provides information about the solutions of a quadratic equation.
discriminante (pág. 478) El valor $b^2 - 4ac$ en una ecuación cuadrática escrita en la forma estándar $ax^2 + bx + c = 0$; da información sobre las soluciones de una ecuación cuadrática.

disjunction (p. 43) Two statements joined by the word *or*.
disyunción (pág. 43) Dos enunciados unidos por la palabra *o*.

domain (p. 71) The set of all input values for a relation or function.
dominio (pág. 71) Conjunto de todos los valores de entrada de una relación o función.

E

elimination method (p. 263) A method for solving a system of equations that involves eliminating variables. Also called the **linear combination method.**
método de eliminación (pág. 263) Método para resolver un sistema de ecuaciones que involucra eliminar variables. También llamado **método de combinación lineal.**

equation (p. 15) A mathematical statement that shows that two expressions are equal.
ecuación (pág. 15) Enunciado matemático que muestra que dos expresiones son iguales.

equilateral (p. 11) A polygon with all sides congruent.
equilátero (pág. 11) Polígono con todos sus lados congruentes.

explicit formula (p. 162) Describes any term in a sequence.
fórmula explícita (pág. 162) Describe cualquier término de una progresión.

exponent (p. 287) The number in an exponential expression that tells how many times to use the base as a factor.
exponente (pág. 287) El número en una expresión exponencial que indica el número de veces para usar la base como un factor.

exponential decay (p. 331) A decrease in a quantity due to multiplying by the same factor during each time period. In a decay function, the constant factor is greater than zero but less than 1.
disminución exponencial (pág. 331) Disminución en una cantidad debido a la multiplicación por el mismo factor durante cada período de tiempo. En una función de disminución, el factor constante es mayor que cero pero menor que 1.

exponential function (p. 326) A function of the form $f(x) = a \cdot b^x$, where a and b are constants, x is the domain, $f(x)$ is the range, and $a \neq 0$, $b > 0$, $b \neq 1$.
función exponencial (pág. 326) Función de la forma $f(x) = a \cdot b^x$, donde a y b son constantes, x es el dominio, $f(x)$ es el rango y $a \neq 0$, $b > 0$, $b \neq 1$.

exponential growth (p. 326) A increase in a quantity due to multiplying by the same factor during each time period. In a growth function, the constant factor is greater than 1.
crecimiento exponencial (pág. 326) Incremento en una cantidad debido a la multiplicación por el mismo factor durante cada período de tiempo. En una función de crecimiento, el factor constante es mayor que 1.

exponential regression (p. 347) A method used to find an exponential function that models a set of data.
regresión exponencial (pág. 347) Un método para encontrar una función exponencial que modela un conjunto de datos.

expression (p. 5) A mathematical phrase that uses numbers, or variables, or both.
expresión (pág. 5) Frase matemática que usa números, o variables, o ambos.

extrema (p. 83) Refers to all maximum and minimum values.
extremidad (pág. 83) Se refiere a todos los valores máximos y mínimos.

F

factor (p. 347) Any of the numbers or symbols that when multiplied together form a product; or the process of finding the factors that form a product.
factor (pág. 347) Cualquiera de los números o símbolos que al multiplicarse entre sí forman un producto; o el proceso de hallar los factores que forman un producto.

factor (p. 386) The process of finding the factors that form a product.
factorizar (pág. 386) Proceso de hallar los factores que forman un producto.

function (p. 68) A relation that pairs each element of the domain with exactly one element of the range.
función (pág. 68) Relación que empareja cada elemento del dominio con un solo elemento del rango.

G

graph of an inequality (p. 35) All the points on a number line or half-plane that make the inequality true.
grafica de una desigualdad (pág. 35) Todos los puntos de una recta numérica o semiplano que hacen que la desigualdad sea verdadera.

geometric sequence (p. 314) A sequence in which the ratio of consecutive terms is a constant.
progresión geométrica (pág. 314) Sucesión en que la razón de los términos consecutivos es constante.

H

half-plane (p. 242) One of the two regions of a coordinate plane created by a line in the plane. A half-plane is closed if its boundary line is included in the region. A half-plane is open if its boundary line is not included in the region.
semiplano (pág. 242) Una de las dos regiones de un plano de coordenadas creadas por una recta sobre el plano. Un semiplano es cerrado si su frontera está incluida en la región. Un semiplano es abierto si su frontera no está incluida en la región.

I

imaginary number (p. 480) A number which is not a real number, such as the square root of -1.
número imaginario (pág. 480) Un número que no es real, como la raíz cuadrada de -1.

independent variable (p. 81) The variable for which input values are substituted in a function.
variable independiente (pág. 81) Variable que es reemplazada por los valores de entrada en una función.

L

leading coefficient (p. 357) The coefficient of the term with the highest degree in a polynomial.
coeficiente líder (pág. 357) Coeficiente del término que tiene el mayor grado en un polinomio.

least common multiple (p. 413) The smallest multiple that two or more numbers have in common.
mínimo común múltiplo (mcm) (pág. 413) El menor múltiplo que dos o más números tienen en común.

like terms (p. 361) Terms that have the same variable(s) raised to the same power(s).
términos semejantes (pág. 361) Términos que tienen las mismas variables elevadas a las mismas potencias.

line of best fit (p. 198) A line on a graph showing the general direction of a group of points.
recta de major ajuste (pág. 198) Una línea de un gráfico que muestra la dirección general de un grupo de puntos.

linear equation (p. 175) An equation in which all the terms have a degree of one or zero.
ecuación lineal (pág. 175) Ecuación en la cual todos los términos tienen grado uno o cero.

linear inequality (p. 240) An inequality in which all the terms have a degree of one or zero.
desigualdad lineal (pág. 240) Desigualdad en la que todos los términos tienen grado uno o cero.

linear regression (p. 198) A method used to find the line of best fit.
regresión lineal (pág. 198) Un método para encontrar la línea de major ajuste.

literal equation (p. 28) A equation containing several different variables.
ecuación literal (pág. 28) Ecuación que contiene varias variables diferentes.

M

monomial (p. 385) A number, a variable, or a product of numbers and variables with whole-number exponents.
monomio (pág. 385) Un número, una variable, or un producto de números y variables con exponents de números enteros.

normal distribution (p. 548) A set of data where the values are grouped symmetrically about the mean, with most of the values occurring near the mean.
distribuciún normal (pág. 548) Conjunto de datos en donde los valores se agrupan simétricamente con respecto a la media, donde la mayor parte de los valores se ubican cerca de la media.

N

negative correlation (p. 197) Two variables are related in a negative association when values for one variable tend to decrease as values for the other variable increase; an increase in x corresponds with a decrease in y.
correlación negativa (pág. 197) Dos variables tienen una correlación negativa cuando los valores de una variable tienden a decrecer cuando los valores de la otra variable aumentan; un incremento en x corresponde a un decremento en y.

normal distribution (p. 548) A set of data where the values are grouped symmetrically about the mean, with most of the values occurring near the mean.
distribución normal (pág. 548) Conjunto de datos en donde los valores se agrupan simétricamente con respecto a la media, donde la mayor parte de los valores se ubican cerca de la media.

O

opposite (p. 364) A number's additive inverse.
opuesto (pág. 364) El inverso aditivo de un número.

outliers (p. 543) Data points in a set of data that do not fit the overall pattern of the data set.
valores atípicos (pág. 543) Puntos de un conjunto de datos que no calzan en el patrón general del conjunto de datos.

P

parabola (p. 427) The graph of a quadratic function.
parábola (pág. 427) Gráfica de una función cuadrática.

parallel lines (p. 264) Lines in the same plane that do not intersect. Parallel lines on a coordinate plane have the same slope.
rectas paralelas (pág. 264) Rectas que están en el mismo plano, pero no se intersecan. Las rectas paralelas sobre un plano de coordenadas tienen la misma pendiente.

parent function (p. 111) The most basic function of a particular type, such as $f(x) = x$ (linear); $f(x) = x^2$ (quadratic); $f(x) = |x|$ (absolute value); and $f(x) = b^x$ (exponential).
función básica (pág. 111) La función más simple de un tipo en particular, como $f(x) = x$ (lineal), $f(x) = x^2$ (cuadrática), $f(x) = |x|$ (valor absoluto) y $f(x) = b^x$ (exponencial).

perfect square trinomials (p. 389) Trinomials that have the form $a^2 + 2ab + b^2$ or $a^2 - 2ab + b^2$ and are the result of squaring binomials of the form $(a + b)^2$ and $(a - b)^2$, respectively.
trinomios de cuadrados perfectos (pág. 389) Trinomios de la forma $a^2 + 2ab + b^2$ o $a^2 - 2ab + b^2$ que resultan de elevar al cuadrado binomios de la forma $(a + b)^2$ y $(a - b)^2$, respectivamente.

piecewise defined function (p. 217) A function that is defined differently for different disjoint intervals in its domain.
función por tramos (pág. 217) Función que se define de manera diferente para diferentes intervalos de su dominio.

point-slope form (p. 179) A linear equation with the form $y - y_1 = m(x - x_1)$, where m is the slope of the line and (x_1, y_1) is a point on the line.
forma punto-pendiente (pág. 179) Ecuación lineal expresada de la forma $y - y_1 = m(x - x_1)$, donde m es la pendiente de la recta y (x_1, y_1) es un punto sobre la recta.

polynomial (p. 356) A single term or the sum of two or more terms.
polinomio (pág. 356) Un único término o la suma de dos o más términos.

positive correlation (p. 197) Two variables are related in a positive association when values for one variable tend to increase as values for the other variable also increase; an increase in x corresponds with an increase in y.
correlación positiva (pág. 197) Dos variables tienen una correlación positiva cuando los valores de una variable tienden a aumentar cuando los valores de la otra variable también aumentan; un aumento en x corresponde a un aumento en y.

power (p. 287) A mathematical expression with two parts, a base and an exponent. For example, in the power 5^3, 5 is the base and 3 is the exponent.
potencia (pág. 287) Expresión matemática con dos partes, una base y un exponente. Por ejemplo, en la potencia 5^3, 5 es la base y 3 es el exponente.

principal square root (p. 300) The positive square root of a number.
raíz cuadrada primaria (pág. 300) La raíz cuadrada positiva de un número.

Q

quadratic function (p. 425) A function in one variable with the form $f(x) = ax^2 + bx + c$, where a, b, and c are real numbers and $a \neq 0$.
función cuadrática (pág. 425) Función en una variable de la forma $f(x) = ax^2 + bx + c$, donde a, b y c son números reales y $a \neq 0$.

quadratic regression (p. 202) A method used to find a quadratic function that models a set of data.
regresion cuadrático (pág. 202) Un método para encontrar una función cuadrática que modela un conjunto de datos.

R

radical expression (p. 300) An expression of the form $\sqrt[n]{a}$.
expresión radical (pág. 300) Expresión de la forma $\sqrt[n]{a}$.

range (p. 71) The set of all output values for a relation or function.
rango (pág. 71) Conjunto de todos los valores de salida de una relación o función.

rate of change (p. 128) The ratio of the change in y to the change in x.
proporcion de cambio (pág. 128) La relación entre el cambio en y para el cambio en x.

rational expression (p. 403) An expression that can be written as the ratio of two polynomials.
expresión racional (pág. 403) Expresión que puede escribirse como razón de dos polinomios.

rationalize the denominator (p. 309) To make the denominator of a fraction rational by multiplying the fraction by an appropriate form of 1.
racionalizar el denominador (pág. 309) Hacer que el denominador de una fracción sea racional, multiplicando la fracción por una forma apropiada de 1.

relation (p. 66) A set of ordered pairs
relación (pág. 66) Cualquier conjunto de pares ordenados.

relative maximum (p. 82) The greatest value of a function over an open interval of the domain.
máximo local (pág. 82) El mayor valor de una función en un intervalo dado del dominio.

relative minimum (p. 82) The least value of a function over an open interval of the domain.
mínimo local (pág. 82) El menor valor de una función en un intervalo dado del dominio.

residual (p. 578) The difference between an actual value and a predicted value.
residual (pág. 578) La diferencia entre un valor real y un valor predicho.

S

scatter plot (p. 194) A graphic display of bivariate data on a coordinate plane that may be used to show a relationship between two variables.
diagrama de dispersión (pág. 194) Representación gráfica de datos bivariados sobre un plano de coordenadas, que puede usarse para mostrar una relación entre dos variables.

sequence (p. 160) A list of items or numbers.
sucesión (pág. 160) Lista de elementos o números.

slope (p. 124) The ratio of the vertical change of a line to the horizontal change of the line.
pendiente (pág. 124) La razón del cambio vertical de una recta al cambio horizontal de la recta.
Two angles whose measures Dos ángulos cuyas

slope-intercept form (p. 175) A linear equation of the form $y = mx + b$ where m is the slope and b is the y-intercept.
forma pendiente-intercepto (pág. 175) Ecuación lineal de la forma $y = mx + b$, donde m es la pendiente y b es el intercepto en el eje de las y.

solution (p. 15) Any value that makes an equation or inequality true when substituted for the variable.
solución (pág. 15) Cualquier valor que hace verdadera una ecuación o desigualdad al reemplazar la variable.

solution region (p. 275) The part of the coordinate plane in which the ordered pairs are solutions to all inequalities in a system.
región solución (pág. 275) La parte del plano de coordenadas en el que los pares ordenados son soluciones a las desigualdades en un sistema.

solution of an inequality (p. 35) An ordered pair or set of ordered pairs that makes an inequality true when substituted for the variables.
solución de una desigualdad (pág. 35) Par ordenado o conjunto de pares ordenados que hacen verdadera una desigualdad al reemplazar las variables.

solutions of the linear inequality (p. 240) An ordered pair or set of ordered pairs that makes an inequality true when substituted for the variables.
solución de una desigualdad lineal (pág. 240) Par ordenado o conjunto de pares ordenados que hacen verdadera una desigualdad al reemplazar las variables.

standard deviation (p. 533) A measure of how much the values in a set of data vary from the mean.
desviación estándar (pág. 533) Medida de la variación de los valores de un conjunto de datos con respecto a la media.

standard form of a linear equation (p. 183) A linear equation of the form $Ax + By = C$, where $A \geq 0$, A and B cannot both be zero, and A, B, and C are integers whose greatest common factor is one.
forma estándar de una ecuación lineal (pág. 183) Ecuación lineal de la forma $Ax + By = C$, donde $A \geq 0$, A y B no pueden ser ambos cero y A, B y C son enteros cuyo máximo común divisor es uno.

standard form of a polynomial (p. 357) A way of writing a polynomial so that the terms are in descending order of degree.
forma estándar de un polinomio (pág. 357) Manera de escribir un polinomio de modo que los términos estén en orden decreciente de grado.

standard form of a quadratic function (p. 425) A function of the form $f(x) = ax^2 + bx + c$, where a, b, and c are real numbers and $a \neq 0$
forma estándar de una función cuadrática (pág. 425) Función de la forma $f(x) = ax^2 + bx + c$, donde a, b y c son números reales y $a \neq 0$.

substitution method for solving a system of linear equations (p. 258) A method that involves solving one of the equations for one of the variables and then substituting that value in the other equation(s).
método de sustitución para resolver un sistema de ecuaciones lineales (pág. 258) Método que involucra resolver una de las ecuaciones para una de las variables y luego sustituir ese valor en la otra ecuación o ecuaciones.

system of linear equations (p. 253) Two or more linear equations using the same variables.
sistema de ecuaciones lineales (pág. 253) Dos o más ecuaciones lineales que usan las mismas variables.

system of linear inequalities (p. 274) Two or more linear inequalities using the same variables.
sistema de desigualdades lineales (pág. 274) Dos o más desigualdades lineales que usan las mismas variables.

T

term (p. 355) A number, a variable, or the product of a number and variable(s).
término (pág. 355) Número, variable, o producto de un número por una o más variables.

transformation (p. 113) A change in the position, size, or shape of a parent graph.
transformación (pág. 113) Cambio en la posición, tamaño o forma de una gráfica original.

translation (p. 113) A transformation that moves each point of a figure the same distance and in the same direction.
traslación (pág. 113) Transformación que mueve cada punto de una figura la misma distancia y en la misma dirección.

tree diagram (p. 313) A graphic organizer for listing the possible outcomes of an experiment.
diagrama de árbol (pág. 313) Organizador gráfico para registrar los resultados posibles de un experimento.

trend line (p. 194) A line drawn on a scatterplot to show the general direction of the association or correlation between two sets of data.
línea de tendencia (pág. 194) Línea que se dibuja en un diagrama de dispersión para mostrar la dirección general de la asociación o correlación entre dos conjuntos de datos.

V

variable (p. 5) A letter or symbol used to represent one or more numbers.
variable (pág. 5) Letra o símbolo que se usa para representar uno o más números.

vertex of a parabola (p. 427) The point at which a maximum or minimum value of a function occurs.
vértice de una parábola (pág. 427) Punto en el que ocurre un valor máximo o mínimo de una función.

vertical line test (p. 81) A visual inspection for checking whether or not a graph represents a function.
prueba de la recta vertical (pág. 81) Inspección visual para comprobar si una gráfica representa o no una función.

X

***x*-intercept** (p. 102) The point where a line crosses the x-axis. Its coordinates will be of the form $(a,0)$, where a is a real number.
intercepto en *x* (pág. 102) Punto donde una línea cruza el eje de las x. Sus coordenadas serán de la forma $(a,0)$, donde a es un número real.

Y

***y*-intercept** (p. 103) The point where a line crosses the y-axis. Its coordinates will be of the form $(0,a)$, where a is a real number.
intercepto en *y* (pág. 103) Punto donde una línea cruza el eje de las y. Sus coordenadas serán de la forma $(0,a)$, donde a es un número real.

Z

***z*-score** (p. 550) A score that indicates by how many standard deviations a value is above or below the mean of the set of data.
puntuación *z* (pág. 550) Puntuación que indica el número de desviaciones estándar que un valor dado está por arriba o por debajo de la media del conjunto de datos.

Verbal & Visual Word Association

Definition in Your Own Words	Important Elements
Visual Representation	**Personal Association**

Academic Vocabulary Word

Word Map

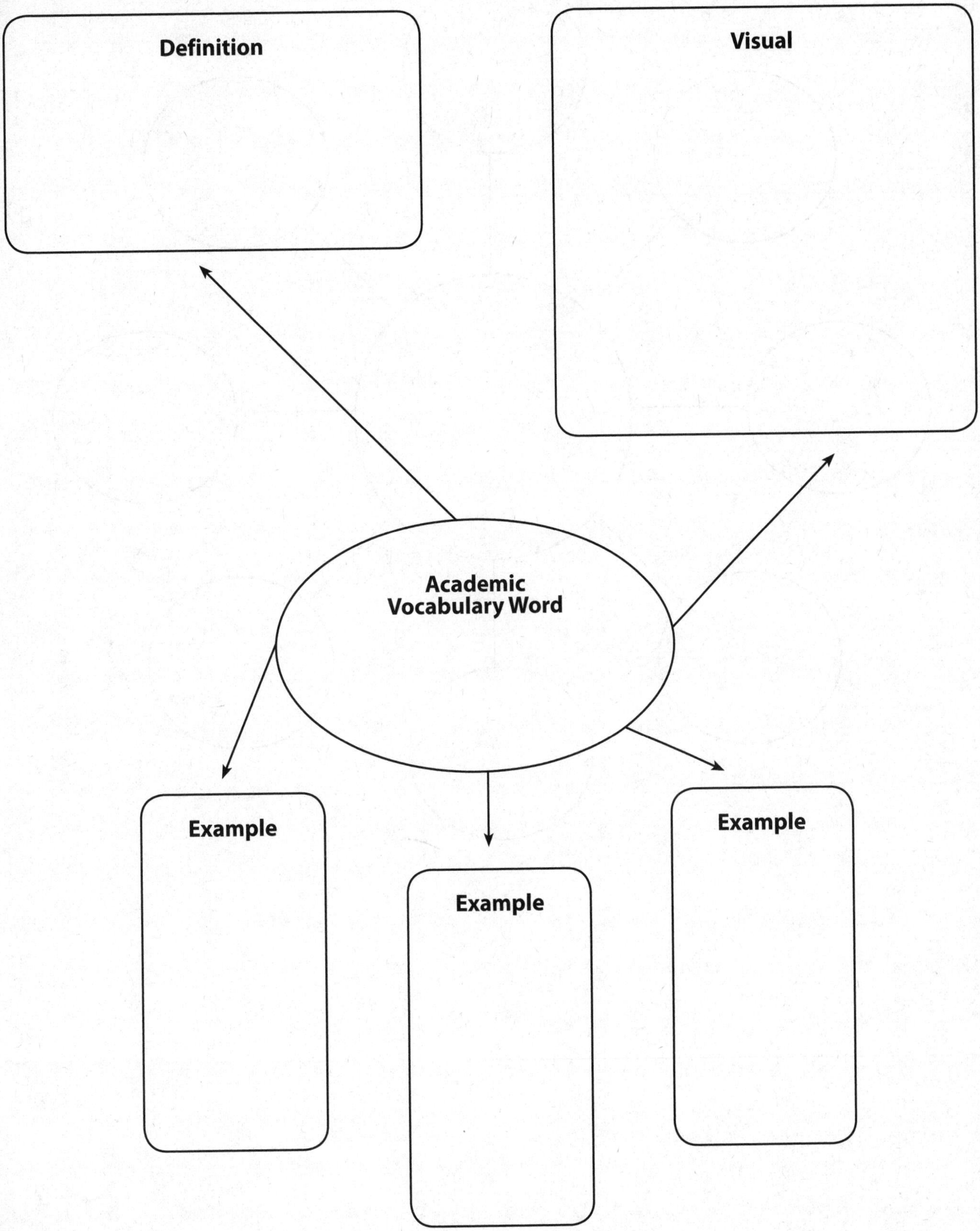
Definition
Visual
Academic Vocabulary Word
Example
Example
Example

Eight Circle Spider

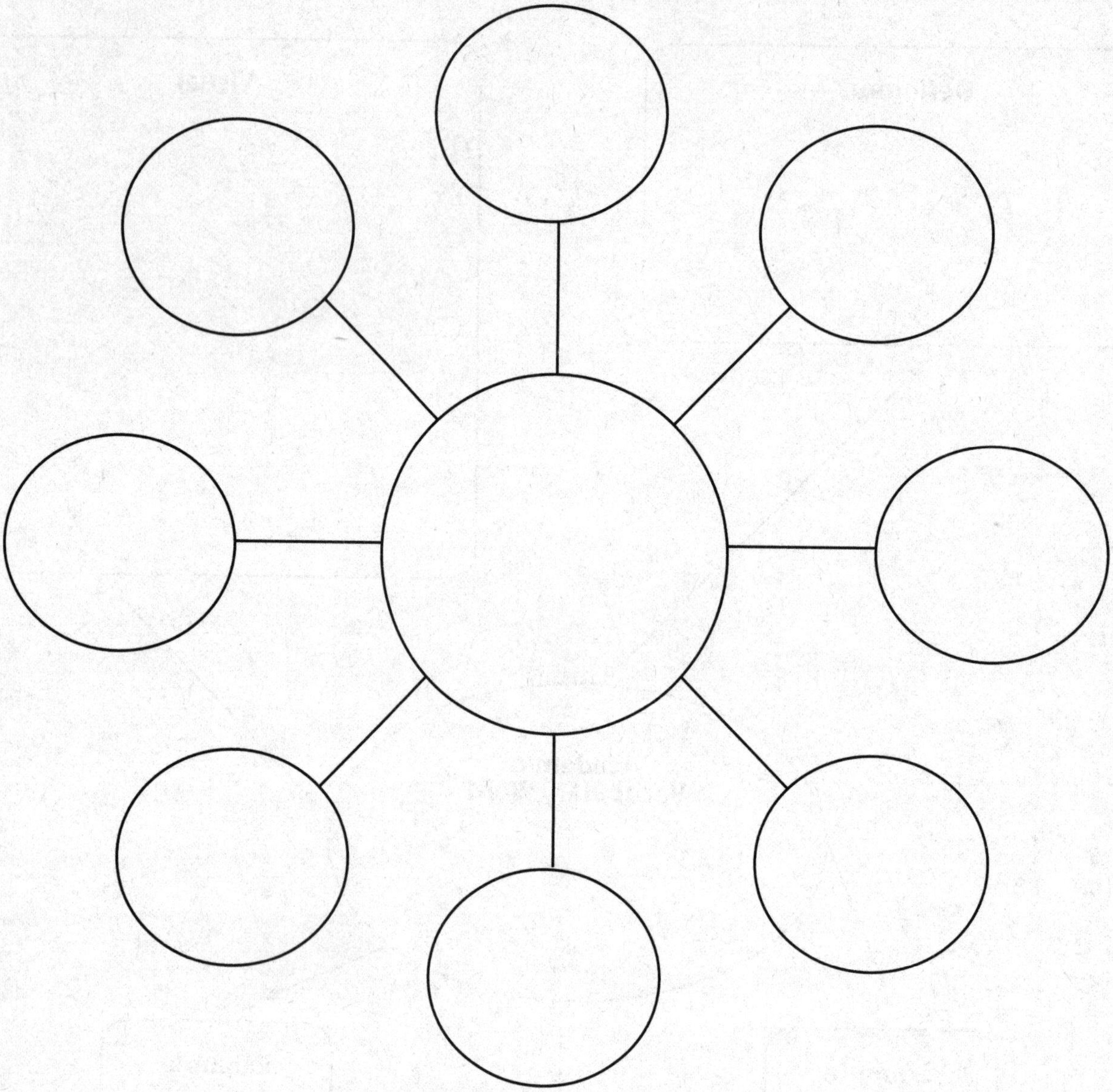

Number Lines

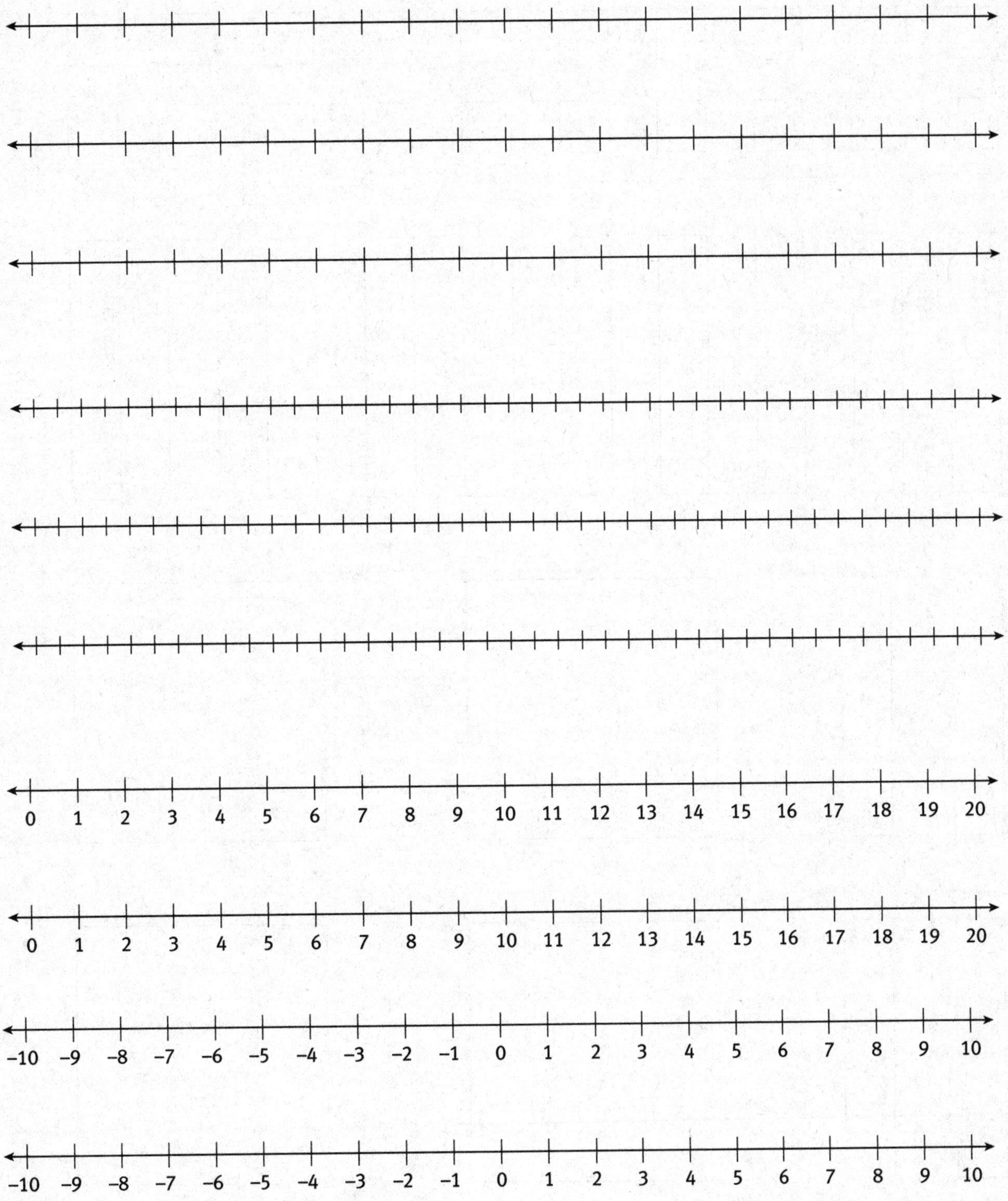

Algebra Tiles

Tables and Coordinate Grids

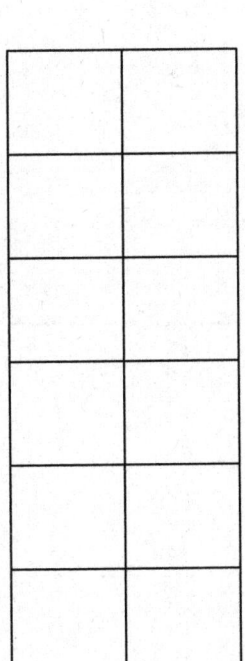

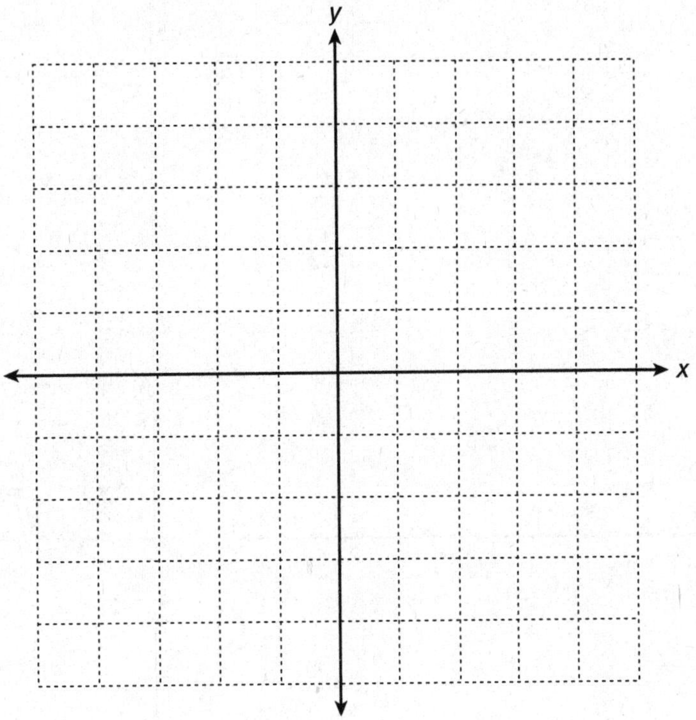

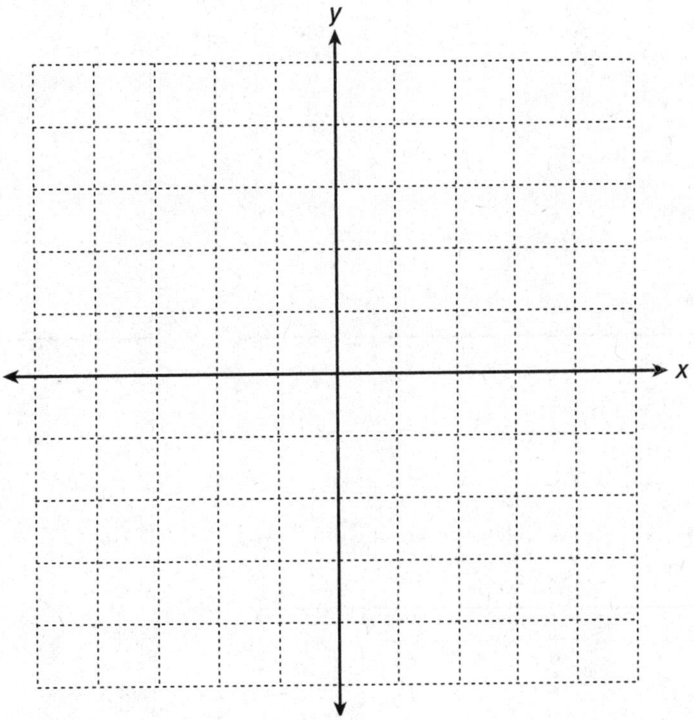

1st Quadrant Grids

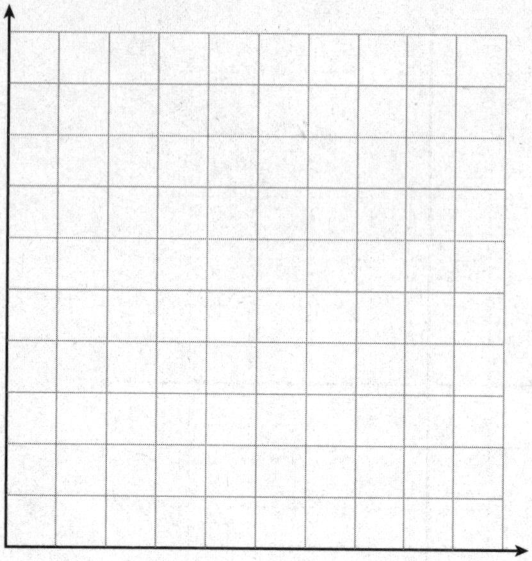

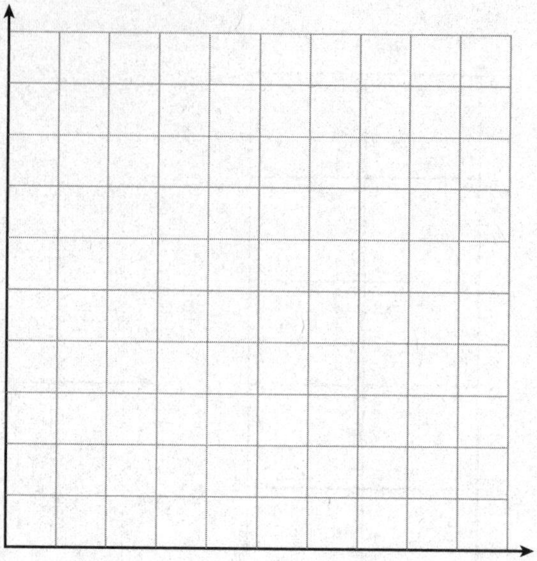

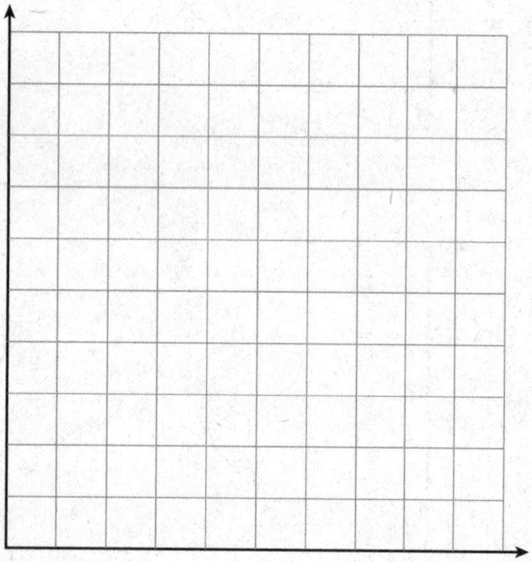

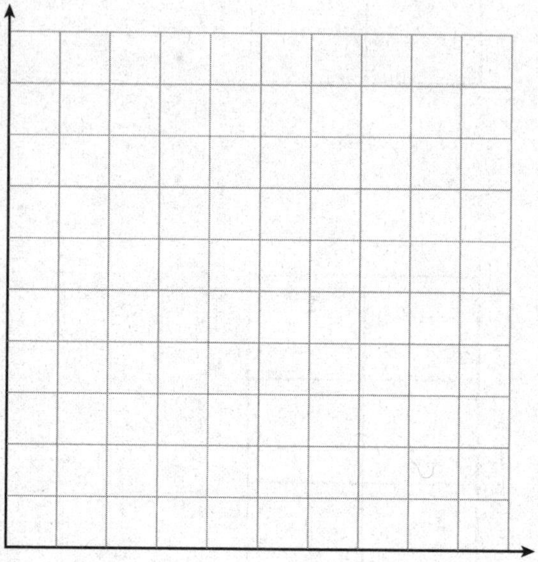

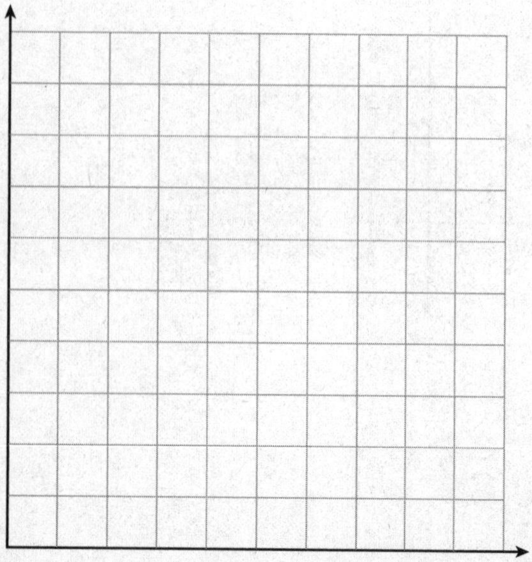

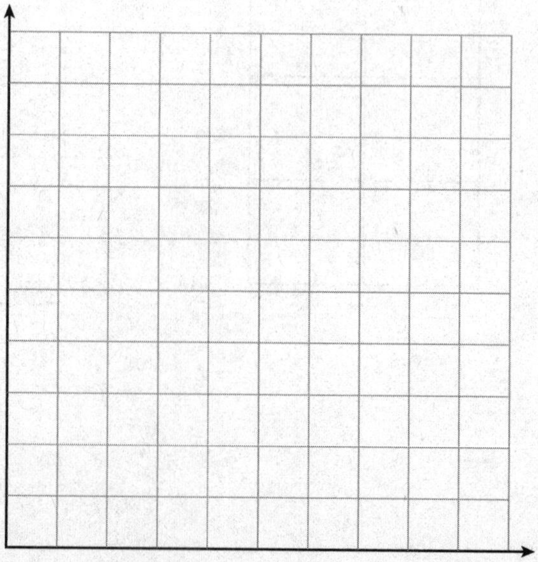

5 by 5 Coordinate Grids

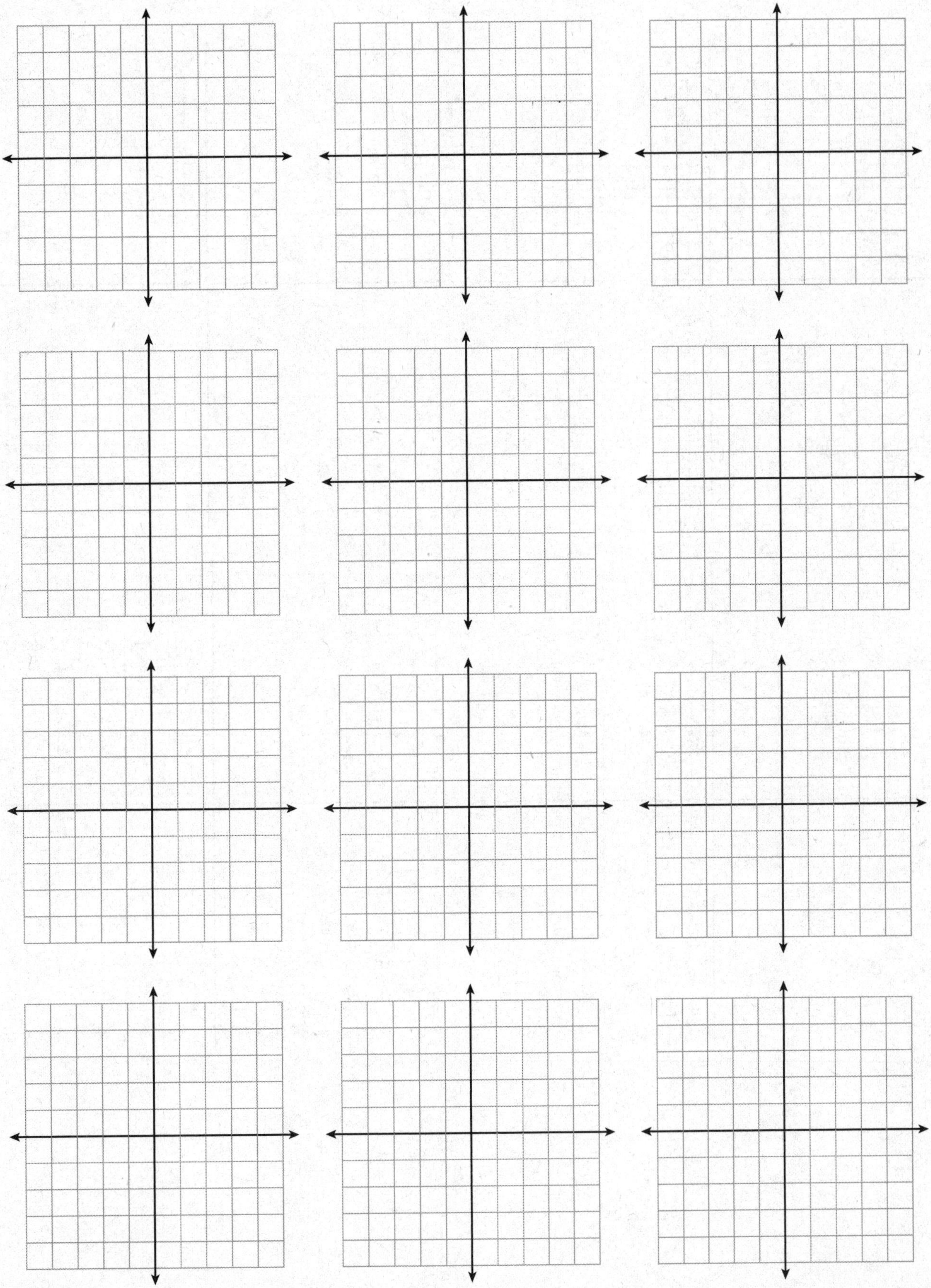

20 × 20 Coordinate Grids

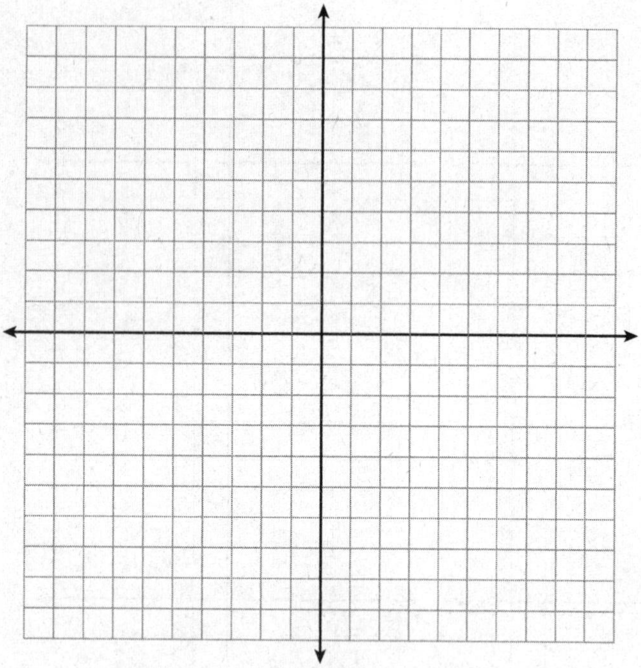

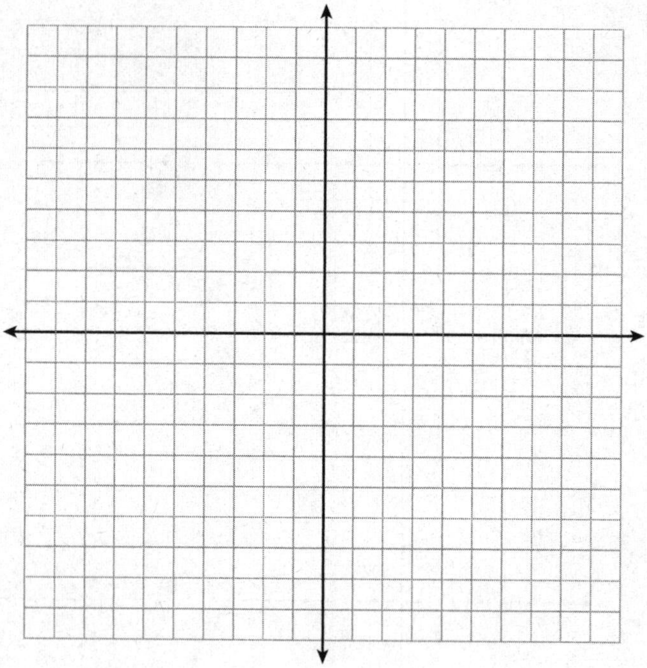

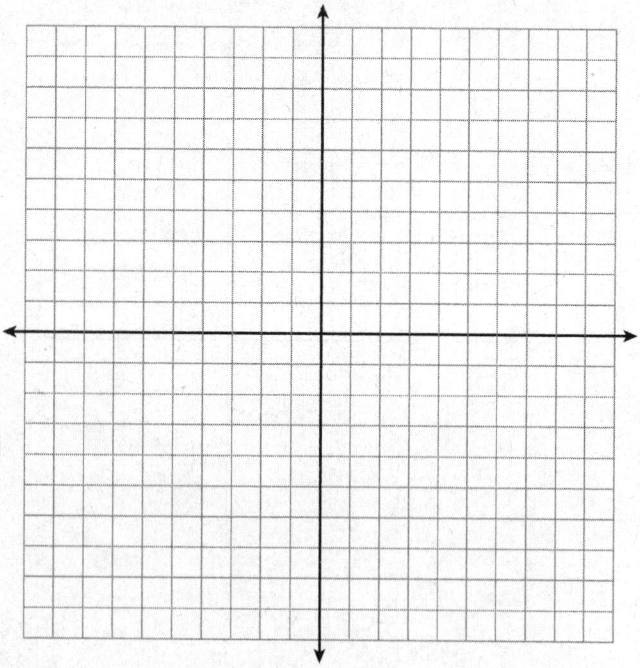

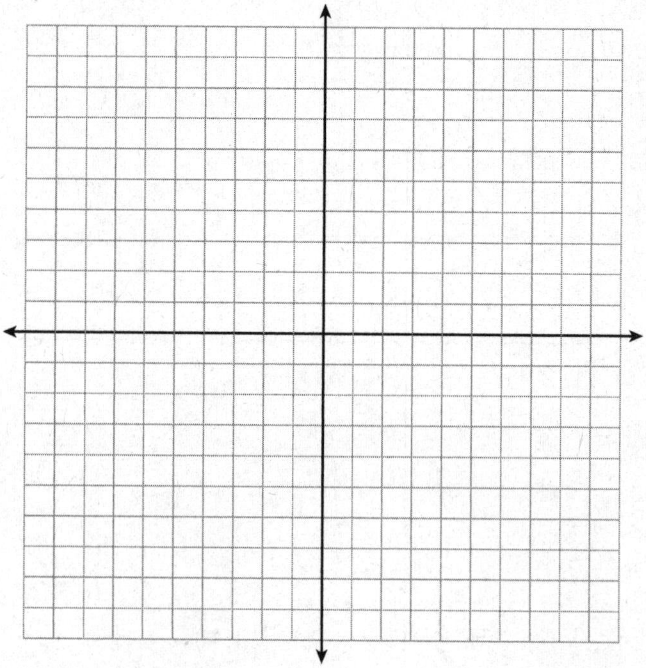

Index

E

F

G

H

I

K

L

M

N

O

P

Q

R

S

T

W

X

Y

Z